WELDING INSPECTION TECHNOLOGY

FOURTH EDITION—2000

Published By
American Welding Society
Education Department

American Welding Society

DISCLAIMER

The American Welding Society, Inc. assumes no responsibility for the information contained in this publication. An independent, substantiating investigation should be made prior to reliance on or use of such information.

International Standard Book Number: 0-87171-467-1
Copyright © 1995 by American Welding Society, Miami, Florida.
Fourth Edition 2000

All rights reserved. No part of this book may be reproduced in any form or by and means, electronic or mechanical, including photocopying, recording, or by and any information storage or retrieval system, without permission in writing from the publisher.

Printed in the United States of America

Table of Contents

Module	Title	Page
1	Welding Inspection and Certification	1-1
2	Safe Practices for Welding Inspectors	2-1
3	Metal Joining and Cutting Processes	3-1
4	Weld Joint Geometry and Welding Symbols	4-1
5	Documents Governing Welding Inspection and Qualification	5-1
6	Metal Properties and Destructive Testing	6-1
7	Metric Practice for Welding Inspection	7-1
8	Welding Metallurgy for the Welding Inspector	8-1
9	Weld and Base Metal Discontinuities	9-1
10	Visual Inspection and Other NDE Methods and Symbols	10-1

REVISION NOTES

The training materials format for the Welding Inspection Technology seminar Workbook has evolved over the past several years, incorporating many upgrades and improvements. Originally, there were 12 Modules in black and white pamphlet form, followed next by a format using a black and white, loose 3-ring binder approach which simplifies the task of revisions and printing format changes. The 1995 revision continued th loose leaf and binder approach, but incorporated the liberal use of color graphics for the first time, as well as improvements in the layout and font types for improved legibility. The current revision simplifies Modules 4 and 7 and corrects the typographical errors that 'crept' into the 1995 version. It is hoped these recent changes make the learning experience more enjoyable, and result in greater retention of the technical material.

The technical content of the last two versions of the training materials has also been changed somewhat to include several new technologies and topics, and to broaden the technical base for the Certified Welding Inspector. However, these latest versions are not entirely new texts, since it required retention of much of the original technical content to comply with the responsibility of meeting the training coverage requirements of the existing CWI Certification Test Question Bank. Future revisions will be necessary, and are anticipated as the welding technology evolves and the usage of the metric system becomes more common within the welding community in the United States.

J. R. Roper, Ph.D., Roper Engineering, authored the new sections on EBW and LBW. Also, thanks to Richard D. Campbell, Welding Solutions, Inc., for his technical and editing input. Special thanks are also given to Richard L. Holdren, Senior Research Engineer, Edison Welding Institute, who authored the 1986 version of the training material, which is retained in large part in the current version.

December 1998

Ted V. Weber
Principal Consultant
Weber & Associates

Module 1
Welding Inspection and Certification

Contents

Introduction 1-2

Who is the Welding Inspector? 1-3

Important Qualities of the Welding Inspector 1-4

Ethical Requirements for the Welding Inspector 1-6

The Welding Inspector as a Communicator 1-7

Personnel Certification Programs 1-10

Key Terms and Definitions 1-13

Module 1—Welding Inspection and Certification

Introduction

In today's world there is increasing emphasis placed on the need for quality, and weld quality is an important part of the overall quality effort. This concern for product quality is due to several factors, including economics, safety, government regulations, global competition, and the use of less conservative designs. While not singularly responsible for the attainment of weld quality, the welding inspector plays a large role in any successful welding quality control program. In reality, many people participate in the creation of a quality welded product. However, the welding inspector is one of the "front line" individuals who must check to see if all of the required manufacturing steps have been completed properly.

To do this job effectively, the welding inspector must have a wide range of knowledge and skills, because it involves more than simply looking at welds. Consequently, this course is specifically designed to provide both experienced and novice welding inspectors a basic background in the more critical job aspects. This does not imply, however, that each welding inspector will use all of this information while working for a particular company. Nor does it mean that the material presented will include all of the information for every welding inspector's situation. Selection of these various topics is based on the general knowledge desirable for an individual to do general welding inspection.

The important thing to realize is that effective welding inspection involves much more than just looking at finished welds. Section 4 of AWS QC1, *Standard for AWS Certification of Welding Inspectors* (see Figure 1.1) outlines the various functions of the welding inspectors. You should become familiar with these various responsibilities because the welding inspector's job is an ongoing

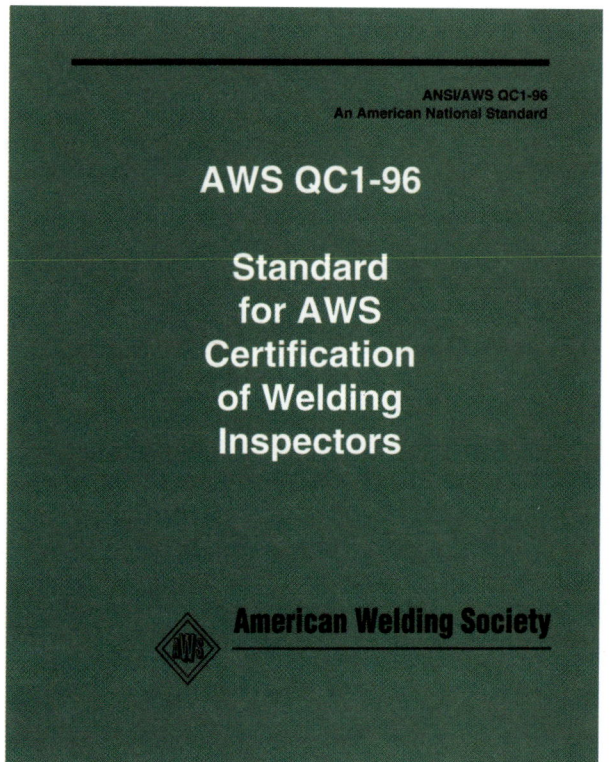

Figure 1.1—AWS QC1, *Standard for AWS Certification of Welding Inspectors*

process. A successful quality control program begins well before the first arc is struck. Therefore, the welding inspector must be familiar with many facets of the fabrication process. Before welding, the inspector will check drawings and specifications to determine such information as the configuration of the component, its specific weld quality requirements, and what degree of inspection is required. This review will also show the need for any special processing during manufacturing. Once welding begins, the welding inspector may observe various processing steps to assure that they are done properly. If all these subsequent steps have been completed satisfactorily, then final inspection should simply confirm the success of those operations.

Another benefit of this course is that it has been designed to provide the welding inspector with the necessary information for the successful completion of the American Welding Society's Certified Welding Inspector (CWI) examination. The ten modules listed below are sources for examination information. The welding inspector must have at least some knowledge in each of these areas. Typically, the information presented will simply be a review, while sometimes it may represent an introduction to a new topic.

Module 1:	Welding Inspection and Certification
Module 2:	Safe Practices for Welding Inspectors
Module 3:	Metal Joining and Cutting Processes
Module 4:	Weld Joint Geometry and Welding Symbols
Module 5:	Documents Governing Welding Inspection and Qualification
Module 6:	Metal Properties and Destructive Testing
Module 7:	Metric Practice for Welding Inspection
Module 8:	Welding Metallurgy for the Welding Inspector
Module 9:	Weld and Base Metal Discontinuities
Module 10:	Visual Inspection and Other NDE Methods and Symbols

Additionally, selected technical references are included in the "Body of Knowledge" required. These include:

- A Selected Code (AWS D1.1, API 1104, etc.)
- Welding Inspection (WI-80)
- AWS A1.1, *Metric Practice Guide for the Welding Industry*
- AWS A2.4, *Standard Symbols for Welding, Brazing, and Nondestructive Examination*
- AWS A3.0, *Standard Welding Terms and Definitions*
- AWS B1.10, *Guide for the Nondestructive Inspection of Welds*
- AWS B1.11, *Guide for the Visual Inspection of Welds*
- ANSI Z49.1, *Safety in Welding, Cutting, and Allied Processes*
- AWS QC1, *Standard for AWS Certification of Welding Inspectors*

Who is the Welding Inspector?

Before turning our discussion to the technical subjects, let us talk about the welding inspector as an individual and the typical responsibilities that accompany the position. The welding inspector is a responsible person, involved in the determination of weld quality **according to applicable codes and/or specifications**. In the performance of inspection tasks, welding inspectors operate in many different circumstances, depending primarily for whom they are working. Thus, there is a special need for job specifications due to the complexity of some components and structures.

The inspection workforce may include destructive testing specialists, nondestructive examination (NDE) specialists, code inspectors, military or government inspectors, owner representatives, in-house inspectors, and others. These individuals may, at times, consider themselves "welding inspectors," since they inspect welds as part of their job responsibility. The three general categories into which the welding inspectors' work-functions can be grouped are:

- Overseer
- Specialist
- Combination Overseer—Specialist

An overseer can be one individual or many individuals whose skills vary such that any amount or type of workmanship may be inspected. Both economics and technical requirements will decide the extent to which these types of inspectors will group themselves and function in various areas of expertise.

The specialist, on the other hand, is an individual who does some specific task(s) in the inspection process. A specialist may or may not act independently of an overseer. The nondestructive examination (NDE) specialist is an example of this category of inspector. This individual has limited responsibilities in the welding inspection process.

It is common to see inspectors serving as both overseer and specialist. Such an individual may be responsible for general weld quality judgments in each of the various fabrication steps, and be required to perform any nondestructive testing that is necessary. Fabricators may employ several overseer type inspectors, each having their own area of general weld inspection responsibility. Because inspection responsibility is divided in these cases, inspectors may have to rely on others for specific aspects of the total inspection program.

For the purposes of this course, we will refer to the welding inspector in general, without regard to how each individual will be used by an employer. It is impractical to address each individual's situation in the scope of this discussion.

To emphasize the differences in job requirements, let's look at some industries using welding inspectors. We see welding inspection being done in the construction of buildings, bridges and other structural units. Energy related applications include power generation facilities, pressure vessels and pipelines, and other distribution equipment requiring pressure containment. The chemical industry also uses welding extensively in the fabrication of pressure-containing processing facilities and equipment. The transportation industry requires assurance of accurate weld quality in such areas as aerospace, automotive, shipbuilding, railroad apparatus and off-road equipment. Finally, the manufacturing of consumer goods often requires specific weld quality requirements. With the diversity shown by this listing, various situations will clearly require different types and degrees of inspection.

Important Qualities of the Welding Inspector

The individual who does welding inspection should possess certain qualities to assure that the job will be done most effectively. Figure 1.2 illustrates these qualities.

The first, and perhaps the most important quality, is a professional attitude. Professional attitude is often the key factor for welding inspector success. Inspector attitude often determines the degree of respect and cooperation received from others during the performance of inspection duties. Included in this category is the ability of the welding inspector to make decisions based on facts so that inspections are fair, impartial and consistent. If decisions are unfair or show partiality or inconsistency, they greatly affect the inspector's credibility. A welding inspector must be completely familiar with the job requirements so that decisions are neither too critical nor too lax. It is a mistake for the inspector to have preconceived ideas as to a component's acceptability. Inspection decisions must be based on facts; the condition of the weld and the acceptance criteria specified in the applicable specification must be the determining factors. Inspectors will often find themselves being "tested" by other personnel on the job, especially when newly assigned to some task. Maintaining a professional attitude helps overcome obstacles to successful job performance.

Next, the welding inspector should be in good physical condition. Since the primary job involves visual inspection, obviously the welding inspector should have good vision, whether natural or corrected. The AWS CWI program requires the inspector to pass an eye examination, with or without corrective lenses, to prove near vision acuity on Jaeger J2 at 12 in., and complete a color perception test. Another aspect of physical condition involves the size of some welded structures. Welds can be located anywhere on very large structures, and inspectors must often go to those areas and make evaluations. Inspectors should be in good enough physical condition to go to any location where the welder has been. This does not imply that inspectors must violate safety regulations just to do their duties. Inspection can often be hampered if not done immediately after welding, because access aids for the welder such as ladders and scaffolding may be removed, making inspection impossible or dangerous. Within safety guidelines, welding inspectors should not let their physical condition prevent them from doing the inspection properly.

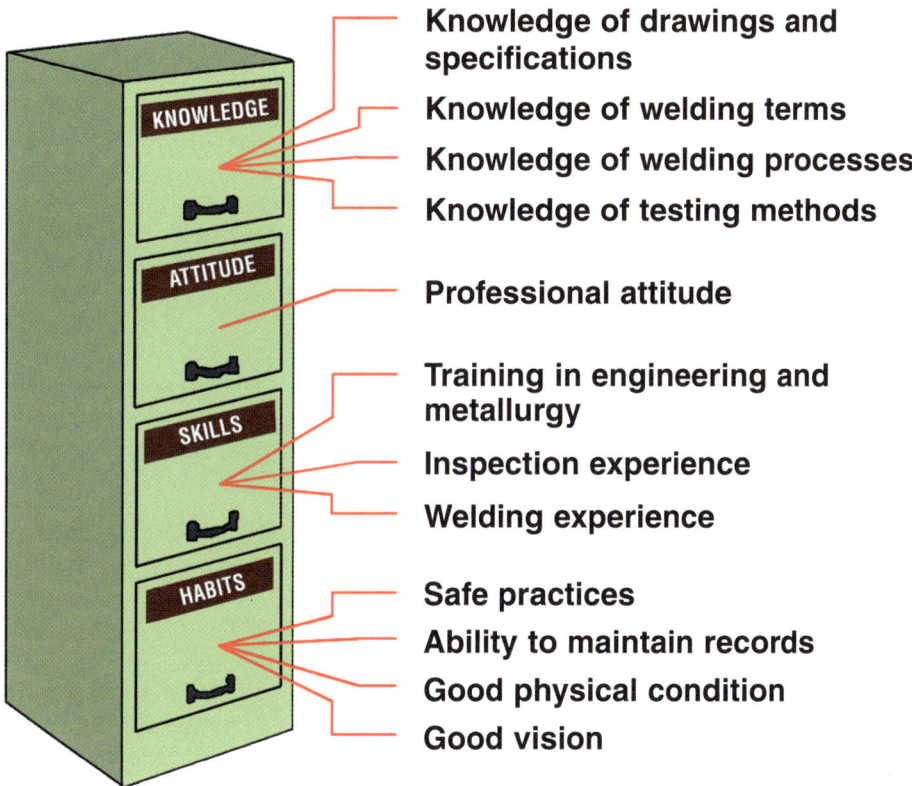

Figure 1.2—The Inspector Possesses a Great Amount of Knowledge, Attitudes, Skills and Habits (KASH)

Another quality the welding inspector should develop is an ability to understand and apply the various documents describing weld requirements. These can include drawings, codes, standards and specifications. Documents provide most of the information regarding what, when, where and how the welding and subsequent inspections are to be done. Therefore, the rules or guidelines under which the welding inspector does the job can be found in these documents. They also state the acceptable quality requirements against which the welding inspector will judge the weld quality. It is important that these documents are reviewed before the start of any work or production because the welding inspector must be aware of the job requirements. Often this pre-job review will reveal required "hold points" for inspections, procedure and welder qualification requirements, special processing steps or design deficiencies such as weld inaccessibility during fabrication. Although welding inspectors should be thorough in their review, this does not mean that the requirements should be memorized. These are reference documents and should be readily available for detailed information any time in the fabrication process. Generally, inspectors are the individuals most familiar with all these documents so they may be called upon by others for information and interpretation regarding the welding.

Most people associated with welding inspection will agree that having inspection experience is very important. Text books and classroom learning cannot teach an inspector all of the things needed to inspect effectively. Experience will aid the welding inspector in becoming more efficient. Better ways of thinking and working will develop with time. Experience will also help the inspector develop the proper attitude and point of view regarding the job. Experience gained working with various codes and specifications improves an inspector's understanding of welding requirements and generally improves job effectiveness. To emphasize the need for inspection experience, we often see a novice inspector paired with an experienced one so the proper techniques can be passed along. Finally, we see that inspector certification programs require some minimum level of experience for qualification.

Another desirable quality of the welding inspector is a basic knowledge of welding and the various welding processes. Because of this, former welders are sometimes selected to be converted into welding inspectors. With a basic knowledge of welding, the inspector is better prepared to understand certain problems that a welder encounters. This aids in gaining respect and cooperation from the welders. Further, this understanding helps the welding inspector to predict what weld discontinuities may be encountered in a specific situation. The welding inspector can then monitor critical welding variables to aid in the prevention of these welding problems. Inspectors experienced in several welding processes, who understand the advantages and limitations of each process, can probably identify potential problems before they occur.

Knowledge of various destructive and nondestructive test methods are also very helpful to the welding inspector. Although inspectors may not necessarily perform these tests, they may from time to time witness the testing or review the test results as they apply to the inspection. Just as with welding processes, the welding inspector is aided by a basic understanding of testing processes. It is important for the inspector to be aware of alternate methods that could be applied to enhance visual inspection. Welding inspectors may not actually perform a given test but they may still be called upon to decide if the results comply with the job requirements.

The ability to be trained is a necessity for the job of welding inspector. Often, an individual is selected for this position based primarily on this attribute. Inspectors do their job most effectively when they receive training in a variety of subjects. By gaining additional knowledge, inspectors become more valuable to their employers.

Another very important responsibility of the welding inspector is safe work habits; good safety habits play a significant role in avoiding injury. Working safely requires a thorough knowledge of the safety hazards, an attitude that all accidents can be avoided, and learning the necessary steps to avoid unsafe exposure. Safety training should be a part of each inspector's training program.

A final attribute, which is not to be taken lightly, is the welding inspector's ability to complete and maintain inspection records. The welding inspector must accurately communicate all aspects of the various inspections, including the results. All records developed should be understandable to anyone familiar with the work. Reports that can only be deciphered by the welding inspector are useless when he or she is absent. Therefore, neatness is important as well. The welding inspector should look at these reports as his or her permanent records should a question arise later. When reports are generated, they should contain information regarding how the inspection was done so, if necessary, it can be duplicated later by someone else with similar results. Once records have been developed, the welding inspector should facilitate easy reference later.

There are a few "rules of etiquette" relating to inspection reports. First, they should be completed in ink, or typewritten. (In today's "age of computers," typing of inspection reports into a computer system is a very effective way of making legible reports, easily retrieved when needed.) If an error is made in a handwritten report, it can be single-lined out and corrected (the error should **not** be totally obliterated). This corrective action should then be initialed and dated. A similar approach is used when the reports are computer generated. The report should also accurately and completely state the job name and inspection location as well as specific test information. The use of sketches and pictures may also help to convey information regarding the inspection results. Then the completed report should be signed and dated by the inspector who did the work.

Ethical Requirements for the Welding Inspector

We have described some of the qualities which are desired of a welding inspector. In addition to those listed above, there are ethical requirements which are dictated by the position. Ethics simply detail what is considered to be common sense and honesty. The position of welding inspector can be very visible to the public if some critical dispute arises and is publicized. Therefore, welding inspectors should live by the rules and report to their supervisors whenever some questionable situation occurs. Simply stated, the welding inspector should act with complete honesty and integrity while doing the job since the inspection function is one of responsibility and importance. If decisions are biased because of associations with dishonest people, or offers of financial gain, then the inspector is not acting with integrity. A welding inspector's decisions should be based totally on available facts without regard to who did the work in question.

The welding inspector's position also carries with it a certain responsibility to the public. The component and/or structure being inspected may be used by others who could be injured should some failure occur. While inspectors may be incapable of discovering every problem, it is their responsibility to report any condition that could result in a safety hazard. When performing an inspection, inspectors should only do those jobs for which they are properly qualified. This reduces the possibility of errors in judgment.

There are situations that occur that may be reported to the public. If the inspector is involved in a dispute regarding the inspection, he or she may be asked to publicly express an opinion. If stated, the opinion should be based totally on facts that the inspector believes to be valid. Probably the best way to deal with public statements, however, is simply to avoid them whenever possible. The inspector should not volunteer information just to gain publicity. However, in situations where a public statement is required, the welding inspector may wish to solicit the advice of a legal representative before speaking.

The ethical requirements of the job carry with them a great deal of responsibility. However, the welding inspector who understands the difference between ethical and unethical behavior should have little difficulty in performing the job with everyone's best interests in mind. Many inspectors are required to make decisions that may have great financial impact on certain parties. In such situations, the inspector may be approached to overlook some feature or reverse a decision for personal financial gain. The welding inspector must recognize such dishonest acts and stand firm on all decisions.

The Welding Inspector as a Communicator

An important aspect of the welding inspector's job is that of communication. The day-to-day inspection effort requires effective communication with many people involved in the fabrication or construction of some item. What must be realized, however, is that communication is not a one way street. The inspector should be able to express thoughts to others, and be ready to listen to a reply. To be effective, this communication sequence must be a continuous loop so that both parties have an opportunity to express their thoughts or interpretations (see Figure 1.3). It is wrong for any individual to think that their ideas will always prevail. Inspectors must be receptive to opinions to which a further response can be made. Often, the best inspector is one who listens well.

As mentioned, the welding inspector has to communicate with several different people involved in the fabrication sequence (see Figure 1.4). In fact, many situations occur where welding inspectors are the central figure of the communication network, since they will constantly be dealing with most of the people involved. Some people that the inspector may communicate with are welders, welding engineers, inspection supervisors, welding supervisors, welding foremen, design engineers, and production supervisors. Each company will dictate exactly how its welding inspectors function.

The communication between the welder and inspector is important to the attainment of quality work. If there is good communication, each individual can do a better job. Welders can discuss problems they encounter, or ask about specific quality requirements. For example, suppose the welders are asked to weld a joint having a root opening which is so tight that a satisfactory weld cannot be accomplished. They may contact the inspector to pass judgment and get the situation corrected right then rather than after the weld is rejected for being made improperly. When effective communication occurs, the welding inspector has the opportunity to supply answers and/or begin corrective action to prevent the occurrence of some problem. The communication between the welder and an inspector is usually improved if the welding inspector has some welding experience. Then the welder has more confidence in the inspector's decisions. If there is poor communication between these two parties, quality can suffer.

Welding engineers rely heavily on welding inspectors to be their "eyes" on the shop floor or construction site. Engineers count on the inspector to spot problems relating to the techniques and

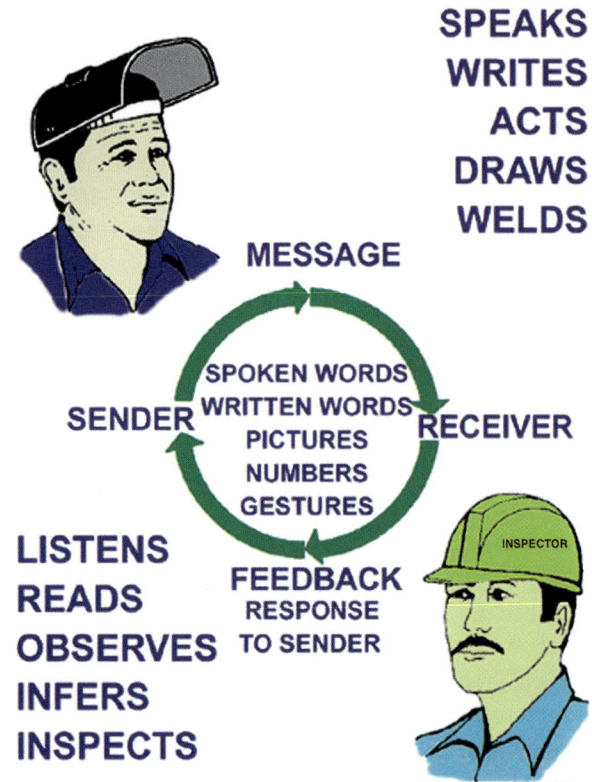

Figure 1.3—The Welding Inspector—A Communicator

processes specified. The welding inspector can also confirm whether specified procedures are being followed. The welding inspector, in turn, can ask the welding engineer about certain aspects of those procedures as well. Often, if a welding procedure is not producing consistent, reliable results, the welding inspector may be the first person to spot the problem. At that point, the welding engineer is notified so that adjustments can be made to alleviate the problem.

The welding inspector will probably work under the direction of some supervisor. This individual is responsible for verifying a welding inspector's qualifications to perform the work. The supervisor should also answer the inspector's questions and aid in the interpretation of quality requirements. In some industry situations, the welding inspector must bring all questions to the supervisor. In turn the supervisor takes that question to someone in engineering, purchasing, etc. The welding inspector must convey a question clearly so it can be described properly by the inspection supervisor to the other party.

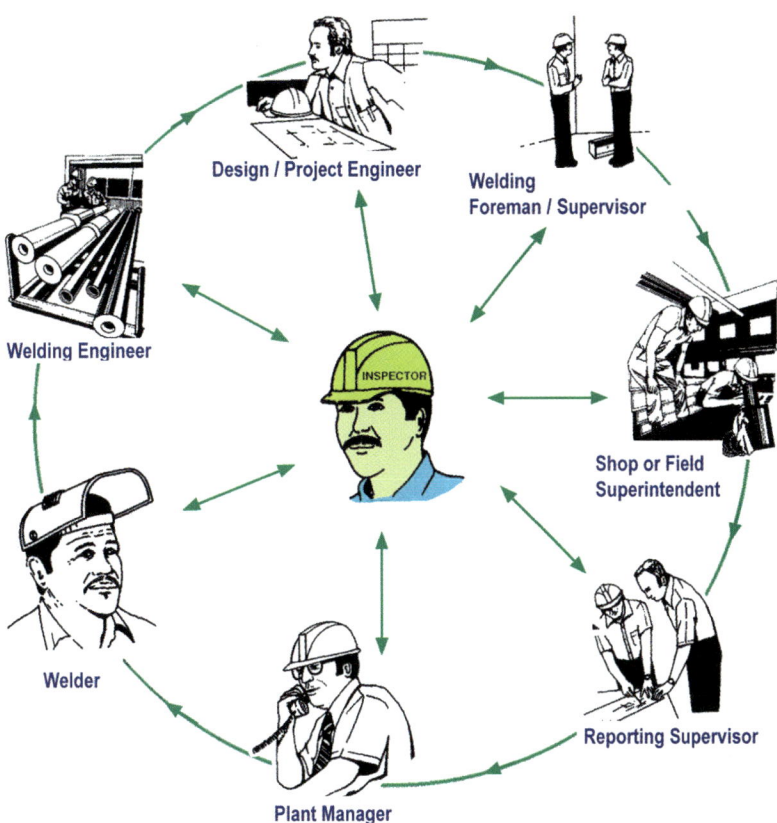

Figure 1.4—Establishing Lines of Communication

During the actual fabrication process, the welding inspector will have opportunities to speak with many other personnel. In some situations, instead of communicating directly with the welders, the welding inspector will deal with the welding supervisor or foreman. This usually involves specific quality requirements or explanations why some aspect of the welding is rejectable.

The welding inspector may also have to gain information from design engineers about the actual weld requirements. During fabrication, other problems may arise which can only be answered by the person who actually designed the structure or component. Another way in which this aspect of communication takes place is through drawings and welding symbols. Although a powerful communication tool, welding and NDE symbols may require clarification by the symbol creator.

Finally, the welding inspector will have some occasion to discuss job scheduling with production personnel. This occurs especially when rejections have been noted which could alter the production schedule. It is important for the welding inspector to keep the production personnel aware of the status of the welding inspection so that schedules can be adjusted if necessary. As indicated, depending on the welding inspector's specific job description, he or she may not deal with all those noted above, or the inspector may communicate with others not discussed here. It is important to realize that all will benefit if some effective communication takes place so surprises do not occur during fabrication.

When we talk of communication, we are not limiting our attention to speaking. There are several ways in which people effectively communicate. They include speaking, writing, drawing, gesturing, and the use of pictures or photographs. Each situation may be dealt with using one or more of these methods. The method is not as important as the fact that communication occurs; messages are sent, received and understood by all concerned.

Personnel Certification Programs

There are several programs presently available to determine whether an individual possesses the necessary experience and knowledge to perform welding inspection effectively. The American Society for Nondestructive Testing has issued guidelines for certification of NDE personnel in ASNT SNT-TC-1A. This document describes the recommended procedures for certifying inspectors performing nondestructive testing. ASNT recognizes three levels of certification, Levels I, II, and III.

AWS has also initiated an NDE Certification Program. Presently, AWS can certify one Level of Radiographic Interpreter (RI). The RI program certifies inspectors for interpreting weld radiographs. Additional NDE methods' certification is available through the joint efforts of AWS and outside training agencies.

For visual inspection of welds, AWS has developed the Certified Welding Inspector program. The front page of the Application form for the Certified Welding Inspector is shown in Figure 1.5.

AWS QC1, *Standard for AWS Certification of Welding Inspectors*, Figure 1.1, establishes the requirements for AWS certification of welding inspection personnel. There are three levels of certification in AWS QC1. The Senior Certified Welding Inspector (SCWI) is a person with at least 15 years experience, including 6 years experience while certified as a Certified Welding Inspector (CWI). The SCWI must pass a separate examination from the CWI examination explained below. Information on the SCWI program and examination are found in a separate course, *Welding Quality Assurance and Inspection Manual—A Guide for the Senior Certified Welding Inspector*. The next certification level is the CWI and the third level is the Certified Associate Welding Inspector (CAWI). Both of these certifications are covered in this course. AWS QC1 describes how personnel are qualified, lists the principles of conduct, and notes the practice by which certification may be maintained. Those major elements will be discussed here.

The first step toward certification is the documentation of relevant educational and work experience. To qualify for the Certified Welding Inspector (CWI) examination, the individual must document his or her educational background. In addition, the candidate's years of welding-related experience according to some code or specification must be documented.

With supporting documentation (e.g., copies of transcripts, reference letters, credited hours of training, quarter hours or semester hours), up to two years of work experience may be substituted by post high school educational experience. Substituted educational experience includes an Associate or higher degree in engineering, physics or physical science and welding technology. Trade and vocational courses can be applied to work experience substitution for completed courses related to welding (up to one year maximum).

Candidates with a high school education, either by diploma or state or military equivalence, must have at least 5 years experience. Individuals with eighth grade schooling are required to have not less than 10 years job experience to qualify for the examination. For individuals with less than eighth grade schooling, not less than 15 years is required.

A subordinate level of qualification is the Certified Associate Welding Inspector (CAWI), which requires fewer years of experience for each educational level. All of the experience noted for both the CWI and CAWI must be work associated with some code or specification to be considered valid.

Individuals who qualify for the Certified Welding Inspector Examination take a three-part examination:

Part A—Fundamentals. The Fundamental examination is a closed book test consisting of 150 multiple choice questions. The topics covered in this portion of the exam include reports and records, destructive tests, welding performance, duties and responsibilities, weld examination, definitions and terminology, safety, welding and nondestructive examination symbols, nondestructive examination methods, welding processes, heat control, metallurgy, mathematical conversions and calculations.

Part B—Practical. The Practical examination consists of 46 questions. It requires measurement of weld replicas with provided measuring tools, and evaluation in accordance with a supplied "Book of Specifications." Not all questions require the use of

American Welding Society
550 N.W. LeJeune Road, Miami, Florida 33126

APPLICATION FOR AWS WELDING INSPECTOR EXAMINATION

Please PRINT or TYPE

AWS USE ONLY
SITE CODE _____
1. Check # _____
2. Date Rec'd _____
3. Amount _____
4. Account # _____

1a. CHECK THE EXAMINATION LEVEL YOU DESIRE
- ☐ A. Certified Welding Inspector (CWI)
- ☐ B. Certified Associate Welding Inspector (CAWI)

1b. Choose ONE of the following for the CODE BOOK test
- ☐ C-1 AWS D1.1
- ☐ C-2 API 1104
- ☐ C-3 ASME B31.1 & ASME Sec IX
- ☐ C-4 ASME Sec VIII & Sec IX

CHARGE MY: VISA   MC   AMEX   DIN
CARD # _____
EXP. DATE _____

SAMPLE

2. PERSONAL

LAST NAME / FIRST / MIDDLE INITIAL

STREET / APT #

CITY / STATE / ZIP

LIST COMPANY NAME ONLY IF ADDRESS ABOVE IS COMPANY ADDRESS

TELEPHONE: AREA CODE AND NUMBER / SOCIAL SECURITY NO. / BIRTH DATE

3. AWS CERTIFICATION STATUS

a. Have you taken a previous AWS QC1 certification examination?   ☐ YES   ☐ NO

a.1. If Yes, give date and location _____

b. Have you ever been certified as an AWS QC1?   ☐ YES   ☐ NO

b.1. If Yes, print your certification no. _____

4. SIC CODES   SOURCE CODE: QC1

Type of Business (Check ONE only)
- A ☐ Contract construction
- B ☐ Chemicals & allied products
- C ☐ Petroleum & coal industries
- D ☐ Primary metal industries
- E ☐ Fabricated metal products
- F ☐ Machinery except elect. (incl. gas welding)
- G ☐ Electrical equip. supplies, electrodes
- H ☐ Transportation equip. - air, aerospace
- I ☐ Transportation equip. - automotive
- J ☐ Transportation equip. - boats, ships
- K ☐ Transportation equip. - railroad
- L ☐ Utilities
- M ☐ Welding distributors & retail trade
- N ☐ Misc. repair services (incl. welding shops)
- O ☐ Educational services (univ. libraries, schools)
- P ☐ Engr. & architectural services (incl. assns.)
- Q ☐ Misc. business services (incl. commercial labs)
- R ☐ Governments (federal, state, local)
- S ☐ other _____

Job Classification (Check ONE only)
- 01 ☐ President, owner, partner, officer
- 02 ☐ Manager, director, superintendent (or assistant)
- 03 ☐ Sales
- 04 ☐ Purchasing
- 05 ☐ Engineer — welding
- 06 ☐ Engineer — other
- 07 ☐ Inspector, tester
- 08 ☐ Supervisor, foreman
- 09 ☐ Welder, welding or cutting operator
- 10 ☐ Architect, designer
- 11 ☐ Consultant
- 12 ☐ Metallurgist
- 13 ☐ Research & development
- 14 ☐ Technician
- 15 ☐ Educator
- 16 ☐ Student
- 17 ☐ Librarian
- 18 ☐ Customer service
- 19 ☐ Other _____

Your Technical Interests (Place a number on line in choice order — 1-2-3, etc.)
- A ___ Ferrous metals
- B ___ Aluminum
- C ___ Non-fer. except aluminum
- D ___ Advanced mat'l/intermetalics
- E ___ Ceramics
- F ___ High energy Processes
- G ___ Arc Welding
- H ___ Brazing & Soldering
- I ___ Resistance Welding
- J ___ Thermal Spray
- K ___ Cutting
- L ___ NDT
- M ___ Safety & Health
- N ___ Pipe & Tubing
- O ___ Pressure Vessels & Tanks
- P ___ Structures
- Q ___ Roll Forming
- R ___ Sheet metal
- S ___ Stamping & punching
- T ___ Bending & shearing
- U ___ Aerospace
- V ___ Automotive
- W ___ Machinery
- X ___ Marine
- Y ___ Other

Major product or service of your company _____

Figure 1.5—AWS Certified Welding Inspector Application Form

the Book of Specifications; some require the individual to answer from practical knowledge. The Practical Test covers welding procedures, welder qualification, mechanical tests and properties, welding inspection and flaws, and nondestructive tests. Test candidates should be familiar with fillet and groove weld gauges, micrometers, dial calipers, and machinist's scales.

Part C—Open Book Code. This portion consists of 46 questions on the code the individual has selected for this part of the examination. The following codes are applicable to this portion of the examination:

- **AWS D1.1.** The AWS D1.1, *Structural Welding Code—Steel* examination covers the following subject areas: general requirements, design of welded connections, prequalification of WPSs, qualification, fabrication, inspection, stud welding and the annexes.

- **API 1104.** The API 1104, *Welding of Pipelines and Related Facilities* examination covers the following subject areas: general, qualification of welding procedures, welder qualification, design and preparation of a joint for production welding, inspection and testing of production welds, standards of acceptability—NDT, repair or removal of defects, radiographic procedure, and automatic welding.

- **AWS D1.5.** The AWS D1.5, *Bridge Welding Code* examination covers the following subject areas: general provisions, design of welded connections, workmanship, technique, qualification, inspection, stud welding, welded steel bridges, fracture control plan for nonredundant members and the annexes.

- **AWS D15.1.** The AWS D15.1, *Railroad Welding Specification—Cars and Locomotives* examination covers welding of metal at least 1/8 in. thick, specific requirements for welding railroad cars, and the requirements for the manufacturing and reconditioning of locomotives and passenger train vehicles.

To successfully complete the examination, individuals must pass all three parts of the test. The passing score in each part for the CWI is 72 percent; the passing for CAWI is 50 percent. Beyond completion of the examination, the test candidate must undergo an eye examination to assure that the individual possesses adequate vision, whether natural or corrected. After all test results are successfully completed, the individual is considered qualified to perform visual inspection of welds. When AWS says that this individual is a Certified Welding Inspector, this simply implies that the person's qualifications are documented with an appropriate certificate. The CWI certificate does not state what code the inspector used on the examination. A CWI is qualified to use and interpret any welding code or standard.

Welding inspectors are a very important part of any effective quality control program. While there are various categories of welding inspectors, in general they are considered to be those individuals responsible for evaluation of the resulting welding. These individuals must possess physical, mental and ethical qualities in order to be effective. The remaining modules will detail those aspects of welding considered important for the welding inspector. In addition, these topics are also considered relevant to the AWS Certified Welding Inspector Examination. Therefore, this text is an appropriate guide for individuals to use in preparation for that series of examinations.

In preparation for that portion of the CWI examination covering welding inspector certification requirements, you are encouraged to read and become familiar with AWS QC1, *Standard for AWS Certification of Welding Inspectors*. Part of the welding inspector's job is the review and interpretation of various documents relating to the welded fabrication. This requires that the individual have a full understanding of the proper terms and definitions that are used. For this reason, included at the end of each module the reader will find, "Key Terms and Definitions" applicable to a module's topic. AWS realizes the need for standardized terms and definitions for use by those involved in the fabrication of welded products. In answer to this need, AWS A3.0, *Standard Welding Terms and Definitions*, was published (see Figure 1.6).

AWS A3.0 was developed by the Committee on Definitions and Symbols to aid in welding information communication. Standard terms and defini-

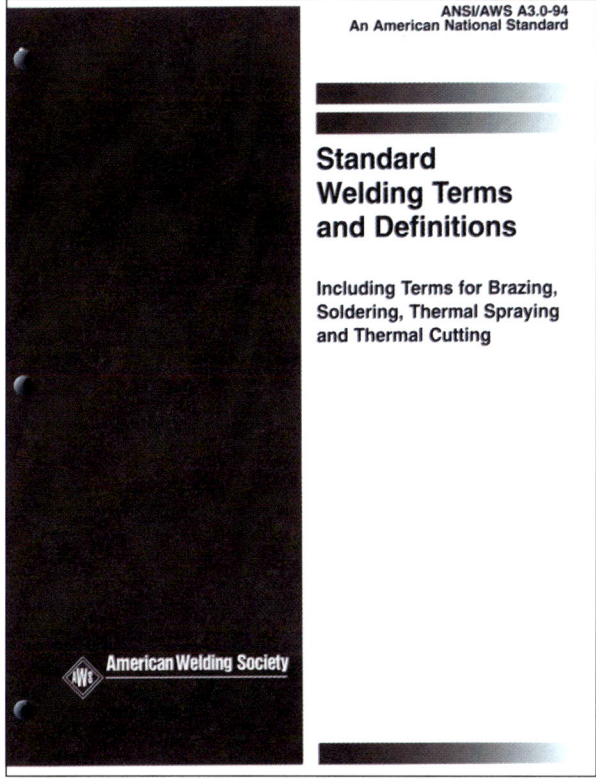

Figure 1.6—AWS A3.0, *Standard Welding Terms and Definitions*

tions published in A3.0 are those that should be used in the oral and written language of welding. While these are the standard, or preferred, terms, they are not the only terms used to describe various situations. The purpose here is to educate, and it is often important to mention some of these common terms, even though they are not preferred terminology. When nonstandard terms are mentioned, they appear in parentheses after the preferred words.

While most of the terms used apply to the actual welding operation, it is important for the welding inspector to understand other definitions which apply to other related operations. Welding inspectors should understand how to describe weld joint configurations and fitup process elements requiring comment. After welding, the inspector may need to describe the location of a weld discontinuity that has been discovered. If a discontinuity requires further attention, it is important that the inspector accurately describe the location of the problem so that the welder will know where the repair is to be made. AWS recommends that standard terminology be used wherever possible, but the inspector must be familiar with nonstandard terms as well.

Key Terms and Definitions

API—American Petroleum Institute. The technical society which provides technical guidance for the petroleum industry.

API 1104—The API Standard, *Welding of Pipelines and Related Facilities*. This standard is often used in construction of cross-country pipelines.

ASME—American Society of Mechanical Engineers. The technical society which provides technical guidance for pressure containing vessels and equipment.

ASNT—American Society for Nondestructive Testing. The technical society which provides technical guidance for NDE.

AWS—American Welding Society. The technical society which provides technical guidance and leadership in all phases of welding.

AWS A3.0—The AWS *Standard Welding Terms and Definitions*. This standard defines welding-related terms with standard definitions.

AWS B5.11—The AWS *Specification for the Qualification of Radiographic Interpreters*.

AWS D1.1—The AWS *Structural Welding Code—Steel*. Used worldwide for construction of buildings and structures.

AWS D1.5—The AWS *Bridge Welding Code* used in the U.S. for construction of bridges.

AWS D15.1—The AWS *Railroad Welding Specification—Cars and Locomotives*. This specification covers welding of railroad cars and locomotives.

CAWI—Certified Associate Welding Inspector.

CWI—Certified Welding Inspector.

KASH—An acronym for Knowledge, Attitude, Skills and Habits, the basic tools of a welding inspector.

NDE—Nondestructive Examination. The act of determining the suitability of some material or component for its intended purpose using techniques that do not affect its serviceability. NDE is the preferred term per ANSI/AWS.

NDI—Nondestructive Inspection. A nonstandard term for **nondestructive examination** (see NDE).

NDT—Nondestructive Testing. A nonstandard term for **nondestructive examination** (see NDE).

QC1—The AWS *Standard for AWS Certification of Welding Inspectors*. Defines the requirements and program for the AWS to certify welding inspectors.

SCWI—Senior Certified Welding Inspector.

SNT-TC-1A—This ASNT Recommended Practice, *Personnel Qualification and Certification in Nondestructive Testing*, outlines the certification program for NDT technicians.

Module 2
Safe Practices for Welding Inspectors

Contents

Introduction	2-2
Eye and Face Protection	2-6
Protective Clothing	2-7
Noise	2-9
Machinery Guards	2-9
Fumes and Gases	2-9
Exposure Factors	2-10
Ventilation	2-11
Handling of Compressed Gases	2-13
Manifolds	2-15
Gases	2-17
Electric Shock	2-18
Key Terms and Definitions	2-20

Module 2—Safe Practices for Welding Inspectors

Introduction

Welding inspectors often work in the same environment as the welder, so they can be exposed to many potential safety hazards. These include electric shock, falling, radiation, eye hazards such as ultraviolet light and particulate matter in the air, smoke and fumes, and falling objects. Safety is not to be taken lightly, although the welding inspector may only be exposed to these conditions momentarily. The welding inspector should strive to observe all safety precautions such as use of safety glasses, hard hats, protective clothing or any other appropriate apparatus for a given situation. For a more detailed look at recommended safety precautions refer to ANSI Z49.1, *Safety in Welding, Cutting, and Allied Processes* (see Figure 2.1).

Safety is an important consideration in all welding, cutting, and related work. No activity is satisfactorily completed if someone is injured. The hazards that may be encountered, and the practices that will reduce personal injury and property damage, are discussed here.

The most important component of an effective safety and health program is leadership support and direction. Management must clearly state objectives and show its commitment to safety and health by consistent support of safe practices. Management must designate approved safe areas for conducting welding and cutting operations. When these operations are done in other than designated areas, management must assure that proper procedures are established and followed to protect personnel and property.

Management must also be certain that only approved welding, cutting, and allied equipment are used. Such equipment includes torches, regulators, welding machines, electrode holders, and personal protection devices (see Figure 2.2). Adequate

Figure 2.1—ANSI Z49.1 Safety in Welding, Cutting, and Allied Processes

supervision must be provided to assure that all equipment is properly used and maintained.

Thorough and effective training is a key aspect of a safety program. Adequate training is mandated under provisions of the *U.S. Occupational Safety and Health Act (OSHA)*, especially those of the *Hazard Communication Standard (29 CFR 1910.1200)*. Welders and other equipment operators work most safely when they are properly trained in the subject.

Figure 2.2—Personal Protective Equipment (PPE)

Proper training includes instruction in the safe use of equipment and processes, and the safety rules that must be followed. Personnel need to know and understand the rules and the consequences of disobeying them. For example, welders must be trained to position themselves while welding or cutting so that their heads are not in the gases or fume plume. A fume plume is a smoke-like cloud containing minute solid particles arising directly from the area of melting metal. The fumes are metallic vapors that have condensed into particulates.

Before work begins, users must always read and understand the manufacturers' instructions on safe practices for the materials and equipment, and the Material Safety Data Sheets (MSDS). Certain AWS specifications call for precautionary labels on consumables and equipment. These labels concerning the safe use of the products should be read and followed (see Figure 2.3).

Manufacturers of welding consumables must, upon request, furnish a Material Safety Data Sheet that identifies materials present in their products that have hazardous properties. The MSDS provides OSHA permissible exposure limits, known as the Threshold Limit Value (TLV), and any other exposure limit used or recommended by the manufacturer. TLV is a registered trademark of the American Conference of Governmental and Industrial Hygienists (ACGIH).

Employers that use consumables must make all applicable MSDS data available to their employees, and also train them to read and understand the contents. The MSDS contain important information about the ingredients contained in welding electrodes, rods, and fluxes. These sheets also show the composition of fumes generated and other hazards

---

WARNING: PROTECT yourself and others. Read and understand this label.
FUMES AND GASES can be dangerous to your health.
ARC RAYS can injure your eyes and burn your skin.
ELECTRIC SHOCK can KILL.

- Before use, read and understand the manufacturer's instructions, Material Safety Data Sheets (MSDS), and your employer's safety practices.
- Keep your head out of fumes.
- Use enough ventilation, exhaust at the arc, or both, to keep fumes and gases from your breathing zone and the general area.
- Wear correct eye, ear, and body protection.
- Do not touch live electrical parts.
- See American National Standard Z49.1, *Safety in Welding, Cutting, and Allied Processes*, published by the American Welding Society, 550 N.W. LeJeune Rd., Miami, Florida 33126; OSHA Safety and Health Standards, available from U.S. Government Printing Office, Washington, DC 20402.

DO NOT REMOVE LABEL

Figure 2.3—Typical Warning Label for Arc Welding Processes and Equipment

that may be caused during use. They also provide methods to be followed to protect the welder and others who might be involved.

Under the *OSHA Hazard Communication Standard, 29 CFR 1910.1200,* employers are responsible for employee hazardous material training in the workplace. Many welding consumables are included in the definition of hazardous materials according to this standard. Welding employers must comply with the communication and training requirements of this standard.

Proper use and maintenance of the equipment must also be taught. For example, defective or worn electrical insulation in arc welding or cutting should not be used. Also, defective or worn hoses used in oxyfuel gas welding and cutting, brazing, or soldering should not be used. Training in equipment operation is fundamental to safe operation.

Personnel must also be trained to recognize safety hazards. If they are to work in an unfamiliar situation or environment, they must be thoroughly briefed on the potential hazards involved. For example, consider a person who must work in confined spaces. If the ventilation is poor and an air-supplied helmet is required, the need and instructions for its proper use must be thoroughly explained to the employee. The consequences of improperly using the equipment must be covered. When employees believe that the safety precautions for a given task are not adequate, or not understood, they should question their supervisor before proceeding.

Good housekeeping is also essential to avoid injuries. A welder's vision is often restricted by necessary eye protection, and personnel passing a welding station must often shield their eyes from the flame or arc radiation. This limited vision makes both the welder and passersby vulnerable to tripping over objects on the floor. Therefore, welders and supervisors must always make sure that the area is clear of tripping hazards. A shop production area should be designed so that gas hoses, cables, mechanical assemblies, and other equipment do not cross walkways or interfere with routine tasks (see Figure 2.4).

When work is above ground or floor level, safety rails or lines must be provided to prevent falls because of restricted vision from eye protection devices. Safety lines and harnesses can be helpful to restrict workers to safe areas, and to restrain them in

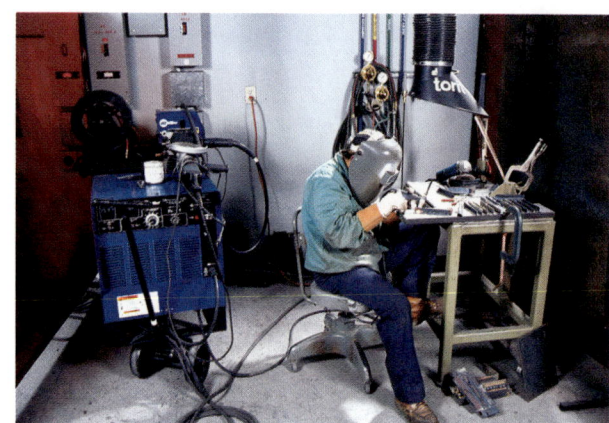

Figure 2.4—Designated Welding Area

case of a fall. Unexpected events, such as fume releases, fire and explosions, do occur in industrial environments. All escape routes should be identified and kept clear so that orderly, rapid, and safe evacuation of an area can take place. Employees must be trained in evacuation procedures. Storage of goods and equipment in evacuation routes must be avoided. If an evacuation route must be temporarily blocked, employees who would normally use that route must be trained to use an alternate route.

Equipment, machines, cables, hoses, and other apparatus should always be placed so that they do not present a hazard to personnel in passageways, on ladders, or on stairways. Warning signs should be posted to identify welding areas, and to specify that eye protection must be worn. Occasionally, a "fire watch" person must be assigned to maintain safety during welding or cutting operations.

Personnel in areas next to welding and cutting must also be protected from radiant energy and hot spatter. This is accomplished with flame-resistant screens or shields, or suitable eye and face protection and protective clothing. Appropriate radiation-protective, semi-transparent materials are permissible. Where operations allow, work stations should be separated by noncombustible screens or shields (see Figure 2.5). Booths and screens should allow circulation of air at floor level and above the screen.

Where arc welding or cutting is regularly performed next to painted walls, the walls should be painted with a finish having low reflectivity of ultraviolet radiation. Paint finish formulated with certain pigments, such as titanium dioxide or zinc

Figure 2.5—Protective Screening Between Workstations

oxide, have low reflectivity to ultraviolet radiation. Color pigments may be added if they do not increase reflectivity. Pigments based on powdered or flaked metals are not recommended because they reflect ultraviolet radiation.

In most welding, cutting, and allied processes, a high-temperature heat source is present. Open flames, electric arcs, hot metal, sparks, and spatter are ready sources of ignition. Many fires are started by sparks, which can travel horizontally up to 35 feet from their source. Sparks can pass through or lodge in cracks, holes, and other small openings in floors and walls.

The risk of fire is increased by combustibles in the work area, or by welding or cutting too close to combustibles that have not been shielded. Materials most commonly ignited are combustible floors, roofs, partitions, and building contents including trash, wood, paper, textiles, plastics, chemicals, and flammable liquids and gases. Outdoors, the most common combustibles are dry grass and brush.

The best protection against fire is to do welding and cutting in specially designated areas or enclosures of noncombustible construction kept free of combustibles. Combustibles should always be removed from the work area or shielded from the operation.

Common combustibles found in welding manufacturing include fuels for both equipment engines and welding or cutting operations. These fuels should be stored and used with care. Equipment manufacturers' instructions should be followed because fuels and their vapors are combustible and can be explosive under some conditions. Acetylene, propane and other flammable gases used in cutting and welding areas require careful handling. Special attention should be given to fuel gas cylinders, hoses, and apparatus to prevent gas leakage.

Combustibles that cannot be removed from the area should be covered with tight fitting, flame-resistant material. These include combustible walls and ceilings. Floors around the work area should be free of combustible materials for a radius of 35 feet. All doorways, windows, cracks, and other openings should be covered with a flame-resistant material. If possible, the work area should be enclosed with portable flame-resistant screens.

Combustibles on the other side of metal walls, ceilings or partitions must be moved to safe locations when welding or cutting is done on or next to the location. If this cannot be done, a fire watch should be stationed near the combustibles. Welding heat can conduct through metal partitions and ignite combustibles on the opposite side. A thorough examination for evidence of fire should be made before leaving the work area. Fire inspection should be continued for at least 30 minutes after the operation is completed.

Welding or cutting should not be done on material having a combustible coating or internal structure, as in walls or ceilings. Hot scrap or slag must not be placed in containers holding combustible materials. Suitable fire extinguishers should always be available nearby, and the fire watch trained in their use.

Welding, brazing, or cutting should not be done on combustible floors or platforms that may readily be ignited by heat from the operation. Welders and inspectors must be alert for traveling vapors from flammable liquids. Vapors are often heavier than air. Vapors from flammable liquid storage areas can travel several hundred feet along floors and in depressions. Light vapors can travel along ceilings to adjacent rooms.

When welding, cutting or similar hot working operations are to be performed in areas not normally assigned for such operations, a "hot work permit" system should be used (see Figure 2.6). The purpose of the hot work permit system is to alert area supervisors to an extraordinary danger of fire that will exist at a particular time. The permit system should include a check list of safety precautions. A

PERMIT NO.______

For electric and acetylene burning and welding with portable equipment in all locations outside of shop.

Date ____________________

Time Started ____________ Finished ____________

Building ____________________

Dept. ____________ Floor ____________

Location on Floor ____________________

Nature of Job ____________________

Operator ____________________

Clock No. ____________________

All precautions have been taken to avoid any possible fire hazard, and permission is given for this work.

Signed ____________________
Foreman

Signed ____________________
Safety supervisor or plant superintendent

PERMIT NO. **0U812**

Date ____________________

Bldg ____________ Floor ______

Nature of Job ____________________

Operator ____________________

INSTRUCTIONS TO OPERATORS

This permit is good only for the location and time shown. Return the permit when work is completed.

PRECAUTIONS AGAINST FIRE

1. Permits should be signed by the foreman of the welder or cutter and by the safety supervisor or plant superintendent.
2. Obtain a written permit before using portable cutting or welding equipment anywhere in the plant except in permanent safe-guarded locations.
3. Make sure sprinkler system is in service.
4. Before starting, sweep floor clean, wet down wooden floors, or cover them with sheet metal or equivalent. In outside work, don't let sparks enter doors or windows.
5. Move combustible material 25 feet away. Cover what can't be moved with asbestos curtain or sheet metal, carefully and completely.
6. Obtain standby fire extinguishers and locate at work site. Instruct helper or fire watcher to extinguish small fires.
7. After completion, watch scene of work a half hour for smoldering fires, and inspect adjoining rooms and floors above and below.
8. Don't use the equipment near flammable liquids, or on closed tanks which have held flammable liquids or other combustibles. Remove inside deposits before working on ducts.
9. Keep cutting and welding equipment in good condition. Carefully follow manufacturer's instructions for its use and maintenance.

Figure 2.6—National Safety Council "Hot Work Permit"

checklist often includes fire extinguisher inspection, establishes the fire watches if necessary, a flammable material search, and area safety instructions for personnel not involved in the hot work. When a hot work permit is issued, the welding inspector must be aware of and adhere to all its requirements.

Flammable gases, vapors, and dust mixed with certain proportions of air or oxygen present explosion and fire dangers. To prevent explosions, avoid all sources of ignition. Welding, brazing, soldering, cutting, or operating equipment that can produce heat or sparks must not be done in atmospheres containing flammable gases, vapors, or dusts. Such flammables must be kept in leak-tight containers or be well removed from the work area. Heat or sparks may cause otherwise low-volatility materials to produce flammable vapors.

Hollow containers must be vented before, and during, any application of heat. Heat must not be applied to a container that has held an unknown material, a combustible substance or a substance that may form flammable vapors without considering the potential hazards. Such containers must first be thoroughly cleaned or filled with an inert gas. Adequate eye and body protection must be worn if the operation involves explosion risks. Burns of the eye or body are serious hazards in the welding industry. Eye, face, and body protection for the operator and others in the work area are required to prevent burns from ultraviolet and infrared radiation, sparks, and spatter.

Eye and Face Protection

Arc Welding and Cutting

Welding helmets or handshields containing appropriate filter plates and cover plates must be used by welders and welding operators and nearby personnel when viewing an arc. Standards for welding helmets, handshields, face shields, goggles, and spectacles are given in ANSI Publication Z87.1, *Practice for Occupational and Educational Eye and Face Protection,* latest edition.

Safety spectacles, goggles, or other suitable eye protection must also be worn during other welding and cutting operations (see Figure 2.7). Such devices must have full conforming side shields when there is danger of exposure to injurious rays or flying particles from grinding or chipping operations. Spectacles and goggles may have clear or colored lenses. Shading depends on the intensity of the radiation that comes from adjacent welding or cutting operations when the welding helmet is raised or removed. Number-2 filter plates are recommended for general purpose protection (see Table 2.1).

Oxyfuel Gas Welding and Cutting, Submerged Arc Welding

Safety goggles with filter plates and full conforming side shields must be worn while performing oxyfuel gas welding and cutting (see Table 2.1). During submerged arc welding, the arc is covered by flux and is not readily visible; therefore, an arc welding helmet is not needed. However, because the arc occasionally flashes through the flux burden, the operator should wear tinted safety glasses.

Torch Brazing and Soldering

Safety spectacles with side shields and appropriate filter plates are recommended for torch brazing and soldering. As with oxyfuel gas welding and cutting, a bright yellow flame may be visible during torch brazing. A filter similar to that used with those processes should be used for torch brazing (see Table 2.1).

Other Brazing Processes and Resistance Welding

Operators and helpers engaged in these processes must wear safety spectacles, goggles, and a face shield to protect their eyes and face from spatter. Filter plates are not necessary but may be used for comfort (refer to Table 2.1).

Protective Clothing

Sturdy shoes or boots, and heavy clothing should be worn to protect the whole body from flying sparks, spatter, and radiation burns. Woolen clothing is preferable to cotton because it is not so readily ignited. Cotton clothing, if used, should be chemically treated to reduce its combustibility. Clothing treated with nondurable flame retardants must be treated again after each washing or cleaning. Clothing or shoes of synthetic or plastic materials, which can melt and cause severe burns, should not be worn. Outer clothing should be kept free of oil and grease, especially in an oxygen-rich atmosphere.

Cuffless pants and covered pockets are recommended to avoid spatter or spark entrapment. Pockets should be emptied of flammable or readily ignitable material before welding because they may be ignited by sparks or weld spatter and result in severe burns. Pants should be worn outside shoes. Protection of the hair with a cap is recommended, especially if a hairpiece is worn. Flammable hair preparations should not be used.

Durable gloves of leather or other suitable material should always be worn. Gloves not only protect the hands from burns and abrasion, but also provide insulation from electrical shock. A variety of special protective clothing is also available for welders. Aprons, leggings, suits, capes, sleeves, and caps, all of durable materials, should be worn when welding overhead or when special circumstances warrant additional protection of the body.

Sparks or hot spatter in the ears can be particularly painful and serious. Properly fitted, flame-resistant ear plugs should be worn whenever operations pose such risks.

Figure 2.7—Eye, Ear, and Face Protective Equipment

Key Words—Eye protection and lens shade AWS F2.2

Table 2.1—Lens Shade Selector
Shade numbers are given as a guide only and may be varied to suit individual needs

Operation	Electrode Size, mm (1/32 in.)	Arc Current (Amperes)	Minimum Protective Shade	Suggested[1] Shade No. (Comfort)
Shielded Metal Arc Welding (SMAW)	less than 2.5 (3)	less than 60	7	—
	2.5–4 (3–5)	60–160	8	10
	4–6.4 (5–8)	>160–250	10	12
	more than 6.4 (8)	>250–500	11	14
Gas Metal Arc Welding and Flux Cored Arc Welding (GMAW and FCAW)		less than 60	7	—
		60–160	10	11
		>160–250	10	12
		>250–500	10	14
Gas Tungsten Arc Welding (GTAW)		Less than 50	8	10
		50–100	8	12
		>150–250	10	14
Air Carbon Arc Cutting (light) (CAC-A) (heavy)		less than 500	10	12
		500–1000	11	14
Plasma Arc Welding (PAW)		less than 20	6	6 to 8
		20–100	8	10
		>100–400	10	12
		>400–800	11	14
Plasma Arc Cutting (PAC)				
Light[2]		less than 300	8	9
Medium		300–400	9	12
Heavy		>400–800	10	14
Torch Brazing (TB)		—	—	3 or 4
Torch Soldering (TS)		—	—	2
Carbon Arc Welding (CAW)		—	—	14

	Plate Thickness		
	mm	in.	
Gas Welding (GW)			
Light[2]	under 3.2	under 1/8	4 or 5
Medium	3.2 to 13	1/8 to 1/2	5 or 6
Heavy	over 13	over 1/2	6 to 8
Oxygen Cutting (OC)			
Light[2]	under 25	under 1	3 or 4
Medium	25 to 150	1 to 6	4 or 5
Heavy	over 150	over 6	5 or 6

1. Shade numbers are given as a rule of thumb. It is recommended to begin with a shade that is too dark to see the weld zone. Then one should go to a lighter shade which gives sufficient view of the weld zone without going below the minimum. In gas welding or oxygen cutting where the torch produces a high yellow light, it is desirable to use a filter lens that adsorbs the yellow or sodium line in the visible light (spectrum) of the operation.
2. These values apply where the actual arc is clearly seen. Experience has shown that lighter filters may be used when the arc is hidden by the workpiece.

Noise

Excessive noise, particularly continuous noise at high levels, can severely damage hearing. It may cause either temporary or permanent hearing loss. U.S. Department of Labor Occupational Safety and Health Administration regulations describe allowable noise exposure levels. Requirements of these regulations may be found in *General Industry Standards, 29 CFR 1910.95*.

In welding, cutting, and allied operations, noise may be generated by the process or the equipment, or both. Hearing protection devices are required for some operations (see Figure 2.7). Additional information is presented in *Arc Welding and Cutting Noise,* American Welding Society, 1979. Air Carbon Arc and Plasma Arc Cutting are processes that have very high noise levels. Engine-driven generators sometimes emit a high noise level, as do some high-frequency, and induction welding power sources.

Machinery Guards

Welders and other workers must also be protected from injury by machinery and equipment they are operating or by other machinery operating in the work area. Moving components and drive belts must be covered by guards to prevent physical contact (see Figure 2.8).

Because welding helmets and dark filter plates restrict the visibility of welders, these people may be even more susceptible than ordinary workers to injury from unseen, unguarded machinery.

Figure 2.8—Machinery Guard

Therefore, special attention to this hazard is required.

When repairing machinery by welding or brazing, the power to the machinery must be disconnected, locked out, tried, and tagged to prevent inadvertent operation and injury. Welders assigned to work on equipment with safety devices removed should fully understand the hazards involved, and the steps to be taken to avoid injury. When the work is completed, the safety devices must be replaced. Rotating and automatic welding machines, fixtures, and welding robots must be equipped with appropriate guards or sensing devices to prevent operation when someone is in the danger area.

Pinch points on welding and other mechanical equipment can also result in serious injury. Examples include resistance welding machines, robots, automatic arc welding machines, jigs, and fixtures. To avoid injury with such equipment, the machine should be equipped so that both of the operator's hands must be at safe locations when the machine is actuated. Otherwise, the pinch points must be suitably guarded mechanically. Metalworking equipment should not be located where a welder could accidentally fall into or against it while welding. During maintenance of the equipment, pinch points should be blocked to prevent them from closing in case of equipment failure. In very hazardous situations, an observer should be stationed to prevent someone from turning the power on until the repair is completed.

Fumes and Gases

Welders, welding operators, and other persons in the area must be protected from over-exposure to fumes and gases produced during welding, brazing, soldering, and cutting. Overexposure is exposure that is hazardous to health, or exceeds the permissible limits specified by a government agency. The *U.S. Department of Labor, Occupational Safety and Health Administration (OSHA), Regulations 29 CFR 1910.1000,* covers this topic. Also, the American Conference of Governmental Industrial Hygienists (ACGIH) lists guidelines in their publication, *Threshold Limit Values for Chemical Substances and Physical Agents in the Workroom Environment*. Persons with special health problems may have unusual sensitivity that requires even more stringent protection.

Fumes and gases are usually a greater concern in arc welding than in oxyfuel gas welding, cutting, or brazing. A welding arc may generate a larger volume of fume and gas, and a greater variety of materials are usually involved. Protection from excess exposure is usually accomplished by ventilation. Where exposure would exceed permissible limits with available ventilation, suitable respiratory protection must be used. Protection must be provided for welding, cutting, and other personnel in the area.

Exposure Factors

Position of the Head

The single most important factor influencing exposure to fumes is the position of the welder's head with respect to the fumes plume. When the head is in such a position that the fumes envelop the face or helmet, exposure levels can be very high. Therefore, welders must be trained to keep their head to one side of the fume plume. Sometimes, the work can be positioned so the fume plume rises to one side.

Types of Ventilation

Ventilation has a significant influence on fume amounts in the work area, and the welder's exposure to them. Ventilation may be local, where the fumes are extracted near the point of welding (see Figure 2.9), or general, where the shop air is changed or filtered. The appropriate type will depend on the welding process, the material being welded, and other shop conditions. Adequate ventilation is necessary to keep the welder's exposure to fumes and gases within safe limits.

Figure 2.9—Movable Fume Extractor Positioned Near the Welding Arc

Work Area

The size of the welding or cutting enclosure is important. It affects the background fume level. Fume exposure inside a tank, pressure vessel, or other confined space will be higher than in a high-bay fabrication area.

Background Fume Level

Background fume levels depend on the number and type of welding stations and the duty cycle for each power source.

Design of Welding Helmet

The extent a helmet curves under the chin toward the chest affects the amount of fume exposure. Close-fitting helmets can be effective in reducing exposure.

Base Metal and Surface Condition

The type of base metal being welded influences fume components and the amount generated. Surface contaminants or coatings may contribute significantly to the hazard potential of the fume. Paints containing lead, and platings containing cadmium, generate dangerous fumes during welding and cutting. Galvanized material creates zinc fumes which are harmful.

Ventilation

The bulk of fumes generated during welding and cutting consists of small particles that remain suspended in the atmosphere for a considerable time. As a result, fume concentration in a closed area can build up over time, as can the concentration of any gases evolved or used in the process. Particles eventually settle on the walls and floor, but the settling rate is low compared to the generation rate of the welding or cutting processes. Therefore, fume concentration must be controlled by ventilation.

Adequate ventilation is the key to control of fumes and gases in the welding environment. Natural, mechanical, or respirator ventilation must be provided for all welding, cutting, brazing, and related operations. The ventilation must ensure that concentrations of hazardous airborne contaminants are maintained below recommended levels.

Many ventilation methods are available. They range from natural drafts to localized devices, such as air-ventilated welding helmets. Examples of ventilation include:

(1) Natural
(2) General area mechanical ventilation
(3) Overhead exhaust hoods
(4) Portable local exhaust devices
(5) Downdraft tables
(6) Crossdraft tables
(7) Extractors built into the welding equipment
(8) Air-ventilated helmets

Welding in Confined Spaces

Special consideration must be given to the safety and health of welders and other workers in confined spaces. See ANSI Publication Z117.1, *Safety Requirements for Working in Tanks and Other Confined Spaces,* latest edition, for further precautions. Gas cylinders must be located outside the confined space to avoid possible contamination of the space with leaking gases or volatiles. Welding power sources should also be located outside to reduce danger of engine exhaust and electric shock. Lighting inside the work area should be low voltage, 12 V, or if 110 V is required, the circuit must be protected by an approved Ground-Fault Circuit-Interrupter (GFCI).

A means for removing persons quickly in case of emergency has to be provided. Safety belts and lifelines, when used, should be attached to the worker's body in a way that avoids the possibility of the person becoming jammed in the exit. A trained helper, a "standby," should be stationed outside the confined space with a preplanned rescue procedure in case of an emergency (including not entering the confined space to aid the first worker without proper breathing apparatus).

Besides keeping airborne contaminants in breathing atmospheres at or below recommended limits, ventilation in confined spaces must also (1) assure adequate oxygen for life support (at least 19.5% by volume) (2) prevent accumulation of an oxygen-enriched atmosphere, (i.e., not over 23.5% by volume) and (3) prevent accumulation of flammable mixtures (see Figure 2.10). Asphyxiation can quickly result in unconsciousness and death without warning if oxygen is not present in sufficient concentration to support life. Air contains approximately 21% oxygen by volume. Confined spaces must not be entered unless well ventilated, or the inspector is

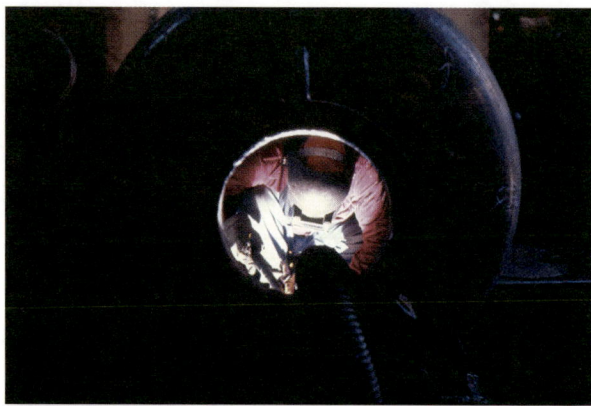
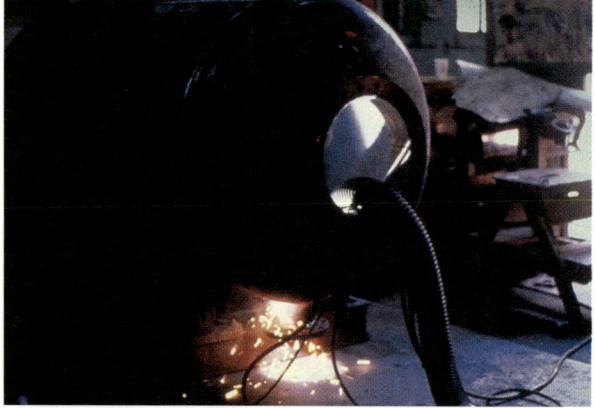

Figure 2.10—Welding in Confined Spaces

wearing an approved air supplied breathing apparatus and has proper training to work in such spaces. A similarly equipped second person must be present as a standby.

Before entering confined spaces, the space should be tested for toxic or flammable gases and vapors, and adequate or excess oxygen. The tests should be made with instruments approved by the U.S. Bureau of Mines. Heavier-than-air gases, such as argon, methylacetylene-propadiene, propane, and carbon dioxide, may accumulate in pits, tank bottoms, low areas, and near floors. Lighter-than-air gases, such as helium and hydrogen, may accumulate in tank tops, high areas, and near ceilings. The precautions for confined spaces also apply to those areas. If practical, a continuous monitoring system with audible alarms should be used for work in a confined space.

Oxygen-enriched atmospheres pose great danger to occupants of confined areas. They are especially hazardous at oxygen concentrations above 25 percent. Materials that burn normally in air may flare up violently in an oxygen-enriched atmosphere. Clothing may burn fiercely; oil or grease soaked clothing or rags may catch fire spontaneously; paper may flare into flame. Very severe and fatal burns can result.

Protection in confined spaces must be provided welders and other personnel in the enclosure. Only clean, respirable air must be used for ventilation. Oxygen, other gases, or mixtures of gases must never be used for ventilation.

Positive pressure self-contained breathing apparatus must be used when welding or cutting related processes are done in confined areas where proper ventilation cannot be provided and there is immediate danger to life and health. It must have an emergency air supply of at least five minutes duration in the event that the main source fails.

Welding of Containers

Welding or cutting outside or inside containers and vessels that have held dangerous substances presents special hazards. Flammable or toxic vapors may be present, or may be generated by the applied heat. The immediate area outside and inside the container should be cleared of all obstacles and hazardous materials. If repairing a container in place, entry of hazardous substances into the container from the outside must be avoided. The required personal and fire protection equipment must be available, serviceable, and in position for immediate use.

When welding or cutting inside vessels that have held dangerous materials, the precautions for confined spaces must also be observed. Gases generated during welding should be discharged in a safe and environmentally acceptable manner according to government rules and regulations. Provisions must be made to prevent pressure buildup inside containers. Testing for gases, fumes, and vapors should be conducted periodically to ensure that recommended limits are maintained during welding.

An alternative method of providing safe welding of containers is to fill them with an inert medium such as water, inert gas, or sand. When using water, the level should be kept to within a few inches of the point where the welding is to be done. The space above the water should be vented to allow the heated air to escape. With inert gas, the percentage of inert gas that must be present in the tank to prevent fire or explosion must be known. How to safely produce and maintain a safe atmosphere during welding must also be known.

Highly Toxic Materials CADMIUM!

Certain materials, which are sometimes present in consumables, base metals, coatings, or atmospheres for welding or cutting operations, have permissible exposure limits of 1.0 mg/m^3 or less. Among these materials are the metals noted in Table 2.2.

Manufacturer's Material Safety Data Sheets should be consulted to find out if any of these materials are present in welding filler metals and fluxes being used. Material Safety Data Sheets should be requested from suppliers. However, welding filler metals and fluxes are not the only source of these materials. They may also be present in base metals, coatings, or other sources in the work area. Radioactive materials under Nuclear Regulatory Commission jurisdiction require special considerations and may also require compliance with state and local regulations. These materials also include x-ray machines and radiographic isotopes.

When toxic materials are encountered as designated constituents in welding, brazing, or cutting operations, special ventilation precautions must be taken. The precautions assure that the levels of these contaminants in the atmosphere are at or below the limits allowed for human exposure. All persons in the immediate vicinity of welding or cutting operations involving these materials must be similarly protected.

Handling of Compressed Gases

Gases used in welding and cutting operations are packaged in containers called cylinders. Only cylinders designed and maintained in accordance with U.S. Department of Transportation (DOT) specifications may be used in the United States. The use of other cylinders may be extremely dangerous and is illegal. Cylinders requiring periodic retest under DOT regulations may not be filled unless the retest is current.

Cylinders may be filled only with the permission of the owner, and should be filled only by recognized gas suppliers or those with the proper training and facilities to do so. Filling one cylinder from another is dangerous and should not be attempted by anyone not qualified to do so. Combustible or incompatible combinations of gases must never be mixed in cylinders.

Welding must not be performed on gas cylinders. Cylinders must not be allowed to become part of an electrical circuit because arcing may result. Cylinders containing shielding gases used in conjunction with arc welding must not be grounded. Electrode holders, welding torches, cables, hoses, and tools should not be stored on gas cylinders to avoid arcing or interference with valve operation. Arc-damaged gas cylinders may rupture and result in injury or death.

Cylinders must not be used as work rests or rollers. They should be protected from bumps, falling objects, weather, and should not be dropped. Cylinders should not be kept in passageways where they might be struck by vehicles. They should be kept in areas where temperatures do not fall below –20°F or exceed 130°F. Any of these exposures, misuses, or abuses could damage them to the extent that they might fail with serious consequences.

Cylinders must not be hoisted using ordinary slings or chains. A proper cradle or cradle sling that securely retains the cylinder should be used. Electromagnets should not be used to handle cylinders.

Cylinders must always be secured by the user against falling during either use or storage (see Figure 2.11). Acetylene and liquefied gas cylinders (dewars) should always be stored and used in the upright position. Other cylinders are preferably stored and used in the upright position, but this is not essential in all circumstances.

Before using gas from a cylinder, the contents should be identified by the label thereon. Contents should not be identified by any other means such as cylinder color, banding, or shape. These may vary

Table 2.2—Toxic Metals

(1)	Antimony	(9)	Lead
(2)	Arsenic	(10)	Manganese
(3)	Barium	(11)	Mercury
(4)	Beryllium	(12)	Nickel
(5)	Cadmium	(13)	Selenium
(6)	Chromium	(14)	Silver
(7)	Cobalt	(15)	Vanadium
(8)	Copper		

Figure 2.11—Inert Gas Cylinders, Attached to Manifold System

Handbook of Compressed Gases published by the Compressed Gas Association.

Many gases in high-pressure cylinders are filled to pressures of 2000 psi or more. Unless the equipment to be used with a gas is designed to operate at full cylinder pressure, an approved pressure-reducing regulator must be used to withdraw gas from a cylinder or manifold. Simple needle valves should never be used. A pressure-relief or safety valve, rated to function at less than the maximum allowable pressure of the welding equipment, should also be employed. Valve functions prevent equipment failure at pressures over working limits if the regulator should fail in service.

Valves on cylinders containing high pressure gas, particularly oxygen, should always be opened slowly to avoid the high temperature of adiabatic recompression. Adiabatic recompression can occur if the valves are opened rapidly. With oxygen, the heat can ignite the valve seat that, in turn, may cause the metal to melt or burn. The cylinder valve outlet should point away from the operator and other persons when opening the valve to avoid injury should a fire occur. The operator should never stand in front of the regulator when opening a cylinder to avoid injury from high pressure release if the regulator fails.

Before connecting a gas cylinder to a pressure regulator or a manifold, the valve outlet should be cleaned. The valve outlet should be wiped clean with a clean, oil-free cloth to remove dirt, moisture, and other foreign matter. Then the valve should be opened momentarily and closed immediately. This is known as "cracking the cylinder valve." Fuel gas cylinders must never be cracked near sources of ignition (i.e., sparks and flames), while smoking, nor in confined spaces.

A regulator should be relieved of gas pressure before connecting it to a cylinder, and also after closing the cylinder valve upon shutdown of operation. The outlet threads on cylinder valves are standardized for specific gases so that only regulators or manifolds with similar threads can be attached (e.g., flammable gas cylinders typically have a left-hand thread while non-flammable gas cylinders have a right-hand thread).

It is preferable not to open valves on low pressure, fuel gas cylinders more than one turn. This usually provides adequate flow and allows rapid

among manufacturers, geographical area, or product line and could be completely misleading. The label on the cylinder is the only proper notice of the contents. If a label is not on a cylinder, the contents should not be used and the cylinder should be returned to the supplier.

A valve protection cap is provided on many cylinders to protect the safety device and the cylinder valve. This cap should always be in place unless the cylinder is in use. The cylinder should never be lifted manually or hoisted by the valve protection cap. The threads that secure these valve protection caps are intended only for that purpose, and may not support full cylinder weight. The caps should always be threaded completely onto the cylinders and hand tightened.

Gas cylinders and other containers must be stored in accordance with all state and local regulations and the appropriate standards of OSHA and the National Fire Protection Association. Safe handling and storage procedures are discussed in the

closure of the cylinder valve in an emergency. High pressure cylinder valves, on the other hand, usually must be opened fully to backseat (seal) the valve to prevent leaks during use.

The cylinder valve should be closed after each use of a cylinder and when an empty cylinder is to be returned to the supplier. This prevents loss of product through leaks that might develop and go undetected while the cylinder is unattended, and avoids hazards that might be caused by leaks. It also prevents backflow of contaminants into the cylinder. It is advisable to return cylinders to the supplier with about 25 psi of contents remaining. This prevents possible contamination by the atmosphere during shipment.

Pressure Relief Devices

Only trained personnel should be allowed to adjust pressure relief devices on cylinders. These devices are intended to provide protection in the event the cylinder is subjected to a hostile environment, usually fire or other source of heat. Such environments may raise the pressure within cylinders. To prevent cylinder pressures from exceeding safe limits, the safety devices are designed to relieve the contents.

A pressure reducing regulator should always be used when withdrawing gas from gas cylinders for welding or cutting operations. Pressure reducing regulators must be used only for the gas and pressure given on the label. They should not be used with other gases or at other pressures although the cylinder valve outlet threads may be the same. The threaded connections to the regulator must not be forced. Improper fit of threads between a gas cylinder and regulator, or between the regulator and hose suggests an improper combination of devices being used.

Use of adapters to change the cylinder connection thread is not recommended because of the danger of using an incorrect regulator or of contaminating the regulator. For example, gases that are oil-contaminated can deposit an oily film on the internal parts of the regulator. This film can contaminate oil-free gas and result in fire or explosion when exposed to pure oxygen.

The threads and connection glands of regulators should be inspected before use for dirt and damage. If a hose or cylinder connection leaks, it should not be forced with excessive torque. Damaged regulators and components should be repaired by properly trained mechanics or returned to the manufacturer for repair.

A suitable valve or flowmeter should be used to control gas flow from a regulator (see Figure 2.12). The internal pressure in a regulator should be released before it is connected to or removed from a gas cylinder or manifold.

Manifolds

A manifold is used when gas is needed without interruption or at a higher delivery rate than can be supplied from a single cylinder. A manifold must be designed for the specific gas and operating pressure, and be leak tight (see Figures 2.13 and 2.14). The manifold components should be approved for

Figure 2.12—Acetylene and Oxygen Regulators and Inert Gas Flowmeters

Figure 2.13—Acetylene Manifold System

Figure 2.14—Oxygen Manifold System

such purpose, and should be used only for the gas and pressure for which they are approved. Oxygen and fuel gas manifolds must meet specific design and safety requirements.

Piping and fittings for acetylene and methylacetylene-propadiene (MPS) manifolds must not be unalloyed copper or alloys containing 70% or more copper. These fuel gases react with copper under certain conditions to form unstable copper acetylide. This compound may detonate under shock or heat.

Manifold piping systems must contain an appropriate overpressure relief valve. Each fuel gas cylinder branch line should incorporate a backflow check valve and flash arrester. Backflow check valves must also be installed in each line at each station outlet where both fuel gas and oxygen are provided for a welding, cutting, or preheating torch. These check valves must be checked periodically for safe operation.

Unless it is known that a piping system is specifically designed and constructed to withstand full cylinder pressure or tank pressure of the compressed gas source supplying it, the piping system must be protected with safety pressure relief devices. The devices must be sufficient to prevent development of pressure in the system beyond the capacity of the weakest element.

Such pressure relief devices may be relief valves or bursting discs. A pressure reducing regulator must never be solely relied upon to prevent over pressurization of the system. A pressure relief device must be located in every section of the system that could be exposed to the full source supply pressure while isolated from other protective relief devices (such as by a closed valve). Some pressure regulators have integral safety relief valves. These valves are designed for the protection of the regulator only, and should not be relied upon to protect the downstream system.

In cryogenic piping systems, relief devices should be located in every section of the system where liquefied gas may become trapped. Upon warming, such liquids vaporize to gas, and in a confined space, the gas pressure can increase dramatically. Pressure relief devices protecting fuel gas piping systems or other hazardous gas systems should be vented to safe locations.

Gases

Oxygen

Oxygen is nonflammable but it supports the combustion of flammable materials. It can initiate combustion and vigorously accelerate it. Therefore, oxygen cylinders and liquid oxygen containers should not be stored near combustibles or with cylinders of fuel gas. Oxygen should never be used as a substitute for compressed air. Pure oxygen supports combustion more vigorously than air, which contains only 21% oxygen. Therefore, the identification of the oxygen and air should be differentiated.

Oil, grease, and combustible dusts may spontaneously ignite on contact with oxygen. All systems and apparatus for oxygen service must be kept free of any combustibles. Valves, piping, or system components that have not been expressly manufactured for oxygen service must be cleaned and approved for this service before use.

Apparatus expressly manufactured for oxygen service, and so labeled, must be kept in the clean condition as originally received. Oxygen valves, regulators, and apparatus should never be lubricated with oil. If lubrication is required, the type of lubricant and the method of applying the lubricant should be specified in the manufacturer's literature. If it is not, then the device should be returned to the manufacturer or authorized representative for service.

Oxygen must never be used to power compressed air tools. These are usually oil lubricated. Similarly, oxygen must not be used to blow dirt from work and clothing because they are often contaminated with oil, or grease, or combustible dust.

Only clean clothing should be worn when working with oxygen systems. Oxygen must not be used to ventilate confined spaces. Severe burns may result from ignition of clothing or hair in an oxygen-rich atmosphere.

Fuel Gases

Fuel gases commonly used in oxyfuel gas welding (OFW) and cutting (OFC) are acetylene, methylacetylene-propadiene (MPS), natural gas, propane, and propylene. Hydrogen is used in a few applications. Gasoline is sometimes used as fuel for oxygen cutting (it vaporizes in the torch). These gases should always be referred to by name.

Acetylene in cylinders is dissolved in a solvent (such as acetone) so that it can be safely stored under pressure. In the free state, acetylene should never be used at pressures over 15 psi (103 kPa) because it can dissociate with explosive violence at higher pressures.

Acetylene and MPS should never be used in contact with silver, mercury, or alloys containing 70 percent or more copper. These gases react with these metals to form unstable compounds that may detonate under shock or heat. Valves on fuel gas cylinders should never be opened to clean the valve outlet near possible sources of flame ignition, or in confined spaces.

When fuel gases are used for a brazing furnace atmosphere, they must be vented to a safe location. Before filling a furnace with fuel gas, the equipment must first be purged with a nonflammable gas. Nitrogen or argon can be used to prevent formation of an explosive air-fuel mixture.

Special attention must be given when using hydrogen. Flames of hydrogen may be difficult to see and parts of the body, clothes, or combustibles may, therefore, unknowingly come in contact with hydrogen flames.

Fuel Gas Fires

The best procedure for avoiding fire from a fuel gas or liquid is to keep it contained within the system, that is, prevent leaks. All fuel systems should be checked carefully for leaks upon assembly and at frequent intervals after that. Fuel gas cylinders should be examined for leaks, especially at fuse plugs, safety devices, and valve packing. One common source of fire in welding and cutting is ignition of leaking fuel by flying sparks or spatter.

In case of a fuel fire, an effective means for controlling the fire is to shut off the fuel valve, if accessible. A fuel gas valve should not be opened beyond the point necessary to provide adequate flow. Opened in this way, it can be shut off quickly in an emergency. Usually, this is less than one turn of the handle. If the immediate valve controlling the burning gas is inaccessible, another upstream valve may cut off the flow of gas.

Most fuel gases in cylinders are in liquid form or dissolved in liquids. Therefore, the cylinders should always be used in the upright position to prevent liquid surges into the system.

A fuel gas cylinder can develop a leak and sometimes result in a fire. In case of fire, the fire alarm should be sounded, and trained fire personnel should be summoned immediately. A small fire near a cylinder valve or a safety device should be extinguished. When possible, extinguish the fire by closing the valve, using water, wet cloths, or fire extinguishers. If the leak cannot be stopped, after the fire is extinguished, the cylinder should be removed by trained fire personnel to a safe outdoor location, and the supplier notified. A warning sign should be posted, and no smoking or other ignition sources should be allowed in the area.

With a large fire at a fuel gas cylinder, the fire alarm should be actuated, and all personnel should be evacuated from the area. The cylinder should be kept wet by fire personnel with a heavy stream of water to keep it cool. It is usually better to allow the fire to continue to burn and consume the issuing gas rather than attempt to extinguish the flame. If the fire is extinguished, there is danger that the escaping gas may reignite with explosive violence.

Shielding Gases

Argon, helium, carbon dioxide (CO_2), and nitrogen are used for shielding with some welding processes. All, except carbon dioxide, are used as brazing atmospheres. They are odorless and colorless and can displace air needed for breathing.

Confined spaces containing these gases must be well ventilated before personnel enter them. If there is any question about the space, it should be checked first for adequate oxygen concentration with an oxygen analyzer. If an analyzer is not available, an air-supplied respirator should be worn by anyone entering the space. Containers of these gases should not be placed in confined spaces, as discussed previously.

Electric Shock

Electric shock can cause sudden death. Injuries and fatalities from electric shock in welding and cutting operations can occur if proper precautionary measures are not followed. Most welding and cutting operations employ some type of electrical equipment. For example, automatic oxyfuel gas cutting machines use electric motor drives, controls and systems. Lightning-caused electrical accidents may not be avoidable. However, all others are avoidable, including those caused by lack of proper training.

Electric shock occurs when an electric current of sufficient amount to create an adverse effect passes through the body. The severity of the shock depends mainly on the amount of current, the duration of flow, the path of flow, and the state of health of the person. The current is caused to flow by the applied voltage. The amount of current depends upon the applied voltage and the resistance of the body path. The frequency of the current may also be a factor when alternating current is involved.

Shock currents greater than about 6 milliamperes (mA) are considered primary because they can cause direct physiological harm. Steady state currents between 0.5 and 6 mA are considered secondary shock currents. Secondary shock currents can cause involuntary muscular reactions without normally causing direct physiological harm. The 0.5 mA level is called the perception threshold because it is the point at which most people just begin to feel the tingle from the current. The level of current sensation varies with the weight of the individual and to some extent between men and women.

Most electrical equipment, if improperly installed, used, or maintained, can be a shock hazard. Shock can occur from lightning-induced voltage surges in power distribution systems. Even earth grounds can attain high potential relative to true ground during severe transient phenomenon. Such circumstances, however, are rare.

In welding and cutting work, most electrical equipment is powered from AC sources of between 115 and 575 V, or by engine-driven generators. Most welding is done with less than 100 arc volts. (Fatalities have resulted with equipment operating at less than 80 volts.) Some arc cutting methods operate at over 400 V, and electron beam welding machines at up to about 150 kV. Most electric shock in the welding industry occurs as the result of accidental contact with bare or poorly insulated conductors operating at such voltages. Therefore, welders must take precautions against contacting bare elements in the welding circuit, and also those in the primary circuits.

Electrical resistance is usually reduced in the presence of water or moisture. Electrical hazards are often more severe under such circumstances. When arc welding or cutting is to be done under damp or wet conditions including heavy perspiration, the inspector must wear dry gloves and clothing in good condition to prevent electric shock. The welding inspector should be protected from electrically conductive surfaces, including the earth. Protection can be afforded by rubber-soled shoes as a minimum, and preferably by an insulating layer such as a rubber mat or a dry wooden board. Similar precautions against accidental contact with bare conducting surfaces must be taken when the welding inspector is required to work in a cramped kneeling, sitting, or lying position. Rings and jewelry should be removed before welding to decrease the possibility of electric shock.

The technology of heart pacemakers and the extent to which they are influenced by other electrical devices is constantly changing. It is impossible to make general statements concerning the possible effects of welding operations on such devices. Wearers of pacemakers or other electronic equipment vital to life should check with the device manufacturer or their doctor to find out whether any hazard exists.

Electric shock hazards are reduced by proper equipment installation and maintenance, good operator practice, proper clothing and body protection, and equipment designed for the job and situation. Equipment should meet applicable NEMA or ANSI standards, such as ANSI/UL 551, *Safety Standard for Transformer Type Arc Welding Machines.*

If significant amounts of welding and cutting are to be done under electrically hazardous conditions, automatic machine controls that safely reduce open circuit voltage are recommended. When special welding and cutting processes require open circuit voltages higher than those specified in ANSI/NEMA Publication EW-1, *Electrical Arc Welding Apparatus,* insulation and operating procedures that are adequate to protect the welder from these higher voltages must be provided.

A good safety training program is essential. Employees must be fully instructed in electrical safety by a competent person before being allowed to commence operations. As a minimum, this training should include the points covered in ANSI Z49.1, *Safety in Welding, Cutting, and Allied Processes* (published by the American Welding Society). Persons should not be allowed to operate electrical equipment until they have been properly trained.

Equipment should be installed in a clean, dry area. When this is not possible, it should be adequately guarded from dirt and moisture. Installation must be done to the requirements of ANSI/NFPA 70, *National Electric Code,* and local codes. This includes disconnects, fusing, and types of incoming power lines.

Terminals for welding leads and power cables must be shielded from accidental contact by personnel or by metal objects, such as vehicles and cranes. Connections between welding leads and power supplies may be guarded using (1) dead front construction and receptacles for plug connections, (2) terminals located in a recessed opening or under a nonremovable hinged cover, (3) insulating sleeves, or (4) other equivalent mechanical means.

The workpiece being welded and the frame or chassis of all electrically powered machines must be connected to a good electrical ground. Grounding can be done by locating the workpiece or machine on a grounded metal floor or platen. The ground can also be connected to a properly grounded building frame or other satisfactory ground. Chains, wire ropes, cranes, hoists, and elevators must not be used as grounding connectors nor to carry welding current.

The work lead is not the grounding lead. The work lead connects the work terminal on the power source to the workpiece. A separate lead is required to ground the workpiece or power source to earth ground.

Care should be taken when connecting the grounding circuit. Otherwise, the welding current may flow through a connection intended only for grounding, and may be of a higher amount than the grounding conductor can safely carry. Special radio-frequency grounding may be necessary for arc welding machines equipped with high-frequency arc initiating devices.

Connections for portable control devices, such as push buttons carried by the operator, must not be connected to circuits with operating voltages above 120 V. Exposed metal parts of portable control devices operating on circuits above 50 V must be grounded by a grounding conductor in the control cable. Controls using intrinsically safe voltages below 30 V are recommended.

Electrical connections must be tight and be checked periodically for tightness. Magnetic work clamps must be free of adherent metal particles and spatter on contact surfaces. Coiled welding leads should be spread out before use to avoid overheating and damage to the insulation. Jobs alternately requiring long and short leads should be equipped with insulated cable connectors so that idle lengths can be disconnected when not needed.

Equipment, cables, fuses, plugs, and receptacles must be used within their current-carrying and duty cycle capacities. Operation of apparatus above the current rating or the duty-cycle results in overheating and rapid deterioration of insulation and other parts. Actual welding current may be higher than that shown by indicators on the welding machine if welding is done with short leads or low voltage, or both. High currents are likely with general purpose welding machines when they are used with processes that use low arc voltage, such as gas tungsten arc welding.

Welding leads should be the flexible type of cable designed especially for the rigors of welding service. Insulation on cables used with high voltages or high-frequency oscillators must provide adequate protection. The recommendations and precautions of the cable manufacturer should always be followed. Cable insulation must be kept in good condition, and cables repaired or replaced promptly when necessary.

Welders should not allow the metal parts of electrodes, electrode holders, or torches to touch their bare skin or any wet covering of the body. Dry gloves in good condition must always be worn. The insulation on electrode holders must be kept in good repair. Electrode holders should not be cooled by immersion in water. If water-cooled welding guns or holders are used, they should be free of water leaks and condensation that would adversely affect the welder's safety. Welders should not drape or coil the welding leads around their bodies.

A welding circuit must be de-energized to avoid electric shock while the electrode, torch, or gun is being changed or adjusted. One exception concerns covered electrodes with shielded metal arc welding. When the circuit is energized, covered electrodes must be changed with dry welding gloves, not with bare hands. De-energization of the circuit is desirable for optimum safety even with covered electrodes.

When a welder has completed the work or has occasion to leave the work station for an appreciable time, the welding machine should be turned off. Similarly, when the machine is to be moved, the input power supply should be electrically disconnected at the source. When equipment is not in use, exposed electrodes should be removed from the holder to eliminate the danger of accidental electrical contact with persons or conducting objects. Also, welding guns of semiautomatic welding equipment should be placed so that the gun switch cannot be operated accidentally.

Fires resulting from electric welding equipment are generally caused by overheating of electrical components. Flying sparks or spatter from the welding or cutting operation, and mishandling fuel in engine driven equipment are among other causes. Most precautions against electrical shock are also applicable to the prevention of fires caused by overheating of equipment. Avoidance of fire from sparks and spatter was covered previously.

The fuel systems of engine driven equipment must be in good condition. Leaks must be repaired promptly. Engine driven machines must be turned off before refueling, and any fuel spills should be wiped up and fumes allowed to dissipate before the engine is restarted. Otherwise, the ignition system, electrical controls, spark producing components, or engine heat may start a fire.

Key Terms and Definitions

ACGIH—American Conference of Governmental and Industrial Hygienists. This group is concerned with the proper, safe levels of exposure to hazardous materials.

adiabatic recompression—the term given to the temperature rise that can occur when some gases at high pressures are released suddenly. (Normal pressure gas releases usually result in a cooling of the gas by the decompression.)

ANSI—American National Standards Institute. An organization promoting technical and safety standards.

ASC Z49.1—Safety in Welding, Cutting, and Allied Processes, a document outlining safe practices for welding and cutting operations.

ANSI Z87.1—*Practice for Occupational and Educational Eye and Face Protection.*

asphyxiation—loss of consciousness as a result of too little oxygen or too much carbon dioxide in the blood.

AWS—American Welding Society. AWS is the technical leader in welding and related issues.

combustibles—any material that can easily catch fire.

cryogenic—very cold service, usually well below zero degrees F.

DOT—Department of Transportation. A federal or state agency covering the transport of materials.

filter lens—in welding, a shaded lens, usually glass, that protects the eyes from radiation from the welding arc and other heat sources. Welding lenses are numbered, with the higher numbers offering the greatest protection. See Table 2.1, Lens Shade Selector, for appropriate lens selection.

fire watch—a person whose primary responsibility is to observe the work operation for the possibility of fires, and to alert the workers if a fire occurs.

flammable—anything that will burn easily or quickly. (Inflammable has the same meaning.)

fume plume—in welding, a smoke-like cloud containing minute solid particles arising directly from molten metal.

fuse plug—a plug filled with a material, usually a metal, that has a very low melting point. Often used as a heat and/or pressure relief device.

fume release—a general term given to the unexpected and undesired release of materials.

galvanized material—any material having a zinc coating on its surface. Common galvanized items are sheet metal and fasteners.

hot work permit—a form designed to insure that all safety precautions have been considered prior to any operation having open flames or high heat.

lock, tag, and try—the phrase noting the physical locking-out of equipment, tagging it for identification, and trying the equipment to make sure it is not operable prior to beginning any repair work.

MSDS—Material Safety Data Sheet. A document that identifies materials present in products that have hazardous or toxic properties.

NEMA—National Electrical Manufacturers Association.

OSHA—Occupational Safety and Health Act. This federal law outlines the requirements for safety in the workplace.

pascal (Pa)—in the metric system, the unit for pressure, or tensile strength. The U.S. customary equivalent is psi, pounds per square inch. One psi equals 6895 Pa.

pinch points—any equipment geometry that can lead to pinching parts of the body, especially the hands or feet, while working on the equipment.

safety glasses—spectacles with hardened and minimum thickness lenses that protect the eyes from flying objects. Improved eye protection occurs when side shields are attached to the safety glasses.

standby—in welding, a person trained and designated to stand by and watch for safety hazards, and to call for help if needed. Most often used for vessel entry safety.

TLV—Threshold Limit Value. The permissible level of exposure limits for hazardous materials.

toxic—poisonous.

vapors—the gaseous form of a substance.

Module 3
Metal Joining and Cutting Processes

Contents

Introduction	3-2
Welding Processes	3-4
Brazing Processes	3-35
Cutting Processes	3-38
Summary	3-44
Key Terms and Definitions	3-44

Module 3—Metal Joining and Cutting Processes

Introduction

Since the welding inspector is primarily concerned with welding, knowledge of the various joining and cutting processes can be very helpful. While it is not mandatory that the inspector be a qualified welder, any hands-on welding experience is beneficial. In fact, many welding inspectors are selected for that position after working as a welder for some time. History has shown that former welders often make good inspectors.

There are certain aspects of the various joining and cutting processes which the successful welding inspector must understand in order to perform most effectively. First, the inspector should realize the important advantages and limitations of each process. The inspector should also be aware of those discontinuities which may result when a particular process is used. Many discontinuities occur regardless of the process used; however, there are others which can occur during the application of a particular process. These will be discussed for each method and referred to as "possible problems."

The welding inspector should also have some knowledge of the equipment requirements for each process, because often discontinuities occur which are the result of equipment deficiencies. The inspector should be somewhat familiar with the various machine controls and what effect their adjustment will have on the resulting weld quality.

When the welding inspector has some understanding of these process fundamentals, he or she is better prepared to perform visual welding inspection. This knowledge will aid in the discovery of problems when they occur rather than later when the cost of correction is greater. The inspector who is capable of spotting problems in-process will be a definite asset to both production and quality control.

Another benefit of having experience with these welding methods is that the production welders will have greater respect for the inspector and resulting decisions. Also, a welder is more likely to bring some problem to the inspector's attention if he or she knows that the inspector understands the practical aspects of the process. Possessing this knowledge will help the inspector gain the cooperation of the welders and others involved with the fabrication operation.

The processes discussed here can be divided into three basic groups: welding, brazing and cutting. Welding and brazing describe methods for joining metals, while cutting results in the removal or separation of material. As each of the joining and cutting processes are discussed, there will be an attempt to describe their important features, including process advantages, process limitations, equipment requirements, electrodes/filler metals, techniques, applications, and possible process problems.

There are numerous joining and cutting processes available for use in the fabrication of metal products. These are shown by the American Welding Society's Master Chart of Welding and Allied Processes, shown in Figure 3.1. This chart separates the joining and cutting methods into various categories, namely, Welding Processes and Allied Processes. The Welding Processes are further divided into seven groups, Arc Welding, Solid-State Welding, Resistance Welding, Oxyfuel Gas Welding, Soldering, Brazing, and Other Welding. The Allied Processes include Thermal Spraying, Adhesive Bonding, and Thermal Cutting (Oxygen, Arc and Other Cutting).

With so many different processes available, it would be difficult to describe each one within the scope of this course. Therefore, the processes selected for discussion include only those which are applicable for the AWS Certified Welding Inspector

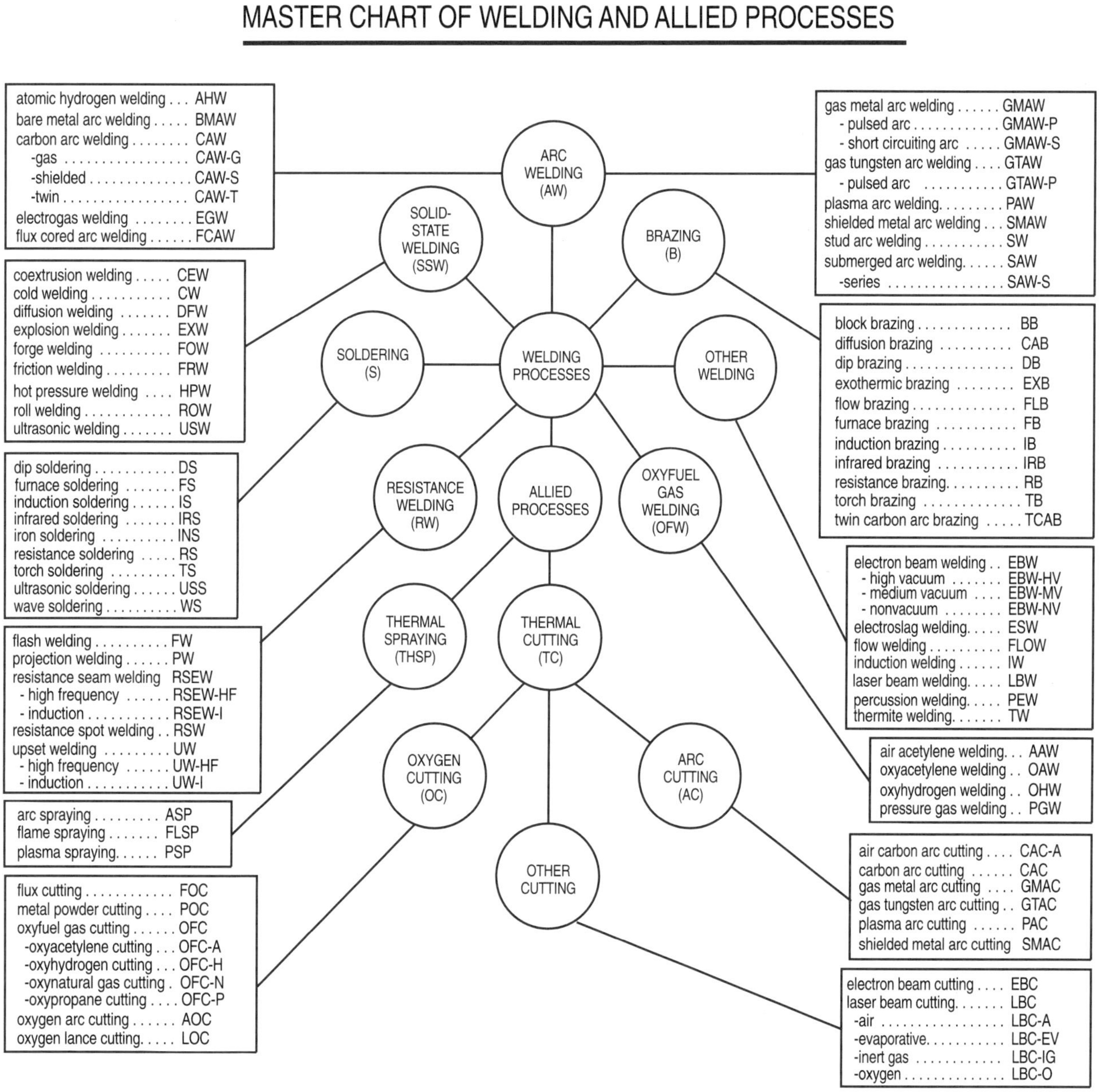

MASTER CHART OF WELDING AND ALLIED PROCESSES

Arc Welding (AW):
- atomic hydrogen welding ... AHW
- bare metal arc welding ..... BMAW
- carbon arc welding .... CAW
  - -gas .... CAW-G
  - -shielded .... CAW-S
  - -twin .... CAW-T
- electrogas welding .... EGW
- flux cored arc welding ..... FCAW
- gas metal arc welding ...... GMAW
  - pulsed arc .... GMAW-P
  - short circuiting arc ..... GMAW-S
- gas tungsten arc welding .... GTAW
  - pulsed arc .... GTAW-P
- plasma arc welding .... PAW
- shielded metal arc welding ... SMAW
- stud arc welding .... SW
- submerged arc welding ...... SAW
  - -series .... SAW-S

Solid-State Welding (SSW):
- coextrusion welding ..... CEW
- cold welding .... CW
- diffusion welding .... DFW
- explosion welding ...... EXW
- forge welding .... FOW
- friction welding .... FRW
- hot pressure welding .... HPW
- roll welding .... ROW
- ultrasonic welding .... USW

Soldering (S):
- dip soldering .... DS
- furnace soldering .... FS
- induction soldering ...... IS
- infrared soldering .... IRS
- iron soldering .... INS
- resistance soldering ..... RS
- torch soldering .... TS
- ultrasonic soldering ..... USS
- wave soldering .... WS

Resistance Welding (RW):
- flash welding .... FW
- projection welding ...... PW
- resistance seam welding RSEW
  - high frequency ...... RSEW-HF
  - induction .... RSEW-I
- resistance spot welding . RSW
- upset welding .... UW
  - high frequency ...... UW-HF
  - induction .... UW-I

Thermal Spraying (THSP):
- arc spraying .... ASP
- flame spraying .... FLSP
- plasma spraying .... PSP

Oxygen Cutting (OC):
- flux cutting .... FOC
- metal powder cutting .... POC
- oxyfuel gas cutting ...... OFC
  - -oxyacetylene cutting ... OFC-A
  - -oxyhydrogen cutting ... OFC-H
  - -oxynatural gas cutting . OFC-N
  - -oxypropane cutting .... OFC-P
- oxygen arc cutting ...... AOC
- oxygen lance cutting ..... LOC

Brazing (B):

Other Welding:
- block brazing .... BB
- diffusion brazing .... CAB
- dip brazing .... DB
- exothermic brazing ...... EXB
- flow brazing .... FLB
- furnace brazing .... FB
- induction brazing .... IB
- infrared brazing .... IRB
- resistance brazing .... RB
- torch brazing .... TB
- twin carbon arc brazing .... TCAB

- electron beam welding .. EBW
  - high vacuum .... EBW-HV
  - medium vacuum .... EBW-MV
  - nonvacuum .... EBW-NV
- electroslag welding ..... ESW
- flow welding .... FLOW
- induction welding ...... IW
- laser beam welding .... LBW
- percussion welding .... PEW
- thermite welding .... TW

Oxyfuel Gas Welding (OFW):
- air acetylene welding ... AAW
- oxyacetylene welding .. OAW
- oxyhydrogen welding .. OHW
- pressure gas welding ... PGW

Arc Cutting (AC):
- air carbon arc cutting .... CAC-A
- carbon arc cutting ...... CAC
- gas metal arc cutting .... GMAC
- gas tungsten arc cutting .. GTAC
- plasma arc cutting ...... PAC
- shielded metal arc cutting SMAC

Other Cutting:
- electron beam cutting .... EBC
- laser beam cutting .... LBC
  - -air .... LBC-A
  - -evaporative .... LBC-EV
  - -inert gas .... LBC-IG
  - -oxygen .... LBC-O

Figure 3.1—Master Chart of Welding and Allied Processes

examination. On that basis, the following processes will be described:

Welding Processes
- Shielded Metal Arc Welding
- Gas Metal Arc Welding
- Flux Cored Arc Welding
- Gas Tungsten Arc Welding
- Submerged Arc Welding
- Plasma Arc Welding
- Electroslag Welding
- Oxyacetylene Welding
- Stud Welding
- Laser Beam Welding
- Electron Beam Welding
- Resistance Welding

Brazing Processes
- Torch Brazing
- Furnace Brazing
- Induction Brazing
- Resistance Brazing
- Dip Brazing
- Infrared Brazing

Cutting Processes
- Oxyfuel Cutting
- Air Carbon Arc Cutting
- Plasma Arc Cutting
- Mechanical Cutting

Welding Processes

Before our discussion of the various welding processes, it is appropriate to define what is meant by the term *welding*. According to AWS, a weld is, "a localized coalescence of metals or nonmetals produced either by heating the materials to the welding temperature, with or without the application of pressure, or by the application of pressure alone and with or without the use of filler metal." Coalescence means "joining together." Therefore welding refers to the operations used to accomplish this joining operation. This section will present important features of some of the more common welding processes, all of which employ the use of heat without pressure.

As each of these welding processes is presented, it is important to note that they all have certain features in common. That is, there are certain elements which must be provided by the welding process in order for it to be capable of producing satisfactory welds. These features include a source of energy to provide heating, a means of shielding the molten metal from the atmosphere, and a filler metal (optional with some processes and joint configurations). The processes differ from one another because they provide these same features in various ways. So, as each process is introduced, be aware of how it satisfies these requirements.

Shielded Metal Arc Welding (SMAW)

The first process to be discussed is shielded metal arc welding. Even though this is the correct name for the process, we more often hear it referred to as "stick welding." This process operates by heating the metal with an electric arc between a covered metal electrode and the metals to be joined. Figure 3.2 shows the "business end" and the various elements of the shielded metal arc welding process.

This illustration shows that the arc is created between the electrode and the workpiece due to the flow of electricity. This arc provides heat, or energy, to melt the base metal, filler metal and electrode coating. As the welding arc progresses to the right, it leaves behind solidified weld metal covered by a layer of converted flux, referred to as slag. This slag tends to float to the outside of the metal since it solidifies after the molten metal has solidified. In doing so, there is less likelihood that it will be trapped inside the weld resulting in a slag inclusion.

Another feature noted in Figure 3.2 is the presence of shielding gas which is produced when the electrode coating is heated and decomposed. These gases assist the flux in the shielding of the molten metal in the arc region.

The primary element of the shielded metal arc welding process is the electrode itself. It is made up of a metal core wire covered with a layer of granular flux held in place by some type of bonding agent. All carbon and low alloy steel electrodes use essentially the same type of steel core wire, a low carbon, rimmed steel. Any alloying is provided from the coating, since it is more economical to achieve alloying in this way.

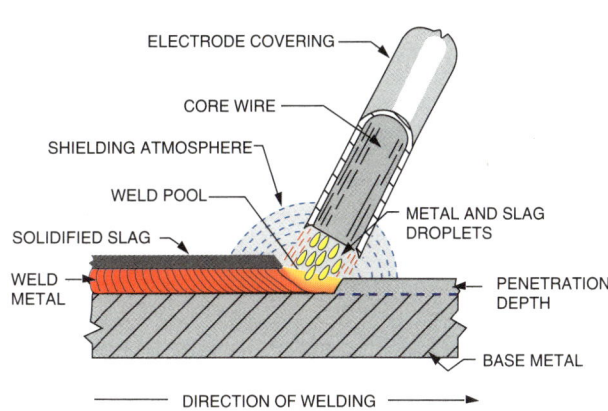

Figure 3.2—Shielded Metal Arc Welding, Including Schematic of Details

The electrode coating is the feature which classifies the various types of electrodes. It actually serves five separate functions:

(1) **Shielding**—the coating decomposes to form a gaseous shield for the molten metal.

(2) **Deoxidation**—the coating provides a fluxing action to remove oxygen and other atmospheric gases.

(3) **Alloying**—the coating provides additional alloying elements for the weld deposit.

(4) **Ionizing**—the coating improves electrical characteristics to increase arc stability.

(5) **Insulating**—the solidified slag provides an insulating blanket to slow down the weld metal cooling rate (minor effect).

Since the electrode is such an important feature of the shielded metal arc welding process, it is necessary to understand how the various types are classified and identified. The American Welding Society has developed a system for the identification of shielded metal arc welding electrodes. Figure 3.3 illustrates the various parts of this system.

American Welding Society Specifications A5.1 and A5.5 describe the requirements for carbon and low alloy steel electrodes, respectively. They describe the various classifications and characteristics of these electrodes.

The identification consists of an "E," which stands for electrode, followed by four or five digits. The first two, or three, numbers refer to the minimum tensile strength of the deposited weld metal. These numbers state the minimum tensile strength in thousands of pounds per square inch. For example, "70" means that the tensile strength of the deposited weld metal is at least 70,000 psi.

The next number refers to the positions in which the electrode can be used. A "1" indicates the electrode is suitable for use in any position. A "2" means that the molten metal is so fluid that the electrode can only be used in the flat position for all welding types and in the horizontal position for fillet welds only. A "4" means the electrode is suitable for welding in a downhill progression. The number "3" is no longer used as a designation.

The last number in the designation describes other characteristics which are determined by the

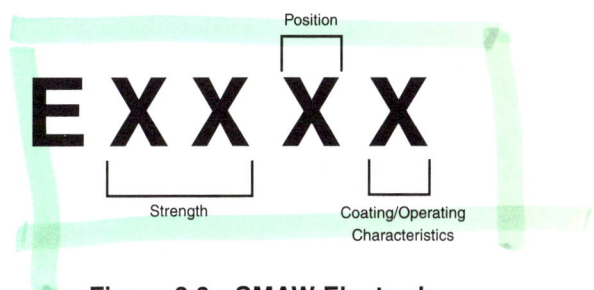

Figure 3.3—SMAW Electrode Identification System

composition of the coating present on the electrode. This coating will determine its operating characteristics and recommended electrical current: AC (alternating current), DCEP (direct current, electrode positive) or DCEN (direct current, electrode negative). Figure 3.4 lists the significance of the last digit of the SMAW electrode identification system.

It is important to note that those electrodes ending in "5," "6," or "8" are classified as "low hydrogen" types. To maintain this low hydrogen (moisture) content, they must be stored in their original factory-sealed metal container or an acceptable storage oven. This oven should be heated electrically and have a temperature control capability in the range of 150° to 350°F. Since this device will assist in the maintenance of a low moisture content (less than 0.2%), it must be suitably vented. Any low hydrogen electrodes which are not to be used immediately should be placed into the holding oven as soon as their airtight container is opened. Most codes require that low hydrogen electrodes be held at a minimum oven temperature of 250°F (120°C) after removal from their sealed container.

However, it is important to note that electrodes other than those mentioned above may be harmed if placed in the oven. Some electrode types are designed to have a certain moisture level. If this moisture is eliminated, the operating characteristics of the electrode will deteriorate significantly.

Those SMAW electrodes used for joining low alloy steels may also have an alpha-numeric suffix which is added to the standard designation after a hyphen. Figure 3.5 shows the significance of these designations.

The equipment for shielded metal arc welding is relatively simple, as can be seen in Figure 3.6. One lead from the welding power source is connected to the piece to be welded and the opposite lead goes to the electrode holder into which the welder places the welding electrode to be consumed. The electrode and base metal are melted by the heat produced from the welding arc created between the end of the electrode and the workpiece when they are brought close together.

The power source for shielded metal arc welding is referred to as a constant current power source, having a "drooping" characteristic. This terminology can be more easily understood by looking at the characteristic volt-ampere (V-A) curve for this type of power source.

As a welder increases arc length, the resistance in the welding circuit increases due to the larger gap the current must cross. As can be seen in the typical volt-amp curve in Figure 3.7(A), this increase in resistance causes a slight decrease (10%) in the current flow across the arc gap. This decrease in current results in a significant increase in voltage (32%) supplied by the power source, which limits the drop in current.

Since heat is a function of voltage, amperage, and time, it can be readily seen that a long arc length (32 volts × 135 amps × 60)/10 IPM = 25,920 J/in.) will result in more heat produced than a short arc length (22 volts × 150 amps × 60)/IPM = 19,800 J/in.).

F-No. Classification	Current	Arc	Penetration	Covering & Slag	Iron Powder
F-3 EXX10	DCEP	Digging	Deep	Cellulose-sodium	0–10%
F-3 EXXX1	AC & DCEP	Digging	Deep	Cellulose-potassium	0%
F-2 EXXX2	AC & DCEN	Medium	Medium	Rutile-sodium	0–10%
F-2 EXXX3	AC & DC	Light	Light	Rutile-potassium	0–10%
F-2 EXXX4	AC & DC	Light	Light	Rutile-iron powder	25–40%
F-4 EXXX5	DCEP	Medium	Medium	Low hydrogen-sodium	0%
F-4 EXXX6	AC or DCEP	Medium	Medium	Low hydrogen-potassium	0%
F-4 EXXX8	AC or DCEP	Medium	Medium	Low hydrogen-iron powder	25–45%
F-1 EXX20	AC or DC	Medium	Medium	Iron oxide-sodium	0%
F-1 EXX24	AC or DC	Light	Light	Rutile-iron powder	50%
F-1 EXX27	AC or DC	Medium	Medium	Iron oxide-iron powder	50%
F-1 EXX28	AC or DCEP	Medium	Medium	Low hydrogen-iron powder	50%

Note: Iron powder percentage is based on weight of the covering.

Figure 3.4—Significance of Last Digit of SMAW Identification

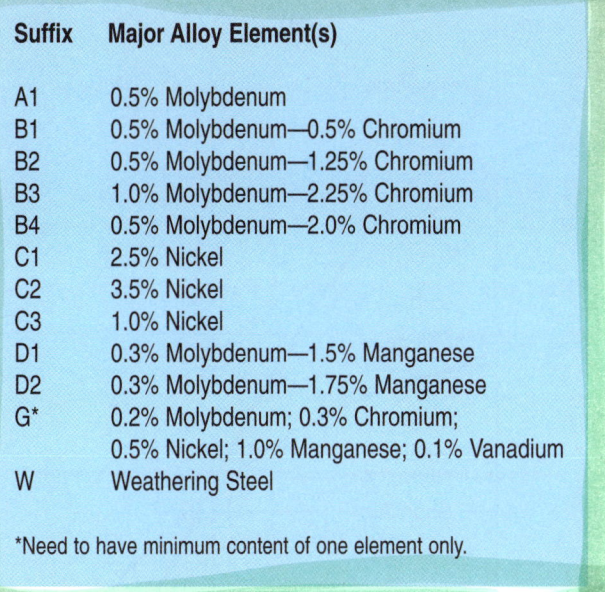

Figure 3.5—Steel Alloy Suffixes for SMAW Electrodes

Suffix	Major Alloy Element(s)
A1	0.5% Molybdenum
B1	0.5% Molybdenum—0.5% Chromium
B2	0.5% Molybdenum—1.25% Chromium
B3	1.0% Molybdenum—2.25% Chromium
B4	0.5% Molybdenum—2.0% Chromium
C1	2.5% Nickel
C2	3.5% Nickel
C3	1.0% Nickel
D1	0.3% Molybdenum—1.5% Manganese
D2	0.3% Molybdenum—1.75% Manganese
G*	0.2% Molybdenum; 0.3% Chromium; 0.5% Nickel; 1.0% Manganese; 0.1% Vanadium
W	Weathering Steel

*Need to have minimum content of one element only.

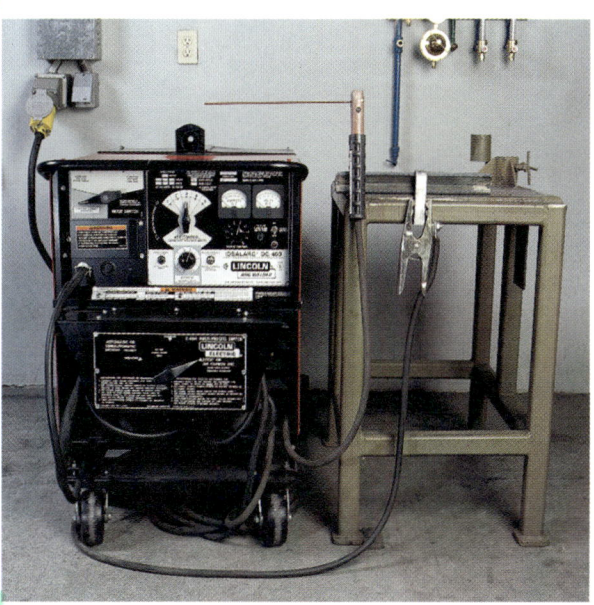

Figure 3.6—Shielded Metal Arc Welding Equipment

This is significant from a process control standpoint, because the welder can increase or decrease the fluidity of the weld pool simply by changing the arc length. However, an extremely long arc length will cause a loss of heat in the weld pool due to loss of arc concentration. An excessively long arc length will also result in a loss of arc stability and weld pool shielding gases.

If equipped with a slope control, the power supply can be adjusted to give the operator more or less control over the fluidity of the weld pool with a slight arc length change. Figure 3.7(B) shows two different slope settings. An experienced welder might choose a flat slope setting to gain more control, while an inexperienced welder might need a steeper slope setting for less change in weld pool characteristics due to an inconsistent art length.

Shielded metal arc welding is used in most industries for numerous applications. It is used for most materials except for some of the more exotic alloys. Even though it is a relatively old method and newer processes have replaced it in some applications, shielded metal arc welding remains as a popular process which will continue to be in great use by the welding industry.

There are several reasons why the process continues to be so popular. First, the equipment is relatively simple and inexpensive. This helps to make the process quite portable. In fact, there are numerous gasoline or diesel engine-driven types which don't rely on electrical input; thus, shielded metal arc welding can be accomplished in remote locations. Also, some of the newer solid state power sources are so small and lightweight that the welder can easily carry them to the work. And, due to the availability of numerous types of electrodes, the process is considered quite versatile. Finally, with the improved equipment and electrodes available today, the resulting weld quality can be consistently high.

One of the limitations of shielded metal arc welding is its speed. The speed is primarily hampered by the fact that the welder must periodically stop welding and replace the consumed electrode with a new one, since they are typically only 9 to 18 in. in length. SMAW has been replaced by other semiautomatic, mechanized and automatic processes in many applications simply because they offer increased productivity when compared to manual shielded metal arc welding.

Another disadvantage, which also affects productivity, is the fact that following welding, there is a layer of solidified slag which must be removed. A further limitation, when low hydrogen type electrodes are used, is that they require storage in an appropriate electrode holding oven which will maintain their low moisture levels.

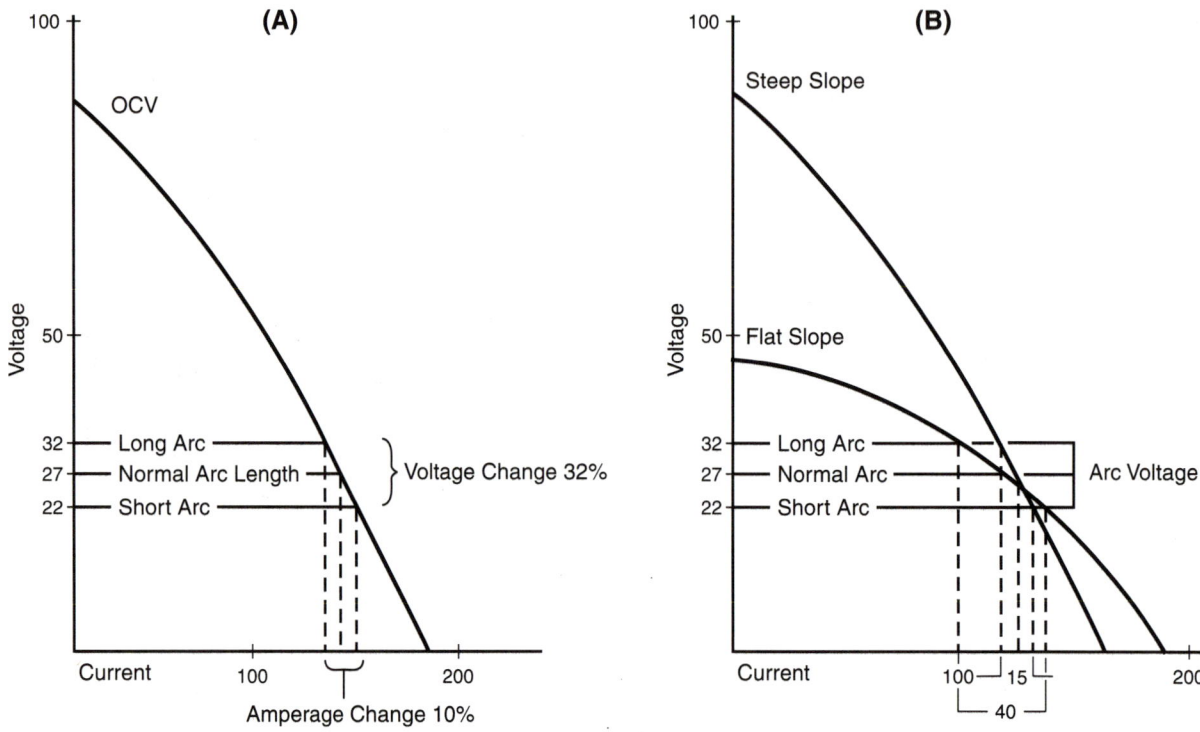

Figure 3.7—Volt-Ampere Curve for Constant Current Power Source

Now that some of the basic principles have been presented, it is appropriate to discuss some of the discontinuities which may result during the shielded metal arc process. While these are not the only discontinuities we can expect, they may result because of the misapplication of this particular process.

One potential problem is the presence of porosity in the finished weld. When porosity is encountered, it is normally the result of the presence of moisture or contamination in the weld region. It could be present in the electrode coating, on the surface of the material, or come from the atmosphere surrounding the welding operation. Porosity can also occur when the welder is using an arc length which is too long. This problem of "long arcing" is especially likely when using low hydrogen electrodes. So, the preferred shorter arc length will aid in the elimination of porosity in the weld metal.

Porosity can also result from the presence of a phenomenon referred to as *arc blow*. While this can occur with any arc welding process, it will be discussed here since it is a common problem which plagues the manual welder.

To understand arc blow, one must first understand that there is a magnetic field developed whenever an electric current is passed through a conductor. This magnetic field is developed in a direction perpendicular to the direction of the electric current, so it can be visualized as a series of concentric circles surrounding the conductor, as shown in Figure 3.8.

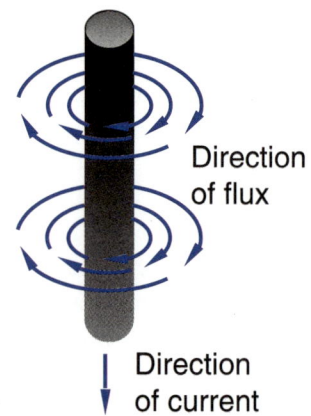

Figure 3.8—Magnetic Field Around Electric Conductor

This magnetic field is strongest when contained entirely within a magnetically permeable material and resists having to travel through the air outside this magnetically permeable material. Consequently, when welding a magnetically permeable material, such as steel, the field can become distorted when the arc approaches the edge of a plate, the end of a weld, or some abrupt change in contour of the part being welded. This is shown in Figure 3.9.

To reduce the effects of arc blow, several alternatives can be tried. They include:

(1) Change from DC to AC.
(2) Hold as short an arc as possible.
(3) Reduce welding current.
(4) Angle the electrode in the direction opposite the arc blow.
(5) Use heavy tack welds at either end of a joint, with intermittent tack welds along the length of the joint.
(6) Weld toward a heavy tack or toward a completed weld.
(7) Use a backstep technique.
(8) Weld away from the ground to reduce back blow; weld toward the ground to reduce forward blow.
(9) Attach the work cable to both ends of the joint to be welded.
(10) Wrap work cable around the workpiece and pass work current through it in such a direction that the magnetic field set up will tend to neutralize the magnetic field causing the arc blow.
(11) Extend the end of the joint by attaching runoff plates.

In addition to porosity, arc blow can also cause spatter, undercut, improper weld contour, and decreased penetration.

Slag inclusions can also occur with SMAW simply because it relies on a flux system for weld protection. With any process incorporating a flux, the possibility of trapping slag within the weld deposit is a definite concern. The welder can reduce this tendency by using techniques which allow the molten slag to flow freely to the surface of the metal. Thorough cleaning of the slag from each weld pass prior to deposition of additional passes will also reduce the occurrence of slag inclusions in multipass welds.

Since shielded metal arc welding is primarily accomplished manually, numerous discontinuities can result from improper manipulation of the electrode. Some of these are incomplete fusion, incomplete joint penetration, cracking, undercut, overlap, incorrect weld size, and improper weld profile.

Gas Metal Arc Welding (GMAW)

The next process to be discussed is gas metal arc welding, GMAW. While gas metal arc welding is the AWS designation for the process, we also hear it commonly referred to as "MIG" welding. It is most commonly employed as a semiautomatic process; however, it is also used in mechanized and automatic applications as well. Therefore, it is very well-suited for robotic welding applications. Gas metal arc welding is characterized by a solid wire electrode which is fed continuously through a welding gun. An arc is created between this wire and the workpiece to heat and melt the base and filler materials. Once molten, the wire becomes deposited in the weld joint. Figure 3.10 depicts the essential elements of the process.

An important feature for GMAW is that all of the shielding for welding is provided by a protective gas atmosphere which is also emitted from the

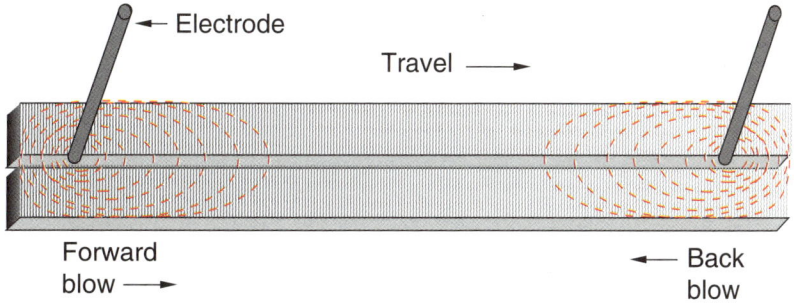

Figure 3.9—Distorted Magnetic Fields at Ends of Welds

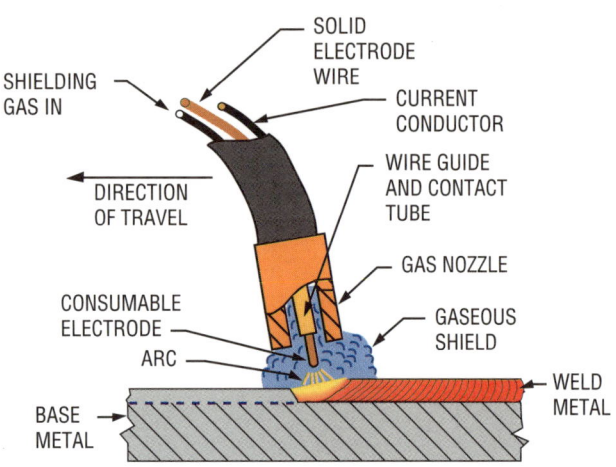

Figure 3.10—Gas Metal Arc Welding, Including Schematic of Details

welding gun from some external source. Gases used include both inert and reactive types. Inert gases such as argon and helium are used for some applications. They can be applied singly, or in combination with each other, or mixed with other reactive gases such as nitrogen, oxygen or carbon dioxide. Many gas metal arc welding applications use carbon dioxide shielding alone, because of its relatively low cost compared to inert gases.

The electrodes used for this process are solid wires which are supplied on spools or reels of various sizes. As is the case for shielded metal arc welding electrodes, there is an approved American Welding Society identification system for gas metal arc welding electrodes. They are denoted by the letters "ER," followed by two or three numbers, the letter "S," a hyphen, and finally, another number, as shown in Figure 3.11.

"ER" designates the wire as being both an electrode and a rod, meaning that it may conduct electricity (electrode), or simply be applied as a filler metal (rod) when used with other welding processes. The next two or three numbers state the minimum tensile strength of the deposited weld metal in thousands of pounds per square inch. So, like the SMAW types, a "70" denotes a filler metal whose tensile strength is at least 70,000 psi. The letter "S" stands for a solid wire. Finally, the number after the hyphen refers to the particular chemistry of the electrode. This will dictate both its operating characteristics as well as what properties are to be expected from the deposited weld. Gas metal arc electrodes typically have increased amounts of deoxidizers such as manganese, silicon, and aluminum to help avoid the formation of porosity.

Even though the wire doesn't have a flux coating, it is still important to store the material properly when not in use. The most critical factor here is that the wire must be kept clean. If allowed to remain out in the open, it may become contaminated with rust, oil, moisture, grinding dust or other materials present in a weld shop environment. So, when idle, the wire should be kept in its original plastic wrapping and/or shipping container. Even when a spool of wire is in place on the wire feeder, it should be covered with a protective covering when not used for prolonged periods of time.

The power supply used for gas metal arc welding is quite different from the type employed for shielded metal arc welding. Instead of a constant current type, gas metal arc welding uses what is referred to as a constant voltage, or constant potential, power source. That is, welding is accomplished using a pre-set value of voltage over the range of welding currents.

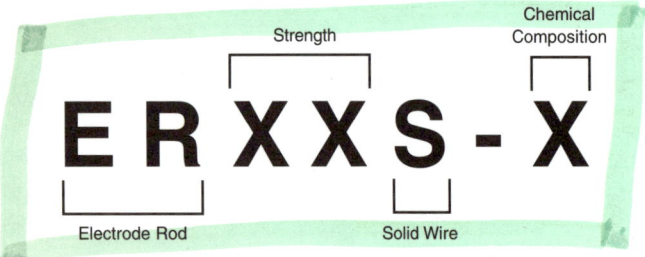

Figure 3.11—GMAW Electrode Identification System

Gas metal arc welding is normally accomplished using direct current, electrode positive (DCEP). When this type of power source is combined with a wire feeder, the result is a welding process which can be either semiautomatic, mechanized or fully automatic. Figure 3.12 shows a typical gas metal arc welding setup.

As can be seen, the equipment is a bit more complex than that used for shielded metal arc welding. A complete setup includes a power source, wire feeder, gas source, and welding gun attached to the wire feeder by a flexible cable through which the electrode and gas travel. To set up for welding, the welder will adjust the voltage at the power source and the wire feed speed at the wire feeder. As the wire feed speed is increased, the welding current increases as well. The melt-off rate of the electrode is proportional to the arc current, so the wire feed speed actually controls this feature as well.

It was mentioned that this power source is a constant potential type; however, a look at a typical V-A curve, Figure 3.13, will show that the line is not flat but actually has a slight slope.

This feature allows the process to function as a semiautomatic type, meaning that the welder does not have to control the feeding of the filler metal as was the case for manual shielded metal arc welding. Another way to describe this system is to call it a "Self-Regulating Constant Potential" system.

This is accomplished because a minor change in the gun's position with respect to the workpiece will result in a substantial increase or decrease in the arc current.

Figure 3.12—Gas Metal Arc Welding Equipment

Looking at Figure 3.13, it can be seen that moving the gun closer to the workpiece will reduce the electrical resistance and produce an instantaneous increase in current. This, in turn, instantaneously burns off the additional electrode to bring the arc length and current back to their original values. This reduces the effect of the operator's manipulation on

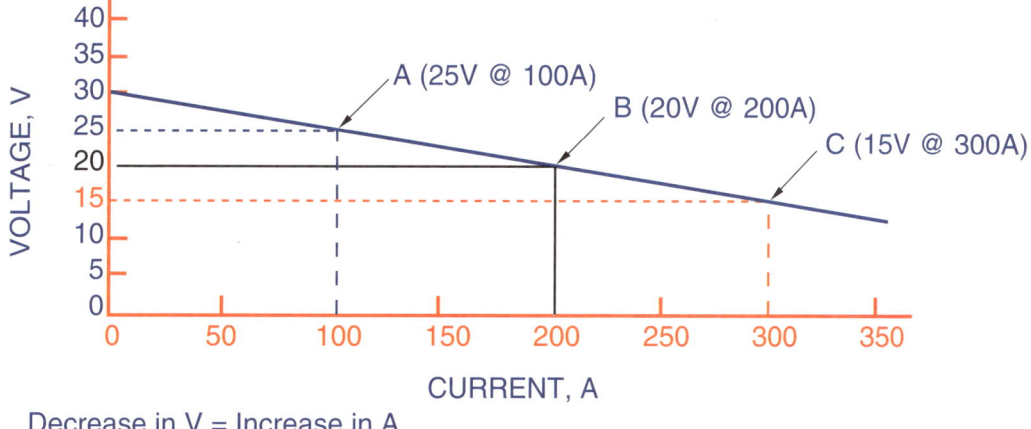

Decrease in V = Increase in A

Figure 3.13—Typical Constant Potential V-A Curve

the welding characteristics, making the process less operator-sensitive and therefore easier to learn the manipulative aspects.

When the machine adjustments are changed, the result is the operating characteristics will be drastically altered. Of primary concern is the manner in which the molten metal is transferred from the end of the electrode, across the arc region, to the base metal. With gas metal arc welding, there are four basic modes of metal transfer. They are spray, globular, pulsed arc, and short circuiting.

Figure 3.14 shows three of the four modes. Their characteristics are so different that it's almost as if there are four separate welding processes. Each specific type has definite advantages and limitations which make it better for some applications than others. The type of metal transfer depends upon several factors, including shielding gas, current and voltage levels, and power source characteristics.

One of the basic ways in which these four types differ is that they provide varying amounts of heat to the workpiece. Spray transfer is considered to be the hottest, followed by pulsed arc, globular, and finally short circuiting. Therefore, spray transfer is the best for heavier sections and full penetration weld joints, as long as they can be positioned in the flat position.

Globular transfer provides almost as much heating and weld metal deposition, but its operating characteristics tend to be less stable, resulting in increased spatter.

Pulsed arc gas metal arc welding requires a welding power source capable of producing a pulsing direct current output which allows the welder to program the exact combination of high and low currents for improved heat control and process flexibility. The welder can set both the amount and duration of the high current pulse. So during operation, the current alternates between a high pulse current and a lower pulse current, both of which can be set with machine controls.

Short circuiting transfer results in the least amount of heating to the base metal, making it an excellent choice for welding of sheet metal and joints having excessive gaps due to poor fitup. The short circuiting transfer method is characteristically colder due to the electrode actually coming in contact with the base metal, creating a short circuit for a portion of the welding cycle. So, the arc is intermittently operating and extinguishing. The brief periods of arc extinction allow for some cooling to occur to aid in reducing the tendency of burning through thin materials. Care must be taken when short circuiting transfer is used for heavier section welding, since incomplete fusion can result from insufficient heating of the base metal.

As mentioned, the shielding gases have a significant effect on the type of metal transfer. Spray transfer can be achieved only when there is at least 80% argon present in the gas mixture. CO_2 is probably the most popular gas for GMAW of carbon steel, primarily due to its relatively low cost and its excellent penetration characteristics. One drawback, however, is that there will be more spatter which may require removal, reducing operator productivity.

The versatility offered by this process has resulted in its use in many industrial applications. GMAW can be effectively used to join or overlay

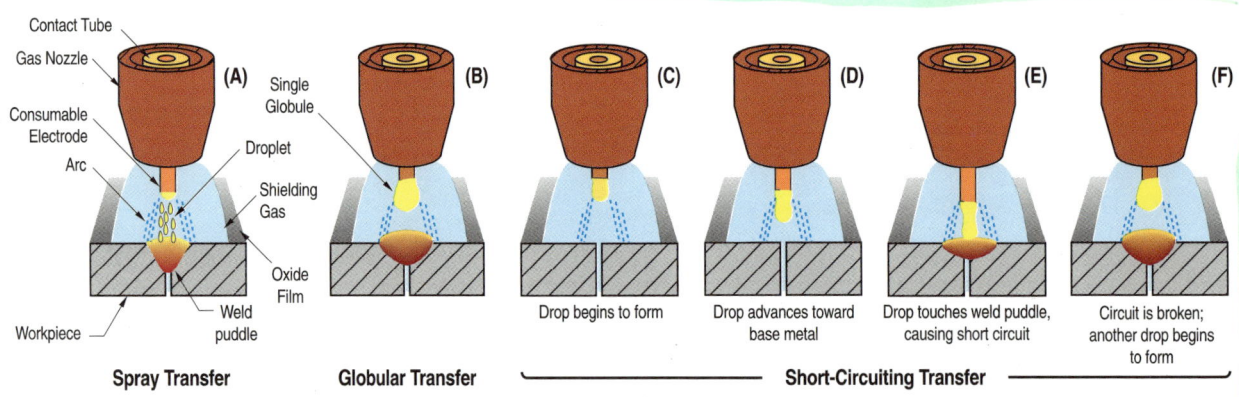

Figure 3.14—Modes of Metal Transfer: (a) Spray, (b) Globular, and (c) Short Circuiting (Pulsed Arc Not Shown)

many types of ferrous and nonferrous metals. The use of gas shielding, instead of a flux which can become contaminated, can reduce the possibility of introducing hydrogen into the weld zone, so GMAW can be used successfully in situations where the presence of hydrogen could cause problems.

Due to the lack of a slag coating which must be removed after welding, GMAW is well suited for automatic and robotic welding or other high production situations. This is one of the major advantages of the process. Since there is little or no cleaning required following welding, the overall operator productivity is greatly improved. This efficiency is further increased by the fact that the continuous spool of wire doesn't require changing nearly as often as the individual electrodes used in SMAW. All of this increases the amount of time in which actual production welding can be accomplished.

The chief advantage of GMAW is lbs/hr of metal deposited, which reduces labor costs. Another benefit of gas metal arc welding is that it is a relatively clean process, primarily due to the fact that there is no flux present. Shops with ventilation problems can find some relief by switching to gas metal arc welding from shielded metal arc welding or flux cored arc welding, because less smoke is generated. With the existence of numerous types of electrodes, and equipment which has become more portable, the versatility of gas metal arc welding continues to improve. One additional benefit relates to the visibility of the process. Since no slag is present, the welder can more easily observe the action of the arc and molten puddle to improve control.

While the use of shielding gas instead of flux does provide some benefits, it can also be thought of as a limitation, since this is the primary way in which the molten metal is protected and cleaned during welding. If the base metal is excessively contaminated, the shielding gas alone may not be sufficient to prevent the occurrence of porosity. GMAW is also very sensitive to drafts or wind which tend to blow the shielding gas away and leave the metal unprotected. For this reason, gas metal arc welding is not well suited for field welding.

It is important to realize that simply increasing the gas flow rate above recommended limits will not necessarily guarantee that adequate shielding will be provided. In fact, high flow rates cause turbulence and may tend to increase the possibility of porosity because these increased flow rates may actually draw atmospheric gases into the weld zone.

Another disadvantage is that the equipment required is more complex than that used for shielded metal arc welding. This increases the possibility that a mechanical problem could cause weld quality problems. Such things as worn gun liners and contact tubes can alter the feeding and electrical characteristics to the point that defective welds can be produced.

The primary inherent problems have already been discussed. They are: porosity due to contamination or loss of shielding, incomplete fusion due to the use of short circuiting transfer on heavy sections, and arc instability caused by worn liners and contact tips. Although such problems could be very harmful to weld quality, they can be alleviated if certain precautions are taken.

To reduce the possibility of porosity, parts should be cleaned prior to welding and the weld zone protected from any excessive wind using enclosures or windbreaks. If porosity persists, the gas supply should be checked to assure that there is not excessive moisture present.

The problem of incomplete fusion is a real one for GMAW, especially when short circuiting transfer is used. This is due in part to the fact that this is an "open arc" process since no flux is used. Without this layer of shielding from the arc heat, the increased heat intensity can lead the welder to believe that there is a tremendous amount of heating of the base metal. Such a feeling can be very deceiving, so the welder must be aware of this condition and assure that the arc is being directed to insure melting of the base metal.

Finally, equipment should be well maintained to alleviate the problems associated with unstable wire feeding. Each time a roll of wire is replaced, the liner should be blown out with clean compressed air to remove particles which could cause jamming. If a wire feed problem persists, the liner should be replaced. The contact tube should be changed periodically as well. When it becomes worn, the point of electrical contact changes so that the electrode extension is increased without the welder knowing it. The electrode extension is also referred to as the contact tube to electrode end distance, as illustrated in Figure 3.15.

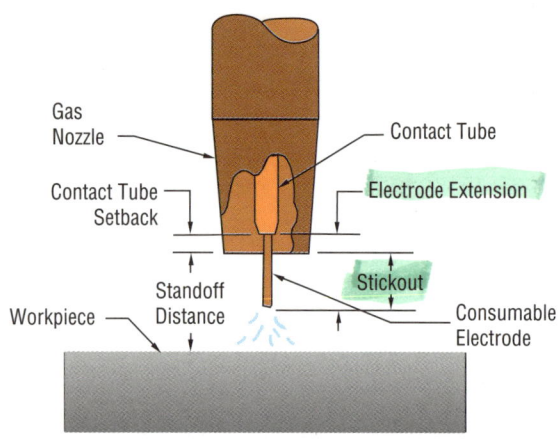

Figure 3.15—Gas Metal Arc Welding Gun Nomenclature

Flux Cored Arc Welding (FCAW)

The next process to be described is flux cored arc welding. This is very similar to gas metal arc welding except that the electrode is tubular and contains a granular flux instead of the solid wire used for gas metal arc welding. The difference can be noted in Figure 3.16 which shows a weldment welded with the self shielded FCAW process and a closeup of the arc region during welding.

It shows the tubular electrode being fed through the contact tube of the welding gun to produce an arc between the electrode and the workpiece. As the welding progresses, a bead of solidified weld metal is deposited. Covering this solidified weld metal is a layer of slag, as occurs for shielded metal arc welding.

With flux cored arc welding, there may or may not be an externally-supplied shielding gas, depending upon what type of electrode is used. Some electrodes are designed to provide all of the necessary shielding from the internal flux, and are referred to as *self-shielding*. Other electrodes require additional shielding from an auxiliary shielding gas. With FCAW, as with other processes, there is a system for identification of the various types of welding electrodes, illustrated in Figure 3.17. A review of each electrode type shows that the designations refer to the polarity, shielding requirements, chemistry and welding position.

An identification begins with the letter "E" which stands for electrode. The first number refers to the minimum tensile strength of the deposited weld metal in ten thousands of pounds per square inch, so a "7" means that the weld metal tensile strength is at least 70,000 psi. The second digit is either a "0" or a "1." A "0" means that the electrode is suitable for use in the flat or horizontal fillet positions only, while a "1" describes an electrode which can be used in any position. Following these numbers is the letter "T" which refers to a tubular electrode. This is followed by a hyphen and then another number which denotes the particular grouping based upon the chemical composition of deposited weld metal, type of current, polarity of operation, whether it requires a shielding gas, and other specific information for the category.

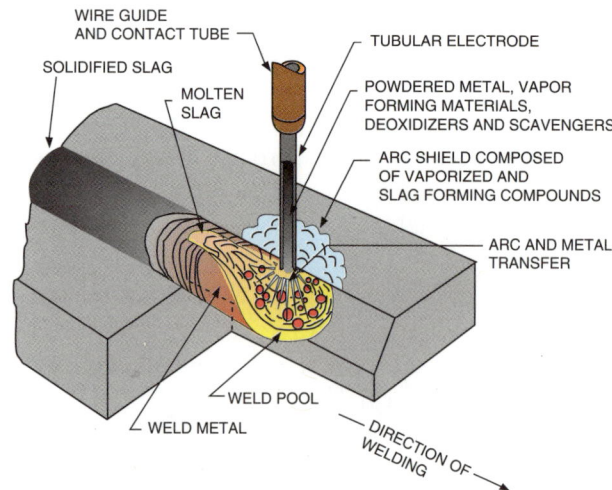

Figure 3.16—Self-Shielded Flux Cored Arc Welding

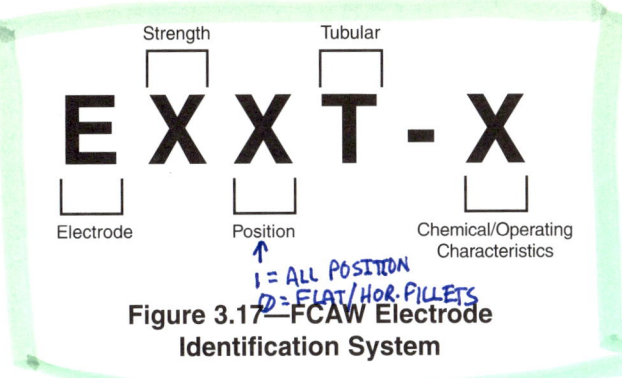

Figure 3.17—FCAW Electrode Identification System

With this identification system, it can be determined whether or not a certain classification of electrode requires an auxiliary shielding gas. This is important to the welding inspector since flux cored arc welding can be performed with or without an external shielding gas. Figure 3.18 shows the two nozzle types.

Some electrodes are formulated to be used without any additional shielding other than that contained within the electrode. They are designated by the suffixes 3, 4, 6, 7, 8, 10, 11, 13, and 14. However, those electrodes having the numerical suffixes 1, 2, 5, 9, or 12 require some external shielding to aid in protecting the molten metal. Both types offer advantages, depending upon the application. Additionally, the suffixes G and GS refer to multiple-pass and single-pass respectively.

For example, the self-shielded types are better suited for field welding where wind could result in a loss of gaseous shielding. Gas shielded types are typically used where the need for improved weld metal properties warrants the additional cost. Gases typically used for flux cored arc welding are CO_2, or 75% Argon—25% CO_2, but other combinations of gases are available.

The equipment used for FCAW is essentially identical to that for GMAW, as shown in Figure 3.19. Some differences might be higher current capacity guns and power sources, lack of gas apparatus for self-shielded electrodes, and knurled wire feed rolls. Like GMAW, FCAW uses a constant

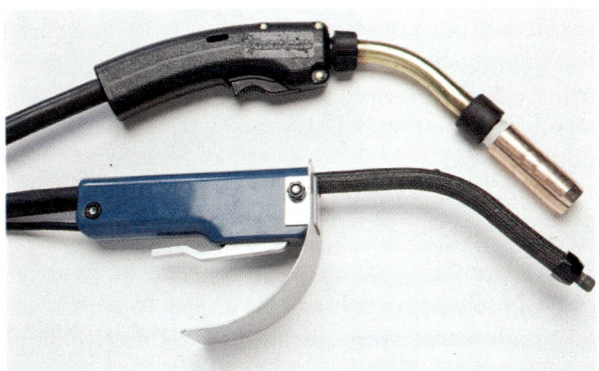

Figure 3.18—FCAW Guns for Gas-Shielding (top) and Self-Shielding Electrodes (bottom)

Figure 3.19—Gas-Shielded (Top) and Self-Shielded (Bottom) Flux Cored Arc Welding Equipment

voltage direct current power supply. Depending on the type of electrode, the operation may be DCEP (1, 2, 3, 4, 6, 9, 12) or DCEN (7, 8, 10, 11, 13, 14) or both DCEN or DCEP (5).

The flux cored arc welding process is rapidly gaining acceptance as the welding process of choice in some industries. Its relatively good performance on contaminated surfaces, and increased deposition rates have helped flux cored arc welding replace SMAW and GMAW for many applications. The process is used in many industries where the predominant materials are ferrous. It can be used with satisfactory results for both shop and field applications. Although the majority of the electrodes produced are ferrous (for both carbon and stainless steels), some nonferrous ones are available as well. Some of the stainless types actually employ a carbon steel sheath surrounding the internal flux which contains the granular alloying elements such as chromium and nickel.

FCAW has gained wide acceptance because of the many advantages it offers. Probably the most significant advantage is that it provides high productivity in terms of the amount of weld metal that can be deposited in a given period of time. It is among the highest for a hand-held process. This is aided by the fact that the electrode comes on continuous reels which increases the "arc time" just as with gas metal arc welding. The process is also characterized by an aggressive, deeply penetrating arc which tends to reduce the possibility of fusion-type discontinuities. Since it is typically used as a semiautomatic process, the skill required for operation is somewhat less than would be the case for a manual process. With the presence of a flux, whether assisted by a gaseous shield or not, FCAW is capable of tolerating a greater degree of base metal contamination than is GMAW. For this same reason, FCAW lends itself well to field situations where the loss of shielding gas due to winds would greatly hamper GMAW quality.

It is important to realize that this process does have certain limitations of which the inspector should be aware. First, since there is a flux present, there is a layer of solidified slag which must be removed before depositing additional layers of weld or before a visual inspection can be made.

Due to the presence of this flux, there is a significant amount of smoke generated during welding. Prolonged exposure in unvented areas could prove to be unhealthy for the welder. This smoke also reduces the welder's visibility to the point where it may be difficult to properly manipulate the arc in the joint. Although smoke extractor systems are available, they tend to add bulk to the gun which increases the weight and decreases visibility. They also may disturb the shielding if an auxiliary gas is being used.

Even though FCAW is considered to be a smoky process, it is not as bad as SMAW in terms of the amount of smoke generated for a given amount of deposited weld metal. The equipment required for FCAW is more complex than that for SMAW, so the initial cost and the possibility of machinery problems may limit its acceptability for some situations.

As with any of the processes, FCAW does have some inherent problems. The first has to do with the flux. Due to its presence, there exists a possibility that the solidified slag could become trapped in the finished weld. This could be due to either improper interpass cleaning or improper technique.

With FCAW, it is critical that the travel speed is fast enough to maintain the arc on the leading edge of the molten puddle. When the travel speed is slow enough to allow the arc to be toward the middle or back of the puddle, molten slag may roll ahead of the puddle and become trapped. Another inherent problem involves wire feeding apparatus. As is the case for GMAW, lack of maintenance can cause wire feeding problems which may affect the quality of the weld. FCAW is also subject to typical discontinuities including incomplete joint penetration, slag inclusions, and porosity.

Gas Tungsten Arc Welding (GTAW)

The next process to be discussed is gas tungsten arc welding, which has several interesting differences when compared to those already discussed. Figure 3.20 shows the basic elements of the process.

The most significant feature of GTAW is that the electrode used is not intended to be consumed during the welding operation. It is made of pure or alloyed tungsten which has the ability to withstand very high temperatures, even those of the welding arc. Therefore, when current is flowing, there is an arc created between the tungsten electrode and the work.

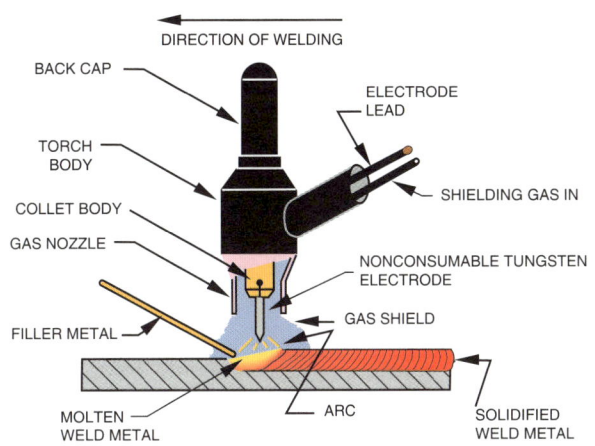

AWS Tungsten Electrode Classifications		
Class	Alloy	Color
EWP	Pure tungsten	Green (Aluminum, AC)
EWCe-2	1.8-2.2% cerium	Orange
EWLa-1	1% lanthanum	Black
EWLa-1.5	1.5% lanthanum	Gold
EWLa-2	2.0% lanthanum	Blue
EWTh-1	0.8 - 1.2% thorium	Yellow
EWTh-2	1.7 - 2.2% thorium	Red (S.S., DC)
EWZr	0.15-0.40% zirconium	Brown

Figure 3.20—Gas Tungsten Arc Welding

When filler metal is required, it must be added externally, usually manually, or with the use of some mechanical wire feed system. All of the arc and metal shielding is achieved through the use of an inert gas which flows out of the nozzle surrounding the tungsten electrode. The deposited weld bead has no slag requiring removal because no flux is used.

As with the other processes, there is a system whereby the various types of tungsten electrodes can be easily identified. The designations consist of a series of letters starting with an "E" which stands for electrode. Next comes a "W" which is the chemical designation for tungsten. These letters are followed by letters and numbers which describe the alloy type. Since there are only five different classifications, they are more commonly differentiated using a color code system. The table shows the classifications and the appropriate color code.

The presence of the thoria or zirconia aids in improving the electrical characteristics by making the tungsten slightly more emissive. This simply means that it is easier to initiate an arc with these thoriated or zirconiated types than is the case for pure tungsten electrodes. Pure tungsten is quite often used for the welding of aluminum because of its ability to form a "ball" end when heated. With a ball end instead of a sharper point, there is a lower concentration of current which reduces the possibility of damaging the tungsten. The EWTh-2 type is most commonly used for the joining of ferrous materials.

The filler metals for GTAW have the ER prefix, followed by chemistry designation. The bare, solid filler wire usually is purchased in 36" lengths, with identification tags on each end.

GTAW can be performed using DCEP, DCEN or AC. The DCEP will result in more heating of the electrode, while the DCEN will tend to heat the base metal more. AC alternately heats the electrode and base metal. AC is typically used for the welding of aluminum because the alternating current will increase the cleaning action to improve weld quality. DCEN is most commonly used for the welding of steels. Figure 3.21 illustrates the effects of these different types of current and polarity in terms of penetration ability, oxide cleaning action, heat balance of arc, and electrode current-carrying capacity.

As mentioned, GTAW uses inert gases for shielding. By inert, we mean that the gases will not combine with the metal, but will protect it from contaminants. Argon and helium are the two inert gases used in GTAW. Some mechanized stainless steel welding applications use a shielding gas con-

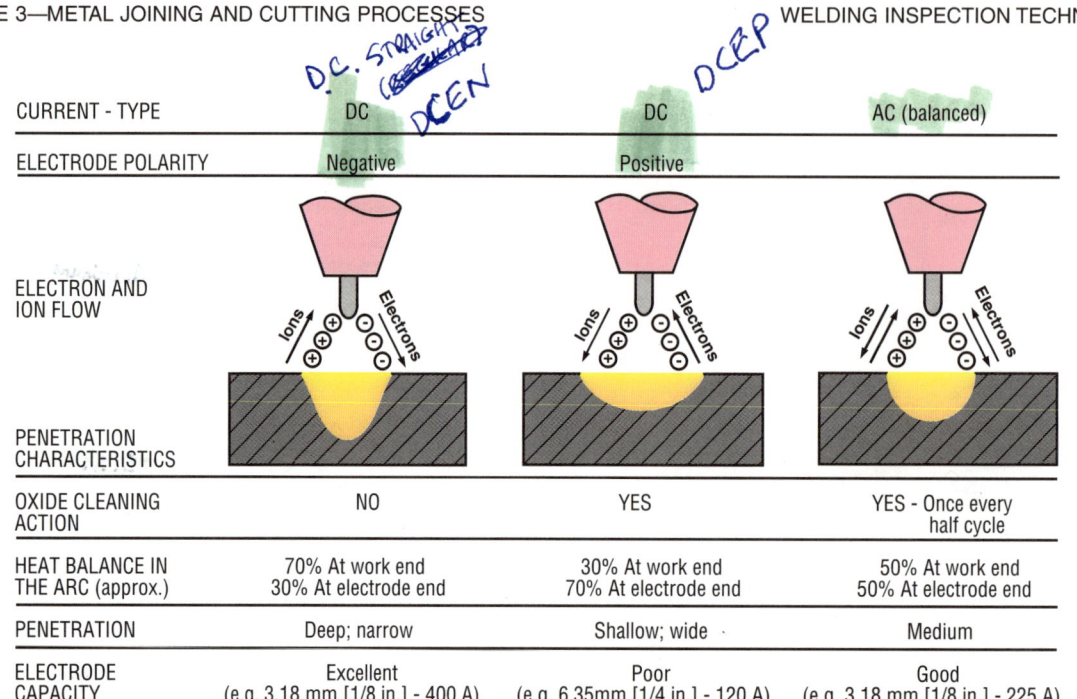

Figure 3.21—Effect of Welding Current Type on Penetration for Gas Tungsten Arc Welding

sisting of argon and a small amount of hydrogen, but this represents a very minor portion of the gas tungsten arc welding which is performed.

The equipment required for GTAW has as its primary element a power source like the one used for SMAW, that is, a constant current type. Since there is a gas present, it is now necessary to have apparatus for its control and transmission. Figure 3.22 shows a typical gas tungsten arc welding setup.

An added feature of this welding system, which is not shown, is a high frequency generator which aids in the initiation of the welding arc. In order to alter the welding heat during the welding operation, a remote current control may also be attached. It can be foot-operated or controlled by some device mounted on the torch itself. This is particularly useful for welding thin materials and open root pipe joints, where instantaneous control is necessary.

There are numerous applications for GTAW in many industries. It is capable of welding virtually all materials, because the electrode is not melted during the welding operation. Its ability to weld at extremely low currents makes gas tungsten arc welding suitable for use on the thinnest (down to 0.005 in.) of metals. Its typically clean and controllable operation causes it to be the perfect choice for extremely critical applications such as those found in the aerospace, food and drug processing, petrochemical, and power piping industries.

The principal advantage of GTAW lies in the fact that it can produce welds of high quality and excellent visual appearance. Also, since no flux is used, the process is quite clean and there is no slag to remove after welding. As mentioned before, extremely thin sections can be welded. Due to the nature of its operation, it is suitable for welding most metals, many of which are not as easily welded using other welding processes. If joint design permits, these materials can be welded without the use of additional filler metal.

When required, numerous types of filler metals exist in wire form for a wide range of metal alloys. In the case where there is no commercially-available wire for a particular metal alloy, it is possible to produce a suitable filler metal by simply shearing a piece of identical base metal to produce a narrow piece which can be hand-fed into the weld zone just as if it were a wire.

Contrasting these advantages are several disadvantages. First, GTAW is among the slowest of the available welding processes. While it produces a clean weld deposit, it is also characterized as having a low tolerance for contamination. Therefore, base and filler metals must be extremely clean prior to welding. When used as a manual process, gas tungsten arc welding requires a high skill level; the welder must coordinate the arc with one hand while feeding the filler metal with the other. GTAW is

Figure 3.22—Gas Tungsten Arc Welding Equipment

Reasons for Tungsten Inclusions

(1) Contact of electrode tip with molten metal;
(2) Contact of filler metal with hot tip of electrode;
(3) Contamination of the electrode tip by spatter;
(4) Exceeding the current limit for a given electrode diameter or type;
(5) Extension of electrodes beyond their normal distances from the collet, resulting in over-heating of the electrode;
(6) Inadequate tightening of the collet;
(7) Inadequate shielding gas flow rates or excessive wind drafts resulting in oxidation of the electrode tip;
8) Defects such as splits or cracks in the electrode;
(9) Use of improper shielding gases; and
(10) Improper grinding of the electrode tip.

normally selected in situations where the need for very high quality warrants additional cost to overcome these limitations.

One of the inherent problems associated with this method has to do with its inability to tolerate contamination. If contamination or moisture is encountered, whether from the base metal, filler metal or shielding gas, the result could be porosity in the deposited weld. When porosity is noted, this is a sign that the process is out of control and some preventive measures are necessary. Checks should be made to determine the source of the contamination so that it can be eliminated.

Another inherent problem which is almost totally confined to the GTAW process is that of tungsten inclusions. As the name implies, this discontinuity occurs when pieces of the tungsten electrode become included in the weld deposit. Tungsten inclusions can occur due to a number of reasons, and several are listed in the following table.

Submerged Arc Welding (SAW)

The last of the more common welding processes to be discussed is submerged arc welding. This method is typically the most efficient one mentioned so far in terms of the rate of weld metal deposition. SAW is characterized by the use of a continuously-fed solid wire electrode which provides an arc that is totally covered by a layer of granular flux; hence the name "submerged" arc. Figure 3.23 shows how a weld is produced using this process.

As mentioned, the wire is fed into the weld zone much the same way as with gas metal arc welding or flux cored arc welding. The major difference, however, is in the method of shielding. With submerged arc welding, a granular flux is distributed ahead of or around the wire electrode to facilitate

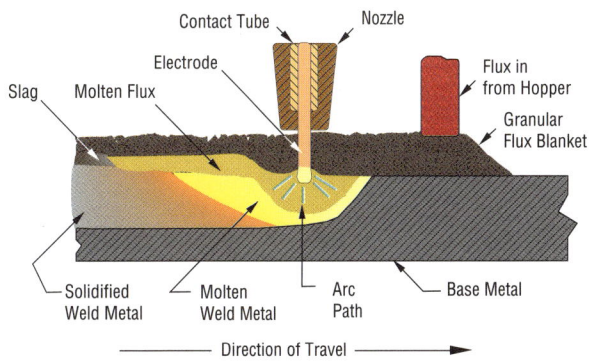

Figure 3.23—Submerged Arc Welding

the protection of the molten metal. As the welding progresses, in addition to the weld bead, there is a layer of formed slag and still-granular flux covering the solidified weld metal. The slag must be removed and is usually discarded, although there are techniques for recombining a portion of it with new flux for reuse in some applications. The still-granular flux can be recovered and reused if care is taken to prevent its contamination. In some cases where the flux must be extremely clean, reuse of the flux may not be advisable.

Since SAW uses a separate electrode and flux, there are numerous combinations available for specific applications. There are two general types of combinations which can be used to provide an alloyed weld deposit: an alloy electrode with a neutral flux, or a mild steel electrode with an alloy flux. Therefore, to properly describe the filler material for SAW, the American Welding Society identification system consists of designations for both the electrode and flux. Figure 3.24 shows what the various parts of the electrode/flux classification system signify, with two examples.

The equipment used for submerged arc welding consists of several components, as shown in Figure 3.25. Since this process can be used as a fully mechanized or semiautomatic method, the equipment used for each is slightly different. In either case, however, some power source is required. Although most submerged arc welding is performed with a

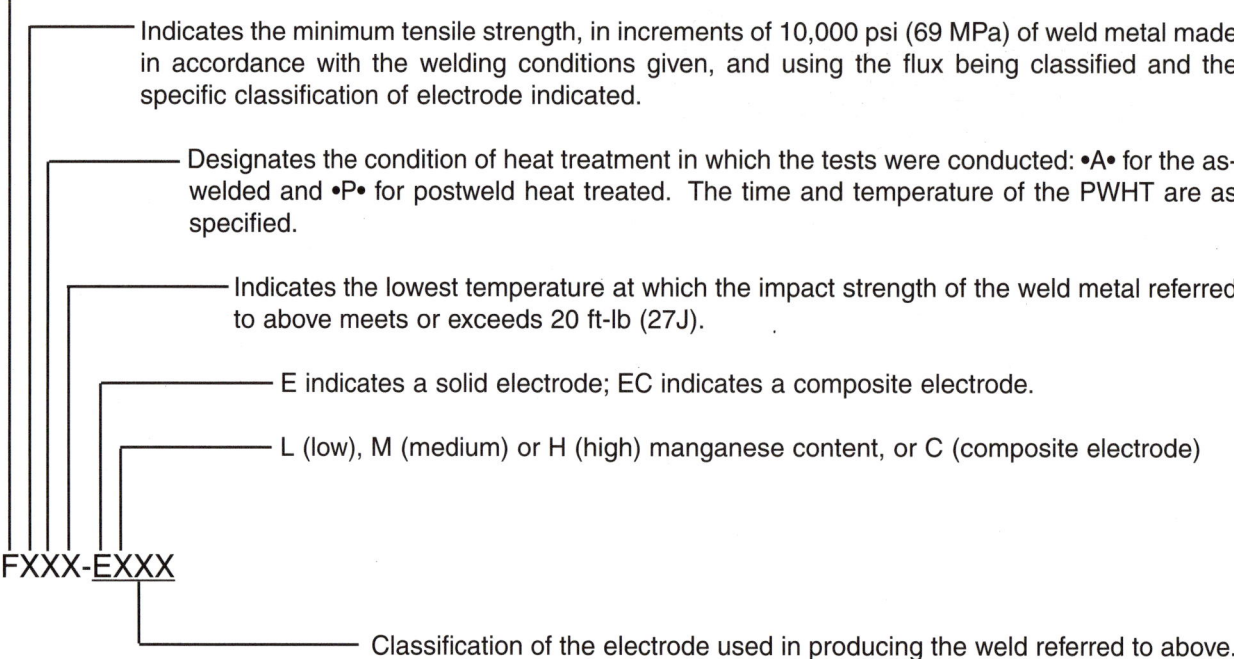

Figure 3.24—SAW Filler Metal Identification System

Figure 3.25—Submerged Arc Welding Equipment

constant voltage power source, there are certain applications where a constant current type is preferred. As with gas metal arc welding and flux cored arc welding, a wire feeder forces the wire through the cable liner to the welding torch.

The flux must be moved to the weld zone; for mechanized systems, the flux is generally placed into a hopper above the welding torch and fed by gravity so that it is distributed either slightly ahead of the arc or around the arc from a nozzle surrounding the contact tip. In the case of semiautomatic submerged arc welding, the flux is forced to the gun using compressed air which "fluidizes" the granular flux, causing it to flow easily, or there is a hopper connected directly to the hand-held gun.

Another equipment variation is the choice of alternating or direct current, either polarity. The type of welding current will affect both penetration and weld bead contour. For some applications, multiple electrodes can be used. The electrodes may be energized by a single power source, or multiple power sources may be necessary. The use of multiple electrodes provides even more versatility for the process.

SAW has found acceptance in many industries, and it can be performed on numerous metals. Due to the high rate of weld metal deposition, it has shown to be quite effective for overlaying or building up material surfaces. In situations where a surface needs improved corrosion or wear resistance, it is often more economical to cover a susceptible base metal with a resistant weld overlay. If this application can be mechanized, submerged arc welding is an excellent choice.

Probably the biggest advantage of SAW is its high deposition rate. It can typically deposit weld metal more efficiently than any of the more common processes. The submerged arc welding process also has high operator appeal because of the lack of a visible arc which allows the operator to control the welding without the need for a filter lens and other heavy protective clothing. The other beneficial feature is that there is less smoke generated than with some of the other processes. Another feature of the process which makes it desirable for many applications is its ability to penetrate deeply.

The major limitation of SAW is that it can only be done in a position where the flux can be supported in the weld joint. When welding in a position other than the normally-used flat or horizontal fillet positions, some device is required to hold the flux in place so it can perform its job. Another limitation is that, like most mechanized processes, there may be a need for extensive fixturing and positioning equipment. As with other processes using a flux, finished welds have a layer of solidified slag which must be removed. If welding parameters are improper, weld contours could be such that this job of slag removal is even more difficult.

The final disadvantage relates to the flux which covers the arc during welding. While it does a good job of protecting the welder from the arc, it also prevents the welder from seeing exactly where the arc is positioned with respect to the joint. With a mechanized setup, it is advisable to track the entire length of the joint without the arc or the flux to check for alignment. If the arc is not properly directed, incomplete fusion can result.

There are some inherent problems related to SAW. The first has to do with the granular flux. Just as with low hydrogen SMAW electrodes, it is necessary to protect the submerged arc welding flux from moisture. It may be necessary to store the flux in heated containers prior to use. If the flux becomes wet, porosity and underbead cracking may result.

Another characteristic problem of SAW is solidification cracking. This results when the welding conditions provide a weld bead having an extreme width-to-depth ratio. That is, if the bead's width is much greater than its depth, or vice versa, centerline shrinkage cracking can occur during solidification. Figure 3.26 shows some conditions which can cause cracking.

Plasma Arc Welding (PAW)

The next process to be discussed is plasma arc welding. A plasma is defined as an ionized gas. With any process using an arc, a plasma is created. However, PAW is so named because of the intensity

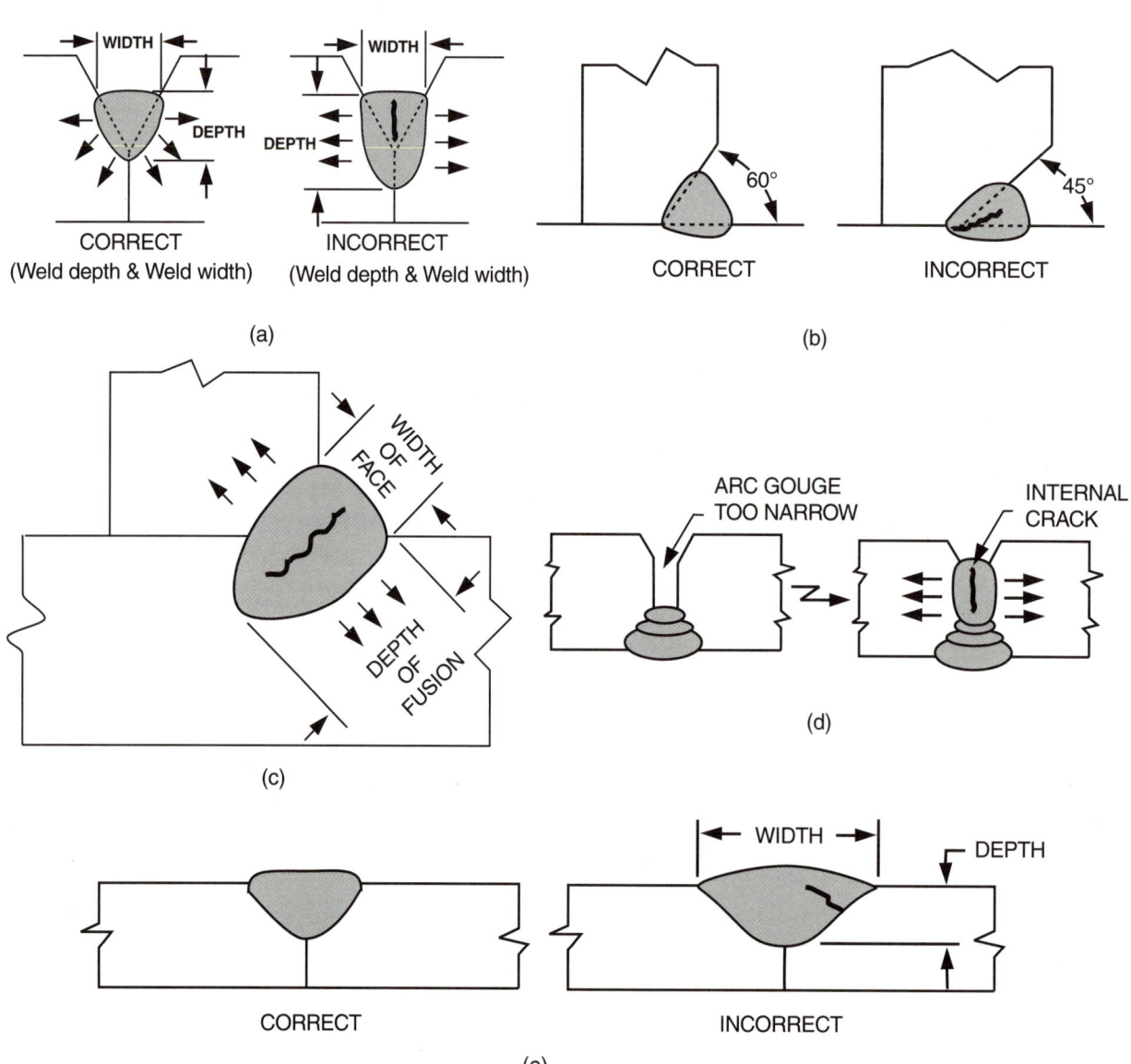

Figure 3.26—Solidification Cracking Because of Weld Profile

Autogenous = To weld w/o filler material

of this plasma region. At first glance, PAW could be easily mistaken for GTAW because the equipment required is quite similar. A typical setup is shown in Figure 3.27.

Both GTAW and PAW use the same type of power source. However, when we look closely at the torch itself, the difference becomes more obvious. Figure 3.28 shows a graphic comparison of the two types of welding torches and the resulting difference in the amount of heating, and therefore penetration, which will occur.

Both the PAW and GTAW torches use a tungsten electrode for the creation of the arc. However, with the PAW torch, there is a copper orifice within the ceramic nozzle. There is a "plasma" gas which is forced through this orifice and past the welding arc resulting in the constriction of the arc.

This constriction, or squeezing, of the arc causes it to be more concentrated, and therefore more intense. One way to illustrate the difference in arc intensity between GTAW and PAW would be to use the analogy of an adjustable water hose nozzle. The GTAW arc would be comparable to the gentle mist setting, while the PAW arc would behave more like the setting which provides a concentrated stream of water having a greater force.

There are two categories of plasma arc operation, the transferred and nontransferred arc. They are shown in Figure 3.29.

With the transferred arc, the arc is created between the tungsten electrode and the workpiece. The nontransferred arc, on the other hand, occurs between the tungsten electrode and the copper orifice. The transferred arc is generally used for both welding and cutting of conductive materials, because it results in the greatest amount of heating of the workpiece. The nontransferred arc is preferred for the cutting of nonconductive materials and for welding of materials when the amount of heating of the workpiece must be minimized.

The similarities between GTAW and PAW extend to the equipment as well. The power sources are identical in most respects. However, as shown in Figure 3.30, there are some additional elements

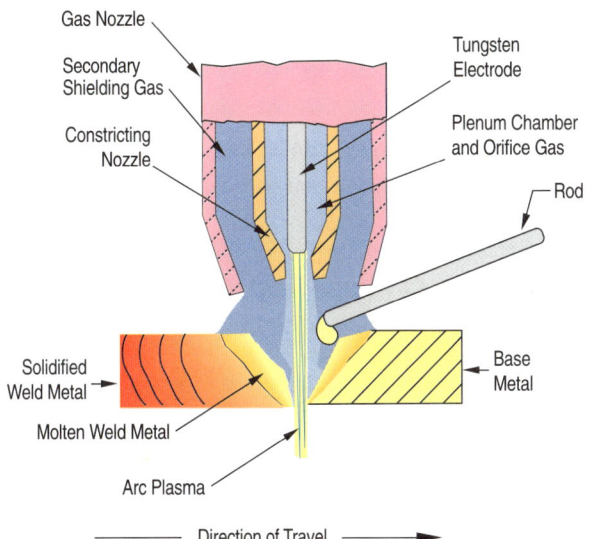

Figure 3.27—Plasma Arc Welding

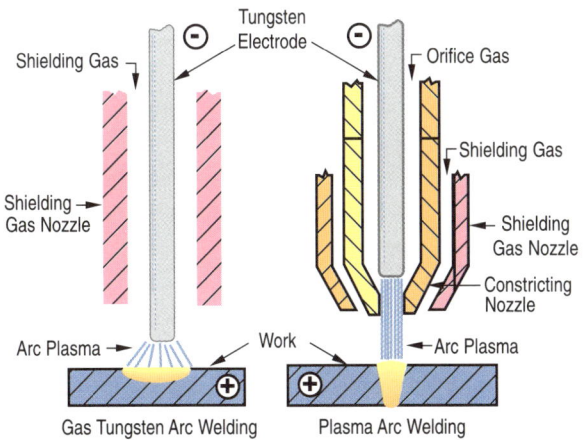

Figure 3.28—Comparison of GTAW and PAW Torches

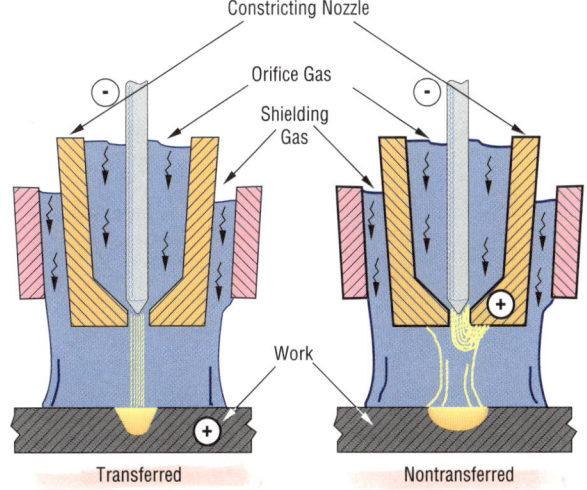

Figure 3.29—Transferred and Nontransferred PAW Comparison

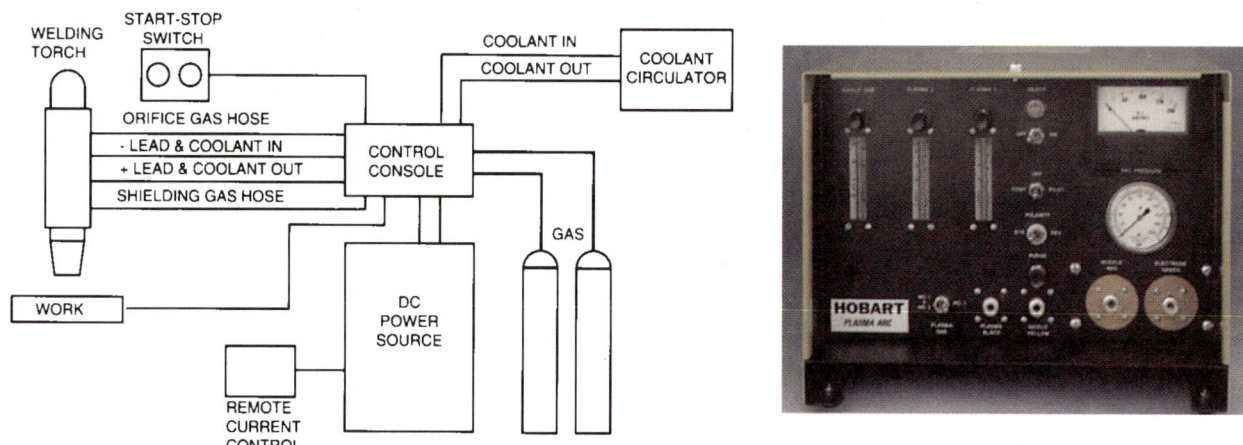

Figure 3.30—Plasma Arc Welding Equipment, Including Control Console

necessary, including the plasma control console and a source of plasma gas.

The torch, as discussed above, does differ slightly; however, a careful check of the internal configuration must be made to be certain. Figure 3.31 illustrates some of the internal structure of a typical PAW torch for manual welding.

As indicated, two separate gases are required: the shielding gas and the orifice (or plasma) gas. Argon is most commonly employed for both types of gas. However, welding of various metals might warrant the use of helium or combinations of argon/helium or argon/hydrogen for one of the other gases.

The primary applications for PAW are similar to those for GTAW. PAW is used for the same materials and thicknesses. PAW becomes the choice where applications warrant the use of a more localized heat source. It is used extensively for full penetration welds in material up to 1/2 in. thick by employing a technique referred to as "keyhole welding." Figure 3.32 shows the typical appearance of a keyhole weld.

Keyhole welding is performed on a square butt joint with no root opening. The concentrated heat of the arc penetrates through the material thickness to form a small keyhole. As welding progresses, the keyhole moves along the joint melting the edges of the base metal which then flow together and solidify after the welding arc passes. This creates a high quality weld, with no elaborate joint preparation and fast travel speeds compared to GTAW.

One advantage of PAW, which was mentioned before, is that it provides a very localized heat source. This allows for faster welding speeds and therefore less distortion. Since the standoff used between the torch end and the workpiece is typically quite long, the welder has better visibility of the weld being made. Also, since the tungsten electrode is recessed within the torch, the welder is less likely to contact the molten metal and produce a tungsten inclusion.

The ability to use this process in a keyhole mode is also desirable. The keyhole is a positive indication of complete penetration and weld uniformity. This weld uniformity is in part due to the fact that plasma arc welding is less sensitive to changes in

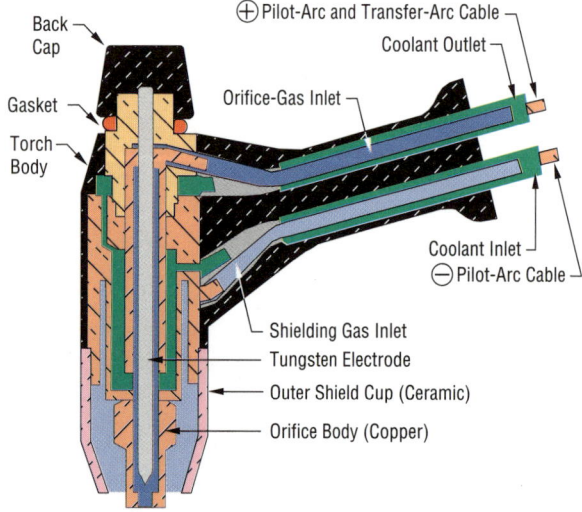

Figure 3.31—Internal Structure of a Typical Manual PAW Torch

Figure 3.32—Keyhole Technique for Plasma Arc Welding (Face – top and Root – bottom)

arc length. The presence of its collimated arc will permit relatively large changes in torch-to-work distance without any change in its melting capacity.

PAW is limited to the effective joining of materials 1 in. or less in thickness. The initial cost of the equipment is slightly greater than that for GTAW, primarily because there is additional apparatus required. Finally, the use of PAW may require greater operator skill than would be the case for GTAW due to the more complex equipment setup.

Among the problems that may be encountered with this process are two types of metal inclusions. Tungsten inclusions may result from too-high current levels; however, the fact that the tungsten is recessed helps to prevent this occurrence. Too-high current could also result in the copper orifice melting and being deposited in the weld metal. Another problem that may be encountered when keyhole welding is being done is referred to as *tunneling*. This occurs when the keyhole is not completely filled at the end of the weld, leaving a cylindrical void which may extend entirely through the throat of the weld. When using the keyhole technique, there is also a possibility of getting incomplete fusion since the arc and joint are so narrow. As a result, even small amounts of mistracking can produce incomplete fusion along the joint.

Electroslag Welding (ESW)

The next process of interest is electroslag welding, but it is not nearly as commonly used as the processes mentioned previously. It typically exhibits the highest deposition rate of any of the welding processes. ESW is characterized by the joining of members which are placed edge to edge so that the joint is vertical. The welding is done in a single pass such that the progression is from the bottom to the top of the joint, without interruption. Even though the welding progresses vertically up the joint, the position of welding is considered flat due to the location of the electrode with respect to the puddle. During welding, the molten metal is supported on two sides by water cooled shoes. See Figure 3.33.

An interesting feature of ESW is that it is not considered to be an arc welding process. It relies on heating from the electrical resistance of the molten flux to melt the base and filler metals. The process does use an arc to initiate the operation; however, that arc is extinguished once there is sufficient flux melted to provide the heating to maintain the weld-

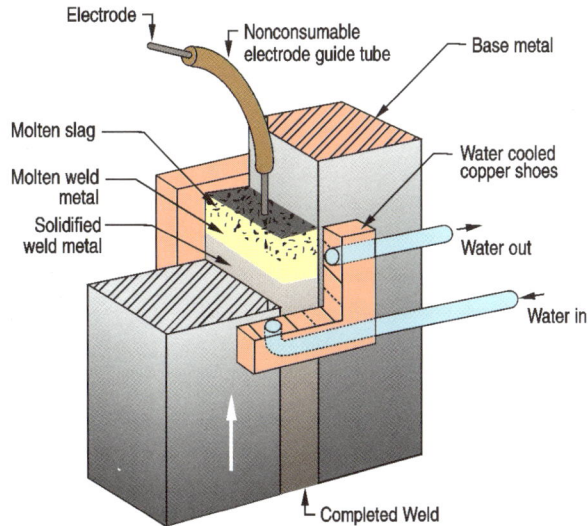

Figure 3.33—Electroslag Welding

ing operation as it progresses upward along the joint.

ESW is used when very heavy sections are being joined. It is essentially limited to the welding of carbon steels in thicknesses greater than 3/4 in. So, only industries dealing with heavy weldments have any real interest in ESW. Figure 3.34 shows an ESW equipment setup.

The major advantage of ESW is its high deposition rate. If single electrode welding is not fast enough, then multiple electrodes can be used. In fact, metal strip can be used instead of wire to increase the deposition rate even more. Another benefit is that there is no special joint preparation required. In fact, a rough, flame cut surface is satisfactory for this method. Since the entire thickness of the joint is fused in a single pass, there is no tendency for any angular distortion to occur during or after welding, so alignment is easily maintained.

The primary limitation of ESW is the extensive time required to set up and get ready to weld. There is a tremendous amount of time and effort required to position the workpieces and guides before any welding takes place. That is why ESW is not economical for thinner sections, even though the deposition rate is quite high.

The ESW process has associated with it several inherent problems. When these problems arise, they can be of major proportions. Gross porosity can occur due to wet flux or the presence of a leak in one of the water cooled shoes. Since electroslag welding resembles a casting process in many respects, there is a possibility of getting centerline cracks due to weld metal shrinkage. Also, due to the tremendous amount of heating, there is a tendency for grain growth in the weld metal. These large grains may result in degradation of the weldment's mechanical properties.

Oxyacetylene Welding (OAW)

The next process is oxyacetylene welding. While the term "oxyfuel welding" is also used, acetylene is the only fuel gas capable of producing high enough temperatures for effective welding. With OAW, the energy for welding is created by a flame, so this process is considered to be a chemical welding method. Just as the heat is provided by a chemical reaction, the shielding for oxyacetylene welding is accomplished by this flame as well. Therefore, no flux or external shielding is necessary. Figure 3.35 illustrates the process being applied with filler metal added from some external source.

The equipment for oxyacetylene welding is relatively simple. A typical setup is shown in Figure 3.36. It consists of several parts: oxygen tank, acetylene tank, pressure regulators, torch, and connecting hoses. The oxygen cylinder is a hollow, high pressure container capable of withstanding a

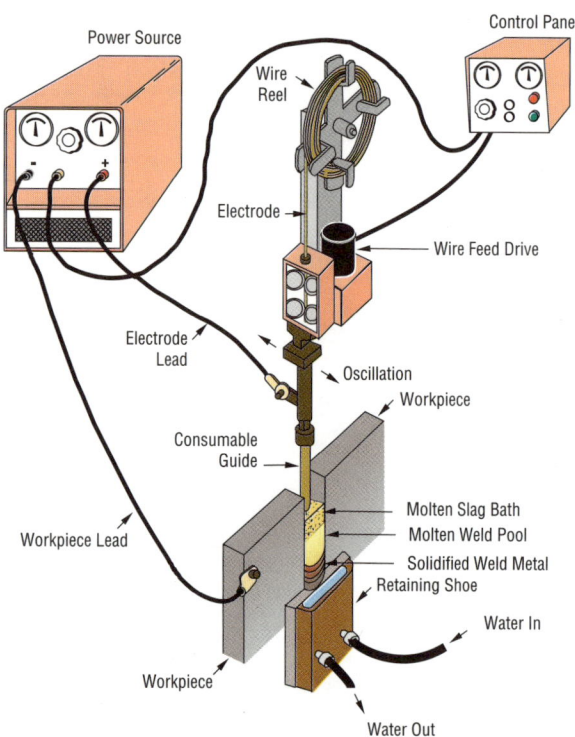

Figure 3.34—Electroslag Welding Equipment

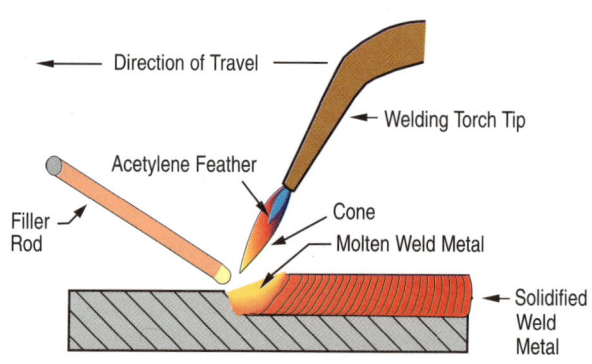

Figure 3.35—Oxyacetylene Welding

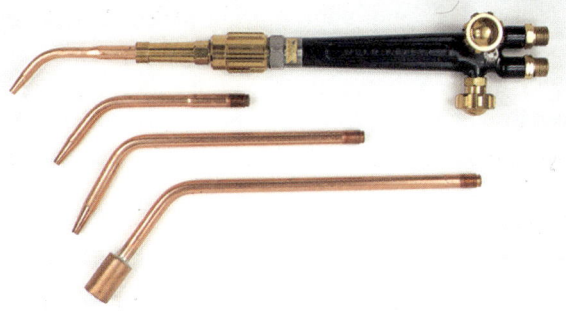

Figure 3.36—Oxyacetylene Welding Equipment

pressure of approximately 2200 psi. The acetylene cylinder on the other hand, is filled with a porous material similar to cement.

Acetylene exists in the cylinder dissolved in liquid acetone. Care must be taken since gaseous acetylene is extremely unstable at pressures exceeding 15 psi and an explosion could occur even without the presence of oxygen. Since the acetylene cylinder contains a liquid it is important that it remains upright to prevent drawing off the liquid.

Each cylinder has attached to its top a pressure regulator which reduces the high internal tank pressure to working pressures. Hoses then connect these regulators to the torch. The torch includes a mixing section where the oxygen and acetylene combine to provide the necessary mixture. The ratio of these two gases can be altered by the adjustment of two separate control valves. Normally, for carbon steel welding, they are adjusted to provide a mixture, which is referred to as a neutral flame. A higher amount of oxygen will create an oxidizing flame and a higher amount of acetylene will produce a carburizing flame. After the gases are mixed, they flow through a detachable tip. Tips are made in a variety of sizes to allow welding of different metal thicknesses.

The filler material used for OAW on steel has a simple identification system. Two examples are RG-45 and RG-60. The "R" designates it as a rod, "G" stands for gas and the 45 and 60 relate to the minimum tensile strength of the weld deposit in thousands of pounds per square inch (psi). So 45 designates a weld deposit having a tensile strength of at least 45,000 psi.

Although not used as extensively as it once was, OAW still sees some usage. Its primary tasks include the welding of thin steel sheet and small diameter steel piping. It is also applied in many maintenance situations as well.

The advantages of OAW include some desirable features of the equipment itself. First, it is relatively inexpensive and can be made very portable. This portability relates not only to the compact size but also to the fact that there is no electrical input required. Care should be taken when moving the equipment so that the primary valves on the cylinders are not damaged. If broken off, a cylinder can turn into a lethal missile. So, whenever transported, the regulators should be removed and the valves covered with special screw-on caps for protection from impact.

The process also has certain limitations. For one, the flame does not provide as concentrated a heat source as can be achieved by an arc. Therefore, if a groove weld is being made, the joint preparation should exhibit a thin "feather edge" to assure that complete fusion is obtained at the root of the joint. This lower heat concentration also results in a relatively slow process, so we typically consider OAW best suited for thin section welding. As with any of the welding processes requiring the filler metal to be fed manually, OAW requires a substantial skill level for best results.

There are certain inherent problems associated with OAW. They are primarily related to either improper manipulation or adjustment of the flame. Since the heat source is not concentrated, care must be taken to direct the flame properly to assure adequate fusion. If the flame is adjusted such that an oxidizing flame or carburizing flame is produced, weld metal properties could be degraded, so it is important to have equipment capable of providing uniform gas flow.

Stud Welding (SW)

The next welding process to be discussed is stud welding. This method is used to weld studs, or attachments, to some metal surface. SW is considered to be an arc welding process because the heat for welding is generated by an arc between the stud and the base metal.

The process is controlled by a mechanical gun which is attached to a power supply through a control panel. So, welding is accomplished very easily and repetitively. The process is performed in four cycles which are timed and sequenced by the control box once the stud is positioned and the trigger is pulled. Figure 3.37 illustrates this sequence.

Sketch (a) shows the stud gun with the stud and ferrule in position, and then in (b) being positioned against the workpiece. In (c), the trigger has been pulled to initiate the current flow, and the gun lifts the stud to maintain the arc. In (d), the arc quickly melts the stud end and a spot on the workpiece beneath the stud. A timer in the gun then cuts off the

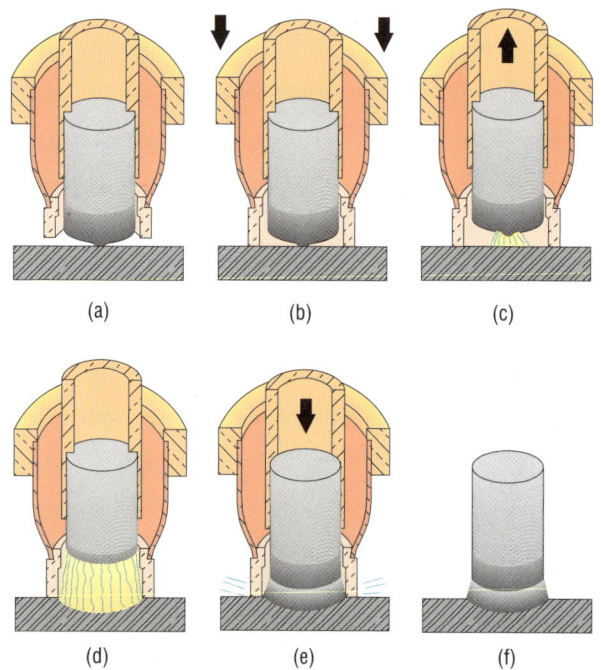

Figure 3.37—Stud Welding Cycle

current and the gun's mainspring plunges the stud into the workpiece (e). The finished stud weld is shown in (f). When properly done, the stud weld should exhibit complete fusion throughout the stud cross-section as well as a reinforcing fillet, or "flash," around the entire circumference of the stud base.

Typical SW equipment is shown in Figure 3.38. Stud welding equipment consists of a DC power source, a control unit, and a stud welding gun. Variations can include automatic stud feeding appa-

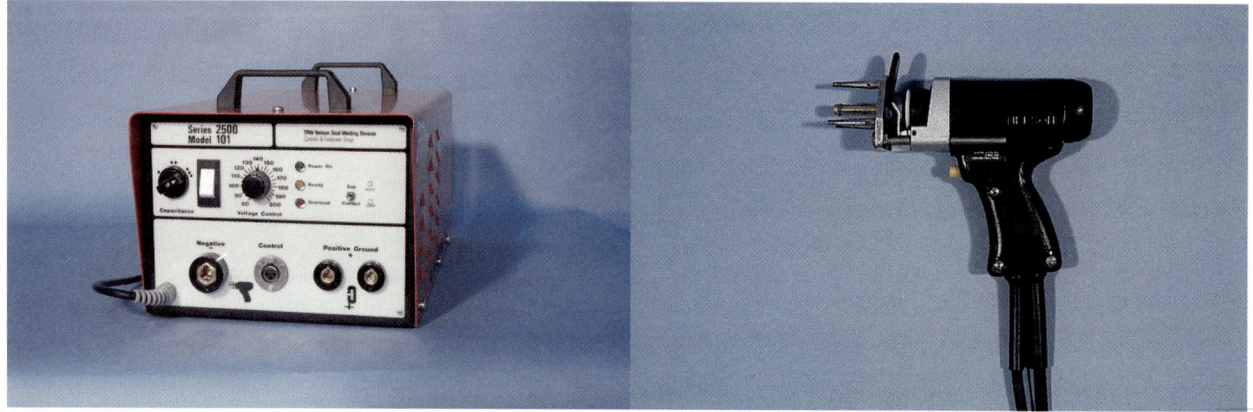

Figure 3.38—Stud Welding Equipment, Including Power Console and Stud Gun

ratus as well as gas shielding for use in the welding of aluminum studs.

Due to the convenience and simplicity offered by SW, it has seen tremendous usage in many industries for a variety of metals. Figure 3.39 shows some of the wide variety of stud shapes and sizes available.

The building and bridge industries use SW extensively as shear connectors to structural steel members. Once the concrete is poured, covering the studs attached to the beams, the mechanical connection obtained allows the steel and concrete to act as a composite unit to enhance the overall strength and rigidity of the structure.

Its wide range of applications is due to the number of advantages which are offered. First, since the process is controlled essentially by the electrical control unit and attached gun, little operator skill is required once the control unit settings are made. Also, SW is a tremendously economical and effective method for welding various attachments to a surface. Its use eliminates the need for hole drilling, tapping or tedious manual welding using some other process. Once welded, a stud can be easily inspected. First a visual examination is made to assure the presence of a 360° flash. Then the stud can be either struck with a hammer or pulled to judge its acceptability. When struck with a hammer, a good stud weld will ring while a poor joint will result in a dull "thud."

Since the process is controlled electrically and mechanically, its predominant limitation relates to this equipment. An electrical or mechanical malfunction could produce poor weld quality. And stud shape is limited to some configuration which can be held in the gun's chuck.

SW has two possible discontinuities. They are lack of 360° flash and incomplete fusion at the interface. Both are caused by improper machine settings or insufficient work connections. Presence of water or heavy rust or mill scale on the base metal surface could also affect the resulting weld quality.

Laser Beam Welding (LBW)

Laser Beam Welding (LBW) is a fusion joining process that produces coalescence of materials with the heat obtained from a concentrated beam of coherent, monochromatic light impinging on the joint to be welded (see Figures 3.40 and 3.41). The high energy of the laser beam causes some of the metal at the joint to vaporize, producing a "keyhole," which is surrounded by molten metal. As the beam is then advanced along the joint (or the part is moved under the beam), molten metal flows from the forward portion of the keyhole around its periphery and solidifies at the rear to form weld metal. The equipment is fairly complex, as seen in Figure 3.43.

The main element for laser welding and cutting equipment is the laser device. Inside the equipment, the laser device is placed between end mirrors. When the laser medium is excited or "pumped," the atoms or molecules in the medium are put into a higher than normal energy state, which generates a source of coherent, monochromatic electromagnetic radiation in the form of light. This light reflects

Figure 3.39—Some Typical Stud and Fastener Configurations Available for Stud Welding

Figure 3.40—Laser Beam Welding Gun

Figure 3.41—Laser Weld Being Made on 1/8 in. (3.2 mm) Thick Type 304 Stainless Steel

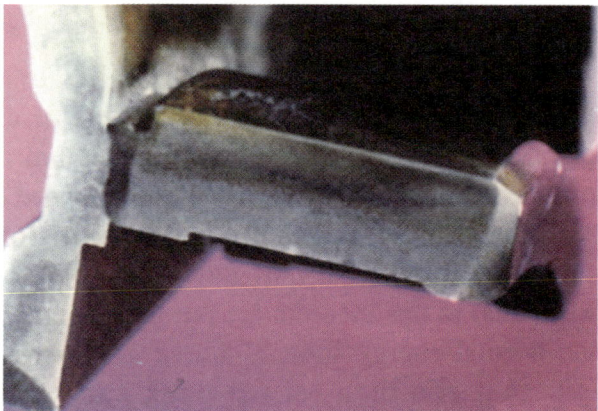

Figure 3.42—Cross Section of a Laser Beam Weld Joining a Boss to a Ring

back and forth between the end mirrors, which increases the energy state even more. This results in the device emitting a beam of laser light. Laser is an acronym for "light amplification by stimulated emission of radiation."

The laser beam has a very small cross section and does not diverge, or broaden, much. Thus, it can be transported over relatively long distances through fiber optics and mirrors. The beam is then focused to a very small spot size at the workpiece through the use of either lenses or reflective-type focusing. This provides the high level of beam power density needed to do a variety of material processing tasks, such as welding, cutting, and heat treating. The small laser beam produces a very narrow and deep weld bead (see Figure 3.42).

The lasers predominantly used for welding are either solid-state or gas lasers. In the solid-state laser, neodymium-doped, yttrium aluminum garnet (Nd-YAG) crystal rods are utilized to produce a continuous monochromatic laser beam output in the 1 to 10 kW power range. In gas lasers for welding, the typical type is the carbon dioxide (CO_2) laser. These are electrically excited and can put out a continuous or a pulsed laser beam, with power levels of up to 25 kW. Such lasers are capable of producing full penetration, single-pass welds in steel up to 1-1/4 in. (32 mm) thick.

LBW is a noncontact process, and thus requires that no pressure be applied. Inert gas shielding is often employed to prevent oxidation of the molten pool, and filler metal is occasionally used.

Major advantages of laser beam welding include the following:

- Low overall heat input results in less grain growth in the heat-affected zone and less workpiece distortion.
- High depth-to-width ratios (on the order of 10:1) are attainable when the weld is made in the keyhole welding mode.
- Single pass laser welds have been made in materials up to 1-1/4 in. (32 mm) thick.
- The laser beam can be focused on a small area, permitting the joining of thin, small, or closely spaced components.
- A wide variety of materials can be welded, including combinations of materials with dissimilar physical properties.
- The laser beam can be readily focused, aligned, and directed by optical elements. Thus, the laser can be located away from the workpiece and the laser beam directed around tooling and obstacles to the workpiece.
- The laser beam is not influenced by the presence of magnetic fields, as in arc and electron beam welding.
- No vacuum or X-ray shielding is required, as in electron beam welding.

Figure 3.43—Production Welding System for Automotive Transmission Components

- The beam can be transmitted to more than one work station using beam switching optics.

Some limitations of the laser beam welding process include:

- Joints must be accurately positioned under the beam.
- Square groove butt joints are required.
- Workpieces must often be forced together.
- The high reflectivity and high thermal conductivity of some materials, such as aluminum and copper alloys, can affect their weldability with lasers.
- The fast cooling rates can produce cracking and embrittlement in the heat-affected zone and can trap porosity in the weld metal.
- With higher power lasers, a plume of vapors is often produced above the weld joint, which interferes with the ability of the laser to reach the joint. A plasma control device is often required, which utilizes an inert gas to blow away the plume of vapors.
- Equipment is expensive, typically in the $100,000 range.

Electron Beam Welding (EBW)

Since electron beam welding (EBW) was initially used as a commercial welding process in the late 1950s, the process has earned a broad acceptance by industry. The process was initially limited strictly to operation in a high vacuum chamber. However, a system was soon developed that required a high vacuum only in the beam generation portion. This permitted the option of welding in either a medium vacuum chamber or a nonvacuum environment. This advancement led to its acceptance by the commercial automotive and consumer product manufacturers. As a consequence, EBW has been employed in a broad range of industries worldwide (see Figures 3.44–3.46).

EBW is a fusion joining process that produces coalescence of materials with heat obtained by impinging a beam of high-energy electrons onto the joint to be welded.

The heart of the electron beam welding process is the electron beam gun/column assembly. Electrons are generated by heating a negatively charged emitting cathode or "filament" to its thermionic emission temperature range, thus causing electrons to "boil off" and be attracted to the positively charged anode (see Figure 3.47). A precisely configured grid or bias cup surrounding the

Figure 3.44—Exterior View of an Electron Beam Vacuum Pump

Figure 3.45—Electron Beam Welding Control Panel

Figure 3.46—Electron Beam Welding Machine Designed for Joining Bimetallic Strip

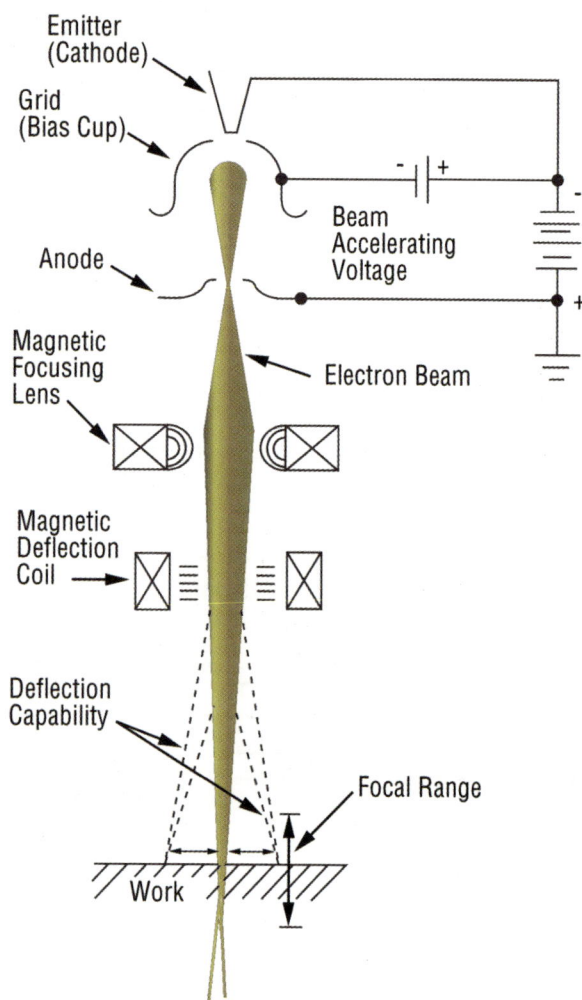

Figure 3.47—Simplified Representation of a Triode Electron Beam Gun Column

emitter helps accelerate and shape the electrons into the beam. The beam then exits the gun through an opening in the anode and continues on toward the workpiece (see Figure 3.48). Once the beam exits from the gun, it will gradually broaden with distance. To counteract this inherent divergence effect, an electro-magnetic lens system is used to converge the beam, which focuses it into a small spot on the workpiece. The beam divergence and convergence angles are relatively small, which gives the concentrated beam a usable focal range, or "depth-of-focus," extending over a distance of an inch or so.

There are four basic welding variables: beam accelerating voltage, beam current, beam focal spot size, and welding travel speed. The basic equipment includes a vacuum chamber, controls and an electron beam gun (see Figures 3.44–3.46). Typical power levels are 30 to 175 kV and 50 to 1000 mA.

The electron beam produces even higher power densities than a laser beam. Like laser beam welding, electron beam welding is usually done in the "keyhole" mode, which produces very deep and narrow weld beads (see Figure 3.49). In most applications, the weld penetration formed is much deeper than it is wide, and the heat-affected zone produced is very narrow. For example, the width of a butt weld in 0.5 in. (13 mm) thick steel plate may be as small as 0.030 in. (0.8 mm) when made in a vacuum. This stands in remarkable contrast to the weld zone produced in arc and gas welded joints, where penetration is achieved primarily through conduction melting.

An electron beam can be readily moved by electromagnetic deflection. In most instances, this deflection is used to adjust the beam-to-joint alignment, or to apply a deflection pattern of a circle, ellipse or other shape. This deflection modifies the average power density into the joint, and results in a change in the weld bead shape. The focal spot can also be adjusted, which will alter the weld shape.

Electron beam welding has unique performance capabilities. The high-power densities and outstanding control solve a wide range of joining problems.

The following are advantages of electron beam welding:

- Low overall heat input results in less grain growth in the heat-affected zone and less workpiece distortion.
- High depth-to-width ratio (on the order of greater than 10:1) are attainable when the weld is made in the keyhole welding mode.
- Single pass electron beam welds have been made in steels up to 4 in. (102 mm) thick.
- A high-purity environment (vacuum) for welding minimizes contamination of the metal by oxygen and nitrogen.
- Rapid travel speeds are possible because of the high melting rates associated with this concentrated heat source.
- Hermetic closures can be welded with the high- or medium-vacuum modes of operation while retaining a vacuum inside the component.
- The beam of electrons can be magnetically deflected to produce various shaped welds and magnetically oscillated to improve weld quality or increase penetration.
- The focused beam of electrons has a relatively long depth of focus, which will accommodate a broad range of work distances.
- Full penetration, single-pass welds with nearly parallel sides, and exhibiting nearly symmetrical shrinkage, can be produced.
- Dissimilar metals and metals with high thermal conductivity such as copper can be welded.

Some of the limitations of electron beam welding are as follows:

- Joints must be accurately positioned under the beam.
- Square groove butt joints are required.

Figure 3.48—Electron Beam Welding a Gear in Medium Vacuum

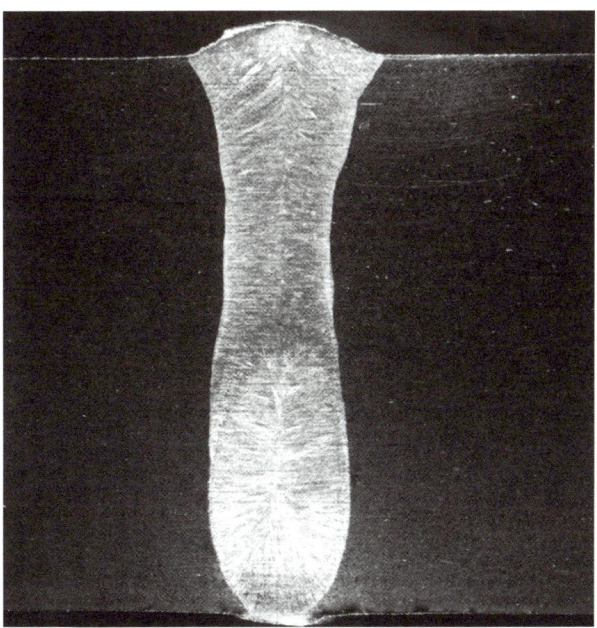

Figure 3.49—Cross Section of a Nonvacuum Electron Beam Weld i 3/4 in. (19mm) Stainless Steel Plate

- Workpieces must often be forced together.
- The fast cooling rates can produce cracking and embrittlement in the heat-affected zone and can trap porosity in the weld metal.
- Equipment is expensive, typically in the $1,000,000 range.
- For high and medium vacuum welding, work chamber size must be large enough to accommodate the assembly operation. The time needed to evacuate the chamber will influence production costs.
- Because the electron beam can be deflected by magnetic fields, nonmagnetic or properly degaussed metals should be used for tooling and fixturing close to the beam path.
- With all modes of EBW, radiation shielding must be maintained to ensure that there is no exposure of personnel to the X-radiation generated by EB welding.
- Adequate ventilation is required with nonvacuum EBW, to ensure proper removal of ozone and other noxious gases formed during this mode of EB welding.

Resistance Welding (RW)

Resistance welding (RW) is a group of welding processes that produces coalescence of the joining surfaces with heat obtained from resistance of the workpieces to the flow of welding current in a circuit of which the workpieces are a part, and by the application of pressure. It is typically used for sheet metal applications, up to about 1/8 in. (3 mm) thick. No filler metals or fluxes are used.

There are three major resistance welding processes: resistance spot welding (RSW), resistance seam welding (RSEW), and projection welding (PW). Electrodes are usually copper alloys, but many different types of electrode materials have been developed for specific purposes, such as for welding of galvanized steel.

The most common of these processes is resistance spot welding (RSW), which is shown in Figure 3.50. The electrodes are typically cylindrical in shape, but can have various configurations. The two electrodes apply a force to hold the two pieces of sheet metal in intimate contact. Current is then passed through the electrodes and the workpieces. Resistance to the flow of current produces heat at the faying surfaces, forming a weld nugget (see

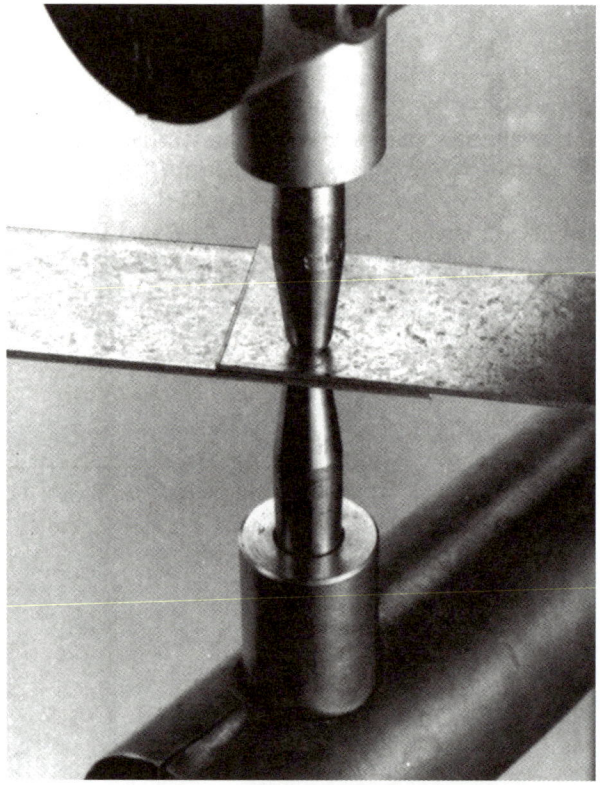

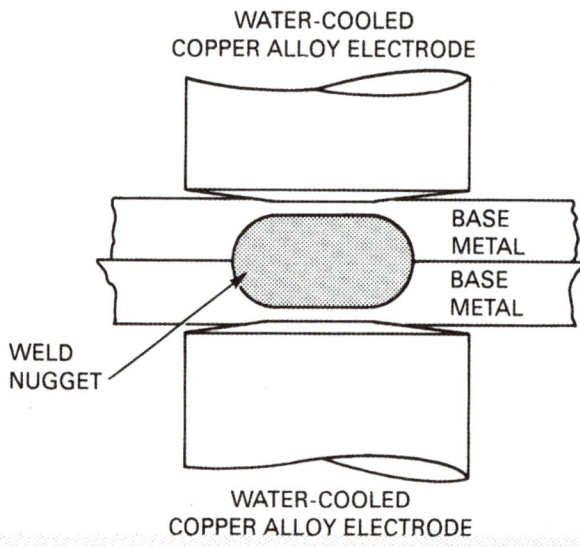

Figure 3.50—Resistance Spot Welding Process

Figure 3.50). The electrodes continue to apply pressure to hold the sheets in intimate contact during welding.

The workpiece surfaces must be very clean to obtain consistent electrical contact and to produce a sound weld nugget. Typically, one spot weld is made at a time.

With projection welding (PW), one sheet has projections or dimples formed in it. When the two sheets are placed together, the current is concentrated to pass through the projections at the faying surfaces. Large, flat electrodes are used on opposite sides of the sheets, and current passes through the projections while the electrodes force the sheets together. This allows several welds to be made during a single welding cycle.

In resistance seam welding (RSEW) a continuous seam weld is made that is actually a series of overlapping spot welds. The electrodes are typically rotating wheels between which the two sheets pass. Current and pressure are applied in a timed manner to produce a continuous seam weld.

Equipment ranges from semiautomatic to fully automatic equipment. With semiautomatic equipment, the operator places the sheets to be welded between the electrodes or places a hand-held gun around the pieces and pushes a switch or foot pedal. The weld is made in a preprogrammed sequence. In automatic equipment, parts are automatically loaded into the machine, welded, then ejected. Robotic resistance spot welding is utilized extensively in the automotive industry.

The main welding variables are welding current, welding time, electrode force, and electrode material and design. Typical welding times for a resistance weld are less than a second with current levels of hundreds to thousands of amperes.

Brazing Processes

Now that the welding processes have been discussed, we will turn our attention to brazing. Brazing differs from welding in that brazing is accomplished without any melting of the base metals. The heat applied is only sufficient for the melting of the filler metal. Another joining process, soldering, is similar in that it too only requires melting of the filler metal to create a bond. Brazing and soldering are differentiated by the temperature at which the filler metal melts. Filler metals melting above 840°F (450°C) are considered braze materials, while those melting below this temperature are used for soldering. Therefore, the common term "silver soldering" is actually incorrect, because silver brazing filler metals melt above 840°F.

Even though the base metals are not melted and there is no fusion between the filler metal and base metals, a bond is created which has substantial strength. When properly applied, the braze joint can develop a strength equal to or greater than the base metal even though the braze material may be much weaker than the base metal. This is possible because of two factors.

First, the braze joint is designed to have a large surface area. Second, the clearance, or gap, between the two pieces to be joined is kept to a minimum. Gaps greater than about 0.010 in. (0.25 mm) may result in a joint having substantially reduced strength. Some typical braze joint configurations are shown in Figure 3.51. As can be seen, all of these joints have relatively large surface areas and tight gaps between parts.

To perform brazing, one of the most important steps is to thoroughly clean the joint surfaces. If the parts are not sufficiently clean, an inadequate joint will result. Once the parts are cleaned and fitted together, heat is applied in some manner. When the parts have been heated to a temperature above the melting point of the braze filler material, the filler metal will be drawn into the joint when placed in contact with the parts as a result of capillary action.

Capillary action is that phenomenon which causes a liquid to be pulled into a tight space between two surfaces. You could observe this occurrence if two pieces of plate glass were held

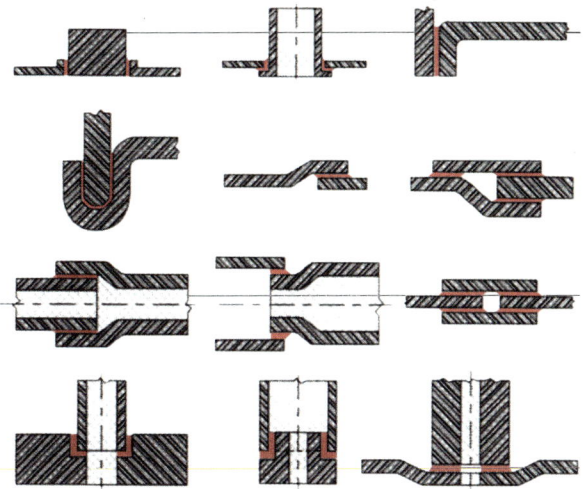

Figure 3.51—Examples of Various Braze Joint Configurations (Red Shows Resulting Braze Zone)

tightly together and stood on edge in a shallow pan of water. Capillary action will cause the liquid between the pieces of glass to rise to a level above that of the water in the pan. Since capillary action is related to surface tension, it is drastically affected by the presence of surface contamination.

So, if the surfaces of the braze joint are not properly cleaned, the ability of the capillary action to occur will be reduced to the point that the braze material will not be drawn sufficiently into the joint. When this occurs, an insufficient bond will result.

The braze filler material is available in a number of configurations and alloy types. The configurations include wire, strip, foil, paste and preforms. Preforms are specially shaped pieces of braze alloy designed for a particular application so that they are preplaced in or near the braze joint during assembly of the parts. Figure 3.52 shows how these braze preforms can be preplaced within a joint prior to the application of the brazing heat. Figure 3.53 then shows how the braze filler metal flows into the joint leaving voids where the preform had been placed.

As with the welding consumables, braze alloys also have American Welding Society designations. Braze alloy designations are preceded by a "B" followed by abbreviations of the most prominent chemical elements included (see Figure 3.54). Within these general groups are types with slightly different properties which are differentiated by individual numbers. The brazing filler metals having an "R" in front of the "B" in their designations denotes their chemistry is identical with Copper and Copper-Alloy Gas Welding Rods.

To maintain the cleanliness of the joint during the application of heat, brazing fluxes are often used. They too carry American Welding Society classifications according to the types of base and filler metals being used. They are designated alphanumerically as shown in Figure 3.55.

There are numerous methods of brazing, with the primary difference being the manner in which the joint is heated. The most familiar method is referred to as torch brazing (TB) where heating is accomplished using an oxyfuel flame. It can be done either manually, mechanically or automatically. Other common heating methods include furnace, induction, resistance, dip and infrared.

Furnace brazing (FB) is performed in a furnace, often with a controlled atmosphere. The braze filler

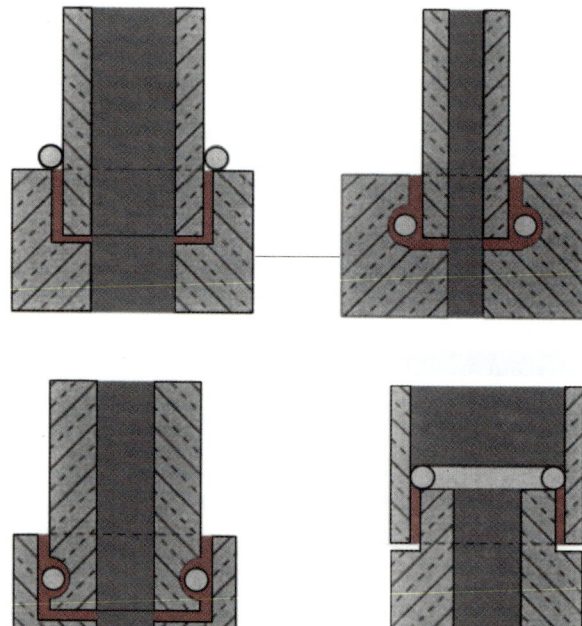

Figure 3.52—Placement of Brazing Preforms in Braze Joints (Red Shows Resulting Braze Zone)

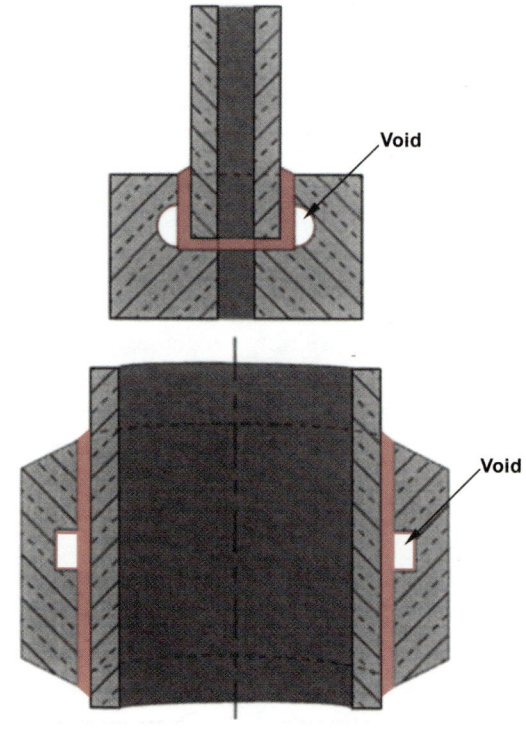

Figure 3.53—Location of Braze Filler Material in Joint After the Application of Heat (Red Shows Resulting Braze Zone)

AWS Brazing Filler Metal Classifications

Designation	Primary Element
BAlSi	Aluminum-Silicon
BCuP	Copper-Phosphorus
BAg	Silver
BAu	Gold
BCu	Copper
RBCuZn	Copper-Zinc
BMg	Magnesium
BNi	Nickel

Figure 3.54—AWS Brazing Filler Metal Identification System (Partial Listing)

material and flux are preplaced at or near the joint and the parts to be joined are then placed in a furnace which heats them in a very controlled manner. FB can be used to produce numerous braze joints simultaneously once the assembly is brought to the brazing temperature.

Induction brazing (IB) relies on the heat produced in a metal when placed within an induction coil. The induction coil is simply a coil through which a high frequency electric current is passed. This flow of electricity will produce substantial heating of a piece of metal placed inside the coil.

Resistance brazing (RB) is accomplished by heating the base metal using its own inherent electrical resistance. When an electric current is passed through the base metals on either side of the braze joint, resistance heating occurs and melts the braze filler metal placed in the joint.

Dip brazing (DB) differs from the other types in that the parts to be joined are immersed in some type of molten bath to provide the necessary heating. This bath can be either molten braze filler metal or some type of molten chemical, such as chemical salts.

Infrared brazing (IRB) relies on heating provided by radiant energy. That is, the joint to be brazed is heated using a high intensity infrared light source.

Brazing is used in many industries, especially aerospace and heating and air conditioning. It can be applied to join virtually all metals, and can also be used to join metals to nonmetals.

One of the biggest advantages of brazing is that it can be used to join dissimilar metals. This is

Classification	Form	Filler Metal Type	Typical Ingredients	Application	Activity Temperature Range °F	°C	Recommended Base Metals
FB1-A	Powder	BAlSi	Fluorides Chlorides	For torch or furnace brazing.	1080–1140	560–615	All brazeable aluminum alloys.
FB2-A	Powder	BMg	Fluorides Chlorides	Because of very limited use of brazing to join magnesium, a detailed classification of brazing fluxes for magnesium is not included.	900–1150	480–620	Magnesium alloys whose designators start with AZ.
FB3-A	Paste	BAg and BCuP	Borates Fluorides	General purpose flux for most ferrous and non-ferrous alloys. (Notable exception Al bronze, etc. See Flux 4A.)	1050–1600	565–870	All brazeable ferrous and non-ferrous metal except those with aluminum or magnesium as a constituent. Also used to braze carbides.
FB4-A	Paste	BAg and BCuP	Chlorides Fluorides Borates	General purpose flux for many alloys containing metals that form refractory oxides.	1100–1600	595–870	Brazeable base metals containing up to 9% aluminum (aluminum brass, aluminum bronze, Monel K500). May also have application when minor amounts of Ti, or other metals are present, which form refractory oxides.

Note: The selection of a flux designation for a specific type of work may be based on the filler metal type and the description above, but the information here is generally not adequate for flux selection.
The above chart represents a partial listing of Table 4.1 Brazing Fluxes from the AWS Brazing Handbook.   Source: AWS Brazing Handbook © 1991

Figure 3.55—AWS Brazing Flux Identification System (Partial Listing)

possible since brazing does not melt the base metals to produce a hybrid alloy which may not have desirable properties. It is also well-suited for the joining of metals which simply don't lend themselves to welding of any type. Another advantage of brazing is the equipment can be relatively inexpensive. Since brazing uses lower temperatures than welding, thin metals are readily joined without as much fear of melt-through or distortion.

The primary limitation is that the parts must be extremely clean prior to brazing. Another limitation is that the joint design must provide sufficient surface area to develop the required strength. Some configurations do not provide such a situation.

There are several inherent problems associated with brazing. The first is the formation of voids or unbonded areas within the joint. These can result from insufficient cleaning or improper heating of the parts. Another problem occurs when too much localized heat is applied to the base metal, resulting in melting and erosion of the base metal. This is normally associated with torch brazing where the combination of the flame's heat and its mechanical action will wear away the base metal adjacent to the braze joint. Another concern is the corrosion of the base metal by some of the extremely reactive fluxes; flux residue must be removed to avoid subsequent corrosion of the joint or base metal.

Cutting Processes

So far the discussion has involved only those methods used for joining metals together. Also of importance in metal fabrication are those processes used to cut or remove metal. These processes are often required prior to welding to produce proper part shapes or make specific joint preparations. During or after welding, some of these same processes can also be employed to remove defective areas of welds or to produce a specific configuration if the as-welded shape is not satisfactory for the intended purpose of the part.

Oxyfuel Gas Cutting (OFC)

The first of these cutting processes is oxyfuel gas cutting. Here, we use an oxyfuel flame to heat the metal to a temperature at which it will readily oxidize, or burn. The temperature needed is referred to as the *kindling* temperature, and for steels, it is about 1700°F (925°C). Once that temperature has been achieved, a high pressure stream of cutting oxygen is directed on the heated surface to produce an oxidation reaction. This stream of oxygen also tends to remove the slag and oxide residue which is produced by this oxidation reaction. Therefore, OFC can be thought of as a type of chemical cutting process. Figure 3.56 shows the basic placement of the cutting torch necessary to achieve a cut in carbon steel.

The equipment used for OFC is essentially the same as that for OAW except that, instead of a welding tip, there is now a cutting attachment which includes an additional lever or valve to turn on the cutting oxygen. Figure 3.57 shows a typical OFC equipment setup found in most welding and fabrication shops.

The cutting operation also requires a special cutting tip which is attached to the end of the torch. It consists of a series of small holes arranged in a circle around the outside edge of the end of the cutting tip. This is where the oxyfuel gas mixture flows to provide the preheat for cutting. Located in the center of these holes is a single cutting-oxygen passage. Cross sectional views of typical cutting tips are shown in Figure 3.58, and torches used for manual and machine cutting are shown in Figure 3.59.

It should be noted that OFC can be accomplished using several different types of fuel gases, such as acetylene, methane (natural gas), propane, gasoline, and methylacetylene-propadiene (MPS). Each provides various degrees of efficiency and may require slightly modified cutting tips. Other factors which should be considered when selecting the proper fuel gas include preheating time

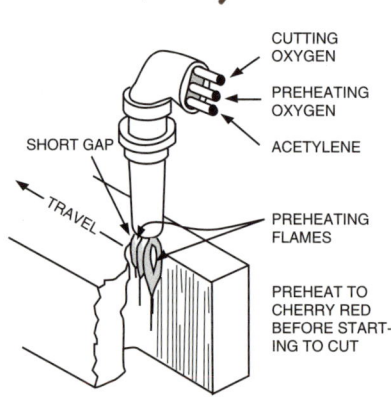

Figure 3.56—Oxyfuel Cutting

Figure 3.57—Oxyfuel Cutting Equipment

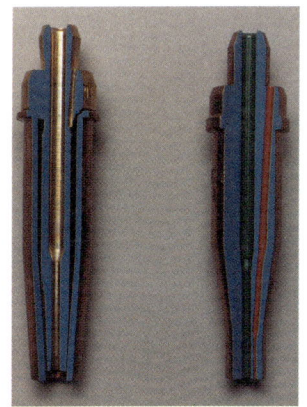 

(A) One Piece Tip   (B) Two Piece Tip

Figure 3.58—Cross Section Through Cutting Tips

required, cutting speeds, cost, availability, amount of oxygen required to burn gas efficiently, and ease and safety of transporting fuel containers.

Cutting is accomplished by applying heat to the part using the preheat flame which is an oxyfuel mixture. Once the metal has been heated to its oxidation temperature, the cutting oxygen is turned on to oxidize the hot metal. The oxidation of the metal

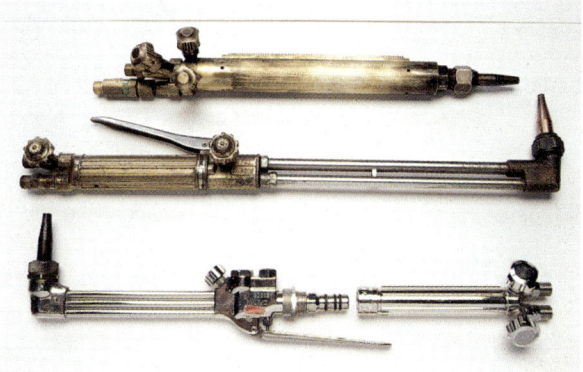

Figure 3.59—OFC Torches for Machine and Manual Oxyfuel Cutting

produces a tremendous amount of heat. This exothermic chemical reaction provides the necessary heat to rapidly melt the metal and simultaneously blow the oxidation products from the joint. The width of the cut produced is referred to as the kerf, as shown in Figure 3.60. Also shown is the drag, which is the amount of offset between the cut entry and exit points, measured along the cut edge.

Although OFC is used extensively by most industries, it is usually limited to the cutting of carbon and low alloy steels. As the amounts of various alloying elements increase, one of two things can happen; either they make the steel more difficult to cut or they may give rise to hardened or heat-checked cut surfaces, or both. The effects of various alloying elements are summarized in Figure 3.61.

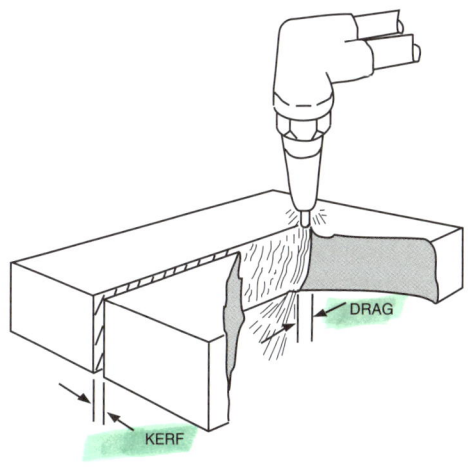

Figure 3.60—Illustration of Kerf and Drag in Oxyfuel Cutting

Element	Effect of element on oxygen cutting
Carbon	Steels up to 0.25% carbon can be cut without difficulty. Higher carbon steels should be preheated to prevent hardening and cracking. Graphite and cementite (Fe2C) are detrimental but cast irons containing 4% carbon can be cut by special techniques.
Manganese	Steels of about 14% manganese and 1.5% carbon are difficult to cut and should be preheated for best results.
Silicon	Silicon, in amounts usually present, has no effect. Transformer irons containing as much as 4% silicon are being cut. Silicon steel containing large amounts of carbon and manganese must be carefully preheated and post-annealed to avoid air hardening and possible surface fissures.
Chromium	Steels up to 5% chromium are cut without much difficulty when the surface is clean. Higher chromium steels, such as 10% chromium steels, require special techniques and the cuts are rough when the usual oxyacetylene cutting process is used. In general, carburizing preheat flames are desirable when cutting this type of steel. The flux injection and iron powder cutting processes enable cuts to be readily made in the usual straight chromium irons and steels as well as stainless steel.
Nickel	Steels containing up to 3% nickel may be cut by the normal oxygen cutting processes; up to about 7% nickel content, cuts are very satisfactory. Cuts of excellent quality may be made in the usual engineering alloys of the stainless steels (18–8 to about 35–15 as the upper limit) by the flux injection or iron powder cutting processes.
Molybdenum	This element affects cutting about the same as chromium. Aircraft quality chrome-molybdenum steel offers no difficulties. High molybdenum-tungsten steels, however, may be cut only by special techniques.
Tungsten	The usual alloys with up to 14% may be cut very readily, but cutting is difficult with a higher percentage of tungsten. The limit seems to be about 20% tungsten.
Copper	In amounts up to about 2%, copper has no effect.
Aluminum	Unless present in large amounts (about 10%) the effect of aluminum is not appreciable.
Phosphorus	This element has no effect in amounts usually tolerated in steel.
Sulfur	Small amounts, such as are present in steels, have no effect. With high percentages of sulfur, the rate of cutting is reduced and sulfur dioxide fumes are noticeable.
Vanadium	In the amounts usually found in steels, this alloy may improve rather than interfere with cutting.

Figure 3.61—Effects of Chemical Elements on Oxyfuel Cutting

As can be seen, in most cases, the addition of certain amounts of alloying elements may prevent conventional OFC. In many cases, these elements are oxidation resistant types. In order for an oxyfuel cut to be effectively accomplished, the material must comply with the following criteria: (1) it must have the capability of burning in a stream of oxygen, (2) its ignition temperature for burning must be lower than its melting temperature, (3) its heat conductivity should be relatively low, (4) the metal oxide produced must melt at some temperature below the melting point of the metal, and (5) the formed slag must be of low viscosity. Therefore, in order to cut cast iron or stainless steel with this process, special techniques involving additional equipment are necessary. These techniques include torch oscillation, use of a waster plate, wire feeding, powder cutting, and flux cutting.

OFC's advantages include its relatively inexpensive and portable equipment, making it feasible for use in both shop and field applications. Cuts can be made on thin or thick sections; ease of cutting usually increases with thickness. When mechanized (Figure 3.62), OFC can produce cuts of reasonable accuracy. When compared to mechanical cutting methods, oxyfuel cutting of steels is more economical. To improve this efficiency even more, multiple torch systems or stack cutting can be used to cut several layers at once.

One of the limitations of OFC is that the finished cut may require additional cleaning or grinding to prepare it for welding. Another important limitation is that since it requires high temperatures, there may be a heat-affected zone produced having a very high hardness. This is especially important if there is a need for machining of this surface. Employment of preheat and postheat will aid in the alleviation of this problem. Also, even though cuts can be reasonably accurate, they still don't compare to the accuracy possible from mechanical cutting methods. Finally, the flame and hot slag produced result in safety hazards for personnel near the cutting operation.

Air Carbon Arc Cutting (CAC-A)

Another very effective cutting process is air carbon arc cutting. This process uses a carbon electrode to create an arc for heating, along with a high pressure stream of compressed air to mechanically remove the molten metal. Figure 3.63 shows the process in use.

The equipment used for CAC-A consists of a special electrode holder which is attached to a constant current power source and a compressed air supply. This special holder, shown in Figure 3.64, grasps the carbon electrode in copper jaws, one of which has a series of holes through which the compressed air passes.

To achieve a cut, the carbon electrode is brought close to the work to create an arc. Once the arc melts the metal, the stream of compressed air is initiated and blows away the molten metal to produce a gouge or cut.

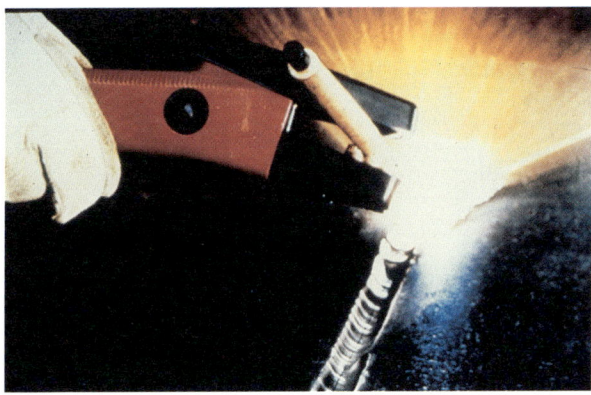

Figure 3.63—Air Carbon Arc Cutting

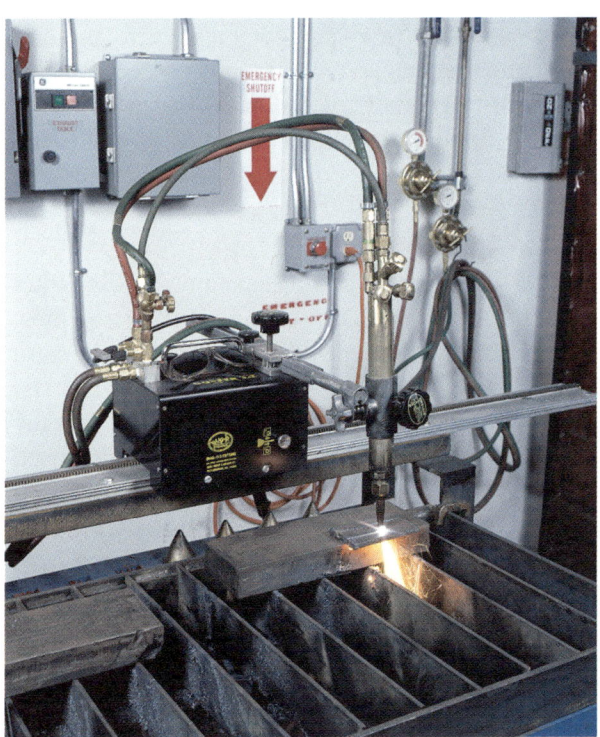

Figure 3.62—Machine OFC Cutting

Figure 3.64—Air Carbon Arc Cutting Electrode Holder

The electrode holder is attached to a power source as well as a source of compressed air. Any nonflammable compressed gas could be used, but compressed air is by far the least expensive, if available. The entire system for air carbon arc cutting is shown in Figure 3.65.

CAC-A has applications in most industries, especially since it can be used to cut any metal. Even though it will cut all metals, there are other considerations that may require other cutting methods for particular alloys. Figure 3.66 shows the current type and polarity for CAC-A cutting of several metals and alloys.

While we tend to think of its application to remove defective areas of the weld or base metal, it is important to realize that it can be used quite effectively as a weld joint preparation tool. For example, two pieces to be butt welded can be aligned with their square-cut edges touching. The CAC-A process can then be employed to produce a uniform U-groove preparation, as shown in Figure 3.67. CAC-A is also used for rough machining of large, complex parts.

Metal	Current Type	Electrode Polarity
Aluminum	DC	Positive
Copper and alloys	AC	NA
Iron, cast, malleable, etc.	DC	Negative
Magnesium	DC	Positive
Nickel and alloys	AC	NA
Carbon steels	DC	Positive
Stainless steels	DC	Positive

Figure 3.66—CAC-A Electrical Requirements for Various Metals

One of the basic advantages of CAC-A is that it is a relatively efficient method for removal of metal. It also has the ability to cut any metal. Since it uses the same power sources as those used for some types of welding, the equipment costs are minimal. All that is necessary is the purchase of the special electrode holder which is attached to an existing power source and a compressed air supply.

The primary disadvantage of the process is safety-related. It is inherently a very noisy and dirty process. Therefore, the operator may elect to use ear protection to reduce the noise level, and breathing filters to eliminate the inhaling of the metal particles produced. A fire watch may also be required to make sure the gouged metal droplets do not create a fire hazard. Another limitation is that the finished cut may require some cleanup prior to additional welding. Carburization of the cut may occur.

Plasma Arc Cutting (PAC)

The final thermal cutting method for discussion is plasma arc cutting. This process is similar in most respects to PAW except that now the purpose is to remove metal rather than join pieces together. The equipment requirements are similar except that the power required may be much higher than that used for welding. The transferred arc type torch is used because of the increased heating of the base metal. Typical PAC torches for manual and machine cutting are shown in Figure 3.68, and a PAC equipment setup is shown in Figure 3.69.

For mechanized PAC cutting, not only is the torch water-cooled internally, but the actual cutting may take place under water or oil to reduce noise and particulate levels.

Figure 3.65—Air Carbon Arc Cutting Equipment

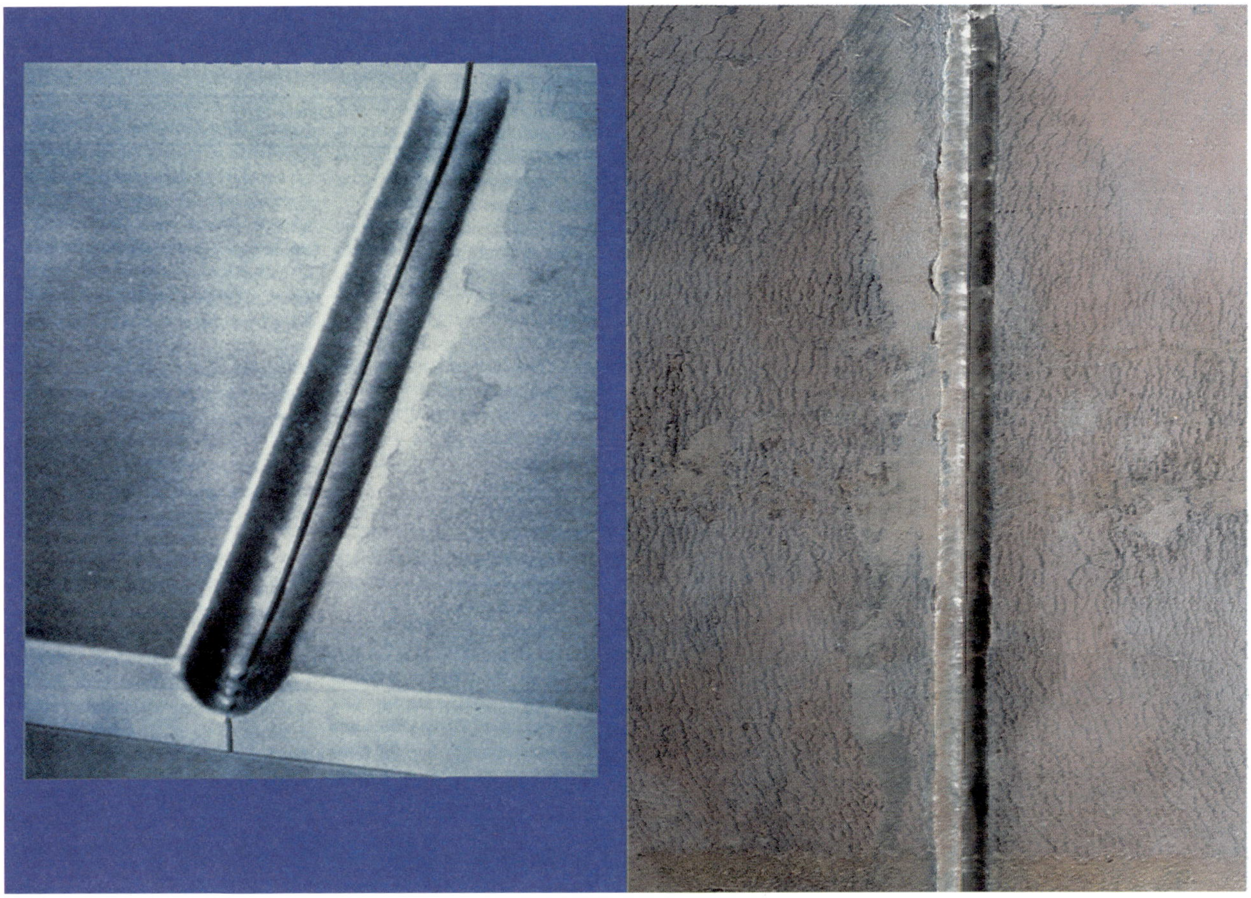

Figure 3.67—Illustration of Joint Preparation Using Mechanized (left) and Manual (right) Air Carbon Arc Cutting

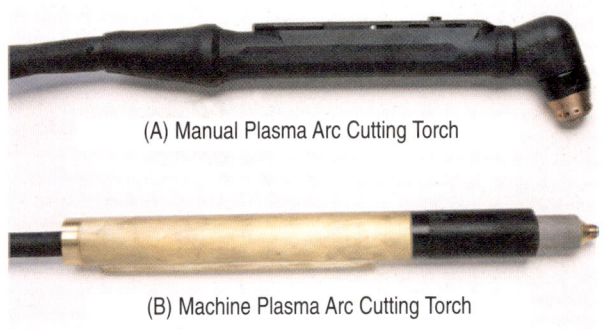

(A) Manual Plasma Arc Cutting Torch

(B) Machine Plasma Arc Cutting Torch

Figure 3.68—Typical Manual and Machine Plasma Arc Cutting Torches

While its primary application is for the cutting of non-ferrous metals, PAC is also useful for the cutting of carbon steels. Advantages include the ability to cut metals which cannot be cut with OFC, the resulting high quality cut, and increased cutting speeds for carbon steel.

One limitation is that the kerf is generally quite large and the cut edges may not be square. Special techniques, such as water injection, can be used to improve this edge configuration if desired. Another limitation is the higher cost of equipment as compared to oxyfuel cutting.

Mechanical Cutting

Finally, brief mention of mechanical cutting methods used in conjunction with welding will be presented. These methods can include shearing, sawing, grinding, milling, turning, shaping, drilling, planing, and chipping. They are used for joint preparation, weld contouring, parts preparation, surface cleaning, and removal of defective welds. See Figure 3.70.

A welding inspector should understand how these methods are used. Their misapplication may have a degrading effect on final weld quality. For

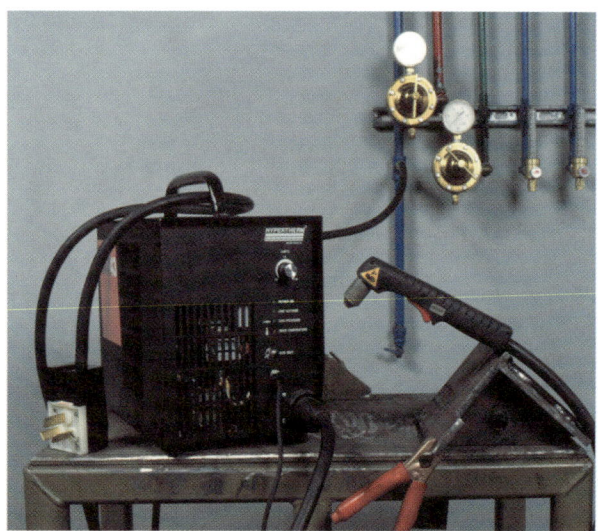

Figure 3.69—Plasma Arc Cutting Equipment

Figure 3.70—Mechanical Grinder

example, many of these methods use cutting fluid to aid their operation. If the fluids are not completely removed from material surfaces prior to welding, problems such as porosity and cracking may result.

Summary

There are numerous joining and cutting processes used in metal fabrication. A welding inspector who understands the fundamentals of various processes can spot problems before or when they occur. Technical understanding combined with information gained from practical experience allows the welding inspector to be better prepared to perform a visual inspection of welds.

Key Terms and Definitions

AC—alternating current; in the U.S., the polarity alternates at 60 cycles per second.

alloy—a substance with metallic properties and composed of two or more elements of which at least one is a metal.

alphanumeric—a combination of numbers and letters used in designations.

ampere—standard unit for measuring the strength of an electric current.

arc blow—the deflection of a welding arc from its normal path because of magnetic forces.

arc length—the distance from the tip of the welding electrode to the adjacent surface of the weld pool.

brazing—joining materials without melting the base metal using a filler metal with a melting point above 840°F (450°C). See soldering.

capillary action—in welding, the force by which a liquid, in contact with a solid, is distributed between closely fitted faying surfaces of the joint to be brazed or soldered.

carbon steel—a mixture of iron and small amounts of carbon.

carburizing—in welding, a term denoting the addition of carbon into the surfaces of hot metals through a solid solution mechanism. May occur during Air Carbon Arc Cutting, CAC-A.

coalescence—joining together of two or more materials.

code—a document adopted by a city, municipality, state or nation, and having legal status.

DC—direct current; constant electric polarity.

DCEN—direct current, electrode negative. Referred to as *straight* polarity.

DCEP—direct current, electrode positive. Referred to as *reverse* polarity.

discontinuity—any interruption of the typical structure of a material; not necessarily a defect.

drag—in OFC and PAC, the amount of the offset between the cut entry and exit points, measured along the cut edge.

electrode—a component of the electrical circuit that terminates at the arc, molten conductive slag, or base metal.

faying surface—the mating surface of a member that is in contact with or in close proximity to another member to which it is to be joined.

ferrous—a term referring to metals that are primarily iron-based, such as steels.

filler metal—the metal or alloy added in making a weld, brazed or soldered joint.

flux—a material used to hinder the formation of oxides and other undesirable substances in molten metal and on solid metal surfaces, and to dissolve or otherwise facilitate the removal of such substances.

inclusion—entrapped foreign solid material, such as slag, flux, tungsten or oxides.

incomplete fusion—a weld discontinuity in which fusion did not occur between weld metal and fusion faces or adjoining weld beads.

incomplete joint penetration—a joint root condition in a groove weld in which weld metal does not extend through the joint thickness.

inert gas—a gas that does not combine chemically with other materials. Argon and helium are most commonly used in welding.

kerf—the width of the cut produced during a cutting process.

keyhole welding—a procedure that produces a hole completely through the workpiece. As the weld progresses, molten metal flows in behind the keyhole to form the weld.

ksi—designation for a thousand pounds per square inch. 70,000 psi is equal to 70 ksi.

low alloy steel—an alloy of iron and carbon, with other elements added for increased strength.

nonferrous—refers to alloys other than the iron-based alloys. Copper, nickel and aluminum alloys are nonferrous.

orifice—in welding, an opening, usually small, that aids in controlling or constricting the flow of materials.

plasma—in welding, an ionized gas stream.

porosity—cavity-type discontinuities formed by gas entrapment during solidification.

position—in welding, the relationship between the weld pool, joint, joint members, and welding heat source during welding. Examples are flat, horizontal, vertical, and overhead.

prefix—an alpha/numeric added at the beginning of an item to modify its meaning.

progression—in welding, the term applied to the direction of vertical welding, uphill or downhill.

psi—pounds per square inch.

reactive gas—a gas that will combine chemically with other materials.

rimmed steel—a steel having a rim, or surface zone having a shallow depth, of extremely low carbon content. Occurs during the steel making practice.

shielding—protecting from contamination.

slag—a nonmetallic product resulting from the mutual dissolution of flux and nonmetallic impurities in some welding and brazing processes.

soldering—joining materials without melting the base metal, using a filler metal having a melting point below 840°F (450°C). See brazing.

solid solution—for metals, one solid dissolving into another solid.

spatter—metal particles expelled during fusion welding that do not form a part of the weld.

suffix—an alpha/numeric following an item which usually changes or modifies its meaning.

tensile strength—usually stated in pounds per square inch (psi); calculated by dividing the maximum load by the cross sectional area. SI units are megapascals (MPa).

undercut—a groove melted into the base metal adjacent to the weld toe or weld root and left unfilled by weld metal.

voltage—electromotive force, or difference in electric potential, expressed in volts.

waster plate—the carbon steel plate placed over austenitic stainless steel plate to permit cutting by the OFC method. CAC-A or PAC are more efficient for cutting these stainless steels.

weld—a localized coalescence of metals or nonmetals produced either by heating the materials to the welding temperature, with or without the application of pressure, or by the application of pressure alone, and with or without the use of filler material.

Module 4
Weld Joint Geometry and Welding Symbols

Contents

Introduction 4-2

Welded Joints 4-2

Welding Symbols 4-28

Supplementary Symbols 4-33

Weld Symbol Dimensioning 4-37

Key Terms and Definitions 4-83

Module 4—Weld Joint Geometry and Welding Symbols

Introduction

The determinations made about welding specifications are part of the design or project engineers' responsibility; and so too are joint design and joint selection. However, it is still the responsibility of fabrication personnel to accurately interpret, and then prepare these joints for fabrication. Knowledge of welded joint terminology is essential in every day job communication. Use of proper terms makes it much easier for welding personnel to relay various fitup and welding problems encountered during the fabrication process to other involved personnel. There is a direct relationship between welded joint terms and supplementary welding symbol data and dimensioning. It is imperative for the welding inspector to master these aspects of communications.

Welded Joints

There are five basic joints used in welded metal fabrication: *butt, corner, T-, lap,* and *edge*. As illustrated in Figure 4.2, certain welds and weld symbols are applicable to the five basic joint designs. A number of different welds can be applied to each joint type depending on the joint design, and these are shown adjacent to each joint type. *Joint design* identifies, "the shape, dimensions, and configuration of the joint."

In the 1994 revision of AWS A3.0, Standard Terms and Definitions, Figure 4.1, additional joint classifications for flanged joints and spliced joints were added. *Flanged joints*, Figure 4.3, are formed into one of the five basic joint types in which at least one of the joint members has a flanged edge shape at the weld joint. A *spliced joint* is "a joint

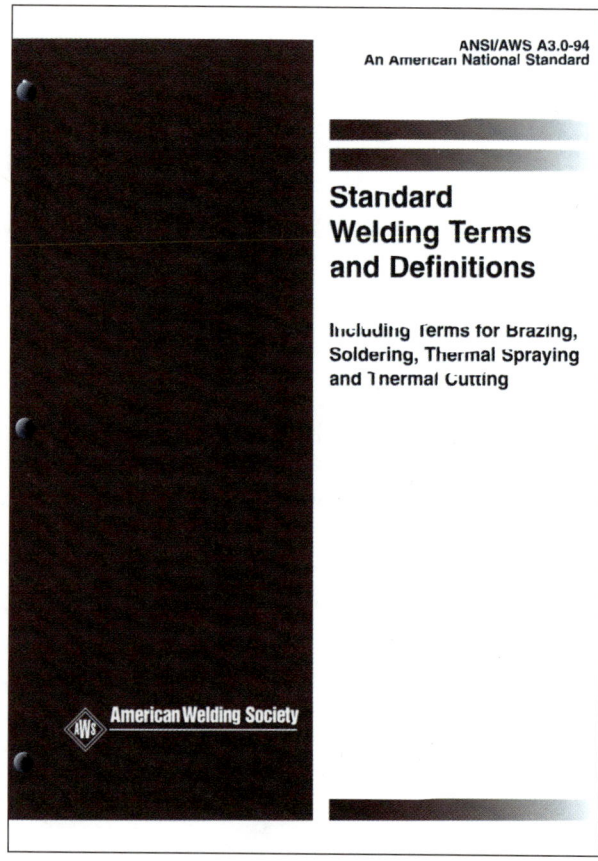

Figure 4.1—AWS A3.0, *Standard Welding Terms and Definitions*

in which an additional workpiece spans the joint and is welded to each joint member" (see Figure 4.4).

The individual workpieces of a joint are called *members*. Members are classified in three ways; butting members, nonbutting members, or splice members. Figures 4.4 and 4.5 provide illustrations of each type of member.

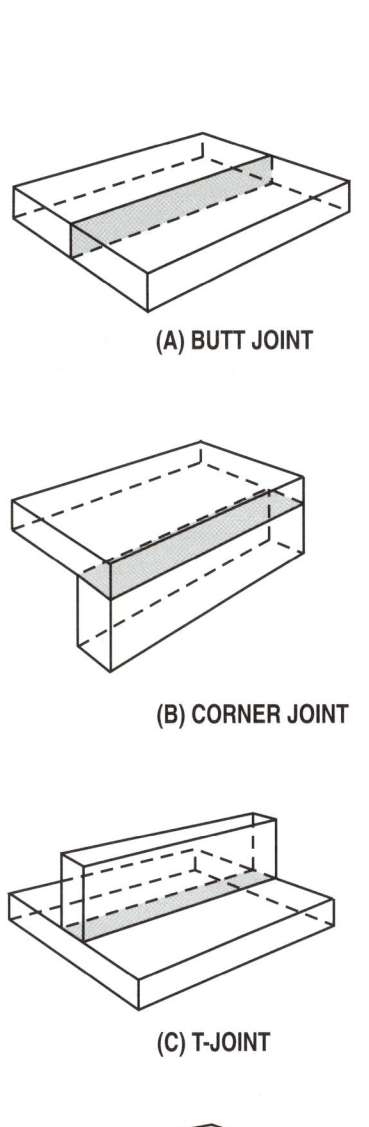

(A) BUTT JOINT

(B) CORNER JOINT

(C) T-JOINT

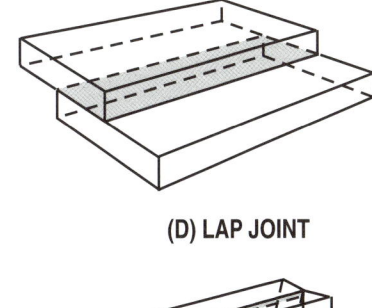

(D) LAP JOINT

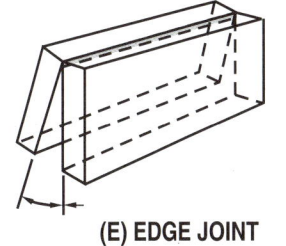

(E) EDGE JOINT

APPLICABLE WELDS and WELD SYMBOL

Symbol	Weld	Symbol	Weld
⊻	Bevel-Groove	⊻	U-Groove
⊐⊏	Flare-Bevel-Groove	⊻	V-Groove
⊃⊂	Flare-V-Groove	⊓	Edge Weld
⊬	J-Groove	∕∕	Scarf (for braze joint)
⊔	Square-Groove		

APPLICABLE WELDS and WELD SYMBOL

Symbol	Weld	Symbol	Weld
⊿	Fillet	⊓	Edge Weld
⊻	Bevel-Groove	⊔	Plug
⊐⊏	Flare-Bevel-Groove	⊔	Slot
⊃⊂	Flare-V-Groove	○	Spot
⊬	J-Groove	⊖	Seam
⊔	Square-Groove	○	Projection
⊻	U-Groove	⊻	V-Groove

APPLICABLE WELDS and WELD SYMBOL

Symbol	Weld	Symbol	Weld
⊿	Fillet	⊔	Slot
⊻	Bevel-Groove	○	Spot
⊃⊂	Flare-V-Groove	⊖	Seam
⊬	J-Groove	○	Projection
⊔	Square-Groove	⊔	Plug

APPLICABLE WELDS and WELD SYMBOL

Symbol	Weld	Symbol	Weld
⊿	Fillet	⊔	Slot
⊻	Bevel-Groove	○	Spot
⊃⊂	Flare-V-Groove	⊖	Seam
⊬	J-Groove	○	Projection
⊔	Square-Groove		*Braze
⊔	Plug		

APPLICABLE WELDS and WELD SYMBOL

Symbol	Weld	Symbol	Weld
⊻	Bevel-Groove	⊻	U-Groove
⊐⊏	Flare-Bevel-Groove	⊻	V-Groove
⊃⊂	Flare-V-Groove	⊓	Edge
⊬	J-Groove	⊖	Seam
⊔	Square-Groove		

Figure 4.2—The Five Basic Types of Joints and Applicable Welds

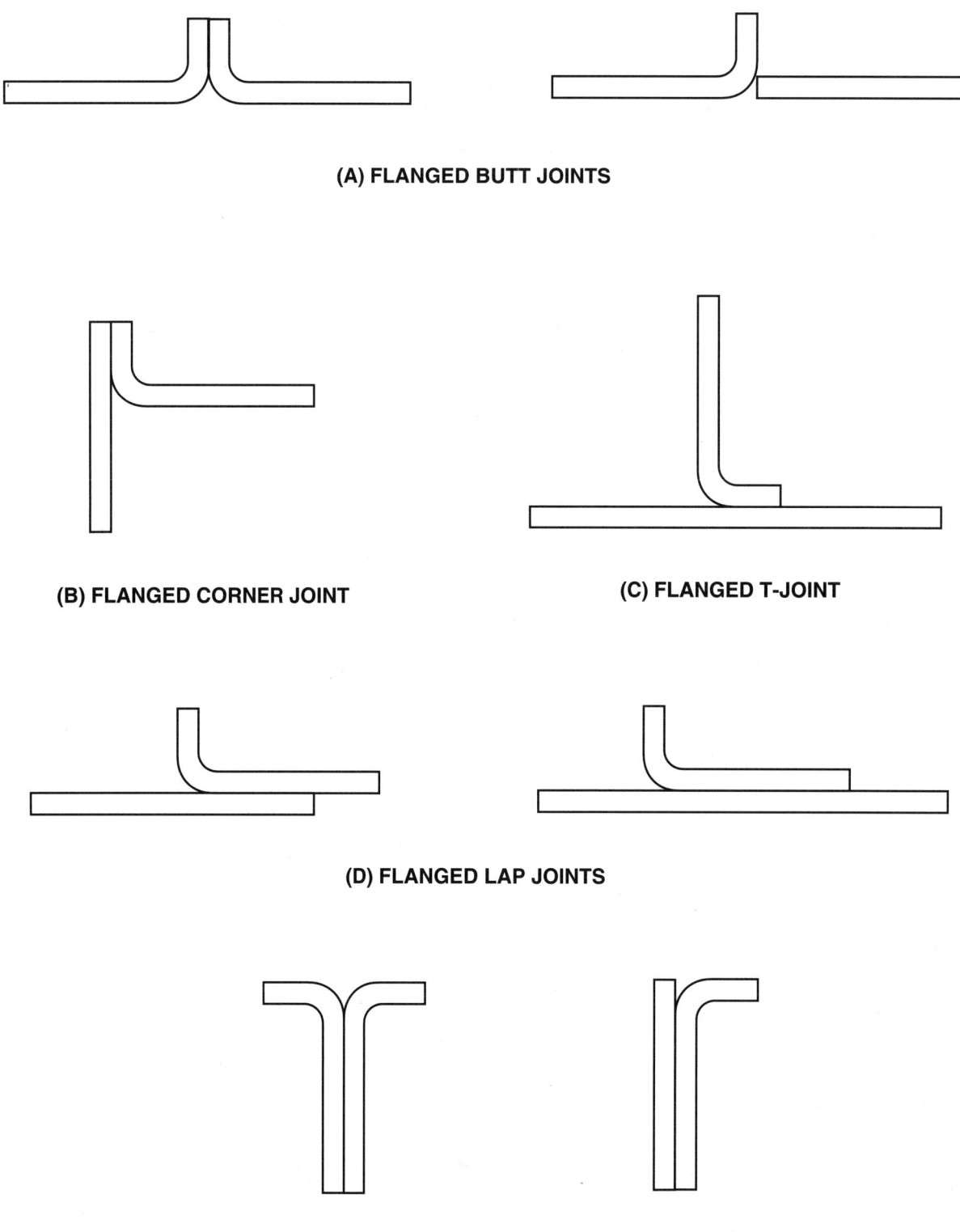

Figure 4.3—Flanged Joints

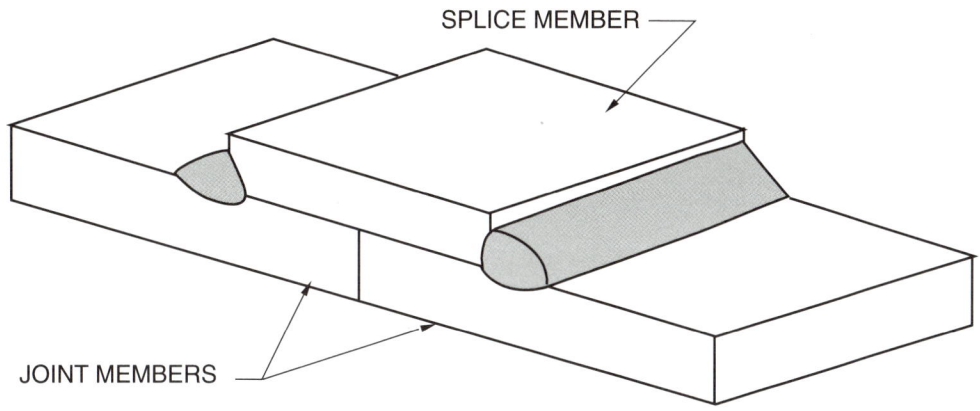

(A) SINGLE-SPLICED BUTT JOINT

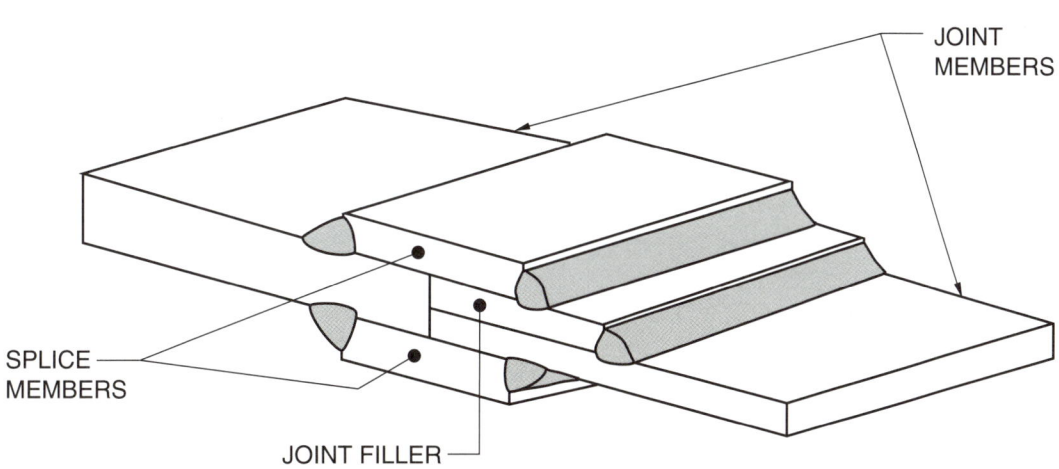

(B) DOUBLE-SPLICED BUTT JOINT WITH JOINT FILLER

Figure 4.4—Spliced Butt Joints

A *butting member* is "a joint member that is prevented by the other member from movement in one direction perpendicular to its thickness dimension." For example, both members of a butt joint, or one member of a T-joint or corner joint are butting members. *A nonbutting member* is "a joint member that is free to move in any direction perpendicular to its thickness dimension." For example, both members of a lap joint, or one member of a T-joint or corner joint are nonbutting members.

A *splice member* is "the workpiece that spans the joint in a spliced joint." In Figure 4.4 two examples are provided for splices used in connection with butt joints.

Identification of the weld type is indicated in the joint geometry. *Joint geometry* is "the shape and dimensions of a joint in cross section prior to welding." When a joint is viewed in cross section, the *edge shape* of each mating member often resembles the weld type and weld symbol specified. Figure 4.6

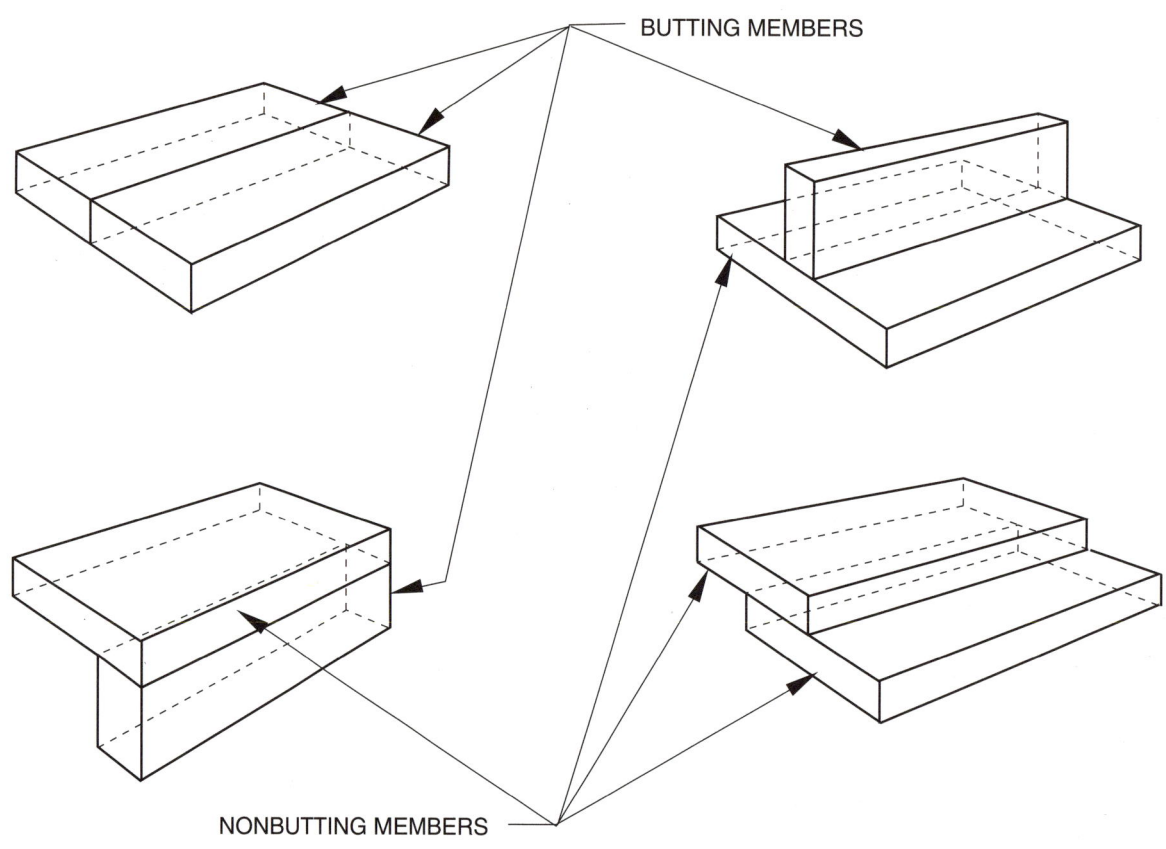

Figure 4.5—Butting and Nonbutting Members

identifies basic edge shapes used in welded metal fabrication and the welds applicable to each. The cross sectional views provided in Figures 4.7 through 4.11, show the relationship between weld symbol appearance and combinations of various edge shapes. These combinations of different edge shapes illustrate a variety of joint configurations for some of the applicable welds identified in the five basic joint arrangements shown in Figure 4.2. Additional weld types and groove designs can be made using various structural or formed shapes when applicable edge or surface preparations are applied to them.

Parts of the Weld Joint

Once the type of joint is identified, it may be necessary to describe the exact joint design required. To do this, welding and inspection personnel must be capable of identifying individual features that make up the joint geometry for a particular joint. The nomenclature associated with these features include:

- joint root
- groove face
- root face
- root edge
- root opening
- bevel
- bevel angle
- groove angle
- groove radius

Depending upon the particular type of joint design, the joint geometry may take on slightly different shapes. One example is the joint root. Joint root is defined as, "that portion of a joint to be welded where the members approach closest to each other. In cross section, the joint root may be either a point, line, or an area." Figure 4.12 illustrates some of the variations in the joint root for several different joint designs. The joint roots are shown as shaded areas in sketches (A)–(D), or a dark line in sketches (E) and (F).

(A) SQUARE EDGE SHAPE

APPLICABLE WELDS
Double-Bevel-Groove   Single-J-Groove
Double-Bevel-Flare-Groove   Square-Groove
Double-J-Groove   Edge
Single-Bevel-Groove   Fillet
Single-Flare-Bevel-Groove

(B) SINGLE-BEVEL EDGE SHAPE

APPLICABLE WELDS
Single-Bevel-Groove
Single-V-Groove

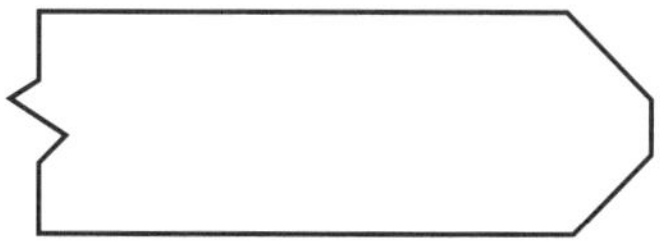

(C) DOUBLE-BEVEL EDGE SHAPE

APPLICABLE WELDS
Double-Bevel-Groove
Double-V-Groove

(D) SINGLE-J-GROOVE EDGE SHAPE

APPLICABLE WELDS
Single-J-Groove
Single-U-Groove

(E) DOUBLE-J-GROOVE EDGE SHAPE

APPLICABLE WELDS
Double-J-Groove
Double-U-Groove

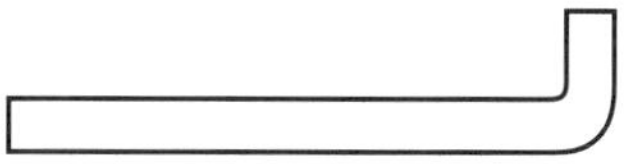

(F) FLANGE EDGE SHAPE

APPLICABLE WELDS
Double-Flare-Bevel-Groove   Projection
Single-Flare-V-Groove   Seam
Edge   Spot
Fillet

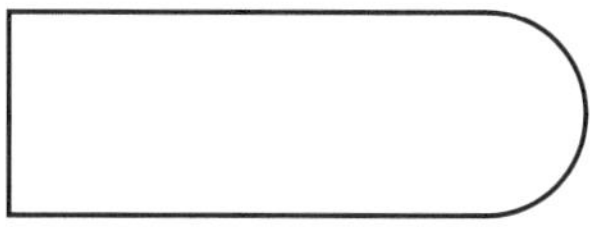

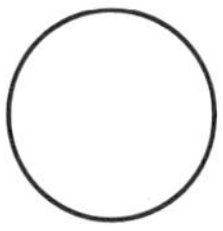

(G) ROUND EDGE SHAPE

APPLICABLE WELDS
Double-Flare-Bevel-Groove
Double-Flare-V-Groove

Figure 4.6—Edge Shapes of Members

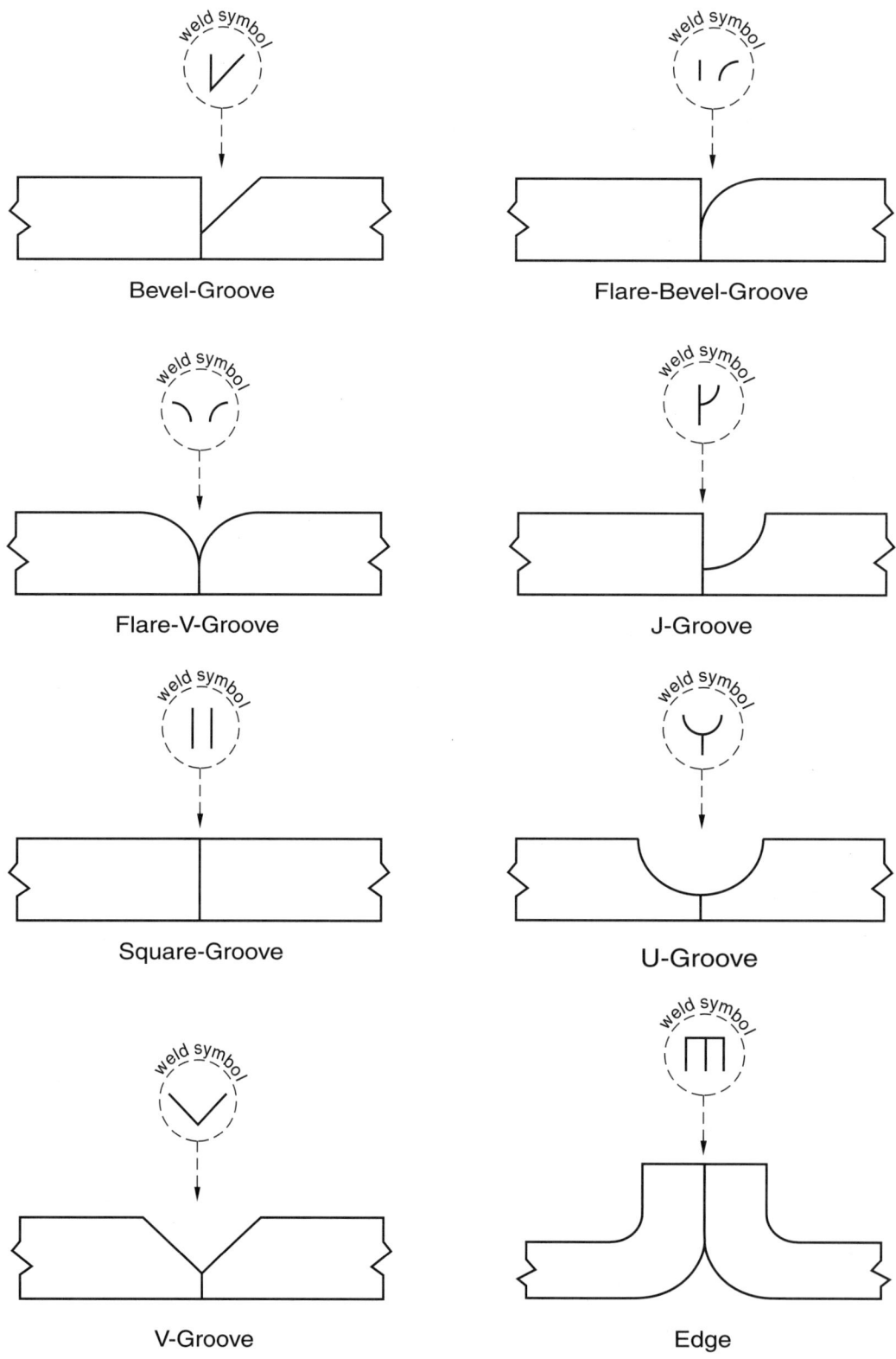

Figure 4.7—Edge Shape Combinations for Butt Joint Variations

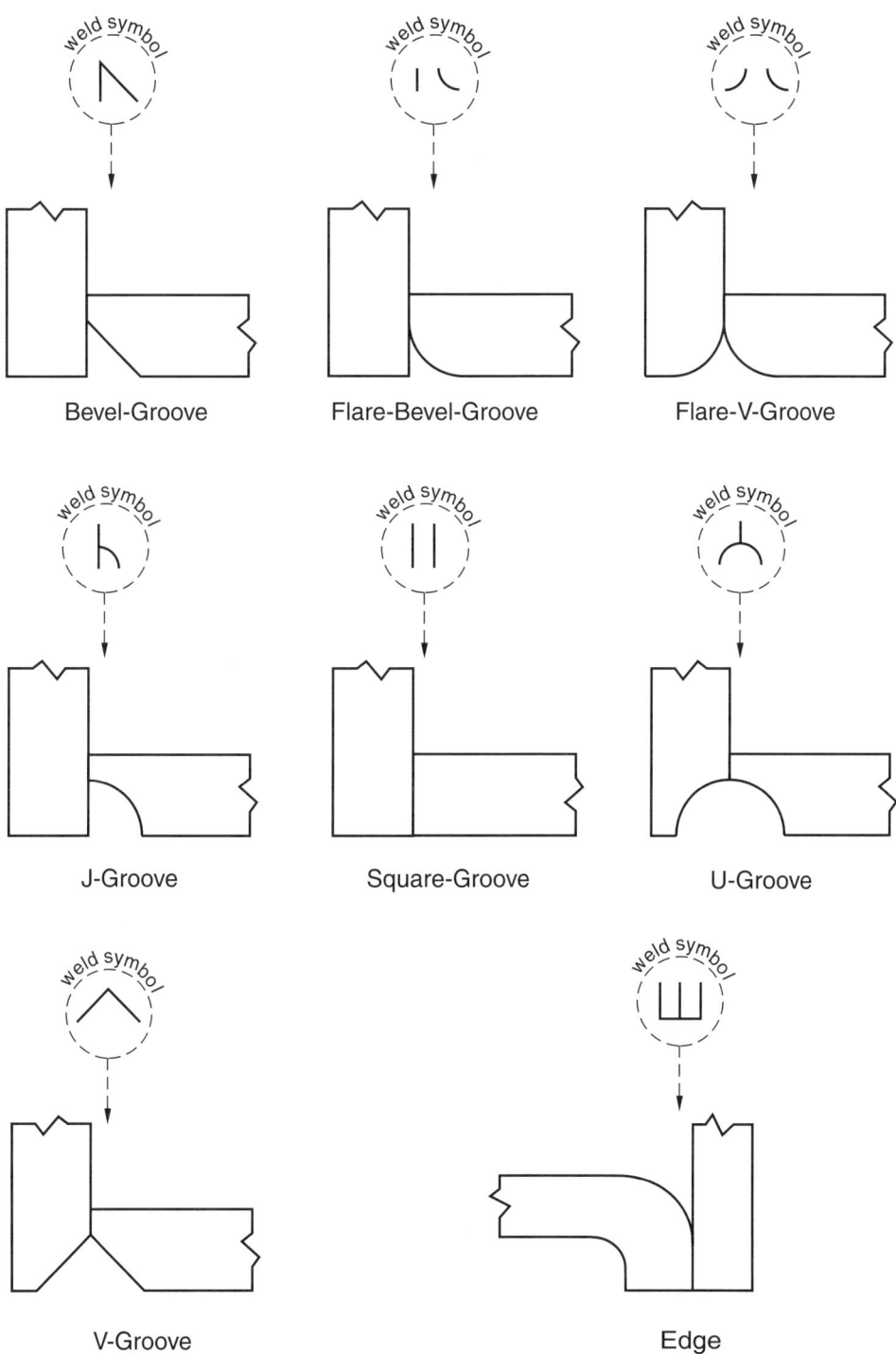

Figure 4.8—Edge Shape Combinations for Corner Joint Variations

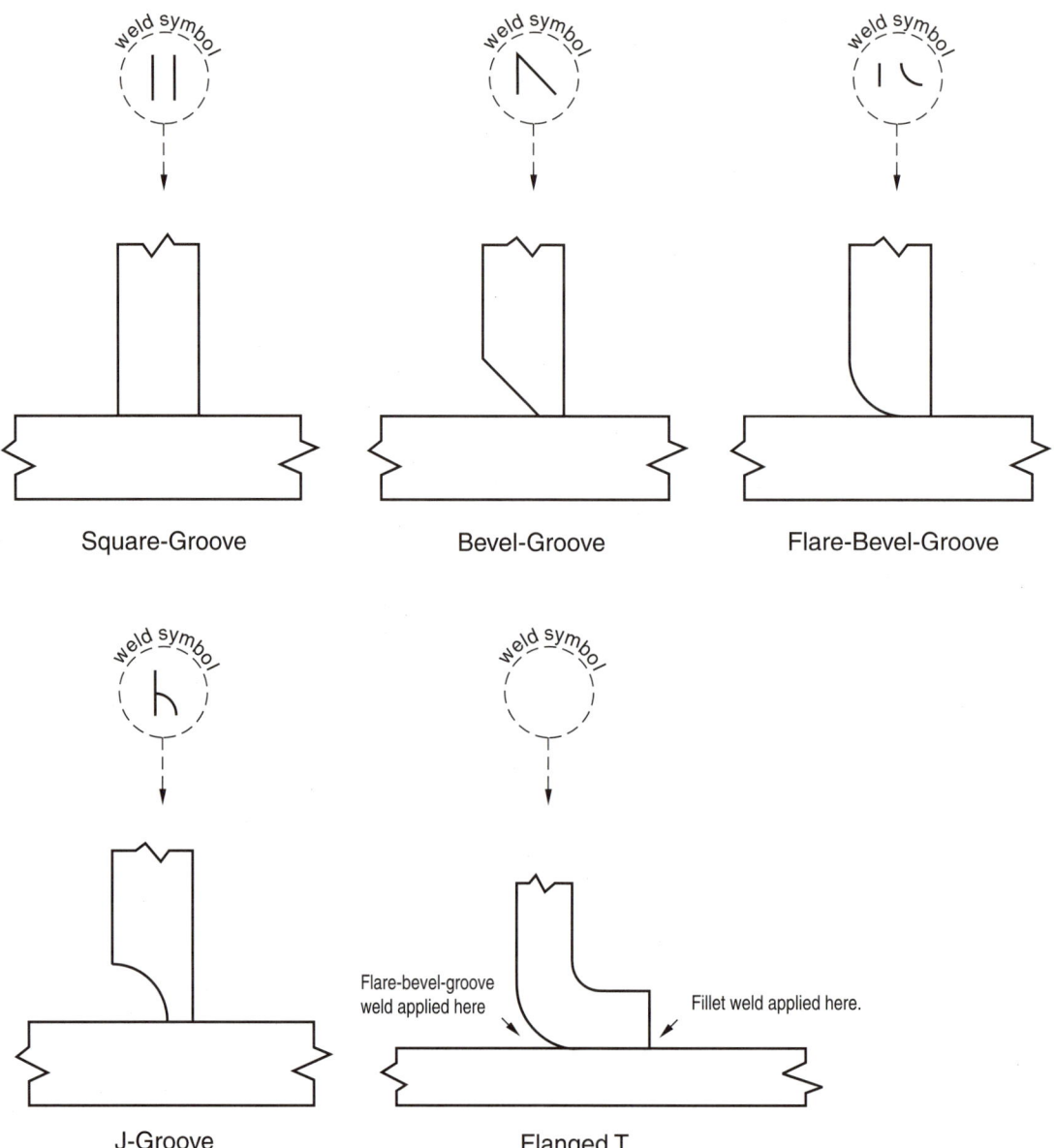

Figure 4.9—Edge Shape Combinations for T-Joint Variations

The nomenclature associated with groove face, root face and root edge is identified in Figure 4.13 *Groove face* is "that surface of a member included in the groove." The *root face* (commonly called the "land") is "that portion of the groove face within the joint root." Finally, *root edge* is defined as "a root face of zero width."

Other features that may require description by welding personnel are shown in Figure 4.14. These elements are often the essential variables in welding procedures, as well as production welding, and welding personnel may be required to actually measure them to determine their compliance with applicable drawings or other documents.

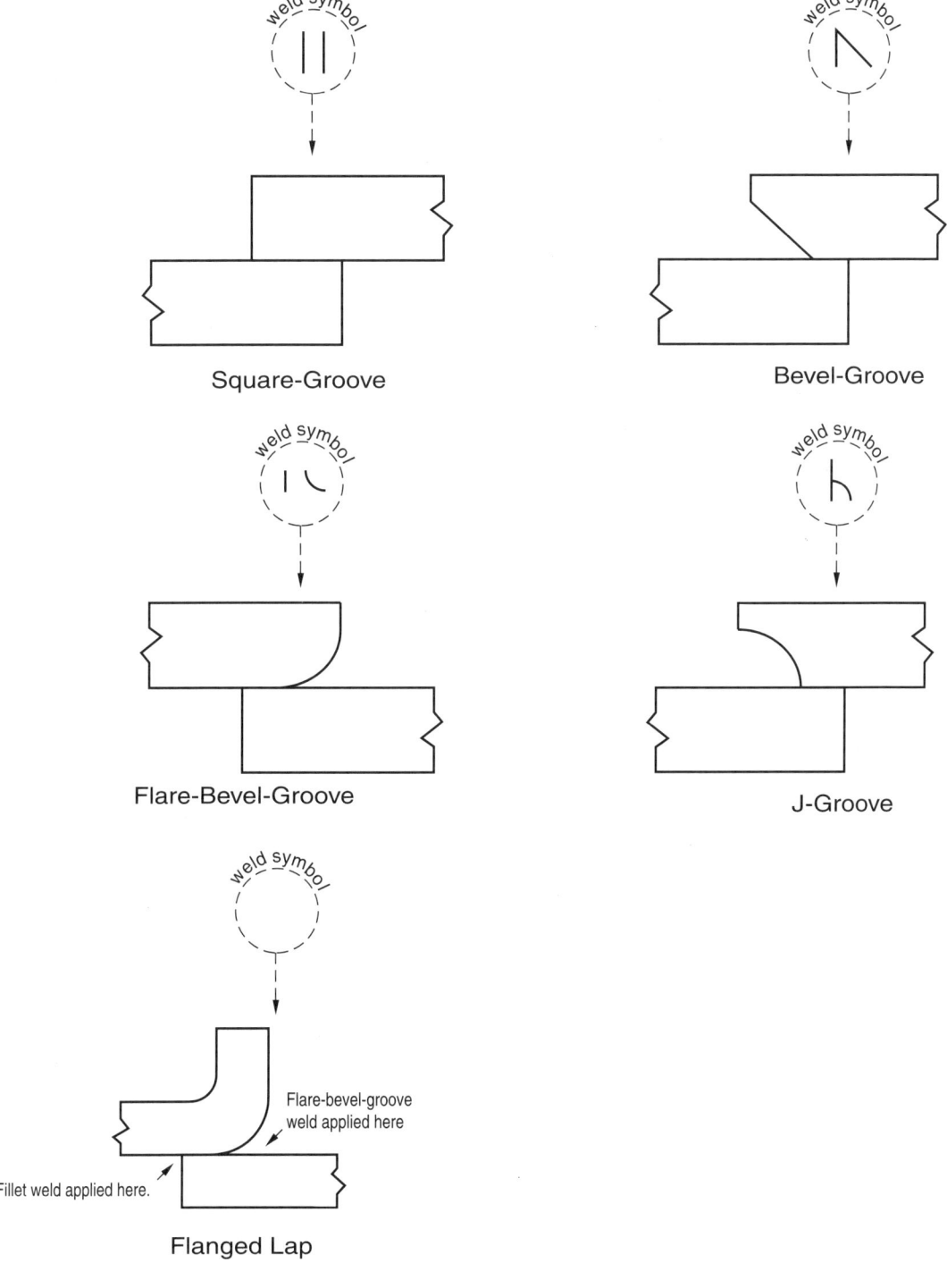

Figure 4.10—Edge Shape Combinations for Lap Joint Variations

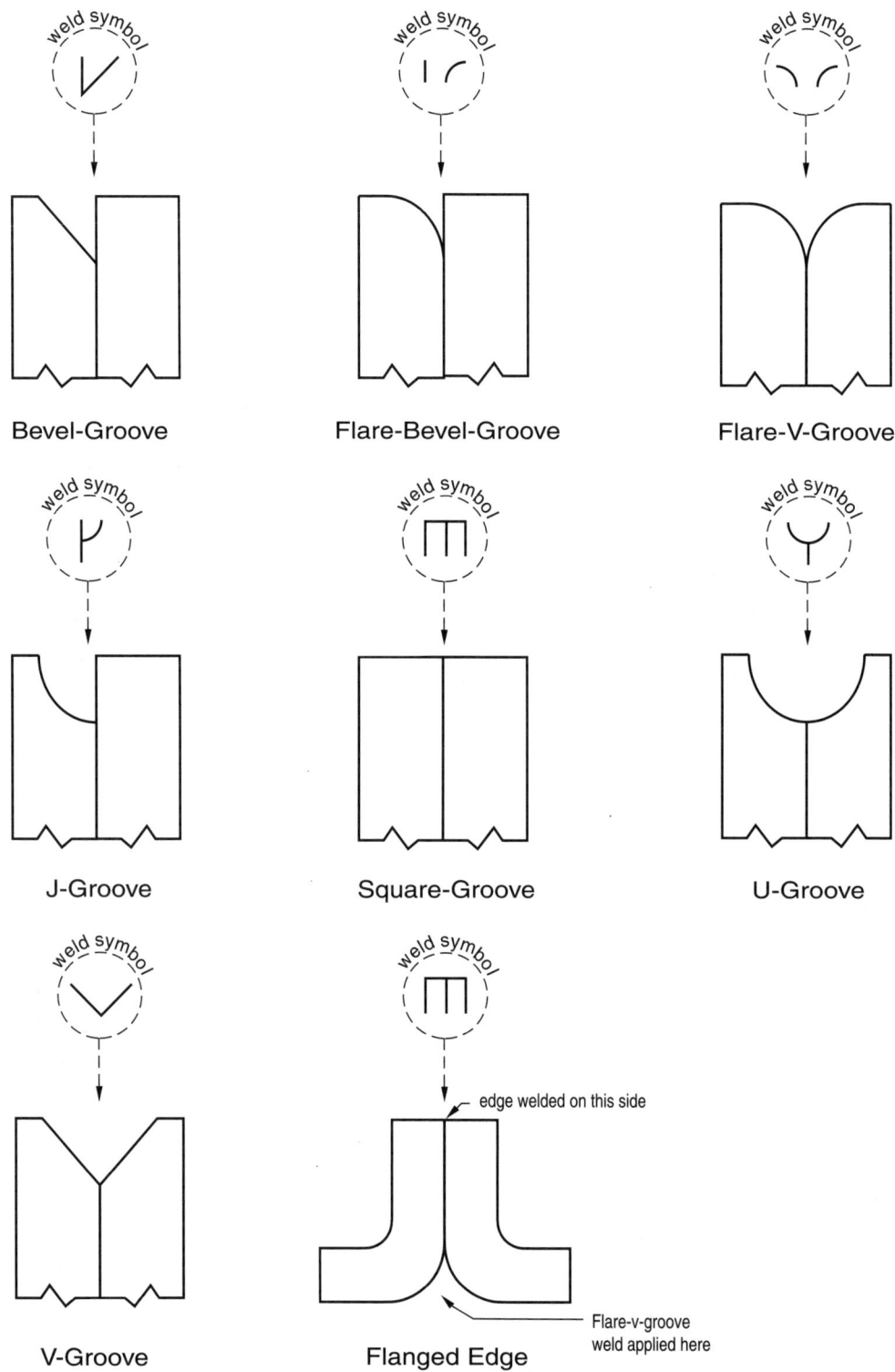

Figure 4.11—Edge Shape Combinations for Edge Joint Variations

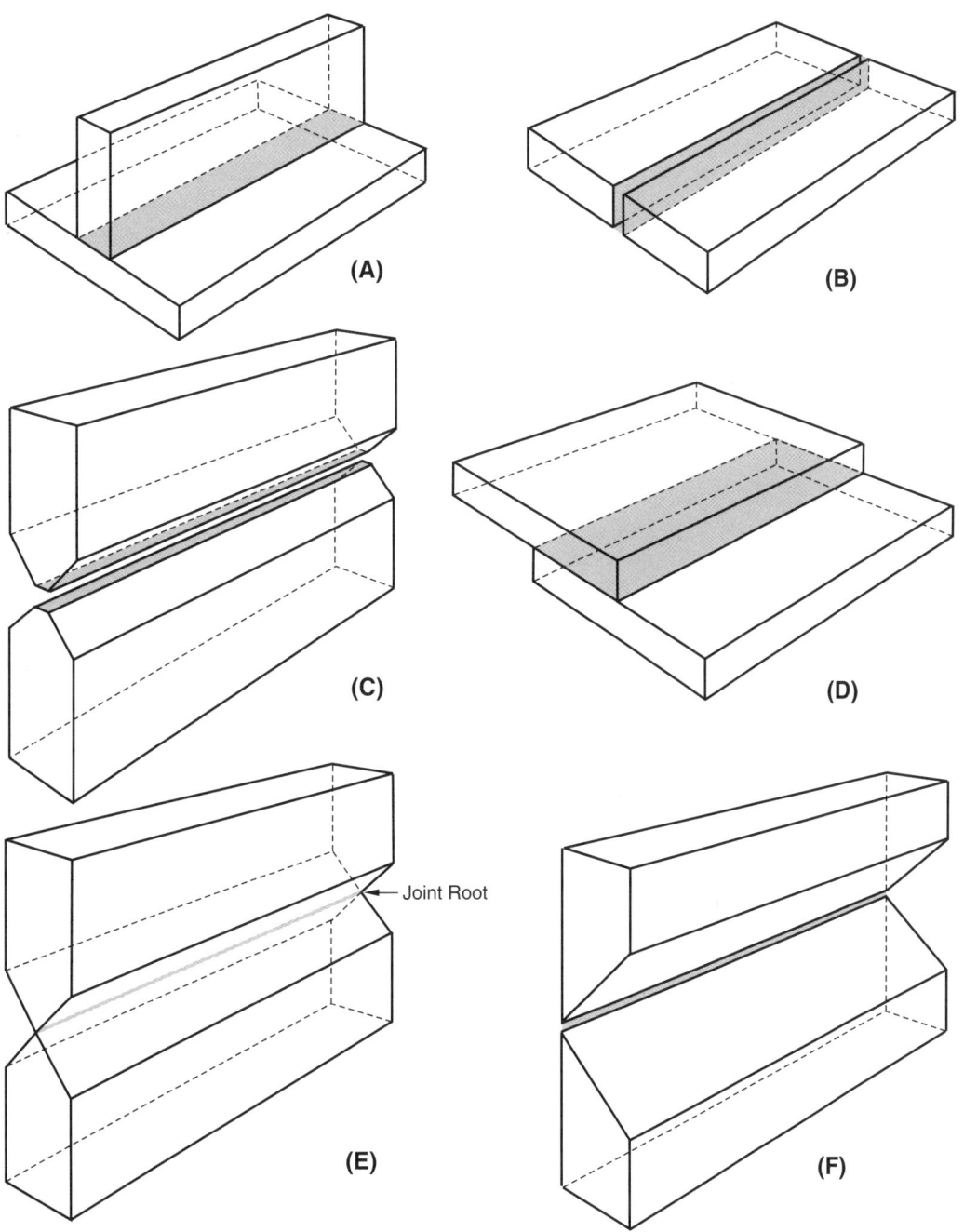

NOTE: Joint root is denoted by shading

Figure 4.12—Joint Roots

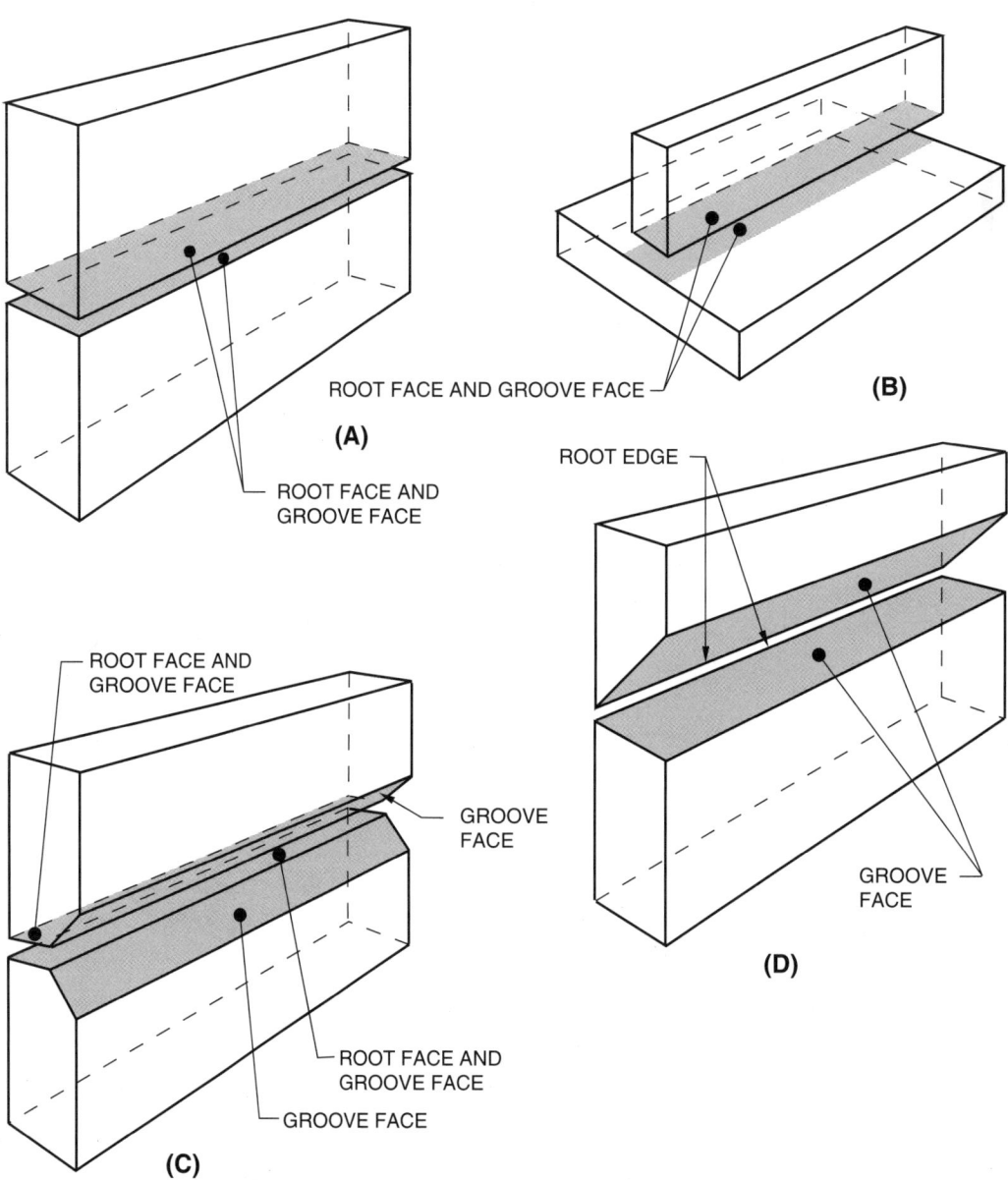

Figure 4.13—Groove Face, Root Face, and Root Edge

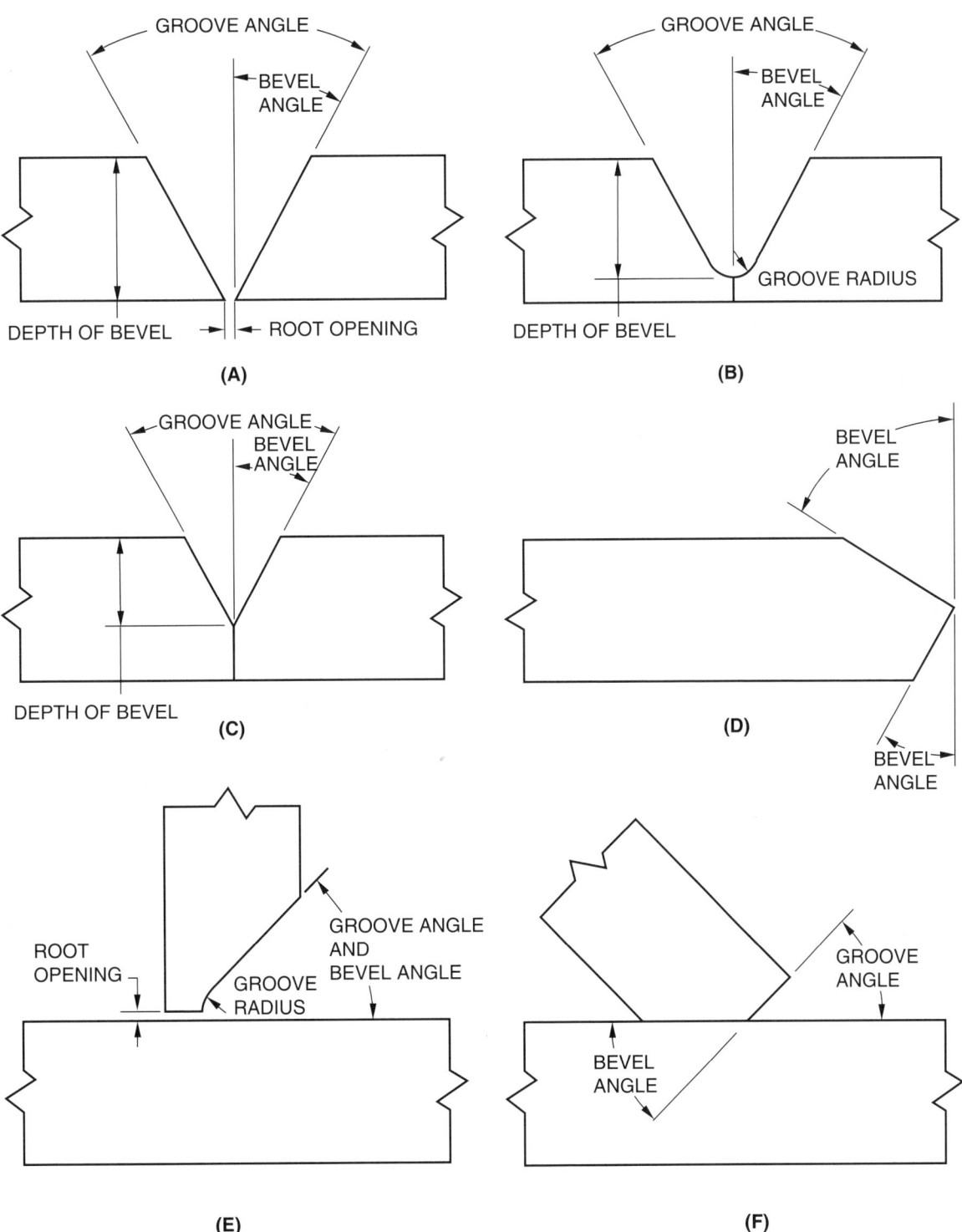

Figure 4.14—Bevel Angle, Depth of Bevel, Groove Angle, Groove Radius, and Root Opening

The *root opening* is described as "the separation between the workpieces at the joint root." The *bevel* (also commonly referred to as the *chamfer*) is "an angular edge preparation." *Bevel angle* is defined as "the angle between the bevel of a joint member and a plane perpendicular to the surface of the member." *Groove angle* is "the total included angle of the groove between workpieces." For a single-bevel-groove-weld, the bevel angle and the groove angle are equal. *Groove radius* applies only to J- and U-groove-welds. It is described as "the radius used to form the shape of a J- or U-groove weld." Normally, a J- or U-groove weld configuration is specified by both a bevel (or groove) angle and a groove radius.

Types of Welds

As was shown in Figure 4.2, numerous welds can be applied to the various types of joints. Using AWS A2.4, *Standard Symbols for Welding, Brazing, and Nondestructive Examination* as a guideline, there are nine categories of welds associated with weld symbols. In each of these categories, certain weld types apply. The categories are:

(1) Groove Welds
(2) Fillet Welds
(3) Plug or Slot Welds
(4) Stud Welds
(5) Spot or Projection Welds
(6) Seam Welds
(7) Back or Backing Welds
(8) Surfacing Welds
(9) Edge Welds

With the variety of joint geometries and weld types available, the welding designer can choose the one which best suits his or her needs. This choice could be based on considerations such as:

- accessibility to the joint for welding
- type of welding process being used
- suitability to the structural design
- cost of welding

Groove Welds

A *groove weld* is "a weld made in a groove between the workpieces." There are eight types of groove welds:

(1) square-groove
(2) scarf
(3) V-groove
(4) bevel-groove
(5) U-groove
(6) J-groove
(7) flare-V-groove
(8) flare-bevel-groove

Their names imply what the actual configurations look like when viewed in cross section. All of these groove weld types can be applied to joints which are welded from a single side or both sides. Figure 4.15 illustrates typical configurations for single and double groove welded joints. As would be expected, a single-welded joint is "a fusion welded joint that is welded from one side only." Similarly, a double-welded joint is "a fusion welded joint that is welded from both sides."

Groove welds of different types are used in many combinations. Selection is influenced by accessibility, economy, adaptation to structural design, expected distortion and the type of welding process used.

Square-groove welds are the most economical to use, but are limited by the thickness of the members. Complete penetration square groove welds, welded from one side, are generally not used for material thicker than one quarter inch.

Thicker material requires the selection of joint geometries that accommodate other types of groove welds. On thicker joints, the particular geometry must provide accessibility for welding, ensure weld soundness and strength, and minimize the amount of metal removed. For economical reasons, these joint designs should be selected with root openings and groove angles that require the least amount of weld metal, but still meet the service conditions of the weldment. Selection of root openings and groove angles are influenced by the metal to be joined, location of the joint within the weldment, and required service conditions.

Welds in J- and U-grooves can be used to minimize weld metal requirements when economic factors outweigh the cost of edge preparation. These types of welds are particularly useful in thicker sections. Single-bevel and J-groove welds are more difficult to weld than V- or U-groove welds because

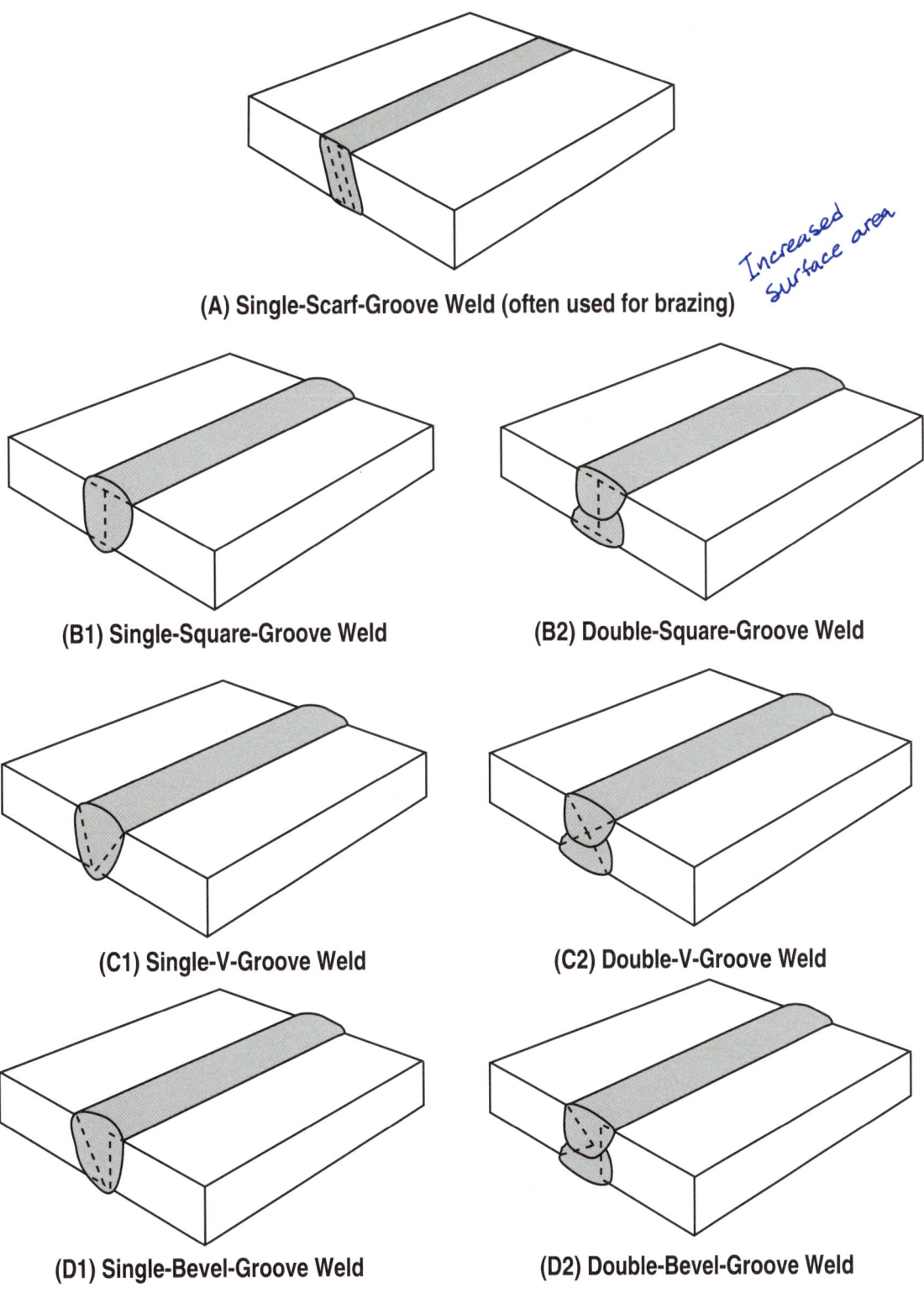

Figure 4.15—Single- and Double-Groove Welds

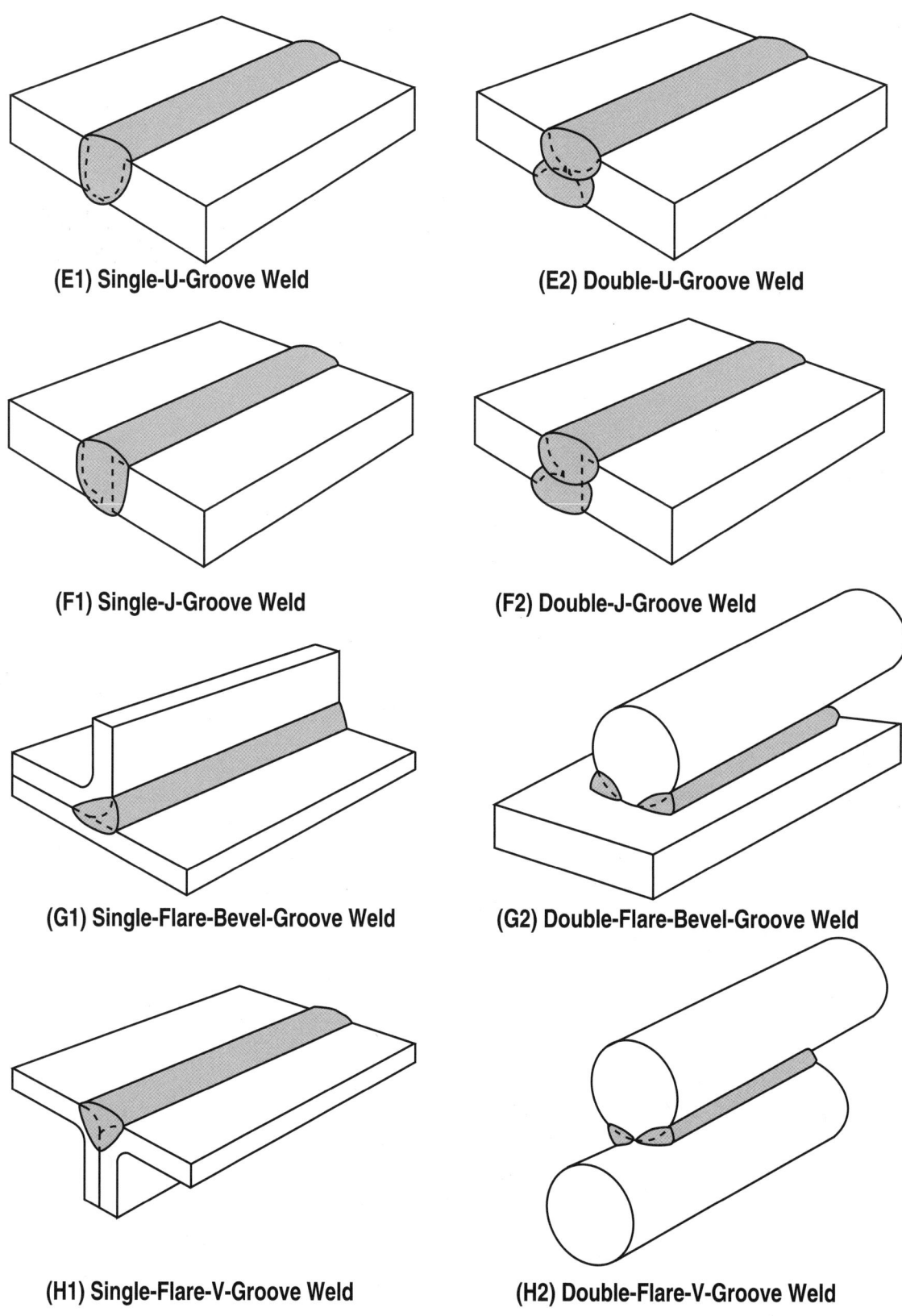

Figure 4.15 (Continued)—Single- and Double-Groove Welds

of the one vertical edge. Flare V-groove and flare-bevel groove welds are used in connection with flanged or round members.

Fillet Welds

AWS A2.4 defines a *fillet weld* as "a weld of approximately triangular cross section joining two surfaces approximately at right angles to each other in a lap joint, T-joint, or corner joint." When the design permits, fillet welds are used in preference to groove welds because of economic reasons. Edge preparations are usually not required for fillet welding, but surface cleaning may be needed. Unlike groove welds, a fillet weld does not take its name from an associated joint geometry; it is a particular type of weld applied to a lap, T-, or corner joint. Fillet welds are sometimes used in combination with groove welds. Figure 4.16 shows typical fillet welds applied to corner, T-, and lap joints.

Fillet welds are made using both single and double sided welds. They are also applied using a single weld pass or multiple weld passes. Examples of both are shown in Figure 4.16.

In addition to continuous pass welds (the complete length of the joint), fillet welds are often intermittently staggered or chained. A *staggered intermittent fillet weld* is "an intermittent weld on both sides of a joint in which the weld increments on one side are alternated with those on the other." A *chain intermittent fillet weld* is "an intermittent weld on both sides of a joint in which the weld increments on one side are approximately opposite those on the other side." Figure 4.16(E) and (F) illustrates both types of intermittent fillet welds.

Plug and Slot Welds

Two types of welds used for joining overlapping members are plug and slot welds. The *plug weld* is "a weld made in a circular hole in one member of a joint fusing that member to another member." A *slot weld* is "a weld made in an elongated hole in one member of a joint fusing that member to another member. The hole may be open at one end." Plug and slot welds require definite depths of filling. Figure 4.17(A) and (B) shows illustrations of plug and slot welds. A fillet weld applied in a circular hole, Figure 4.16(D), or slot is not considered a plug or slot weld.

Stud Welds

Figure 4.17(C) provides an example of a stud weld.

The most common stud materials welded with the arc stud welding process are low carbon steel, stainless steel, and aluminum. Other materials are used for studs on a special application basis.

Most stud weld bases are round. However, there are many applications which use a square or rectangular shaped stud. Applications of stud welds include attaching wood floors to steel decks or framework; fastening linings or insulation in tanks, boxcars, and other containers; mounting machine accessories; securing tubing and wire harnesses; and welding shear connectors and concrete anchors to structures.

Spot and Projection Welds

A *spot weld* is "a weld made between and upon overlapping members in which coalescence (the act of combining or uniting) may start and occur on the faying surfaces or may begin from the outer surface of one member." A *faying surface* is defined as, "the mating surface of a member that is in contact with or in close proximity to another member to which it is to be joined." Spot welds are commonly associated with resistance welding. However, a very effective way to join a lap joint shape on thin metals is with an arc spot weld. In arc spot welding, the weld forms by melting through the top member using an arc welding process, and fusion occurs between it and the overlapping member." Figure 4.18(A) and (B) illustrates resistance and arc spot welds.

Projection welds are made using the resistance welding process. The weld is formed by the heat obtained from the resistance to the flow of the welding current. The resulting welds are localized at predetermined points by projections, embossments, or intersections. Figure 4.18(C) shows cross-sectional views of an embossed member of a lap joint to be projection welded, and the desired weld upon completion.

Seam Welds

A *seam weld* is "a continuous weld made between or upon overlapping members, in which coalescence may start and occur on the faying surfaces, or may have proceeded from the outer surface of one member. The continuous weld may consist

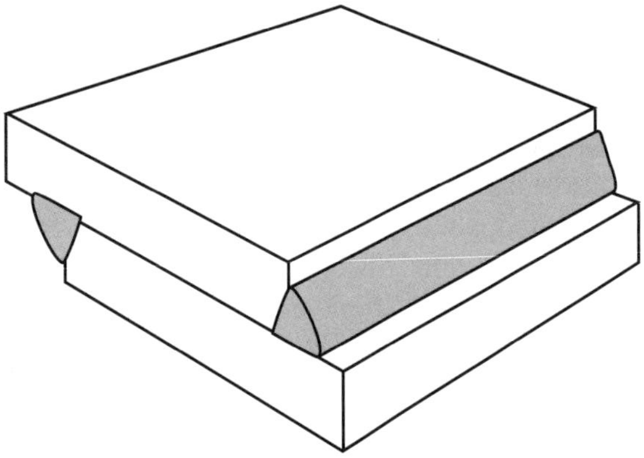

(A) Double Sided - Single Pass Fillet Welds on a Lap Joint

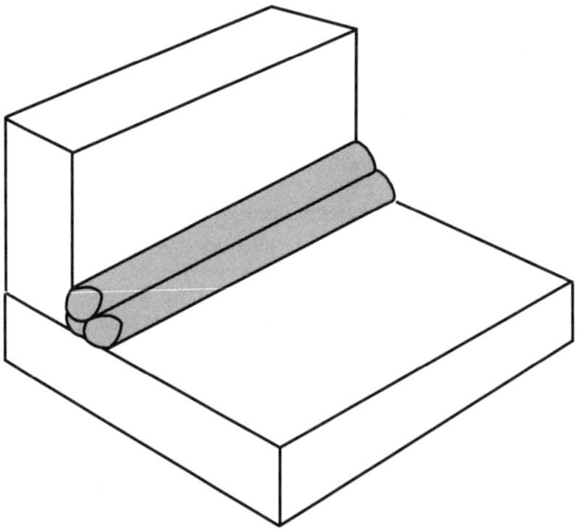

(B) Single Sided - Multiple Pass Fillet Welds on a Corner Joint

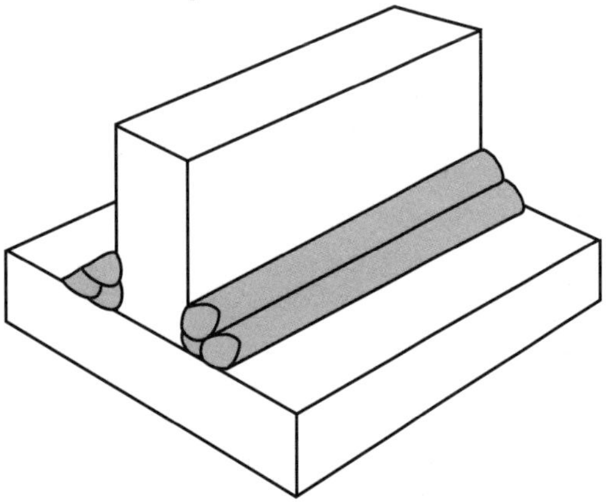

(C) Double Sided - Multiple Pass Fillet Welds on a T-Joint

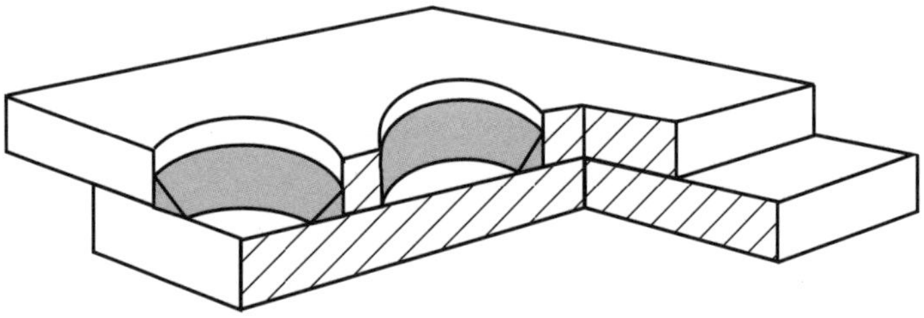

(D) Fillet Welds around the diameter of a hole

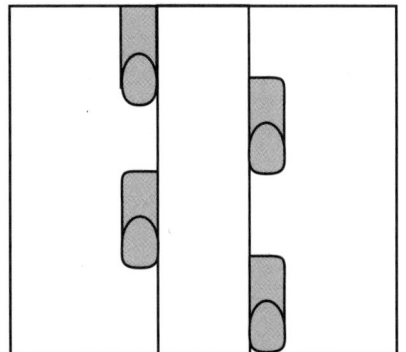

(E) Staggered Intermittent Fillet Welds Top View

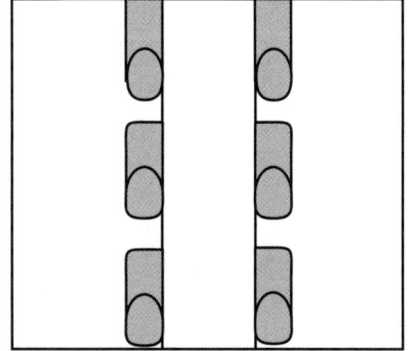

(F) Chain Intermittent Fillet Welds Top View

Figure 4.16—Fillet Weld Applications

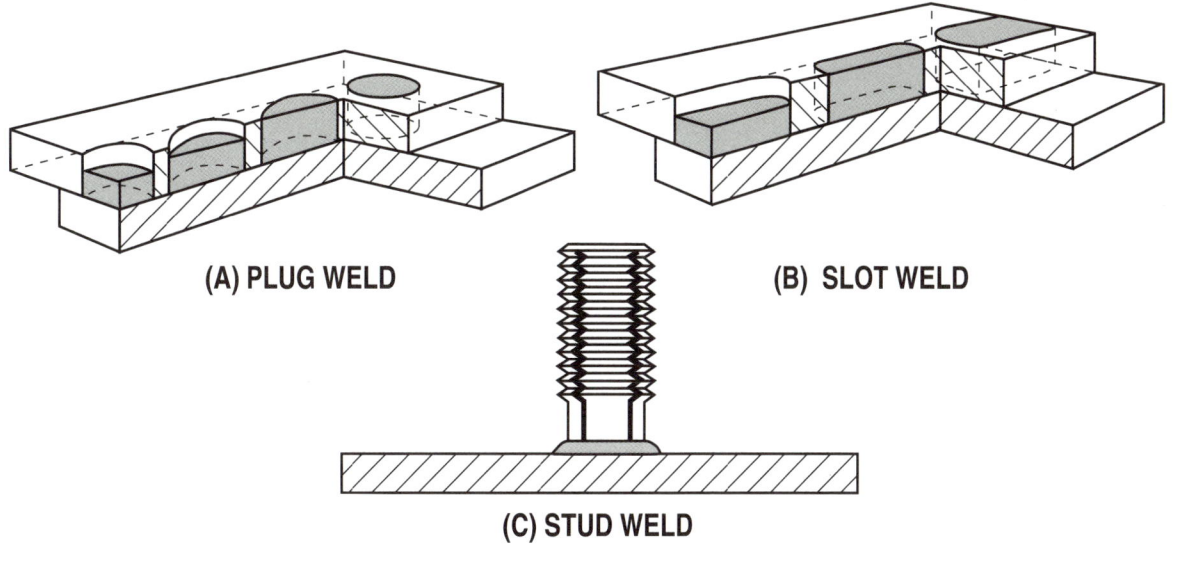

Figure 4.17—Plug, Slot, and Stud Welds

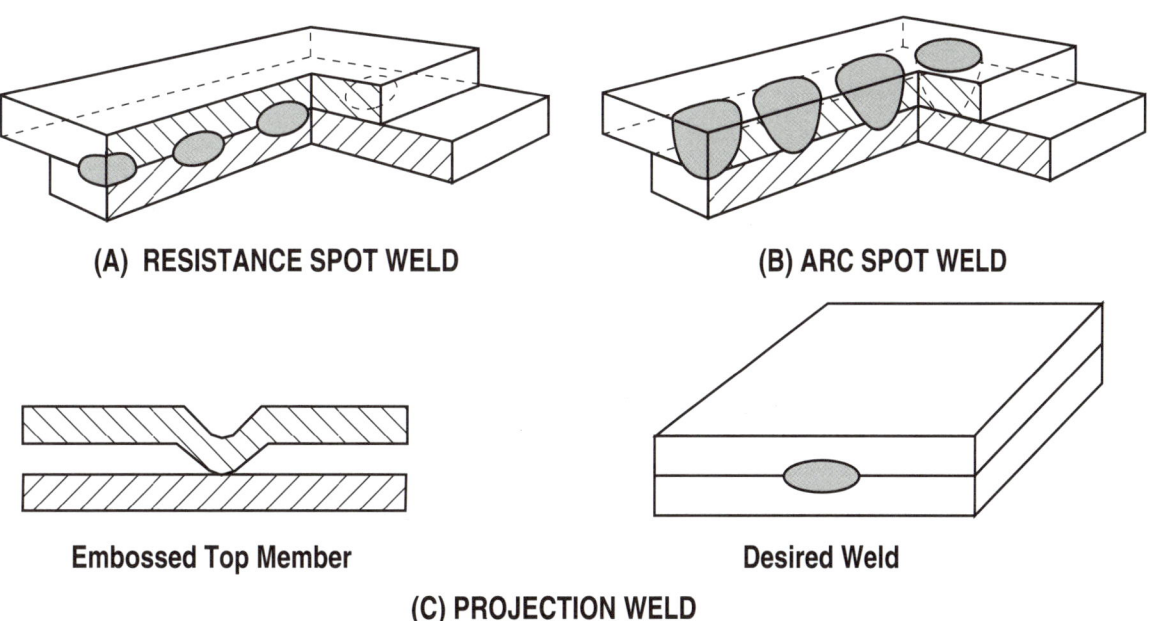

Figure 4.18—Spot and Projection Welds

of a single weld bead or a series of overlapping spot welds." Some means must be provided to move the welding head along the seam during welding, or to move the workpiece beneath the welding head. This type of weld is associated with arc and resistance welding. Seam welds for both welding processes are illustrated in Figure 4.19(A)–(D).

Back and Backing Welds

As their names imply, these welds are made on the back side of a weld joint. Although they apply to the same location, they are deposited differently. AWS A3.0 describes a *back weld* as "a weld made at the back of a single groove weld." A *backing weld* is "backing in the form of a weld." A back

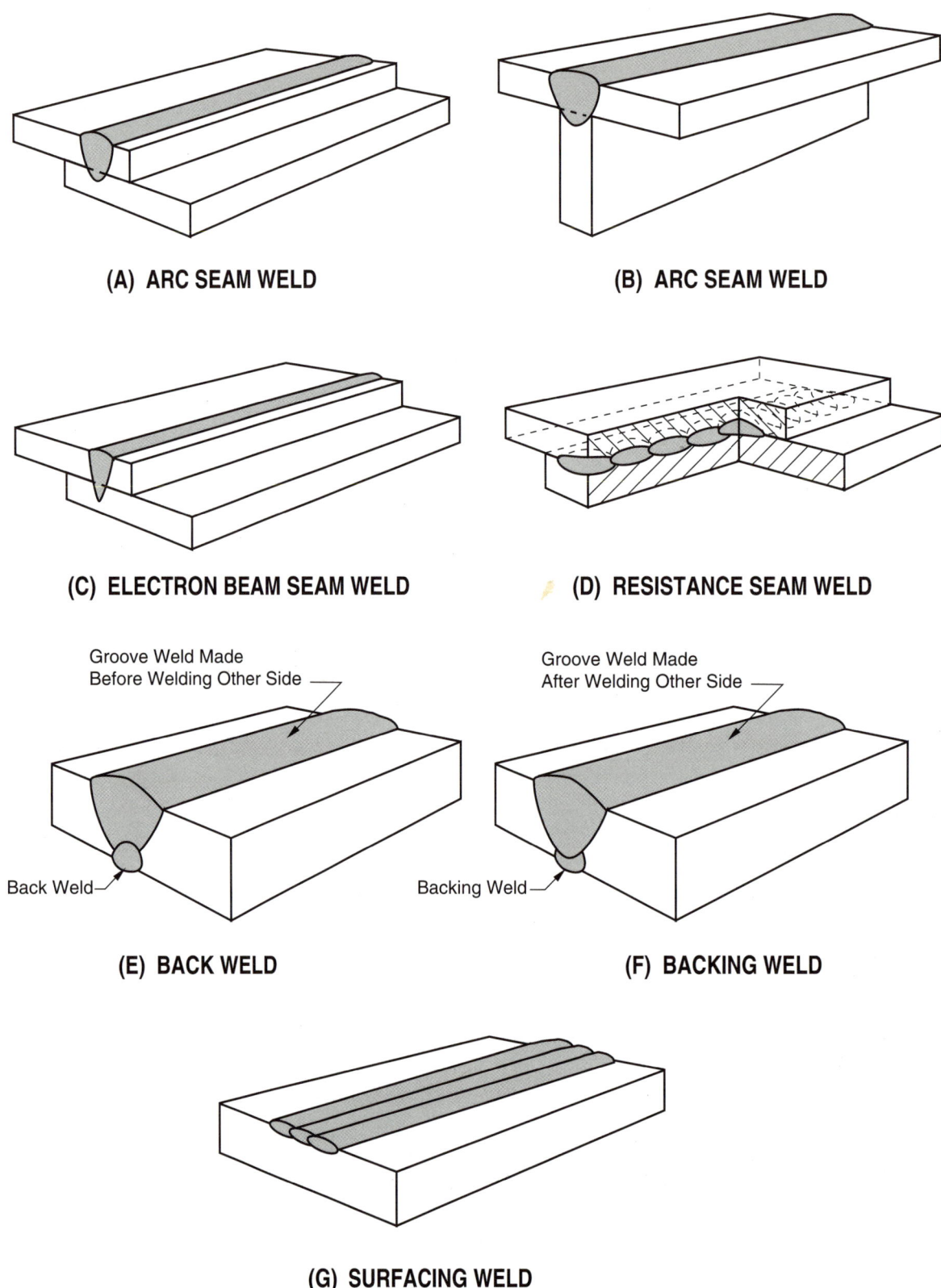

Figure 4.19—Seam, Back, Backing, and Surfacing Welds

weld is applied *after* the front side is welded, while the backing weld is deposited *before* welding the front side. Figure 4.19(E) and (F) illustrates the application of both back and backing welds.

Surfacing Welds

As the name implies, this type of weld is applied to the surface of a metal. A *surfacing weld* is defined as "a weld applied to a surface, as opposed to making a joint, to obtain desired properties or dimensions." Figure 4.19(G) illustrates a typical surfacing weld. Other terms associated with surfacing are:

- *Buildup,* "a surfacing variation in which surfacing material is deposited to achieve the required dimensions."

- *Buttering,* "a surfacing variation that deposits surfacing metal on one or more surfaces to provide metallurgically compatible weld metal for the subsequent completion of a weld."

- *Cladding,* "a surfacing variation that deposits or applies surfacing material, usually to improve corrosion or heat resistance."

- *Hardfacing,* "a surfacing variation in which surfacing material is deposited to reduce wear."

Edge Welds

An *edge weld* is "a weld in an edge joint, a flanged butt joint, or a flanged corner joint in which the full thickness of the members are fused." An edge-flange joint has two flanged members, while a corner-flange joint has only one of the members flanged. Figure 4.20 illustrates edge weld placements on corner and edge flange joints.

Completed Welds

Welding and inspection personnel should be aware of terms associated with certain conditions or features of completed welds. Knowledge of these terms aids in the communication process and enhances one's ability to interpret welding symbol information and locate areas of a weld that may require additional post-weld cleaning or detailing.

Terms related to groove welds (see Figure 4.21) consist of:

- Weld Face
- Weld Toe
- Weld Root
- Root Surface
- Face Reinforcement
- Root Reinforcement

The *weld face* is "the exposed surface of a weld on the side from which welding was done." *Weld toe* is "the junctions of the weld face and the base metal." Opposite the weld toe is the weld root. *Weld root* is "the points, shown in cross section, at which the root surface intersects the base metal surfaces." Similar to weld face is the *root surface,* or "the exposed surface of a weld opposite the side from which welding was done." In other words the root surface is bounded by the weld root on either side.

Additional terminology associated with groove welds relates to weld reinforcement; *weld reinforcement* is "weld metal in excess of the quantity required to fill a joint." The *face reinforcement* (commonly referred to as the *crown* or *cap*), refers to "weld reinforcement on the side of the joint from which welding was done."

Conversely, the *root reinforcement* is "weld reinforcement opposite the side from which welding was done." Root reinforcement is used only in the case of a single welded joint, meaning, welding performed from one side (see Figure 4.21(C)). When a double sided weld is made, the term face reinforcement is applied to the amount of reinforcement present on both sides. This point is illustrated in Figure 4.21(A) where a back weld is used.

Standard terminology also exists for parts of fillet welds. As with the groove weld, the surface of the fillet weld is referred to as the *weld face*. The junctions of the weld face with the base metal are the *weld toes*. The furthest penetration of the weld metal into the joint is the *weld root*. "The distance from the beginning of the *joint* root to the toe of the fillet weld" is called the fillet weld leg. Figure 4.22 identifies various parts of a fillet weld.

Three other dimensional features of fillet welds are concavity, convexity, and throat. Concavity and convexity are the amount of curvature of the weld face, and throat is the length through the weld cross section. The method for measuring these is shown in Figure 4.27.

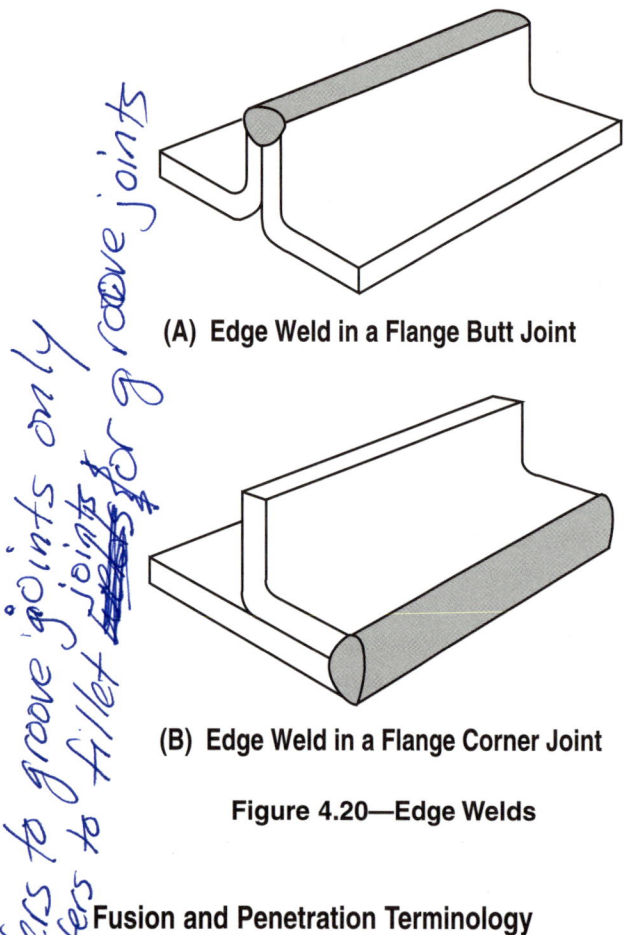

(A) Edge Weld in a Flange Butt Joint

(B) Edge Weld in a Flange Corner Joint

Figure 4.20—Edge Welds

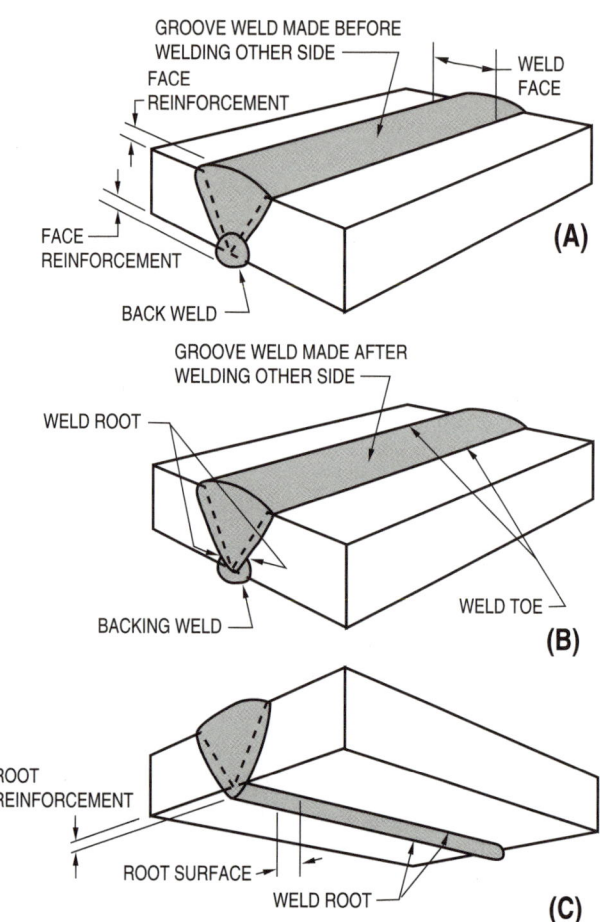

Figure 4.21—Completed Groove Weld Terms

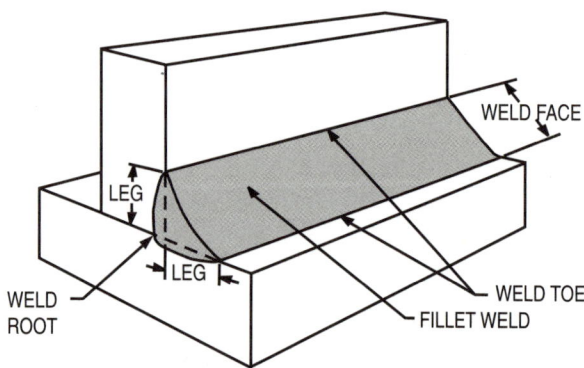

Figure 4.22—Completed Fillet Weld Terms

Fusion and Penetration Terminology

In general, fusion refers to the actual melting together of the filler metal and base metal, or of the base metal only. Penetration is a term which relates to the distance that the weld metal has progressed into the joint. The degree of penetration achieved has a direct effect on the strength of the joint and is therefore related to the weld size.

Numerous terms exist that describe the degree or location of either fusion or penetration. During the welding operation, the original groove face is melted so that the final limits of the weld metal are deeper than the original surfaces. The groove face (before welding) is now referred to as the *fusion face,* since it will be melted during welding. The boundary between the weld metal and base metal is referred to as the *weld interface.* The *depth of fusion* is "the distance from the fusion face to the weld interface." The depth of fusion is always measured perpendicular to the fusion face. The *fusion zone* is "the area of base metal melted as determined on the cross section of a weld. These terms are applied similarly for other types of welds such as fillet and surfacing welds. Figure 4.23 illustrates various terms associated with fusion.

As shown in Figure 4.24, there are also several terms which refer to penetration of the weld. *Root penetration* is "the distance that the weld metal has

Interface: the point at which fusion stops.

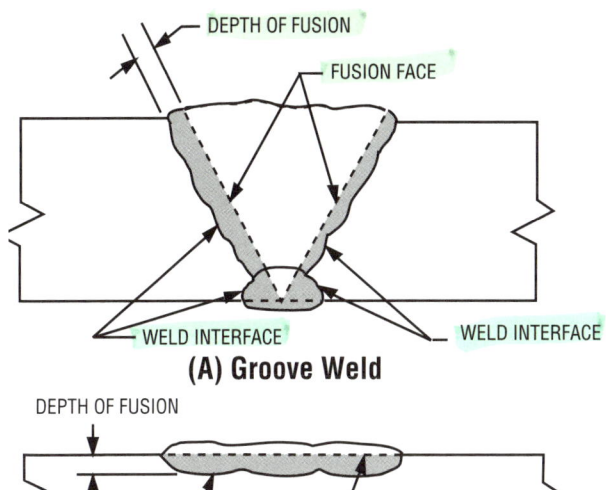

(A) Groove Weld

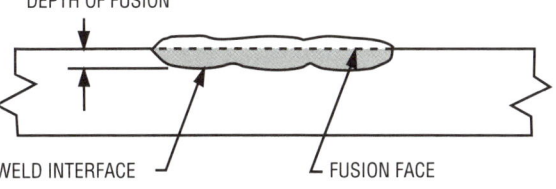

(B) Surfacing Weld

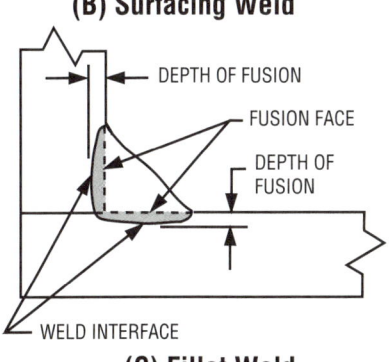

(C) Fillet Weld

Note: Fusion zones indicated by shading.

Figure 4.23—Fusion Terminology

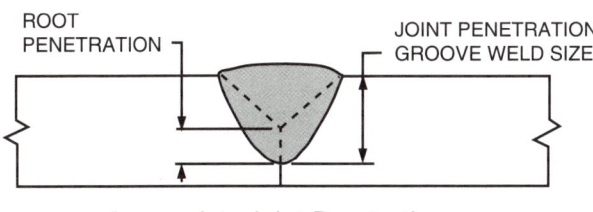

Incomplete Joint Penetration

Figure 4.24—Penetration Terminology

melted into the joint beyond the joint root." The *joint penetration* is "the distance from the furthest extension of the weld into the joint to the weld face, excluding any weld reinforcement which may be present." For groove welds, this same length is also referred to as the weld size.

Another related term is *heat-affected zone.* This region, shown in Figure 4.25, is defined as "that portion of the base metal that has not been melted, but whose mechanical properties or microstructure have been altered by the heat of welding, brazing, soldering, or cutting."

Weld Size Terminology

The previous discussion describes joint penetration, and the relationship to weld size, for single-groove weld configurations. For a double-groove weld configuration where the joint penetration is less than complete, the weld size is equal to the sum of the joint penetrations from both sides (see Figure 4.26(A)).

In a complete penetration groove weld, the weld size is equal to the thickness of the thinner of the two members joined, since there is no credit given

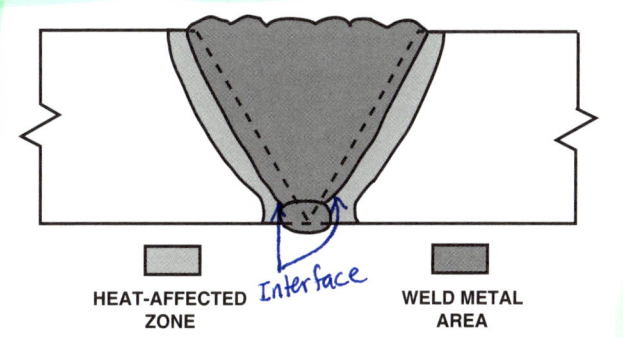

Figure 4.25—Heat Affected Zone

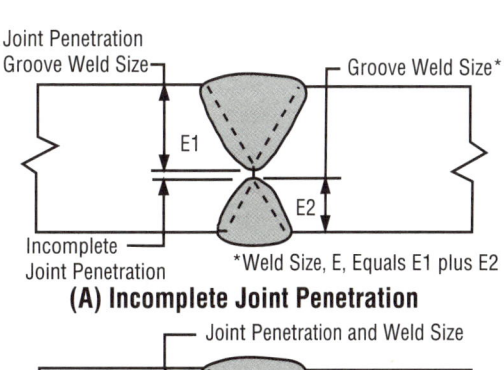

(A) Incomplete Joint Penetration

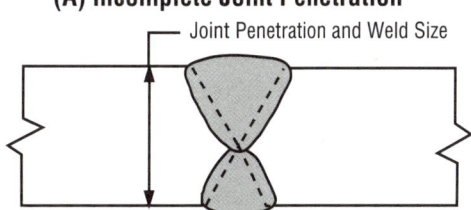

(B) Complete Joint Penetration

Figure 4.26—Penetration and Weld Size

for any weld reinforcement that is present (see Figure 4.26(B)).

To determine the size of a fillet weld, you must first know whether the final weld configuration is *convex* or *concave*. Convex means that the weld face exhibits some buildup causing it to appear curved slightly outward. This is referred to as the amount of convexity. *Convexity* in a fillet weld is synonymous with weld reinforcement in a groove weld. If a weld has a concave profile, this means that its face is "dished in."

For either configuration, the fillet weld size for *equal leg fillet welds* is described as "the leg lengths of the largest isosceles (two legs of equal length) right triangle which can be inscribed within the fillet weld cross section."

These inscribed isosceles right triangles are shown with dotted lines in the two illustrations of Figure 4.27. So, for the convex fillet weld, the leg and size are equal. However, the size of a concave fillet weld is slightly less than its leg length.

For *unequal leg fillet welds,* the fillet weld size is defined as, "the leg lengths of the largest right triangle that can be inscribed within the fillet weld cross section." Figure 4.28 shows this.

It can be noted that there are additional notations on the illustrations in Figure 4.27 which refer to fillet weld throats. There are really three different types of *weld throats*. The first is the theoretical throat, or "the minimum amount of weld that the designer counts on when originally specifying a weld size."

The *theoretical throat* is described as "the distance from the beginning of the joint root perpendicular to the hypotenuse (side of the triangle opposite the right angle) of the largest right triangle that can be inscribed within the cross section of a fillet weld. This dimension is based on the assumption that the root opening is equal to zero."

Effective throat takes into account any additional joint penetration that may be present. So, the *effective throat* can be defined as "the minimum distance minus any convexity between the weld root and the face of a fillet weld." The final throat dimension, the actual throat, takes into account both the joint penetration as well as any additional convexity present at the weld face.

Actual throat is "the shortest distance between the weld root and the face of a fillet weld." For a

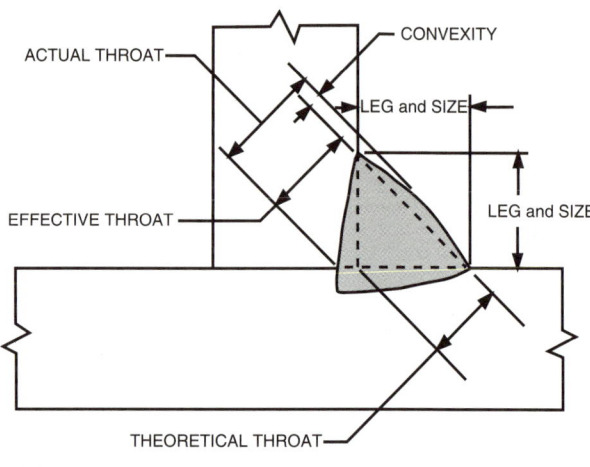

Figure 4.27—Fillet Weld Size

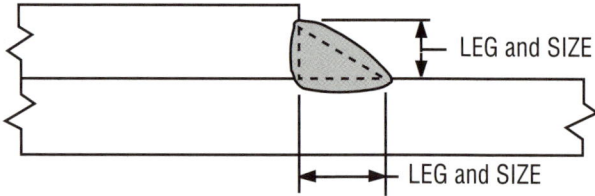

Figure 4.28—Unequal Leg Fillet Weld

concave fillet weld, the effective throat and actual throat are equal, since there is no convexity present.

Inspection personnel may also be asked to determine the sizes of other types of welds. One example might be a spot or seam weld, where the weld size is equal to the diameter of the weld metal in the plane of the faying surfaces as shown in Figure 4.29. A second example is for an edge or flange

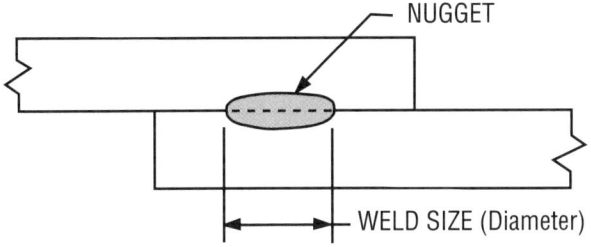

Figure 4.29—Size of Seam or Spot Weld

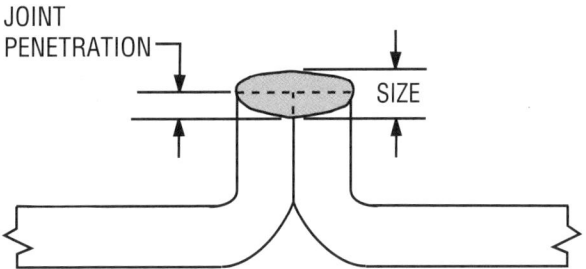

Figure 4.30—Edge Weld Size

weld as shown in Figure 4.30; the weld size is equal to the total thickness of the weld from the weld root to the weld face.

Weld Application Terminology

To complete this discussion of welding terms, it is appropriate to mention additional terminology associated with the actual application of welds. Often welding procedures will refer to these details, so inspection personnel should be familiar with their meanings. The first aspect is the difference among the terms weld pass, weld bead and weld layer. A *weld pass* is a single progression of welding along a joint. The *weld bead* is that weld which results from a weld pass. A *weld layer* is a single level of weld within a multiple-pass weld. A weld layer may consist of a single bead or multiple beads (see Figure 4.31).

When a weld bead is deposited, it could have a different name, depending upon the technique the welder uses. If the welder progresses along the joint with little or no side-to-side motion (oscillation), the resulting weld bead is referred to as a *stringer bead*. A *weave bead* results when the welder manipulates the electrode laterally, or side to side, as the weld is deposited along the joint. The weave bead is typically wider than the stringer bead. Due to the amount of lateral motion used, the travel speed, as measured along the longitudinal axis of the weld, is less than would be the case for a stringer bead. Examples of these are shown in Figure 4.32.

When fillet welds are required, there will be some cases where the design does not warrant the use of continuous welds. The designer may therefore specify intermittent fillet welds. If there are intermittent fillet welds specified on both sides of a particular joint, they can be detailed as either chain intermittent or staggered intermittent fillet welds.

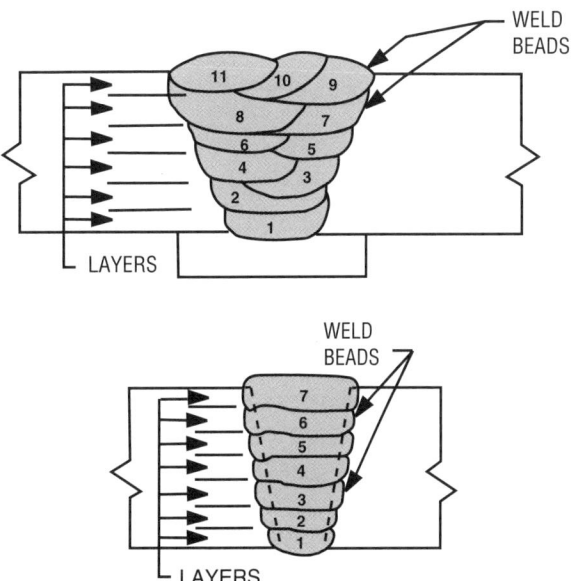

CROSS-SECTIONAL WELDING SEQUENCE

Figure 4.31—Weld Pass, Bead, and Layer

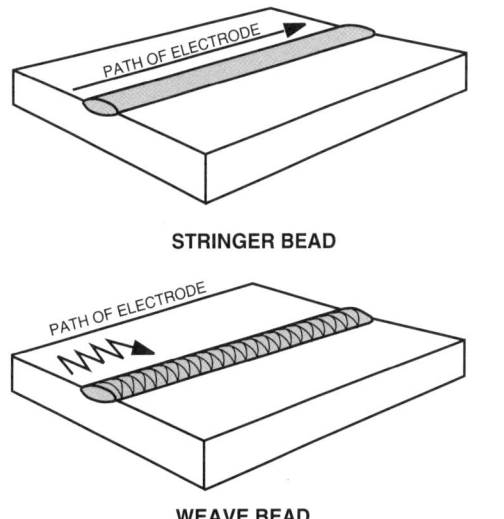

Figure 4.32—Stringer and Weave Beads

The *chain intermittent fillet weld* has the increments on either side of the joint directly opposite each other. Similarly, a *staggered intermittent fillet weld* is an intermittent fillet weld on both sides of a joint in which the weld increments on one side are alternated with respect to those on the other side. Both types of the intermittent fillet welds are shown in Figure 4.33.

Another term related to the actual welding operation is boxing (commonly referred to as an end return). *Boxing* is defined as "the continuation of a fillet weld around a corner of a member as an extension of the principal weld" (see Figure 4.34).

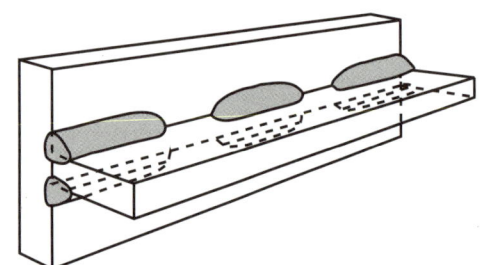

CHAIN INTERMITTENT FILLET WELD

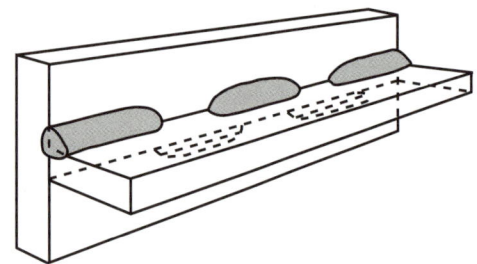

STAGGERED INTERMITTENT FILLET WELD

Figure 4.33—Intermittent Fillet Welds

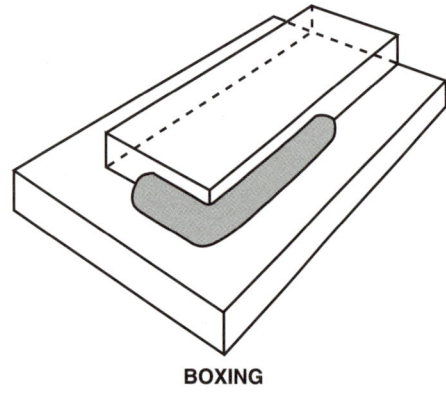

BOXING

Figure 4.34—Boxing Technique

Finally, there are several terms that describe the actual sequence in which the welding is to be done. This is commonly done to reduce the amount of distortion caused by welding. Three common techniques are backstep sequence, block sequence, and cascade sequence (see Figure 4.35). The *backstep sequence* is a technique where each individual weld pass is deposited in the direction opposite that of the overall progression of welding.

A *block sequence* is defined as "a combined longitudinal and cross sectional sequence for a continuous multiple pass weld in which separated increments are completely or partially welded before intervening increments are welded." With the block sequence, it is important that each subsequent layer is slightly shorter than the previous one so that the end of the block has a gentle slope. This will provide the best chance of obtaining adequate fusion when the adjacent block is filled in later.

A *cascade sequence* is described as "a combined longitudinal and cross sectional sequence in which weld passes are made in overlapping layers." This method differs from the block sequence in that each subsequent pass is longer than the previous one.

Welding Symbols

Welding symbols provide a system for placing complete welding information on drawings. They quickly indicate to the designer, draftsman, supervisors, and welding personnel, including welding inspectors, which welding technique is needed for each joint to satisfy the requirements for material strength and service conditions.

For layout and fitting personnel, welding symbols often relay information that affect the finished size of a prepared part. For example, root opening changes will cause a change in the actual size of a member when only design size dimensions are provided on a drawing. The inspector must be aware of these requirements, and the effects of changes in the specified parameters.

Fitters and layout personnel must be aware of the placement and size of tack welds. Oversized tack welds, and tack welds placed outside the designated weld area, cause extra steps in the finishing phases of a project. One example of this is the placement of tack welds outside the designated area of intermittent fillet welds. Another example is the

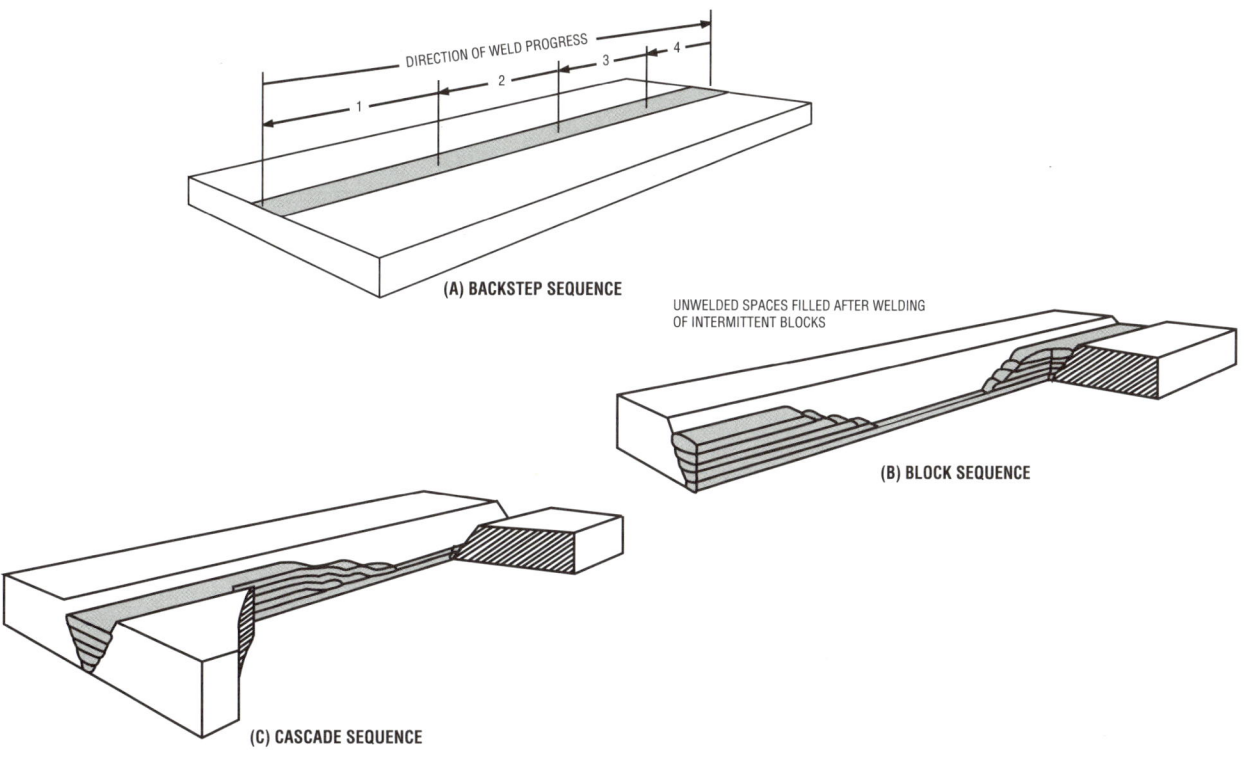

Figure 4.35—Backstep, Block, and Cascade Welding Sequences

appearance of the weld face after the welder has fused an oversized tack weld into the required weld. These examples usually fall under the requirement of inspection responsibilities and the welding inspector must be familiar with the meaning of welding symbols to complete the inspection duties.

In this section, the welding inspector will be provided a basic understanding of the information which can appear in a welding symbol, the identification of basic weld symbols, the use of supplementary symbols, and an understanding of the terminology associated with the basic welding symbol.

A detailed reference regarding weld and welding symbols and the associated terminology is found in the current edition of AWS A2.4, *Standard Symbols for Welding, Brazing, and Nondestructive Examination.* This document is shown in Figure 4.36 and is published by the American Welding Society.

Weld Symbol vs. Welding Symbol

AWS makes a distinction between the terms weld symbol and welding symbol. The *weld symbol* (see Figure 4.38) identifies each specific type of weld and is only part of the total information contained in the welding symbol. Weld symbols are drawn above and below the reference line of the welding symbol. The *welding symbol* (Figure 4.39) indicates the total symbol, including all information applied to it, to specify the weld(s) required. *All welding symbols require a reference line and an arrow,* and these are shown in Figure 4.37.

Welding Symbol Elements

Except for the reference line and arrow, not all elements need be used unless required for clarity. A welding symbol may include the following elements:

- Reference Line (required element)
- Arrow (required element)
- Tail
- Basic weld symbol
- Dimensions and other data
- Supplementary symbols
- Finish symbols
- Specification, process, or other reference

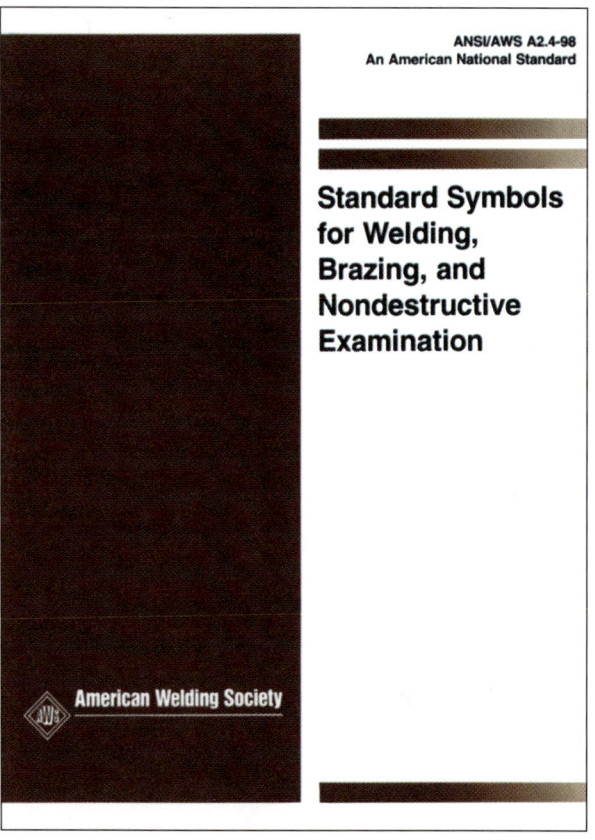

Figure 4.36—AWS A2.4, *Standard Symbols for Welding, Brazing, and Nondestructive Examination*

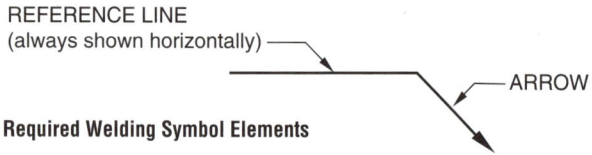

Figure 4.37—Reference Line and Arrow

The *reference line* is always drawn horizontally. It is used to apply weld symbols and other data, and has a particular significance that remains the same regardless of any elements added to it. The lower side of the reference line is termed the *arrow side* and the upper side of the reference line is termed the *other side*. This convention is shown in Figure 4.40. The direction of the arrow creates no change in the significance of the reference line. Multiple reference lines may also be used with the basic weld symbols and are shown in Figure 4.41.

The *arrow* connects the reference line to the weld joint or area to be welded. It may be shown with or without a break, or with multiple arrows. *When the arrow is shown with a break, the broken arrow **always** breaks toward the member of the joint that is to be prepared or shaped,* and is shown in Figure 4.42. Figure 4.42 also shows that multiple arrows may be added to show the same weld required in several different locations. Additional examples of multiple arrows are shown in Figure 4.60.

Arrows point to a line or lines on the drawing which clearly identify the proposed joint or weld area. When possible the arrow should point to a solid line (object line, visible line) but the arrow may point to a dashed line (hidden line).

The *tail* of the welding symbol is used to indicate the welding and cutting processes, as well as the welding specifications, procedures, or the supplementary information to be used in making the weld. When the welding process, specification, procedure, or supplementary information is not necessary to identify the welding information, the tail is omitted from the welding symbol. Figure 4.43 illustrates the tail.

Process, references, specifications, codes, drawing notes, or any other applicable documents pertaining to the welding may be specified by placing the reference in the tail of the welding symbol. The information contained in the referenced documents does not have to be repeated in the welding symbol (see Figure 4.44).

Repetitions of identical welding symbols on the drawing are avoided by designating a single welding symbol as *typical* or abbreviated as "TYP," and pointing the arrow to the representative joint (see Figure 4.45). Typical designations must clearly identify all applicable joints, e.g., "TYP @ 4 stiffeners." See Figure 4.62 for "typical" weld applications.

Weld Symbol Locations

Regardless which way the arrow points, when weld symbols are placed below the reference line the weld must be made on the arrow side of the joint. Weld symbols placed above the reference line require the weld to be made on the other side of the joint. Weld symbols placed on both sides of the reference line indicate the weld is to be made on both sides of the joint. The both sides designation does

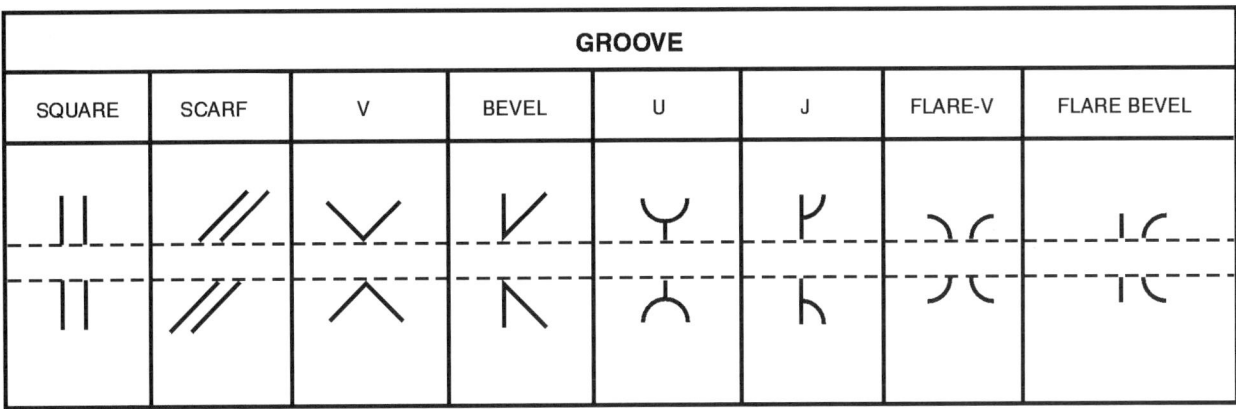

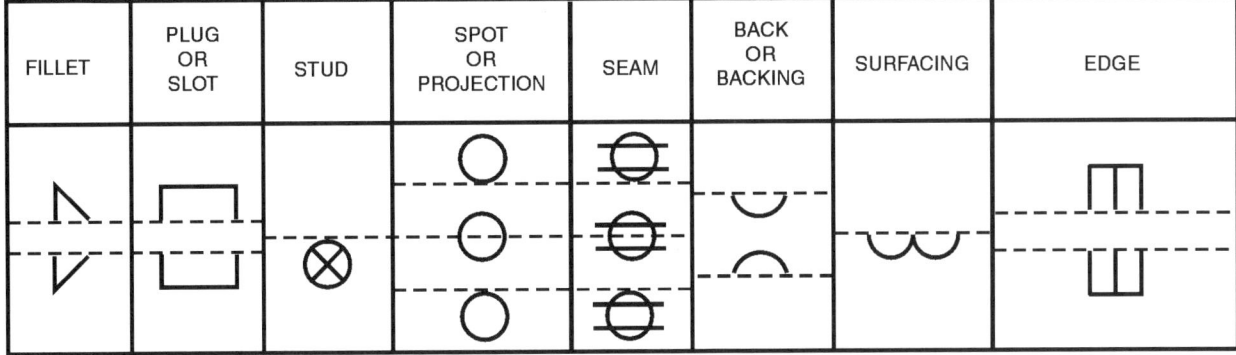

NOTE: THE REFERENCE LINE IS SHOWN DASHED (- - - -) FOR ILLUSTRATIVE PURPOSES.

DEFINITION:
  weld symbol. A graphical character connected to the welding symbol indicating the type of weld.

Figure 4.38—Weld Symbols

not apply to all weld symbols. Some symbols have no arrow side or other side significance, although supplementary symbols used in conjunction with them may (see Figure 4.46).

For fillet and groove symbols, the arrow always connects the welding symbol reference line to one side of the joint. That side is considered the arrow side of the joint, with the opposite side considered the other side of the joint. In addition, the perpendicular leg for fillet, bevel-groove, J-groove, and flare-bevel-groove symbols is always drawn to the left, as shown in Figure 4.47.

With plug, slot, spot, projection and seam weld symbols, the arrow connects the weld symbol reference line to the outer surface of one of the joint members, at the center line of the desired weld. The member toward which the arrow points is considered the arrow side member. The opposite member is considered to be the other side member. This is shown in Figure 4.48.

When only one member of a joint is to be prepared, such as for a bevel-groove, the arrow will have a break and point toward the member that must be prepared. Such joints will always be shown with a broken arrow when no joint details are given. If it is obvious which member is to be prepared, the arrow does not need to be broken. Figure 4.49 illustrates the broken arrow use.

Combined Weld Symbols

Some welded joints require more than one type of weld. This is a common occurrence in groove

DEFINITION:
welding symbol. A graphical representation of a weld.

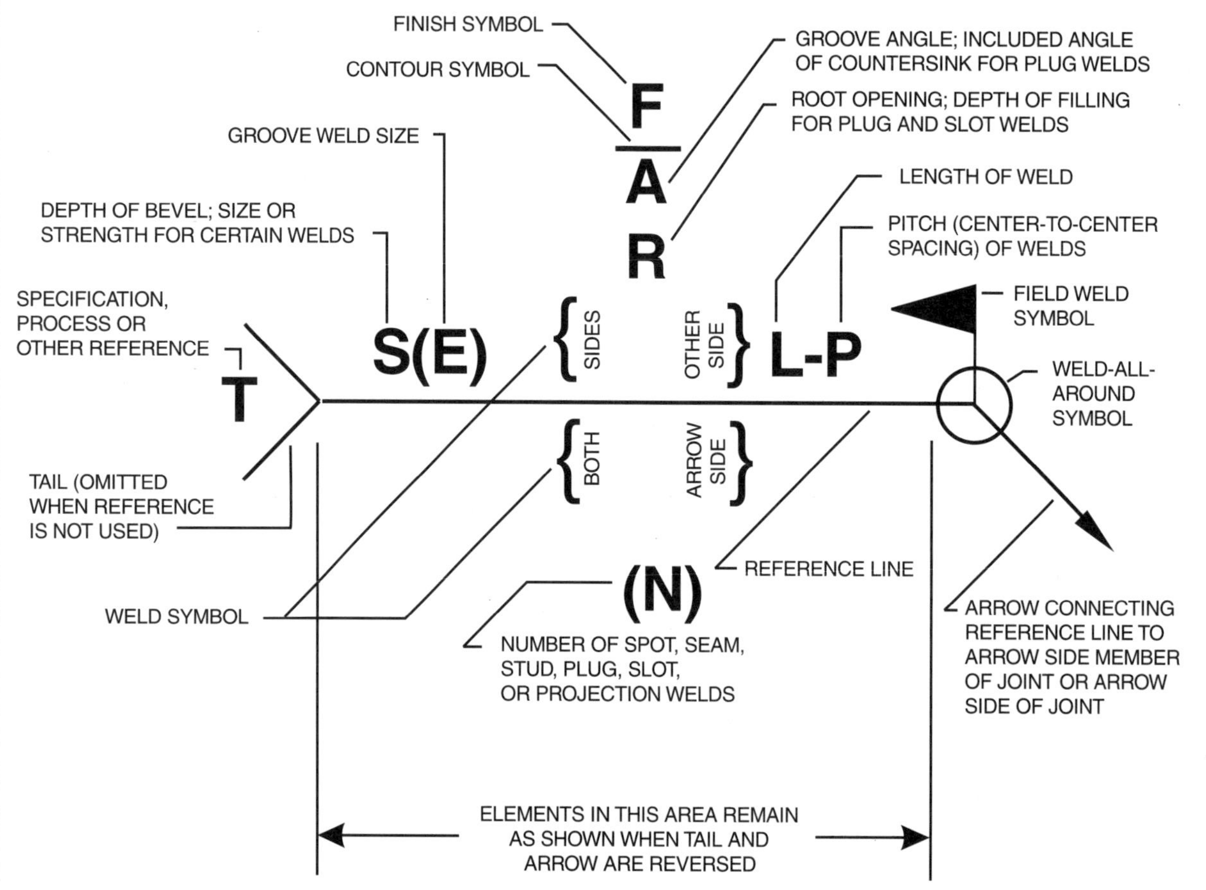

Figure 4.39—Standard Location of Elements of Welding Symbol

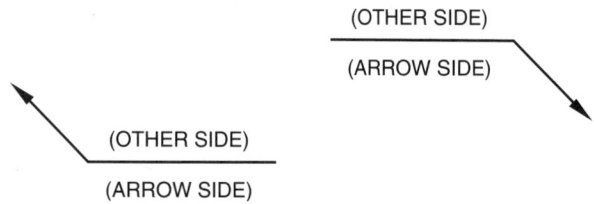

Figure 4.40—Arrow Side—Other Side Positions

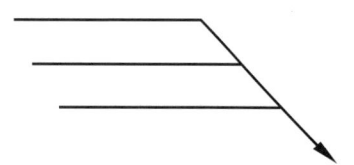

Figure 4.41—Multiple Reference Lines

welded joints for structural fabrication. Often the groove weld is finished with a fillet weld. As shown in Figure 4.50, a number of different combinations may apply to welded joints.

Multiple Reference Lines

The addition of two or more reference lines to the welding symbol are applied for several reasons. First, they are used to show the *sequence of operations.* That is, the first operation (shown on the reference line closest to the arrow) must be completed before the next operation can be performed and so on. Second, the addition of extra reference lines are also used when supplementary data applicable to each weld must be included, either in combination

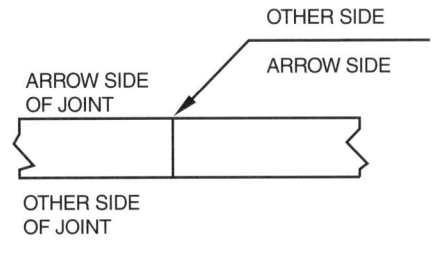

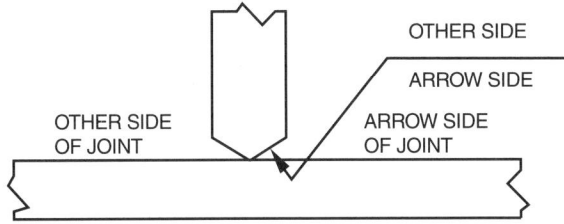

Note: Break in the arrow pointing to member to be prepared.

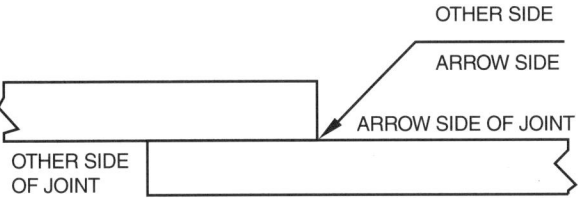

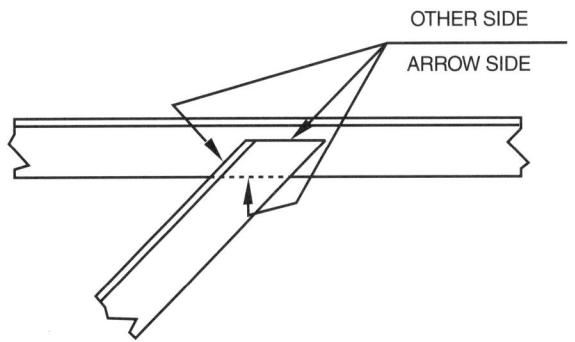

Note: Multiple arrows pointing to areas to be welded.

Figure 4.42—Placement and Location Significance of Arrow

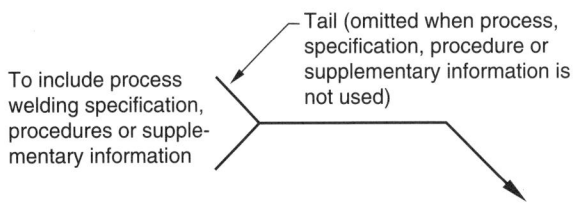

Figure 4.43—Conventions for use of a Tail

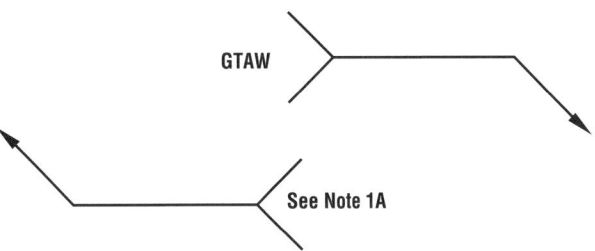

Figure 4.44—Examples of use of a Tail

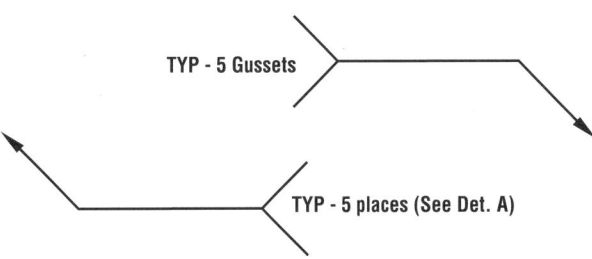

Figure 4.45—Use of "Typical" Notes

with the symbol or in the tail. Figure 4.51 illustrates this usage.

Supplementary Symbols

Supplementary symbols are used in connection with the weld symbol and can indicate extent of welding, weld appearance, material included in the preparation of the welded joint, or to indicate welding which is performed in some place other than in the shop. Certain supplementary symbols are used in combination with the basic weld symbols; others will appear on the reference line. Figure 4.52 identifies these supplementary symbols.

Finish Methods

Supplementary symbols specifying contour are included with the weld symbol when the finished face of the weld is to appear flat, flush, convex or concave. Certain mechanical finishing symbols may be added to indicate the type of method used to obtain the desired contour. These mechanical methods are indicated by the use of a letter designation, shown below, that signifies the finishing method required, but not the degree of finish. The letter "U" may be used when finish is necessary but the

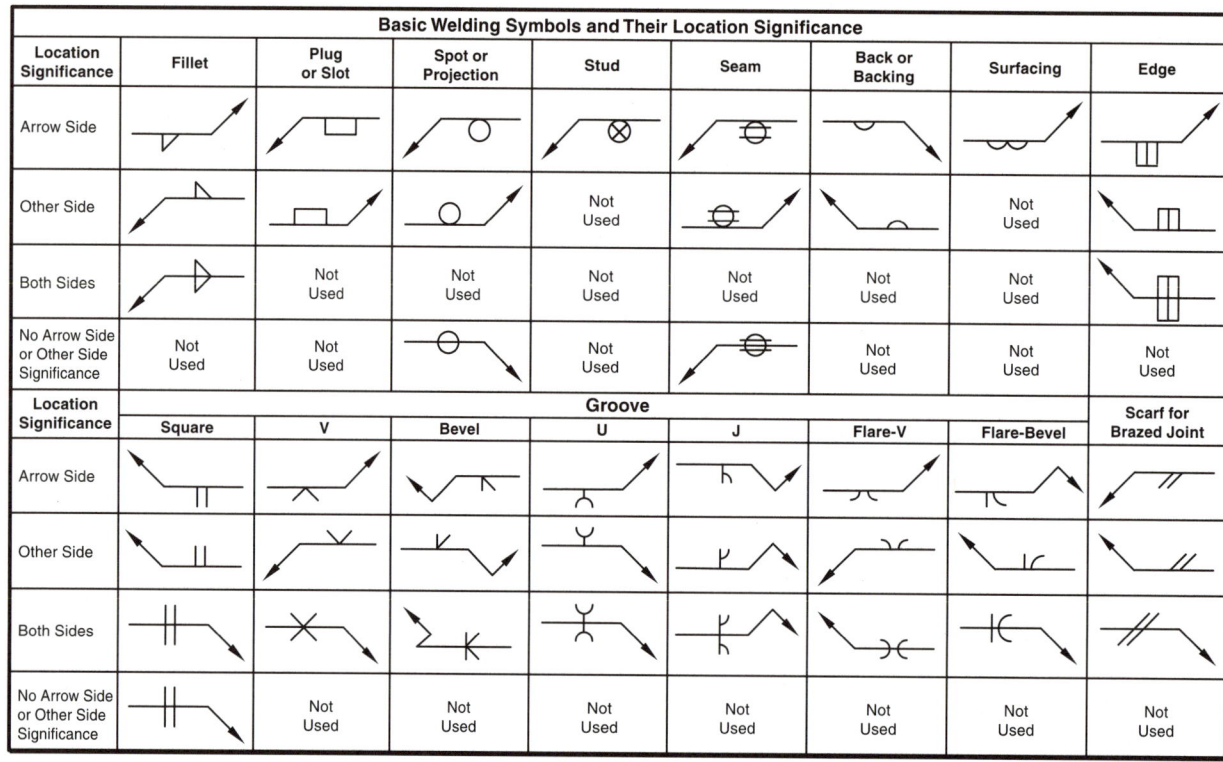

Figure 4.46—Reference Line Location for Basic Weld Symbols

method has not been determined. Figure 4.53 illustrates the use of the contour and finish supplementary symbols in the upper two sketches.

Mechanical Methods:

C = Chipping
G = Grinding
H = Hammering
M = Machining
R = Rolling
U = Unspecified

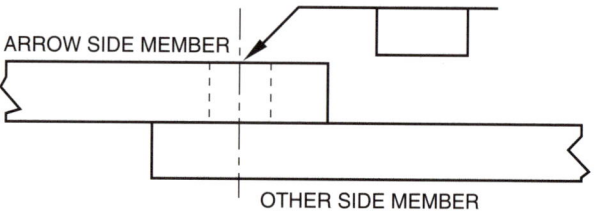

Figure 4.48—Example of Plug Weld Side

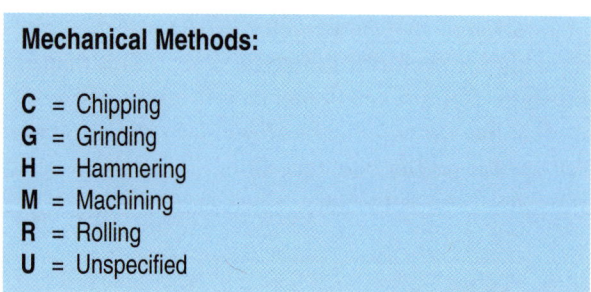

Figure 4.47—Perpendicular Leg of Weld Symbol

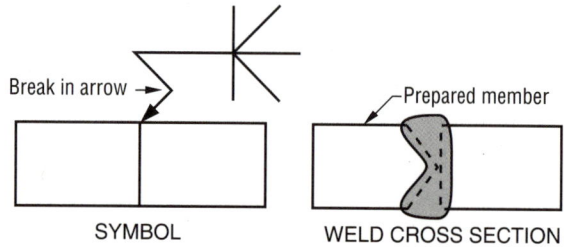

Figure 4.49—Use of Broken Arrow

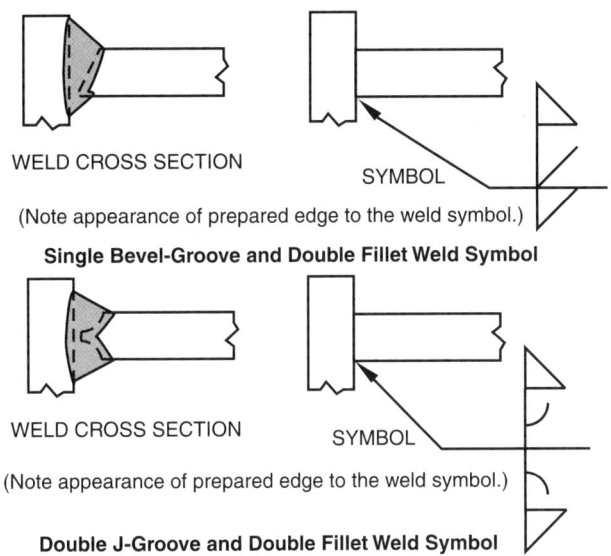

Figure 4.50—Combinations of Weld Symbols

Field Weld Symbols

Field welds are welds not made in a shop or at the place where the parts or assemblies were initially constructed. The symbol known as a *field flag* is placed either above or below, and at a right angle to, the reference line at the junction of the arrow. It has no side significance in regards to the weld required. The flag may point in either direction, either opposite or in the same direction as the arrow. Figure 4.53, bottom sketch, shows a number of welding symbols used in combination with the field weld symbol.

Melt-Through Symbols

The melt-through symbol is used only when complete root penetration plus visible root reinforcement is required in welds made from one side. The symbol is placed on the side of the reference line opposite the weld symbol. The height requirement of root reinforcement is specified by placing the required dimension to the left of the melt-through symbol. Examples are shown in Figure 4.54. The height of root reinforcement may be unspecified.

Melt-through symbols used with an edge welding symbol is also placed on the opposite side of the reference line and the symbol remains the same whether the joint is detailed or not detailed on the print. When the melt-through symbol is used in connection with a corner-flange welding symbol it is also placed on the opposite side of the reference line; however, the arrow will have a break pointing to the member that is flanged when no detail is given (see Figure 4.55). See Figure 4.61 for examples of melt-through applications.

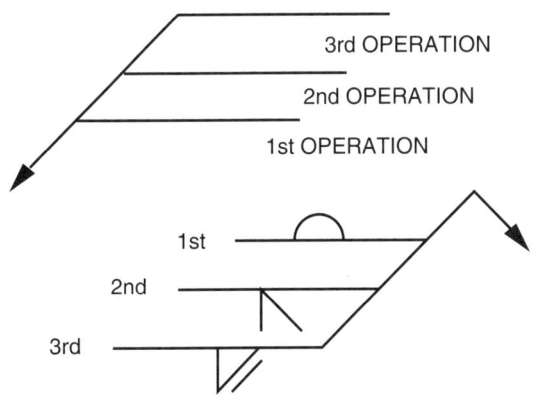

Figure 4.51—Use of Multiple Reference Line to Signify Sequence of Operations

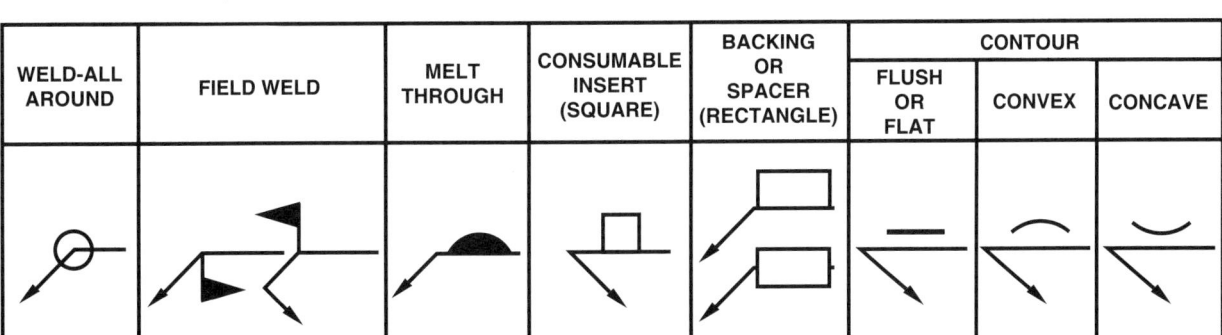

Figure 4.52—Supplementary Symbols

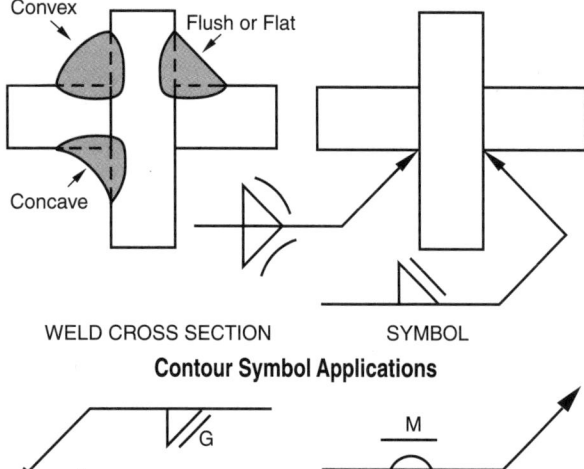

Contour Symbol Applications

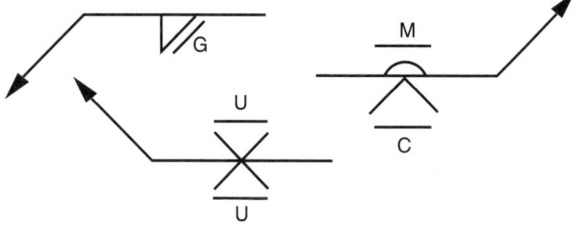

Finish Methods Specified and Unspecified

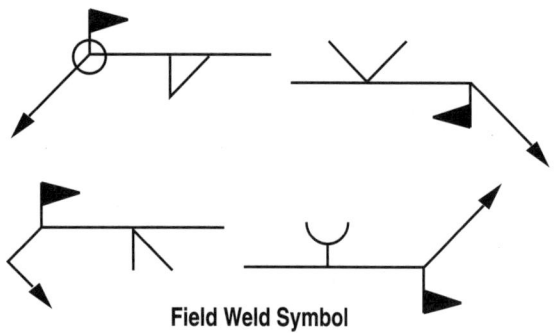

Field Weld Symbol

Figure 4.53—Contour, Finish, and Field Weld Symbols

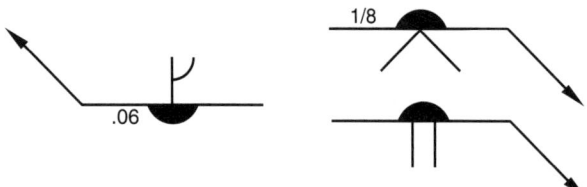

Figure 4.54—Melt-Through Symbol Use

Backing and Spacer Symbols

Joints with *backing* are specified by placing the backing symbol on the side of the reference line opposite the groove weld symbol. If the backing is to be removed after welding, an "R" is placed within the backing symbol (see Figure 4.56). The material and dimensions of the backing are specified in the tail of the symbol or in a note placed on the drawing near the weld joint. The *backing symbol* is distinct from the *back weld* and *backing weld* symbol. *Backing* is material or a device placed on the back side of the groove. *Backing welds* and *back welds* are welds applied to the backside of the joint. The backing symbol, although similar to a plug or slot weld symbol is distinguished by the presence of groove weld symbols with which it is used.

Joints that require *spacers* are specified by modifying the groove symbol to show a rectangle within it. Spacers are illustrated in Figure 4.57. The spacer is applied to joints welded from both sides and is generally centered on the root faces of the prepared members. It may be used to hold critical root openings in position. The spacer may be removed after welding from one side has been completed, or may remain as part of the welded joint. When used in connection with multiple reference lines, the symbol appears on the line closest to the arrow. Material and dimensions of the spacer are shown in the tail of the symbol or noted on the drawing near the welded joint. The spacer symbol is centered (straddled) on the reference line, and is similar to the appearance of the welded joint; centering it on the reference line distinguishes it from the backing symbol.

Consumable Insert Symbols

Consumable inserts are strips or rings of filler metal, added to the weld joint, that completely fuse to the joining members. The insert may have a special composition of filler metal to prevent porosity and enable the weld metal to meet specific requirements. Inserts are used on certain groove welded plate and pipe joints to improve the likelihood of full penetration. Generally, consumable inserts are welded with the GTAW welding process. The symbol is specified by placing the symbol on the opposite side of the groove weld symbol. The AWS consumable insert class (type) is placed in the tail of the symbol; the insert symbols are shown in Figure 4.58.

Weld-All-Around Symbol

This symbol, shown in Figure 4.59, is used to show weld applications made completely around

Edge-Flange Weld with Melt-Through Symbol Applied

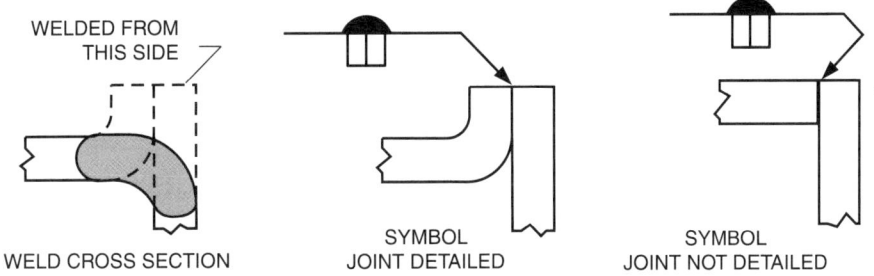

Corner-Flange Weld with Melt-Through Symbol Applied

Figure 4.55—Corner-Flange Weld with Melt-Through Symbol

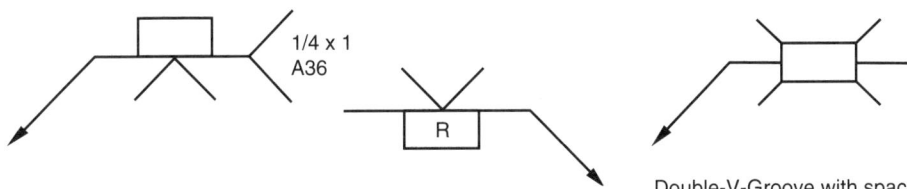

Figure 4.56—Backing Symbol Use

the joints specified. The symbol can be used with combination or single type welds. The series of joints may involve different directions and may be on more than one plane. The symbol is centered on the junction between the reference and the arrow line. Weld-all-around symbols are not used for circumferential welds made around pipe.

Weld Symbol Dimensioning

It has been previously noted that each basic weld symbol is a miniature detail of the weld to be placed at the welded joint. Therefore, if a specific set of dimensions are applied to the welding symbol, and notations, specifications, or references are placed in the tail of the welding symbol, the need for an enlarged, detailed view may be eliminated on the drawing.

Double-V-Groove with spacer

Double-U-Groove with spacer

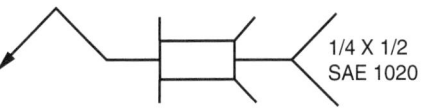

Double-Bevel-Groove with spacer

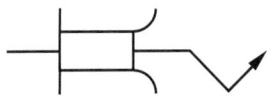

Double-J-Groove with spacer

Figure 4.57—Groove Weld Symbol with Spacer

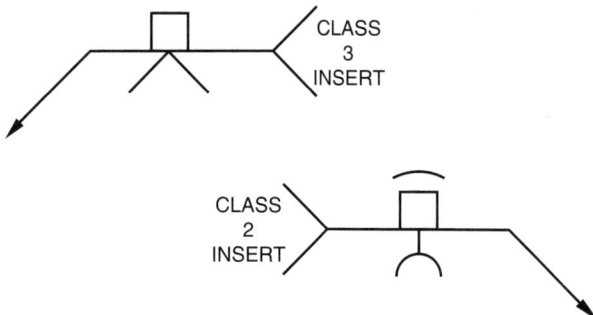

Figure 4.58—Consumable Inserts

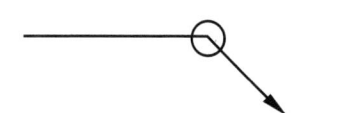

Figure 4.59—Weld-All-Around Symbol

There are certain specific and definite locations on the welding symbol that designate weld dimensions. The weld's size or strength, length, pitch, or number can be specified. In addition, dimension information regarding root openings, depth of fill, depth of preparation, and groove angle preparation can also be included.

Each element of the welding symbol becomes an important tool for welding personnel and the welding inspector. Your ability as an inspector to accurately interpret welding symbols is extremely important since welding symbol information definitely affects part or assembly preparations. The data gathered while interpreting a drawing must include the information specified for joint or weld preparation. This section examines in detail the dimensioning aspects of welding symbols for each type of weld.

Fillet Welds

Fillet welds are dimensioned according to size, length, and pitch when required. Dimensions of fillet welds are placed on the same side of the reference line as the weld symbol. Double fillet dimensions are placed on both sides of the reference line whether they are different or identical. Fillet dimensions specified in drawing notes need not be repeated on the symbol. Figures 4.63–4.65 illustrate these fillet weld dimensioning aspects.

Fillet size is located to the left of the weld symbol, and is not enclosed in parentheses as is the case for groove welds. Unequal leg fillet sizes are also placed to the left of the weld symbol. Dimensioning data will not indicate which size applies to either leg; this must be shown by a drawing detail or note.

Fillet length is placed to the right of the symbol. Length dimensions do not appear when the weld is made for the full length of the joint. The extent of fillet weld length may be represented graphically with the use of cross hatching in conjunction with drawing objects and dimensions. The welding symbol for specific sizes and locations can also be made in conjunction with the drawing dimension lines. *Pitch* dimensions (center to center spacing) of welds are placed to the right of the length dimension and separated by a hyphen.

Chain intermittent fillet weld dimensions are placed on both sides of the reference line; the welds are made opposite of each other on both joints. *Staggered intermittent fillet welds* are dimensioned in the same manner, with welds placed on opposite sides of the joint, but not opposite of each other; instead, they are spaced symmetrically. Figure 4.66 shows the length and intermittent convention for fillet welds, and Figures 4.70 through 4.73 show additional examples of fillet weld dimensioning.

Plug and Slot Welds

Plug and slot welds are identified by the same weld symbol; the weld symbol location for both types can be placed on either side of the reference line. Three dimensioning elements distinguish plug welds from slot welds; first, plug weld diameter is measured as size, while slot welds are measured for width. Plug weld size is indicated through the use of a diameter symbol—Ø. This symbol is omitted in the width specification for slot welds. Second, a length dimension is used for slot welds. The spacing (pitch) dimension for a plug weld is located in the same position as slot weld length. Third, the location and orientation of slots must be shown on the drawing (see Figures 4.67 and 4.69).

Plug welds are dimensioned according to weld size, angle of countersink, depth of filling, pitch and the number of welds required. Plug weld information is placed on the side of the reference line where the weld symbol appears. The arrow of the welding symbol must connect the welding symbol reference line to the outer surface of one of the joint members at the centerline of the desired weld (see Figure 4.68).

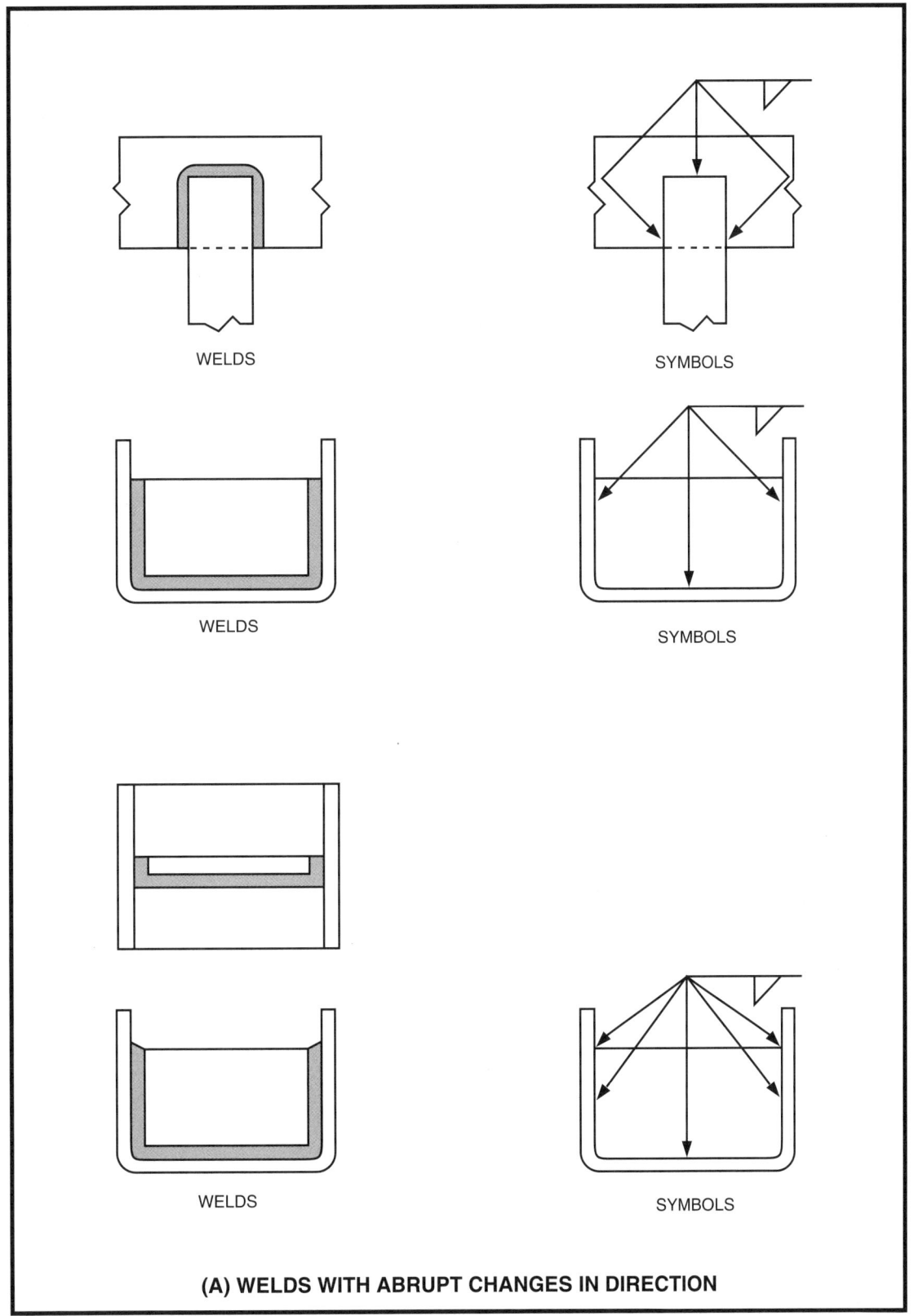

Figure 4.60—Specification of Extent of Welding

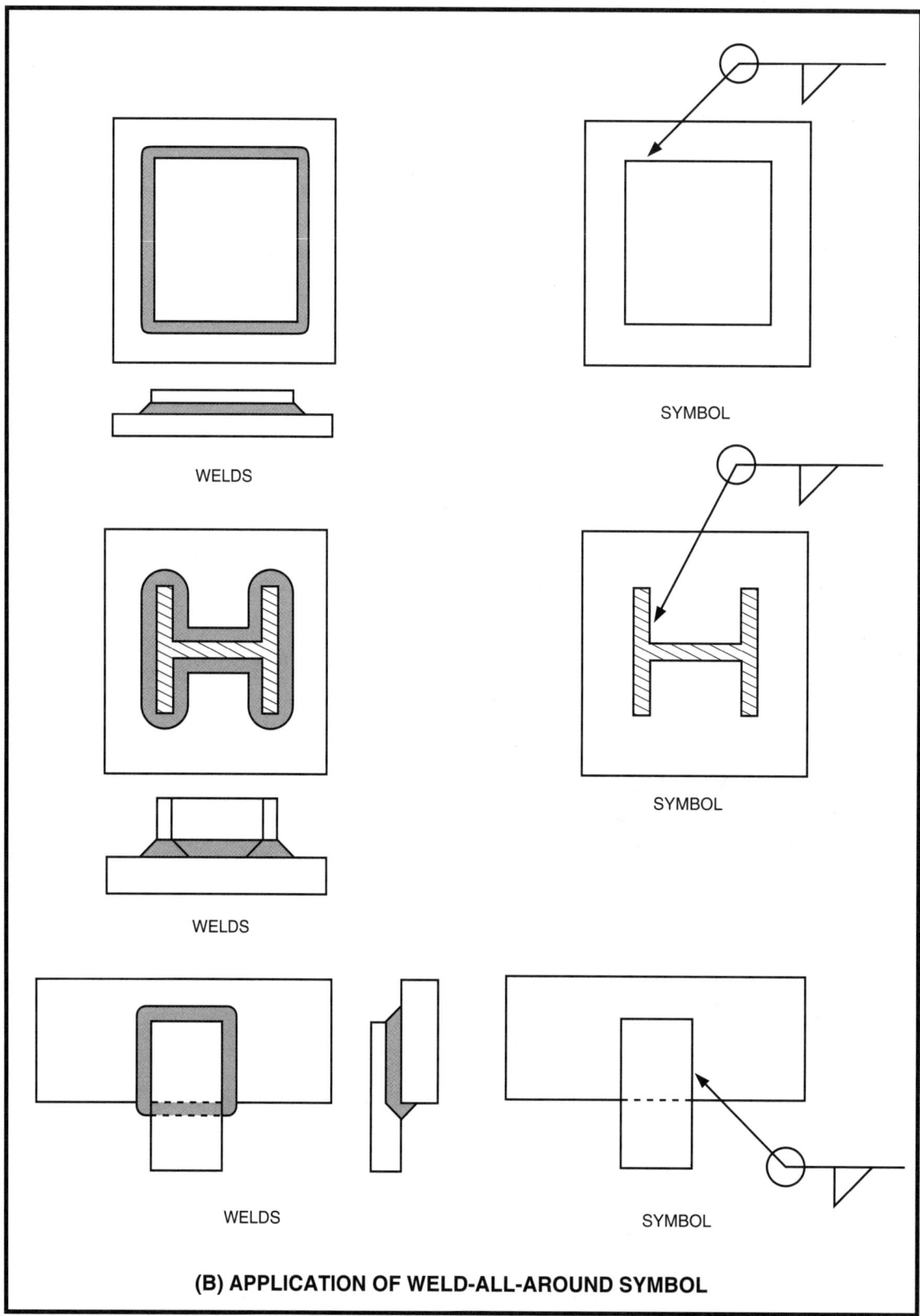

Figure 4.60 (Continued)—Specification of Extent of Welding

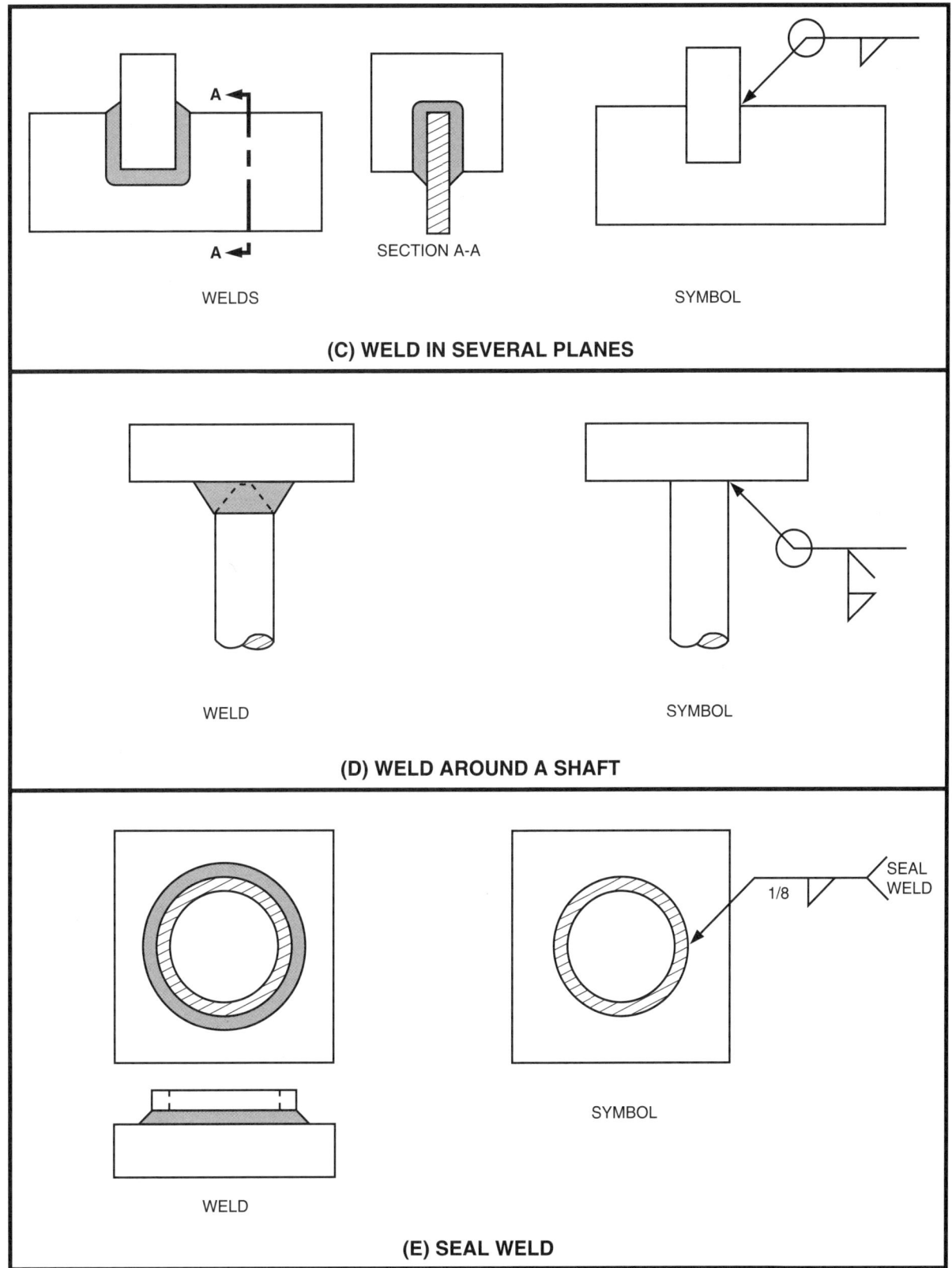

Figure 4.60 (Continued)—Specification of Extent of Welding

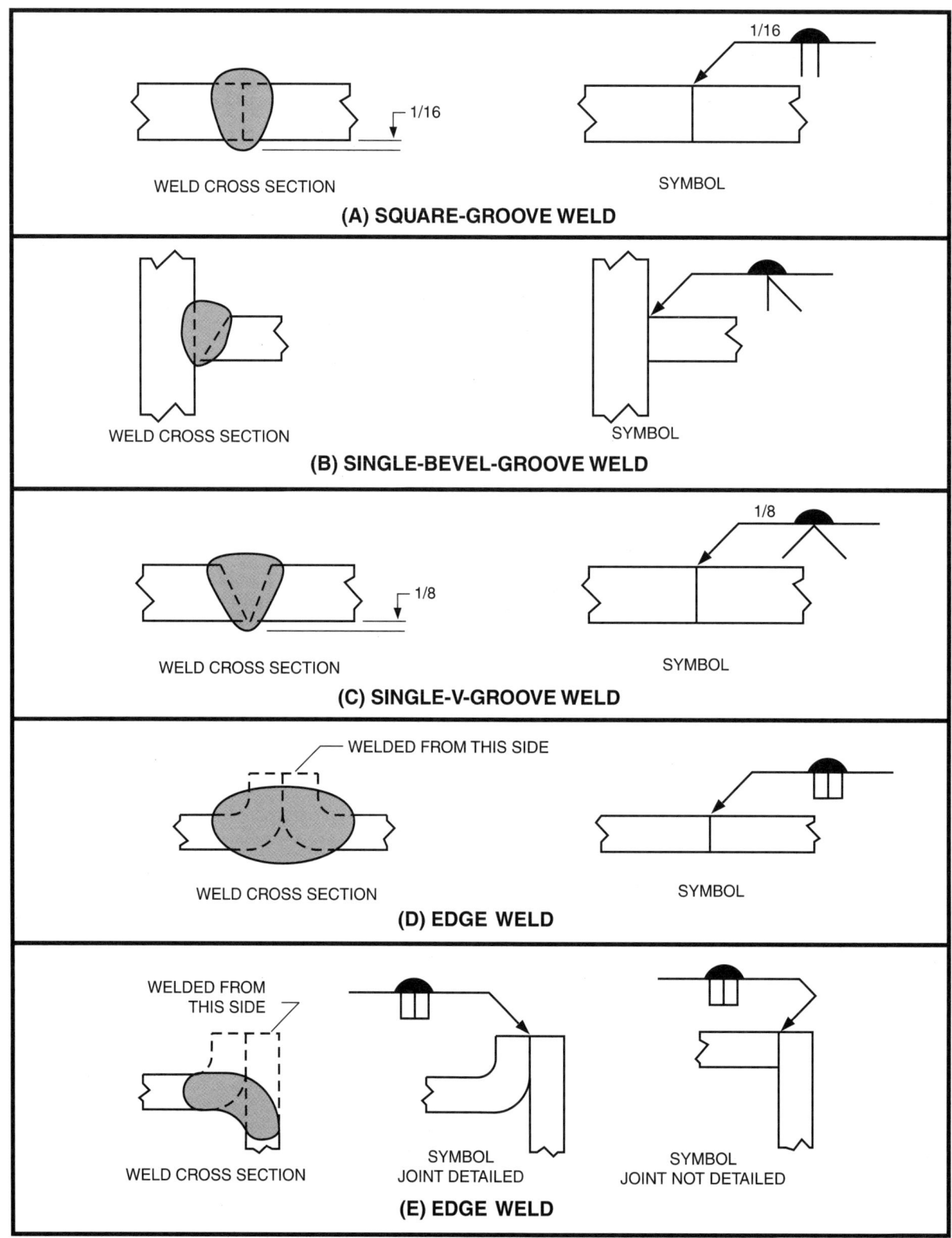

Figure 4.61—Applications of Melt-Through Symbol

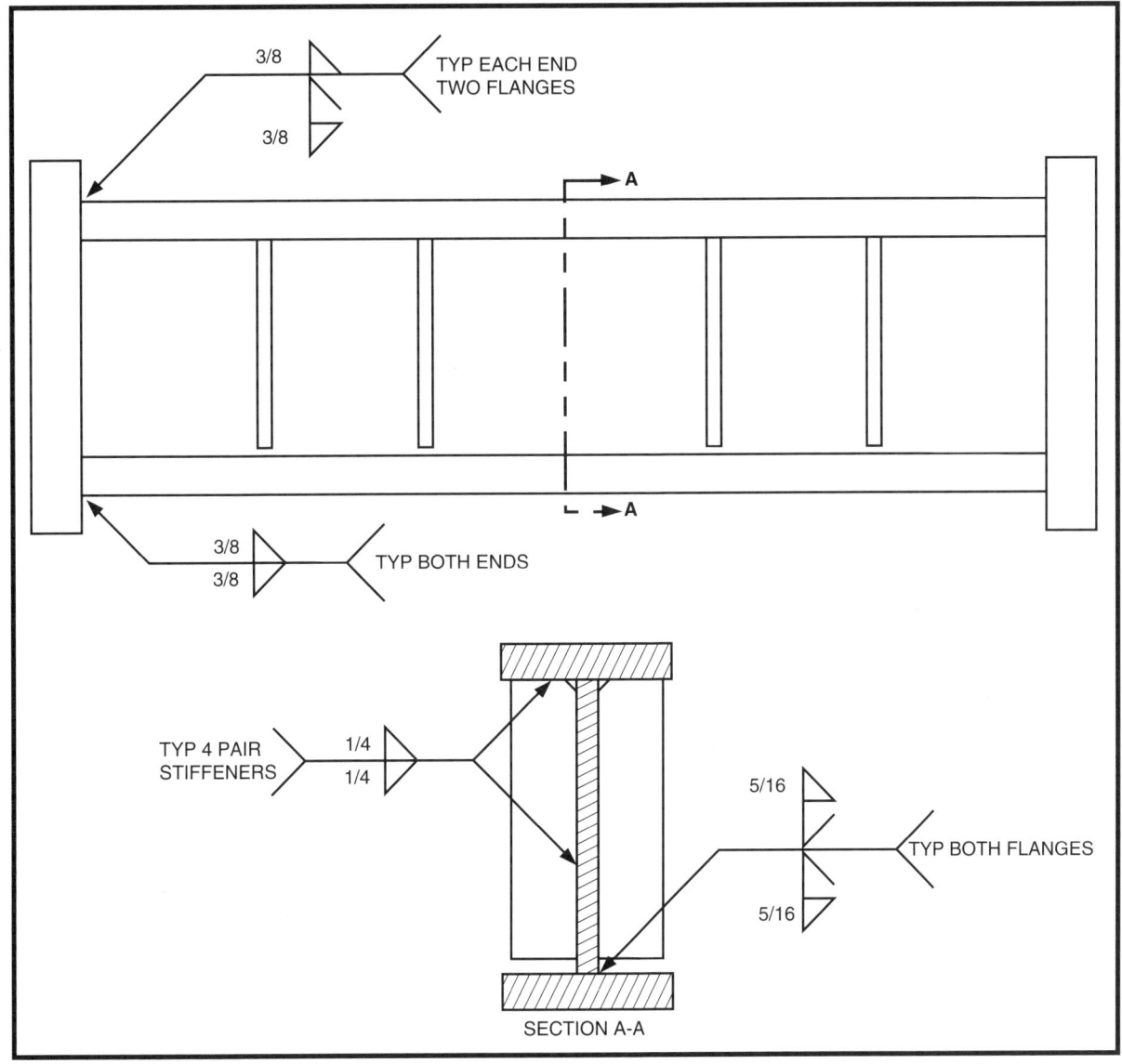

Figure 4.62—Applications of "Typical" Welding Symbols

Plug weld size is located to the left of the symbol, regardless of arrow or other side significance. Size is determined by the diameter of the hole at the faying surface (the point where the surfaces of the members come in contact with each other).

Angle of countersink for plug welds is located above or below the weld symbol depending upon symbol location on the reference line. Angle of countersink is the included angle for tapered holes.

Depth of filling is indicated by placing the filling dimension inside the plug weld symbol for filling that is less than complete. If the dimension has been omitted, this indicates that the hole is to be completely filled.

Spacing or *pitch* is placed to the right of the weld symbol. Plug weld spacing in any configuration other than a straight line must be dimensioned on the drawing.

Number of plug welds. When definite numbers of plug welds are required, the desired number is specified in parentheses on the same side of the reference line as the weld symbol. This dimension is located above or below the weld symbol depending upon symbol placement on the reference line.

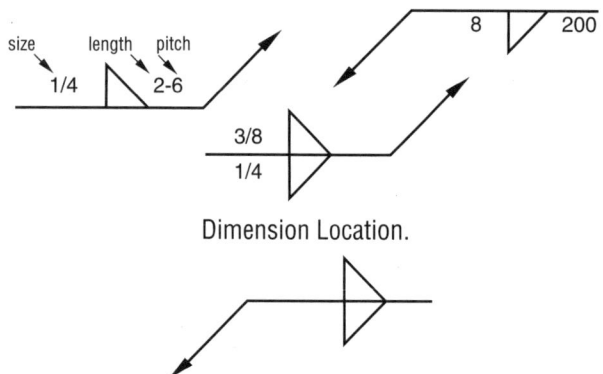

Figure 4.63—Fillet Weld Dimensions

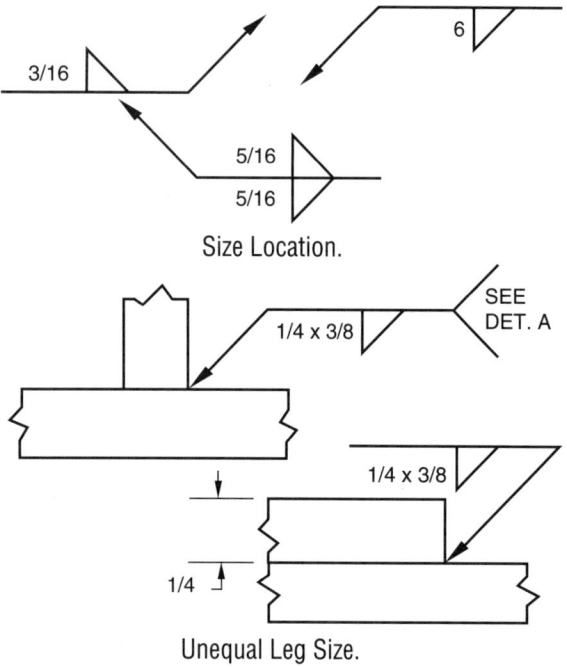

Figure 4.64—Size—Unequal Leg Fillet Welds

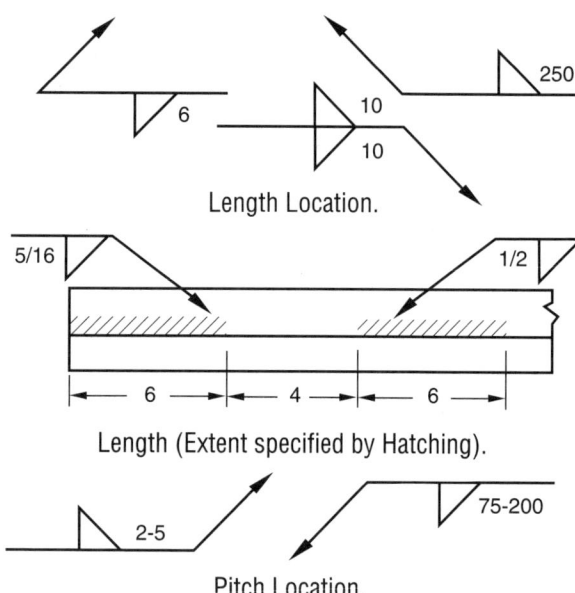

Figure 4.65—Length, Pitch Fillet Welds

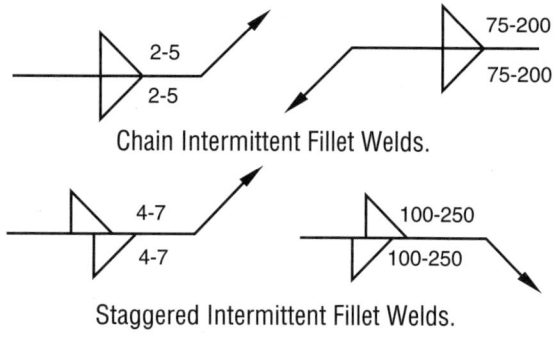

Figure 4.66—Intermittent Fillet Welds

Plug weld contours that are obtained by welding, will have face appearances which are approximately flush or convex. When post weld finishing is specified, the appropriate letter is applied above the contour symbol.

At times, the type of weld specified for holes will require a fillet weld. In these instances, the plug weld symbol will not be specified; instead, the fillet weld symbol will be used, and a weld-all-around symbol is usually included to complete the required weld configuration. See Figure 4.73(A) previously noted, and Figure 4.74 for other examples of plug weld dimensioning.

Slot Welds

Slot welds are dimensioned according to width, length, angle of countersink, depth of filling, pitch and the number of welds required. Slot weld information is placed on the side of the reference line where the weld symbol appears. The arrow of the welding symbol must connect the welding symbol reference line to the outer surface of one of the joint members at the centerline of the desired weld. In addition, the location and orientation of the slots must be specified on the drawing (see Figure 4.69).

Slot weld width is located to the left of the symbol, regardless of arrow or other side significance. Width is the dimension of the slot, measured in the direction of the minor axis at the faying surface.

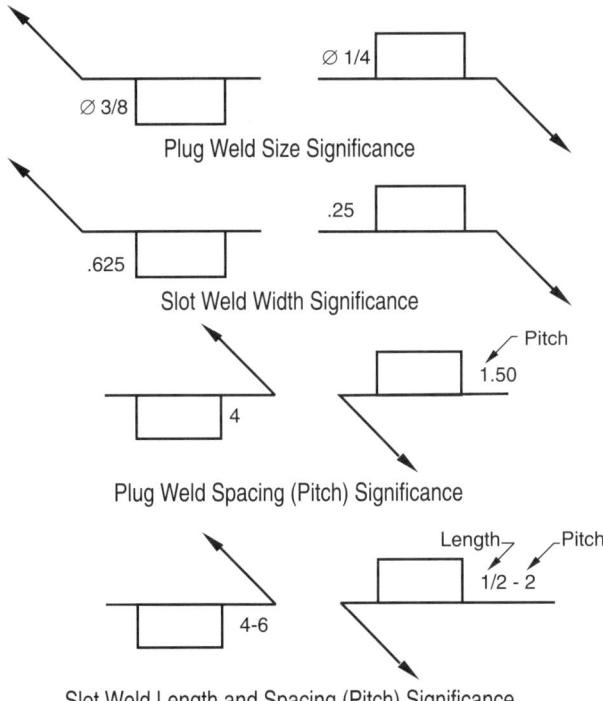

Figure 4.67—Differences in Plug and Slot Welds

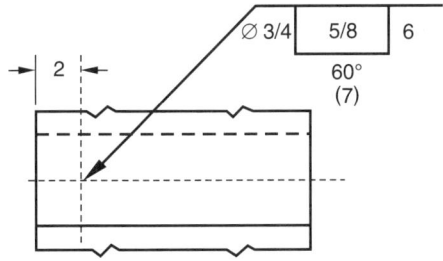

Figure 4.68—Plug Weld Dimensions

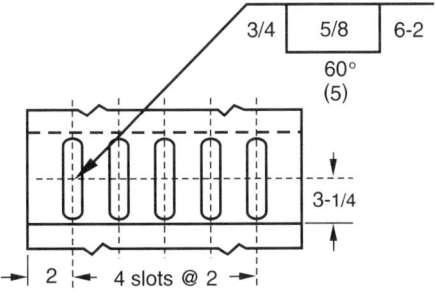

Figure 4.69—Slot Weld Dimensions

Slot weld length is located to the right of the symbol, regardless of arrow or other side significance. Length is the dimension of the slot, measured in the direction of the major axis at the faying surface.

Angle of countersink for slot welds is located above or below the weld symbol depending upon symbol location on the reference line. Angle of countersink is the included angle for tapered slots.

Depth of filling is indicated by placing the dimension inside the slot weld symbol for filling that is less than complete. If the dimension has been omitted, this indicates that the slot is to be completely filled.

Spacing or *pitch* (the center-to-center distance of two or more slot welds), is placed to the right of the length dimension, separated by a hyphen.

Number of slot welds. When definite numbers of slot welds are required, the desired number is specified in parentheses on the same side of the reference line as the weld symbol. This dimension is located above or below the weld symbol depending upon symbol placement on the reference line. If the angle of countersink is included in the welding symbol, then the number of slots required is placed in parentheses above or below the angle of countersink as appropriate. See Figure 4.75 for examples of slot weld dimensioning.

Slot weld contours that are obtained by welding, will have face appearances which are approximately flush or convex. When post weld finishing (contour obtained after welding) is specified, the appropriate letter is applied above the contour symbol. This signifies the method used for obtaining the desired contour, but does not specify the degree of finish. The degree of finish is indicated by a drawing note, or detail.

At times the type of weld specified for a slot will require a fillet weld. In these instances, the slot weld symbol will not be specified; instead the fillet weld symbol will be applied and a weld-all-around symbol included.

Spot and Projection Welds

Spot welds and projection welds share the same symbol, a circle placed above, below, or straddling the reference line. They can be differentiated by differences in the welding process, joint design, detailing on the drawing, and the reference placed in the tail.

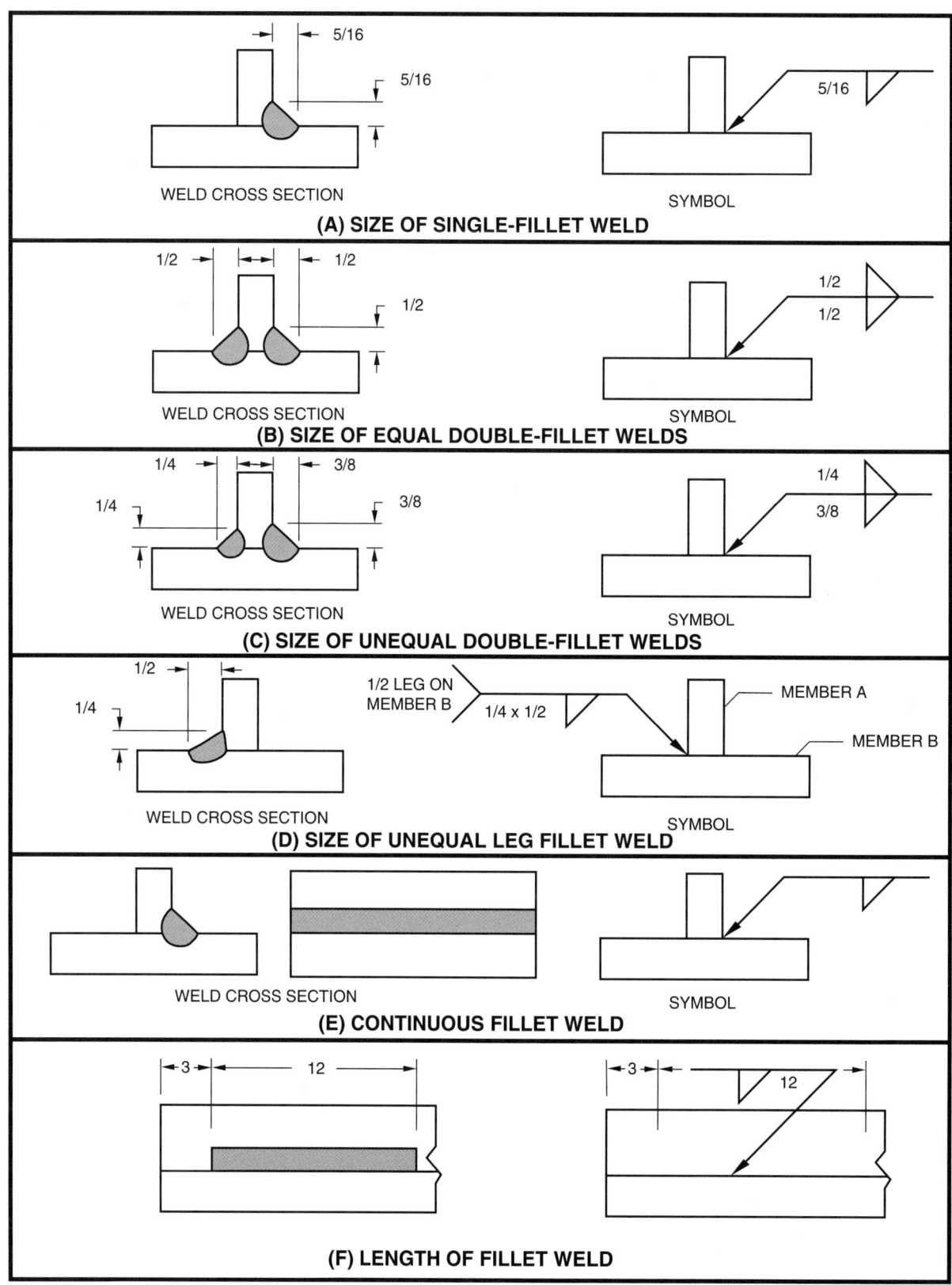

Figure 4.70—Specification of Size and Length of Fillet Welds

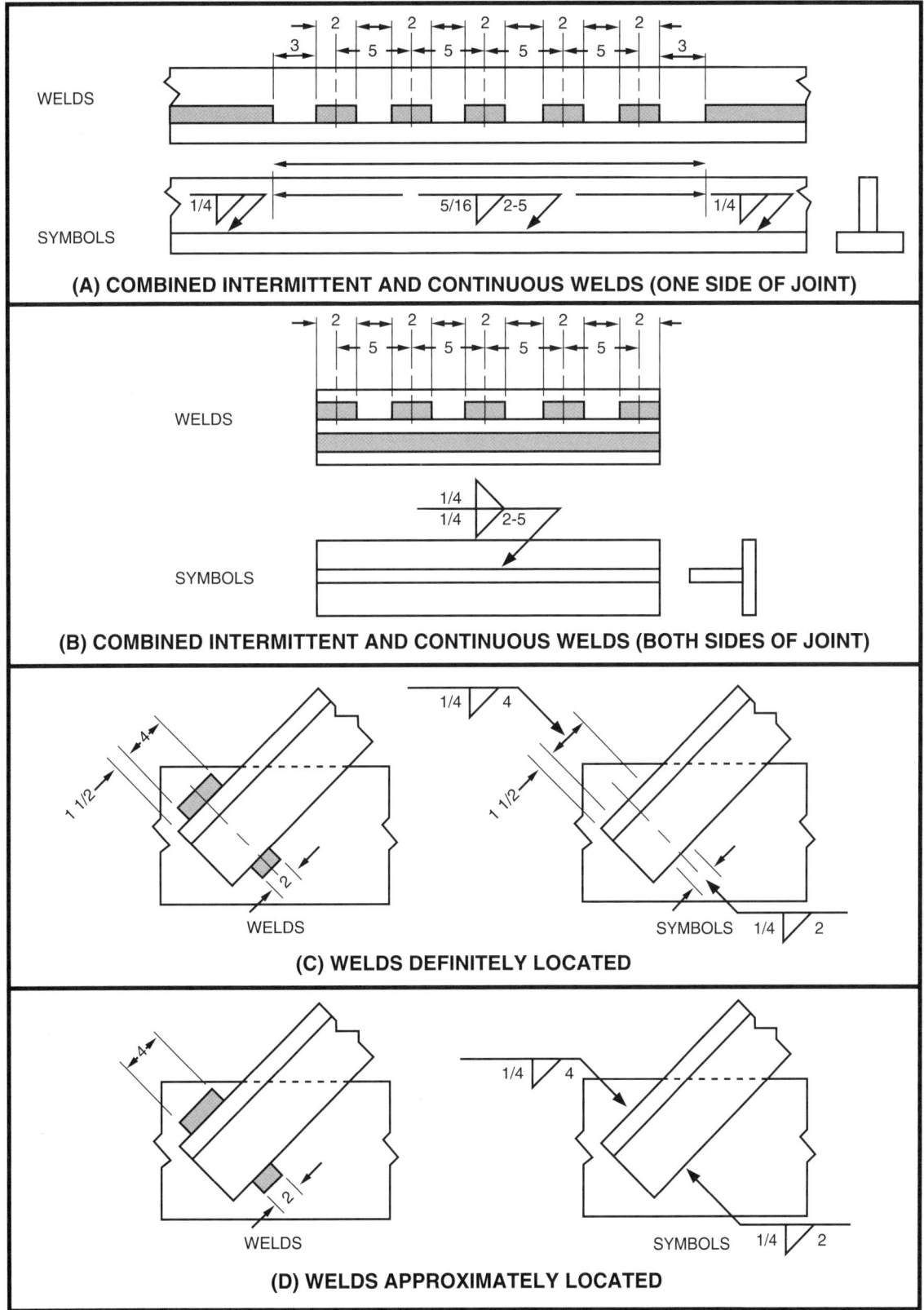

Figure 4.71—Specification of Location and Extent of Fillet Welds

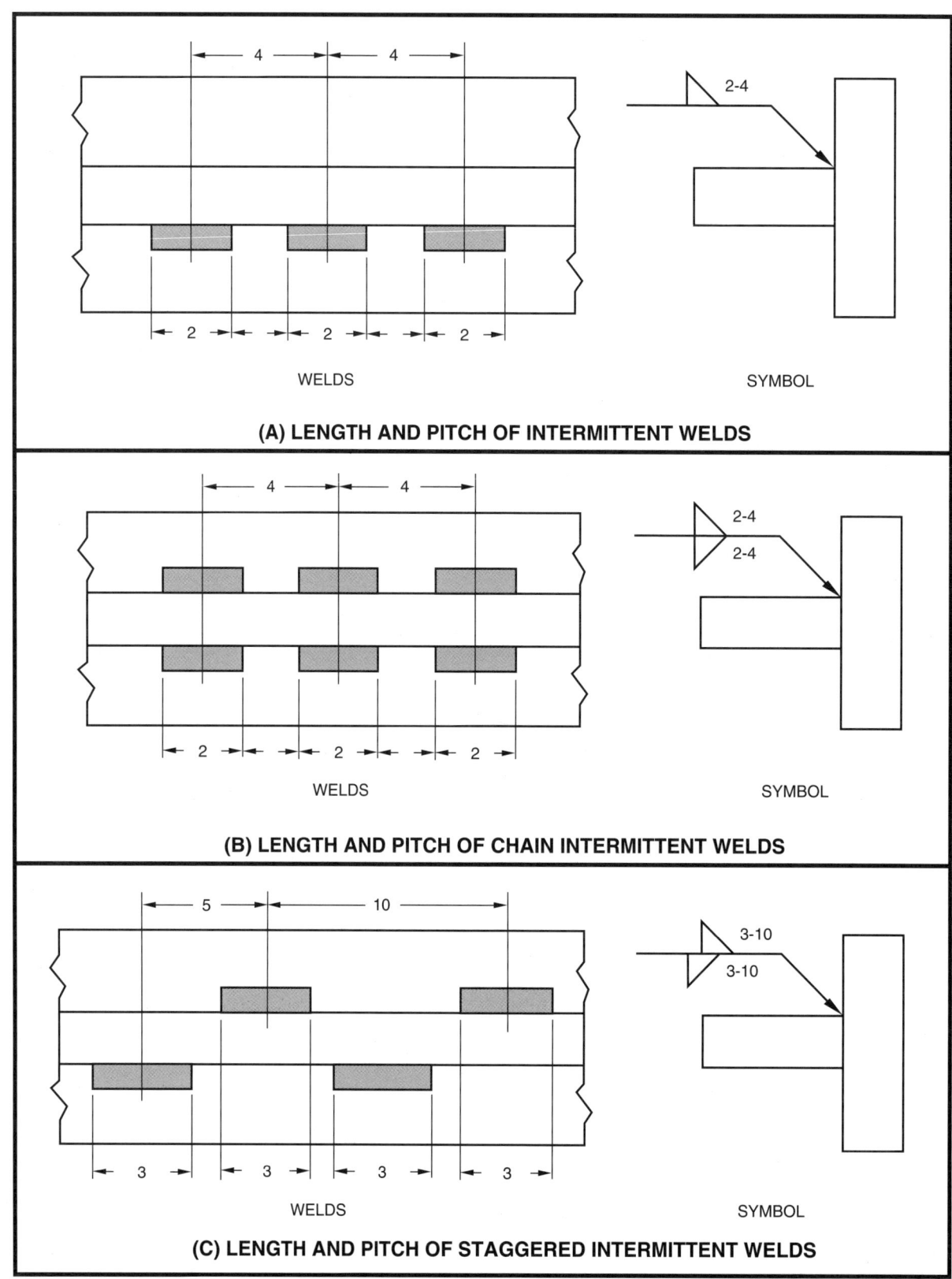

Figure 4.72—Applications of Fillet Weld Symbols

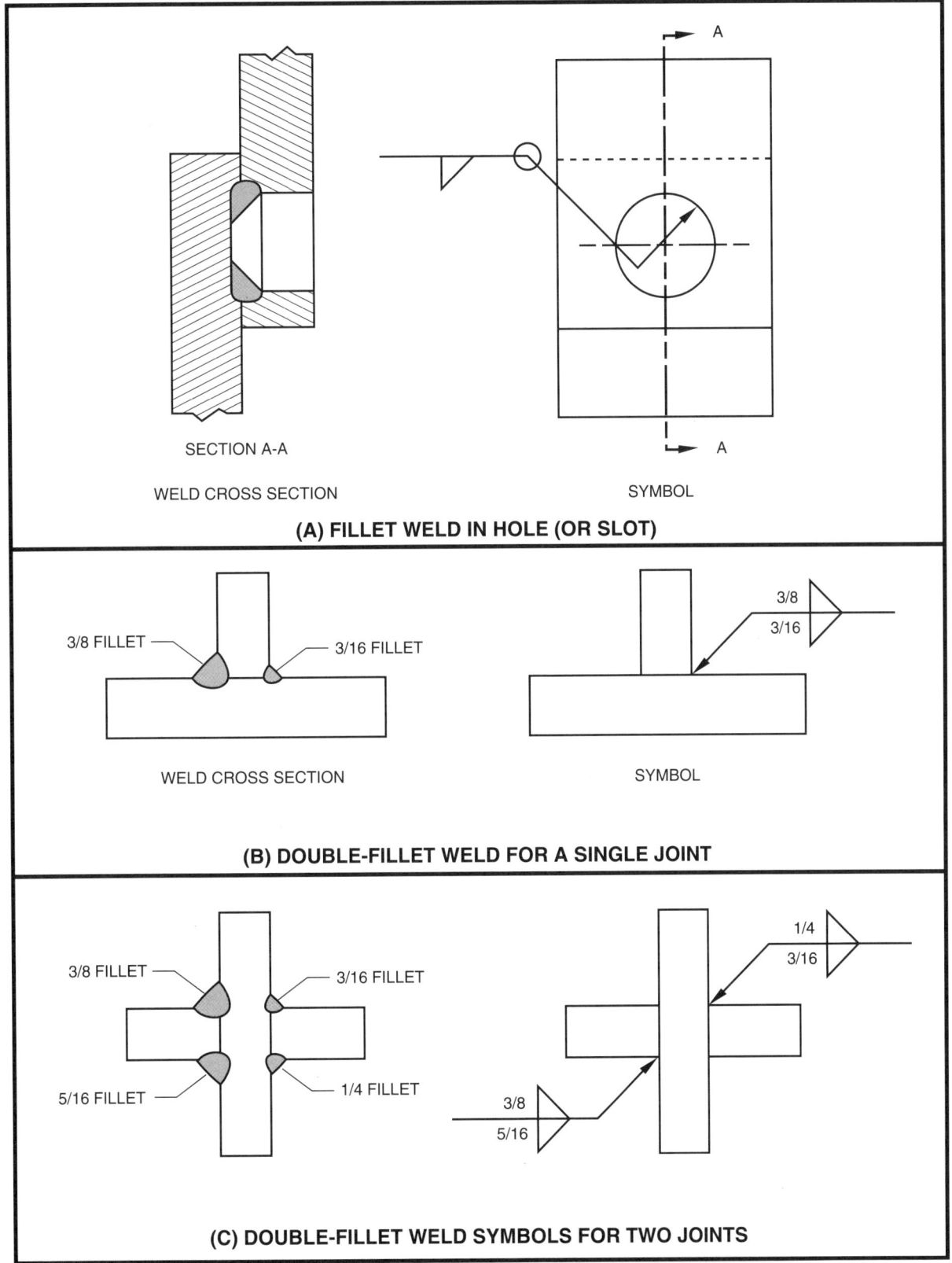

Figure 4.73—Applications of Fillet Weld Symbol

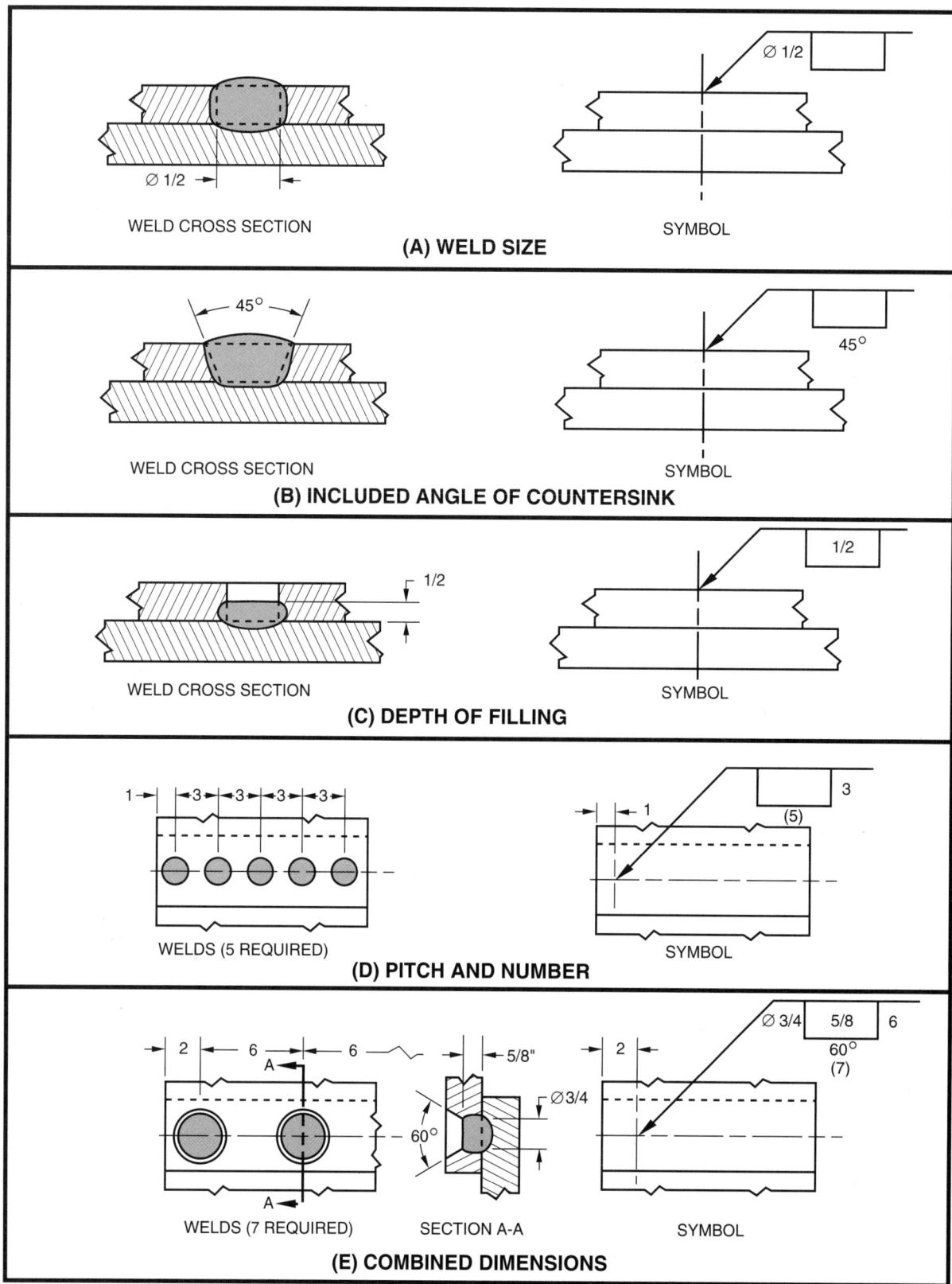

Figure 4.74—Applications of Plug Weld Dimensions

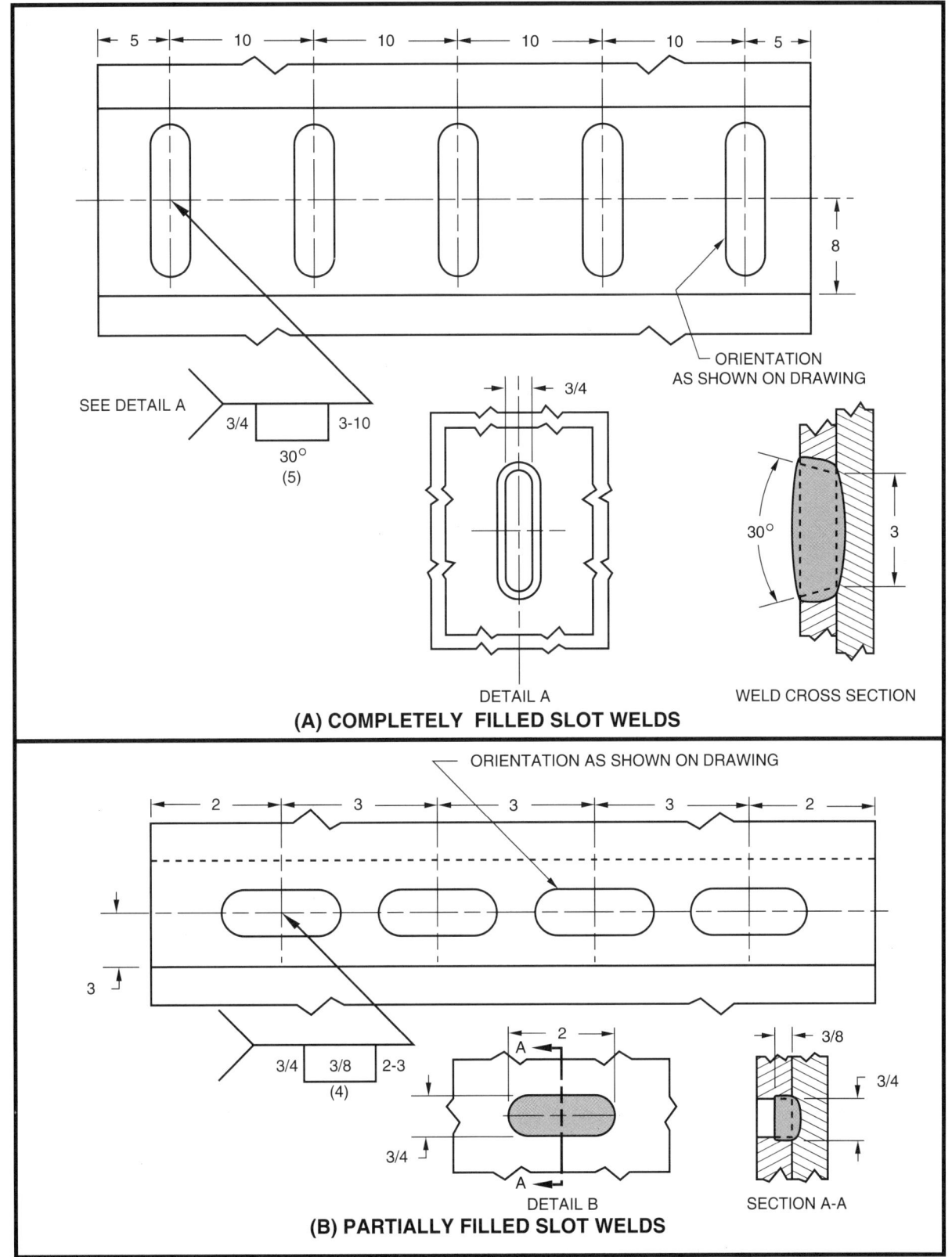

Figure 4.75—Applications of Slot Weld Dimensions

Spot Welds

A spot weld can be made using resistance welding, gas tungsten arc welding (GTAW), electron beam welding, ultrasonic, and many other welding processes. It has limited applications when used with the gas metal arc welding (GMAW), and shielded metal arc welding (SMAW) processes. Depending upon the welding process specified, the spot welding symbol is placed below, above, or centered on the reference line.

Welding symbol dimensions for spot welds include: weld size or strength, spacing and number of spots required. The welding process is always indicated in the tail of the symbol. Dimensions are placed on the same side of the reference line as the symbol, or on either side in the case of no arrow or other side significance. When spot welds are grouped, or the extent of spot welding is specified, dimensioning and location will be clearly marked on the drawing.

Spot weld size or *strength* is placed to the left side of the symbol. Weld size is measured by the diameter of the weld at the point of contact between the faying surfaces of the members. Either size or strength, but not both dimensions, will appear in conjunction with the welding symbol.

Spacing or *pitch* of two or more spot welds made in a straight line are indicated by placing the appropriate distance dimension to the right of the spot weld symbol.

The *number of spot welds* required is placed above or below the symbol, depending upon symbol placement, and is specified in parentheses.

Grouped spot welds may be represented by the use of intersecting center lines on the drawing. If this is the case, multiple arrows connecting the reference line of the welding symbol will point to at least one of the centerlines passing through each weld location. Should the spot welds be grouped randomly, the area where the welds will be applied must be clearly indicated on the drawing.

Extent of spot welding. At times, spot welds may extend less than the distance between abrupt changes in the direction of welding, or less than the full length of the joint. In situations where this occurs, the desired extent of welding must be dimensioned on the drawing.

Spot weld contours that are obtained by welding, will have face appearances which are approximately flush or convex. When post weld finishing is specified, the appropriate letter is applied above the contour symbol. This signifies the method used for obtaining the desired contour, but does not specify the degree of finish. The degree of finish is indicated by a drawing note. Examples of spot weld dimensioning are shown in Figures 4.77 and 4.78.

Projection Welds

The projection weld symbol is placed either above or below the reference line because of the joint design and welding process used (resistance type welding). The symbol for this weld type is never placed to straddle the reference line. When projection welding is used, the welding process, will always be identified in the tail of the welding symbol. The side designation of the projection weld symbol indicates which member is embossed (see Figure 4.76).

Seam Welds

The seam weld symbol, dependent upon its location on the reference line and the welding process used, may or may not have arrow-side or other side significance. When the symbol is placed centered on the line, it does not indicate a both sides designation; rather, it specifies no arrow or other side significance.

Seam welds are dimensioned according to size or strength, length and/or pitch and the number of welds required. The welding process used will be specified in the tail of the welding symbol.

Size or *strength* dimensions are placed to the left of the symbol on the same side as the weld symbol location, or to the left on either side in the case of no side significance. Seam weld size is measured according to the width of the weld at the faying surfaces of the members. Strength is specified by pounds per linear inch or in newtons per millimeter for metric measurements. Size and strength designations are not specified at the same time.

Length and pitch of seam welds. The length dimension of a seam weld is placed on the right side of the weld symbol. Length dimensioning is omitted if the seam extends for the full length of the weld joint, or the full distance between abrupt changes in the direction of welding.

Sometimes seam welds are made intermittently. In these instances a *pitch* dimension will be placed

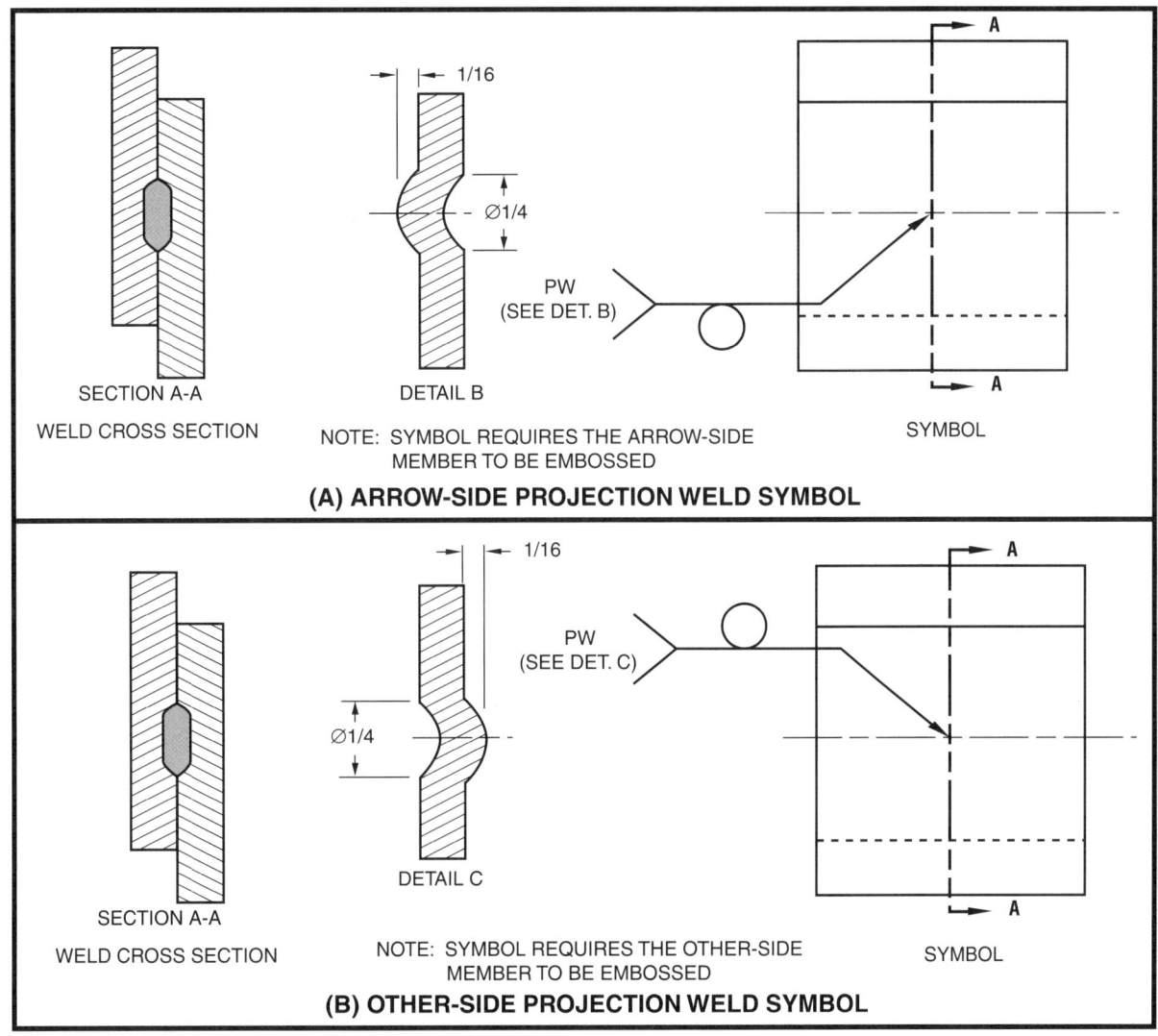

Figure 4.76—Projection Weld Dimensioning

to the right of the length dimension separated by a hyphen (-). If two or more seam welds are applied intermittently, length and pitch are understood to measure parallel to the weld axis. When the orientation is other than parallel to the weld axis, a detailed drawing is used to clarify the specific orientation.

The *number of seam welds* required is placed above or below the symbol (depending upon symbol placement) and is specified in parentheses.

Seam weld contours that are obtained by welding, will have face appearances which are approximately flush or convex. When post weld finishing is specified, the appropriate letter is applied above the contour symbol. This signifies the method used for obtaining the desired contour, but does not specify the degree of finish. The degree of finish is indicated by a drawing note, or detail. See Figures 4.79 and 4.80 for seam weld dimensioning examples.

Stud Welds

The stud welding symbol is a new category of weld symbol. In the ordinary sense, the stud weld symbol does not indicate the welding of a joint. For this reason it has no arrow-side or other-side significance. The symbol is always placed ***below*** the reference line and points directly to the surface where the studs are welded. Studs are dimensioned

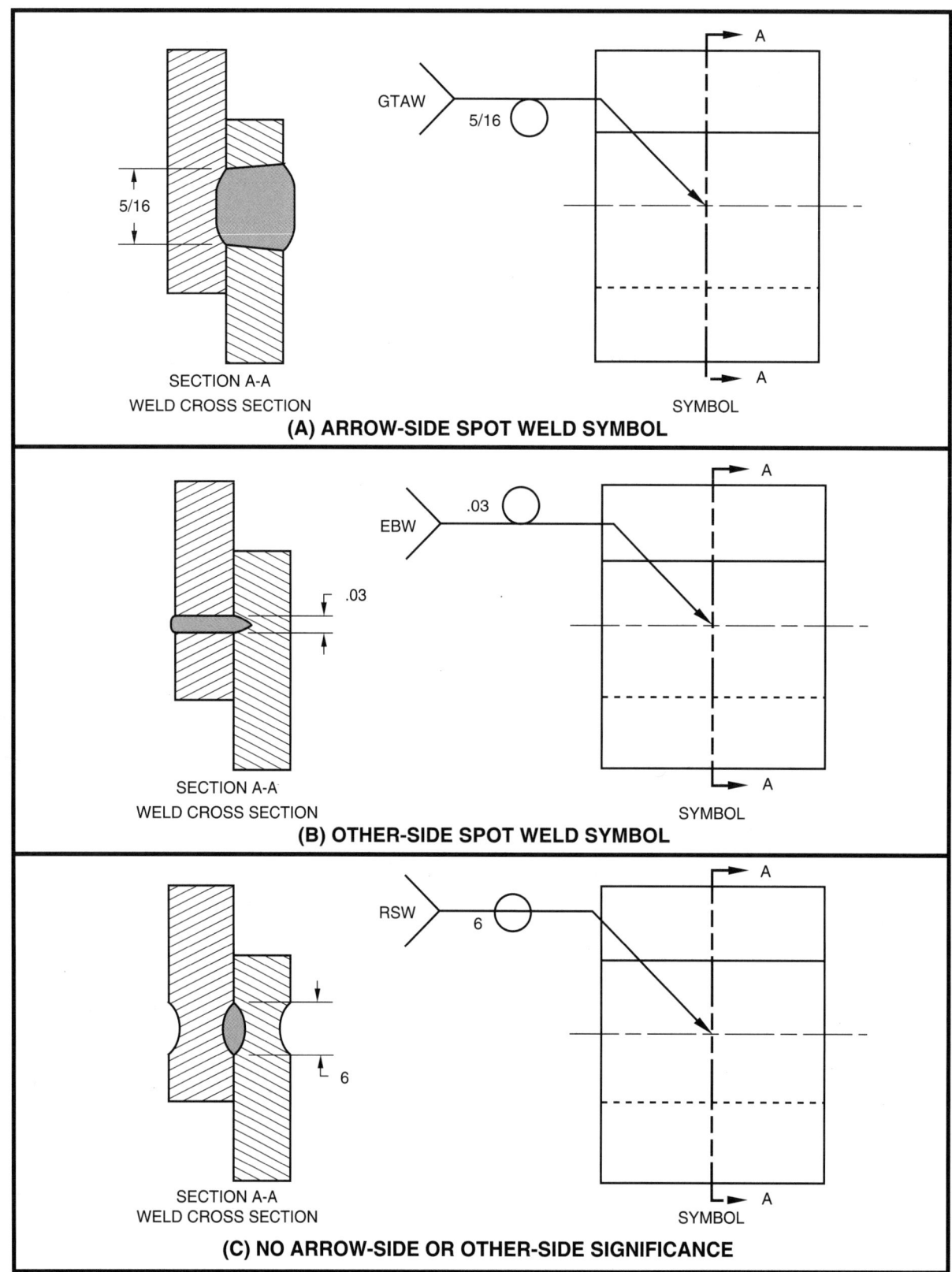

Figure 4.77—Applications of Spot Weld Symbol

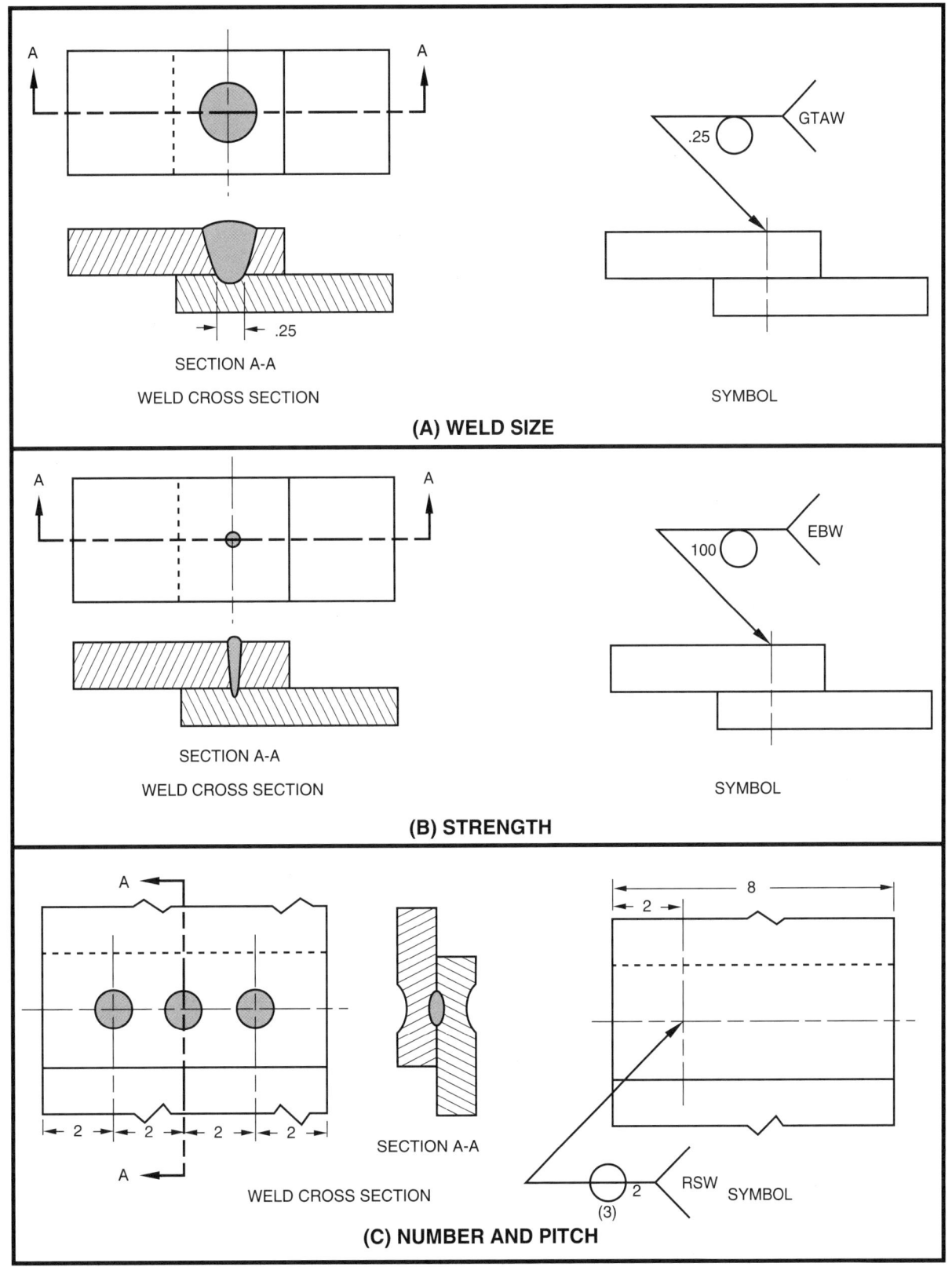

Figure 4.78—Spot Weld Dimensioning

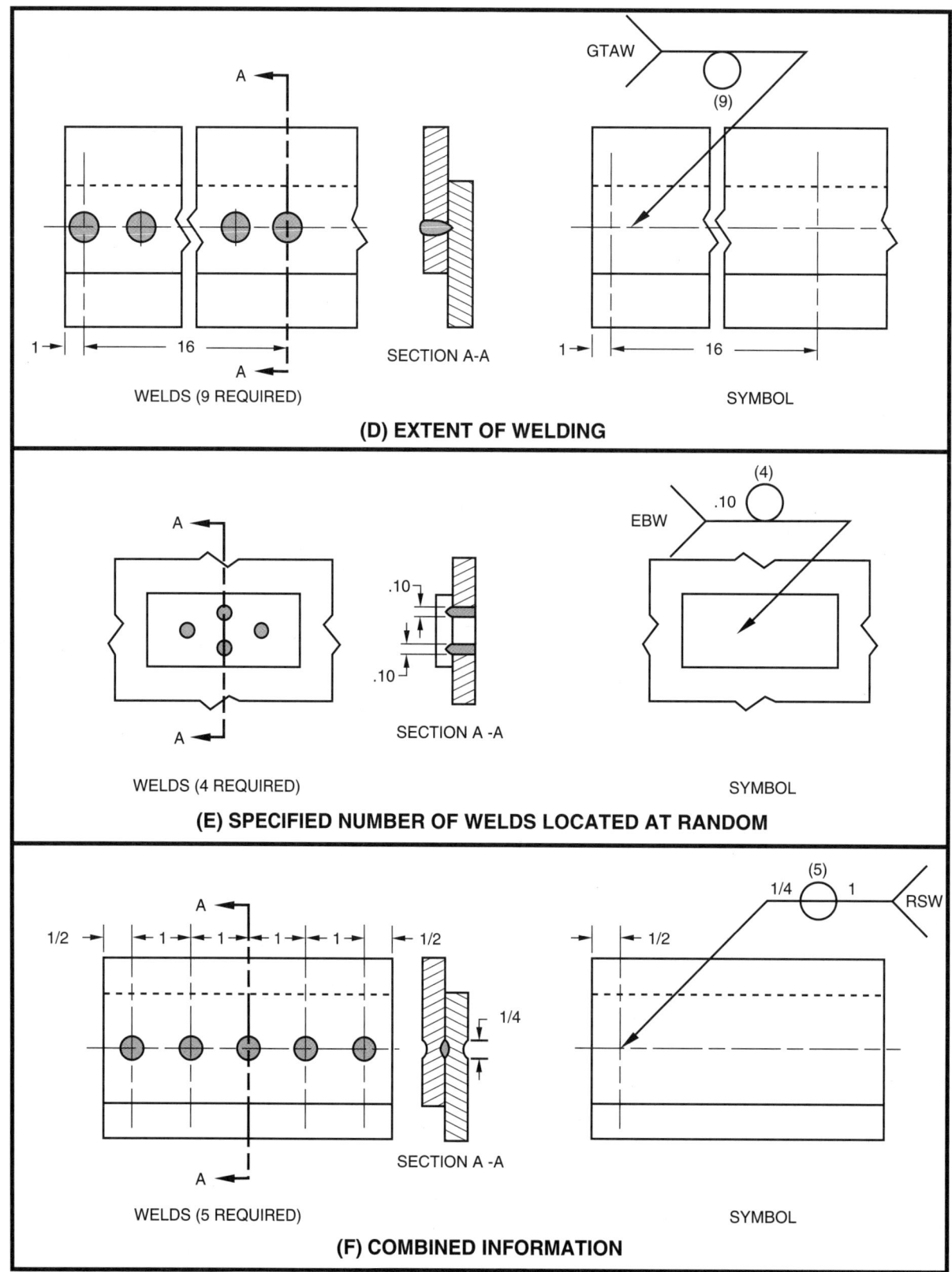

Figure 4.78 (Continued)—Spot Weld Dimensioning

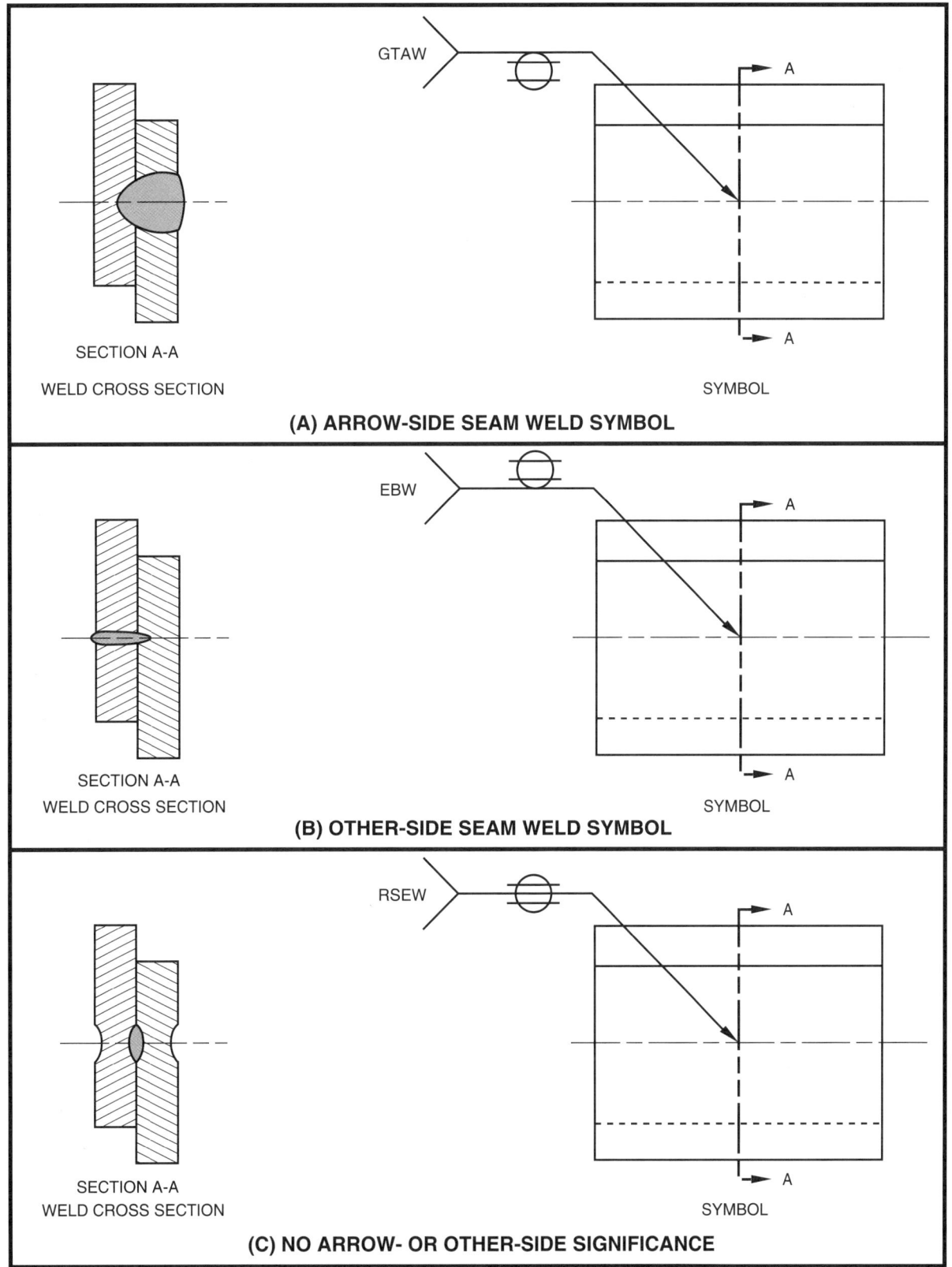

Figure 4.79—Applications of Seam Weld Symbol

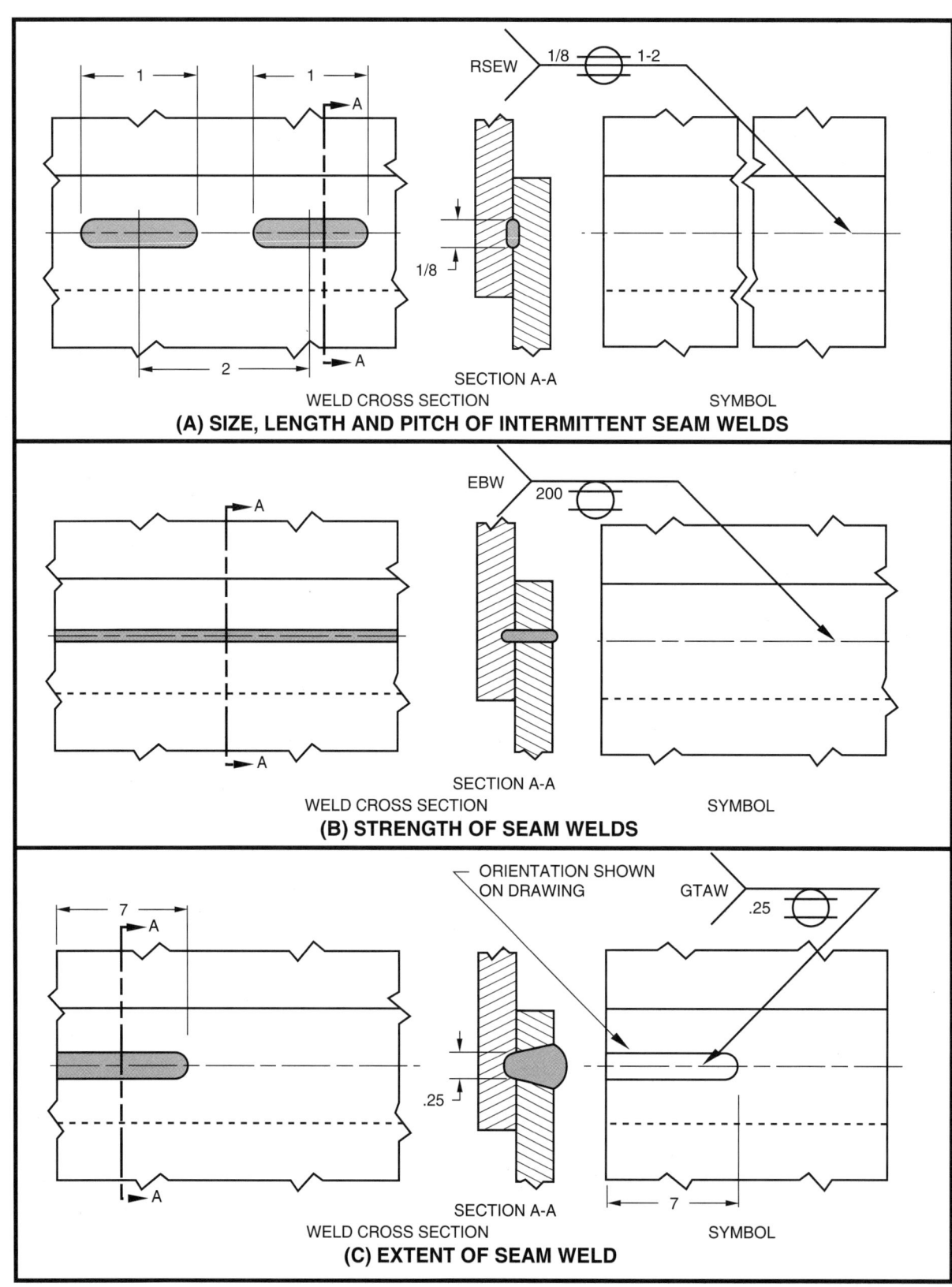

Figure 4.80—Seam Weld Dimensioning

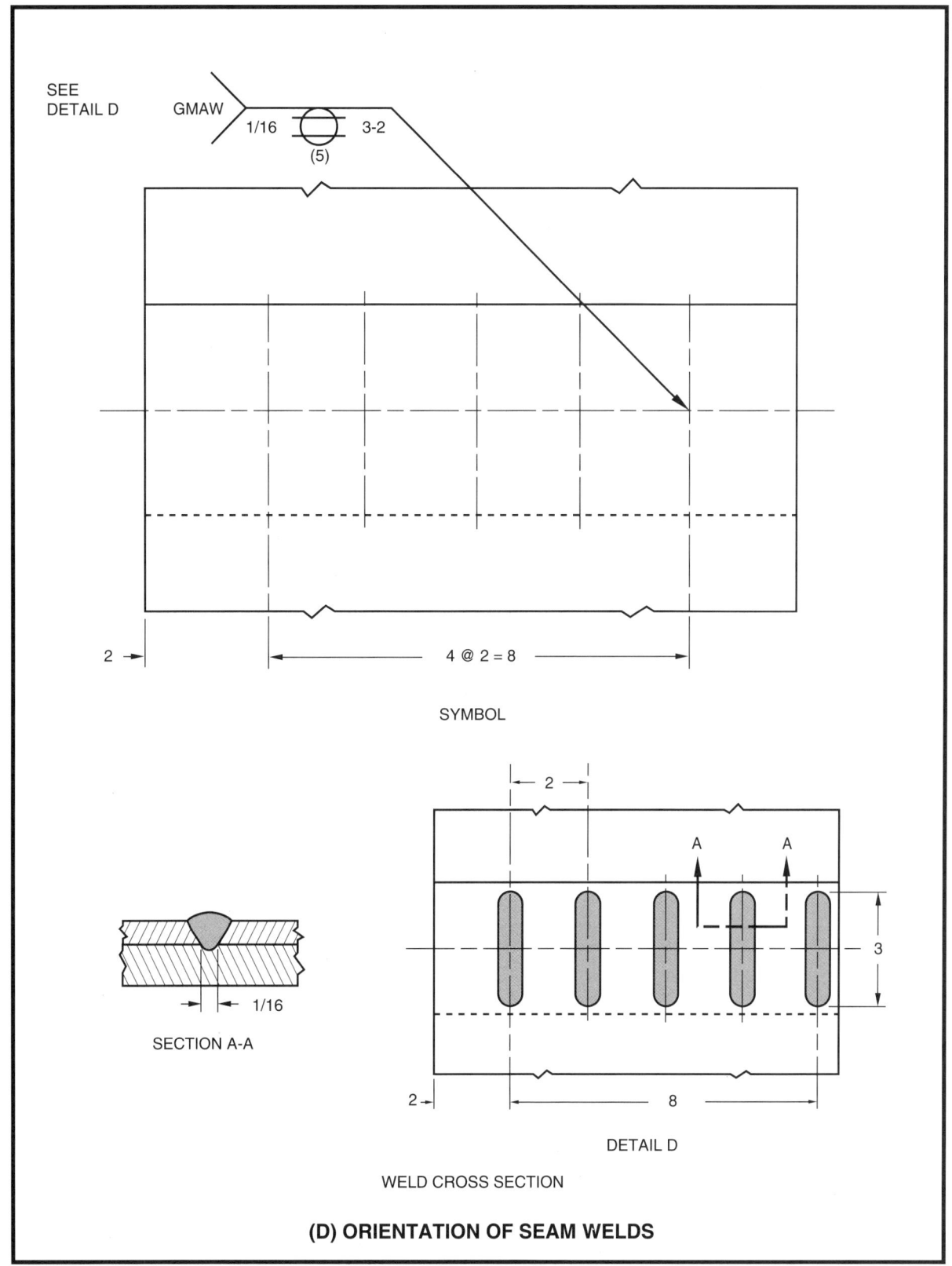

Figure 4.80 (Continued)—Seam Weld Dimensioning

according to *stud size* (left of the symbol), *spacing* (right of the symbol) and *number of studs* required (placed below the symbol in parenthesis). The location of the first and last stud placed on a single line will be dimensioned on the drawing, and the arrow will point directly to the start of each line of studs. In the case of multiple lines of studs, multiple arrows will point to each line (see Figure 4.81).

Surfacing Welds

Many times welders will be called upon to place layers of welds (buildup) on metal surfaces, or make cross hatched surfacing patterns on the outside of heavy equipment. In fabrication shops that include machine shops, the welder may be called upon to build up shafts or other pieces of materials so the machinist can turn the part down to sound base metal and achieve a desired dimension or diameter. Surfacing welds may also be employed to correct dimensions of parts.

Surfacing is also used to provide corrosion or heat resistant surfaces (cladding). In certain surfacing applications, "butter passes" are welded to the surfaces of existing members, before the prepared parts are installed. "Buttering" is done to keep the weld metal uniform (metallurgically compatible),

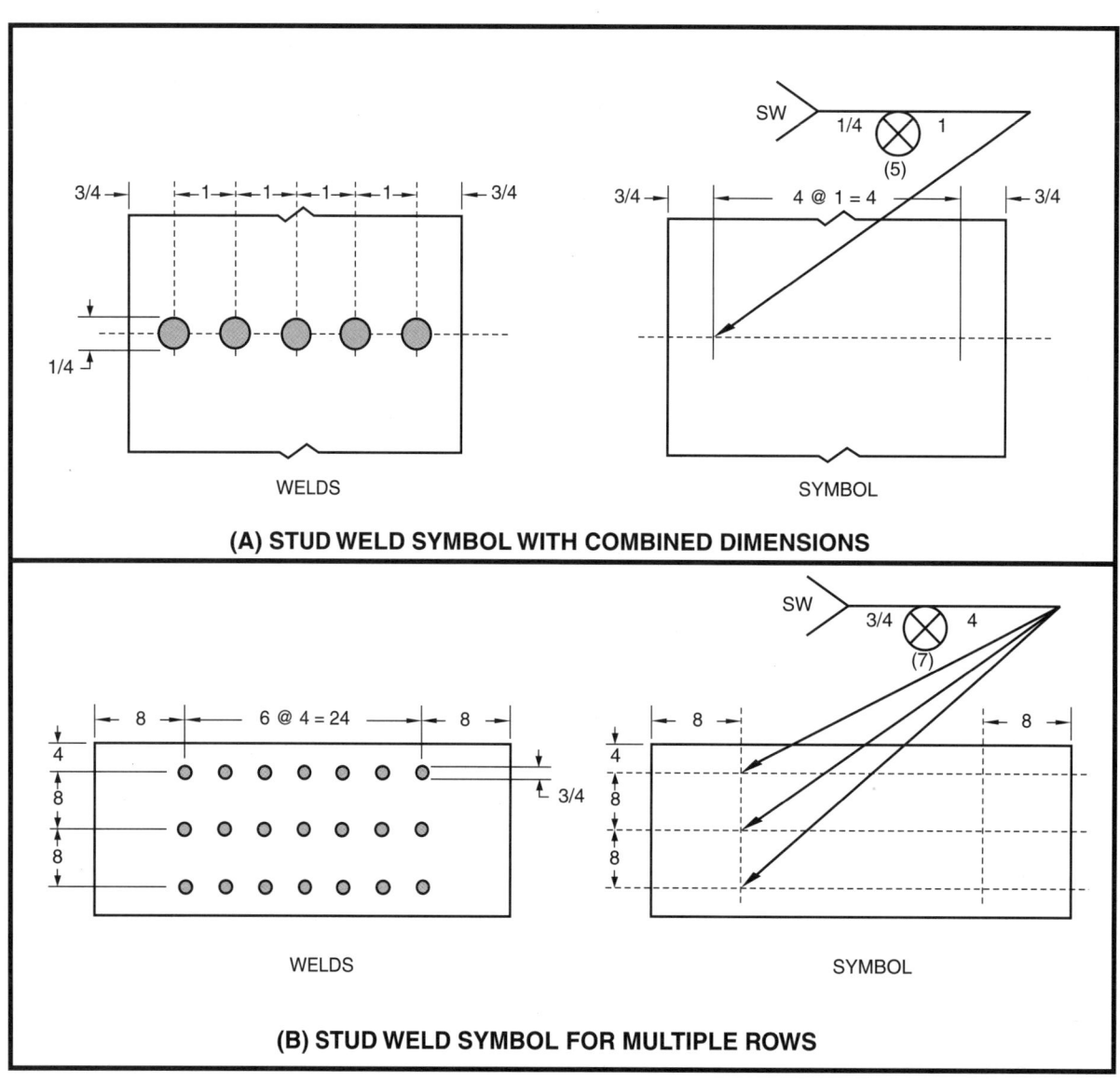

Figure 4.81—Applications of Stud Weld Symbols

and in some applications, make a transition between two dissimilar weld metals. Surface welds can be applied using single or multiple pass welds, and can include single or multiple layers.

Surfacing weld symbols do not indicate welded joints; therefore, there is no arrow side or other side significance. The symbol always appears *below* the reference line. Dimensions are placed on the same side of the reference line as the weld symbol. The arrow of the welding symbol points clearly to the area to be built up by the surface welds.

Size (minimum thickness) dimensions are placed to the left of the weld symbol. Welding direction is placed in the tail of the welding symbol. The direction can also be specified on the drawing.

In the case of multiple layer surfacing welds, the use of multiple reference lines can be used and these can show the required size (thickness) of each layer and direction of welding in the tail of the symbol or on the drawing.

When the entire area of the surface is to be built up, no dimension other than thickness is necessary on the welding symbol. In cases where only a portion of the area will be built-up by surface welds, the extent of weld, location, and orientation will be shown on the drawing. See Figure 4.82 for surfacing weld dimensioning.

Back or Backing Weld Symbols

Back and backing weld symbols are identical. The term *back weld* or *backing weld* is specified in the tail of the welding symbol and provides an indication of the welding sequence when used in a combined weld symbol having a single reference line.

Backing welds are made to the opposite side of a groove *before* the groove weld is applied. When shown in conjunction with a welding symbol using multiple reference lines, the backing symbol will be located on the reference line closest to the arrow.

Back welds are made after the groove has been welded, usually *after* some type of back gouge operation has been performed to ensure that the first weld root is sound. When used with a welding symbol having multiple reference lines, the symbol will appear on the line after the one containing the groove welding symbol. The symbol will always appear on the opposite side of the welded groove.

Back or backing weld *contours* that are obtained by welding, will have face appearances which are approximately flush or convex. When post weld finishing is specified, the appropriate letter is applied above the contour symbol. This signifies the method used for obtaining the desired contour, but does not specify the degree of finish. The degree of finish is indicated by a drawing note, or detail. See Figure 4.83 for examples of back and backing weld symbol use.

Groove Welds

Previously, a statement was made that the weld symbol is actually a miniature detail of the part or surface which it points to. Groove welds usually require some edge preparation at the joint, and the root opening of all grooved joints affects part preparation when a separation of members is specified. Eight types of groove weld symbols have been developed according to AWS A2.4 standards and are illustrated as shown in Figure 4.84.

All groove weld symbols have an arrow-side, other-side, and both-sides significance. The square groove weld symbol may have no arrow side or other-side significance, meaning the weld can be started from either side. As with other weld symbols, location significance is determined by the side of the reference line on which the symbol is placed.

Broken arrows are used with the bevel-groove weld symbol, the J-groove weld symbol and the flare bevel groove weld symbol. Use of a broken arrow for these three symbols identifies the joint member that must be prepared. A broken arrow need not be used if the joint is detailed on the drawing.

Single-groove dimensions are placed on the same side of the reference line as the symbol. Double-groove dimensions are placed on both sides of the reference line for each groove, except for the root opening which appears only once (see Figure 4.88(B)).

Dimensions that are common to all groove welds include, depth of bevel, groove weld size, root opening, and groove angle. Additional dimensioning applicable to J- and U-groove welds include a radius and root face. Radius is also used in the size specification for flare-bevel and flare-V-groove welds (see Figures 4.88–4.93).

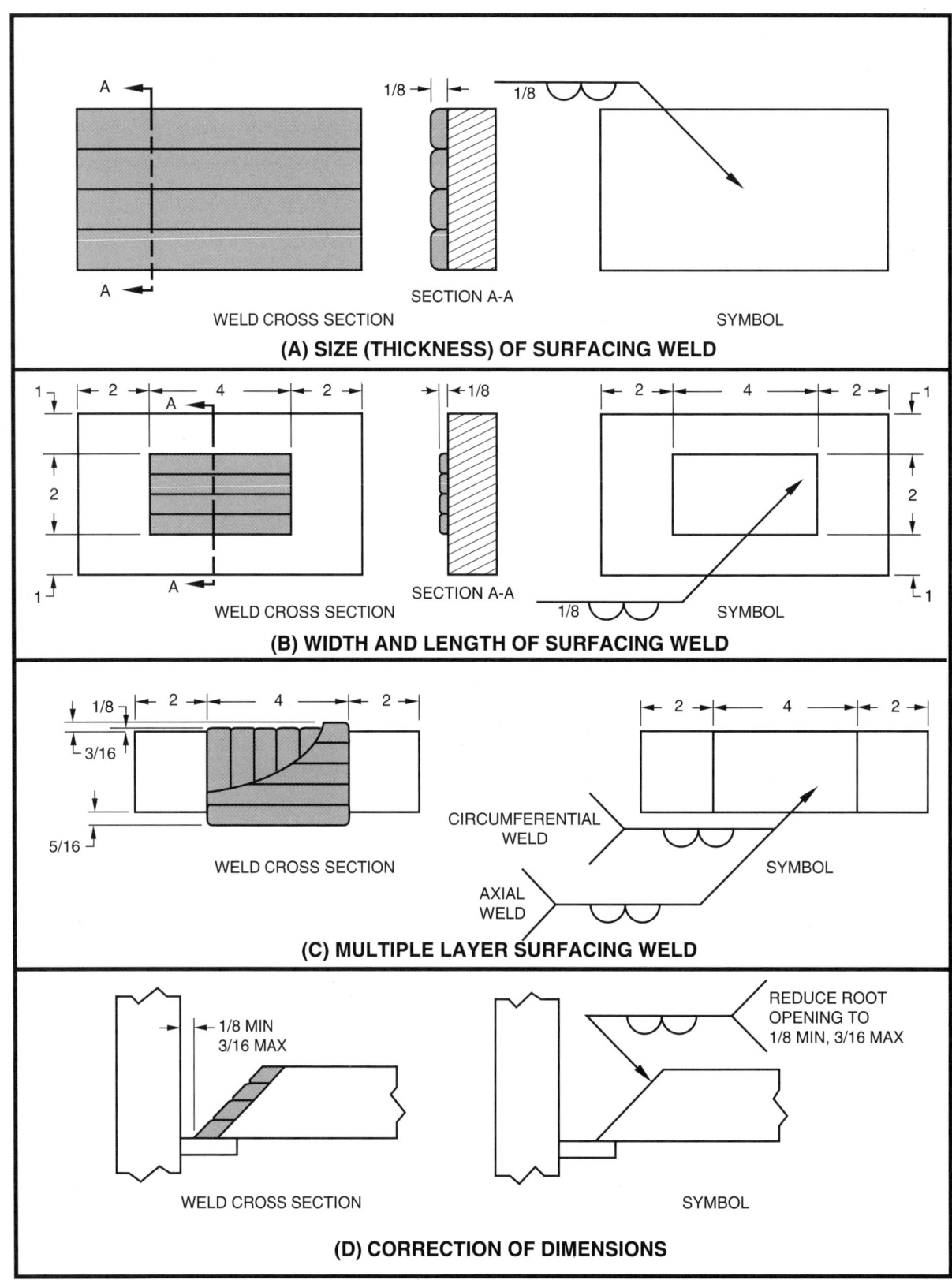

Figure 4.82—Surfacing Weld Dimensioning

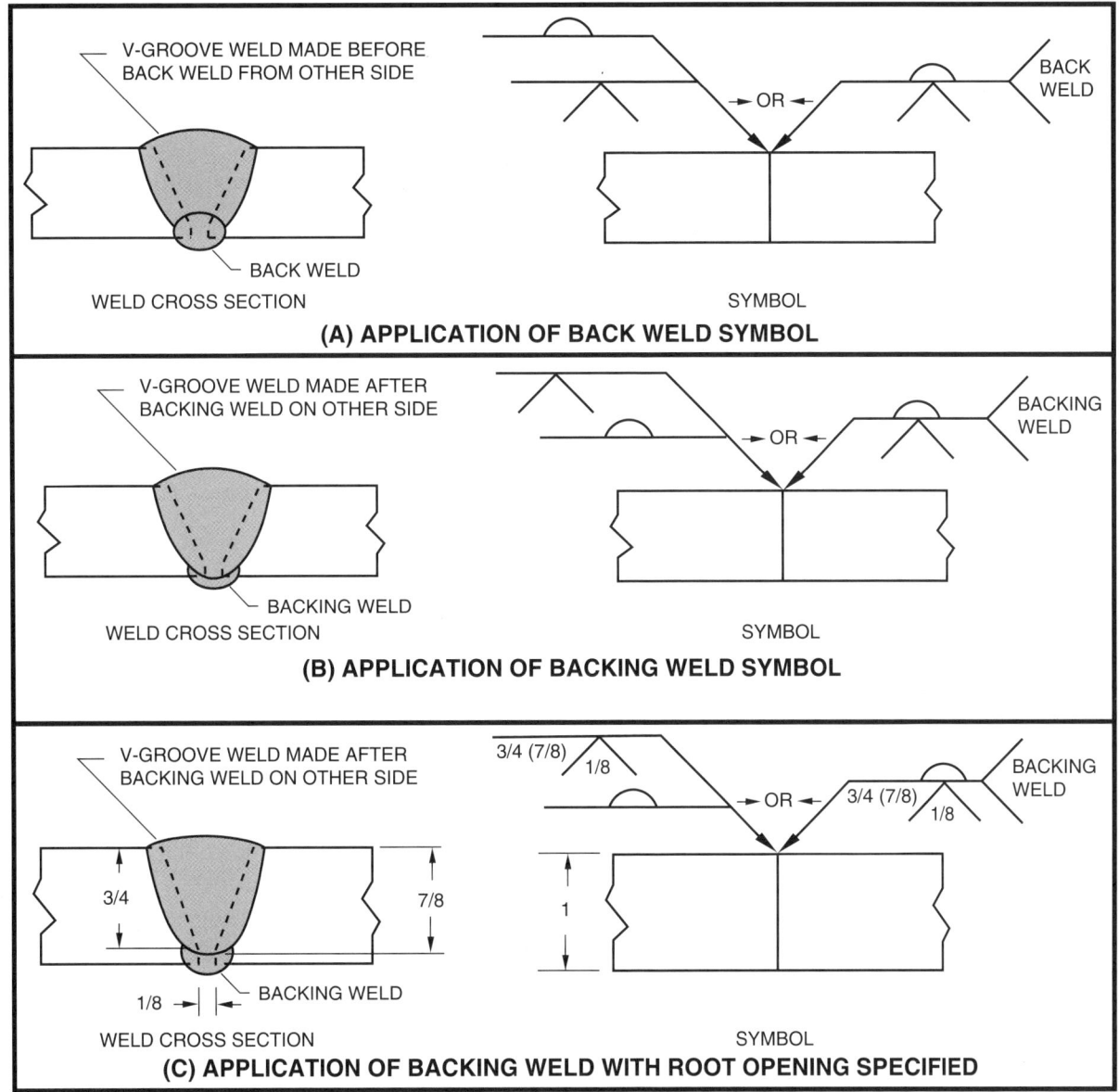

Figure 4.83—Application of Back and Backing Weld Symbols

Depth of bevel dimensioning is placed to the left of the weld symbol represented by "S" in Figure 4.85. *Depth of bevel* is defined as "the perpendicular distance from the base metal surface to the root edge or the beginning of the root face" (see Figure 4.90(A), (B), and (C)).

Groove weld size is "the joint penetration of the weld placed into a groove." Penetration may include the fusion of the base metal at or beyond the depth of bevel, the groove face and/or the root face, represented by "(E)" in Figure 4.85. Groove weld size dimensions are placed in parentheses between the dimension for depth of bevel and the weld symbol (see Figures 4.88–4.93).

Except for square-groove welds, groove weld size "(E)," in relation to the depth of bevel "S," is shown as "S(E)" to the left of the weld symbol. Because of the joint geometry of square edge shapes, only weld size "(E)" is shown for a square-groove weld (see Figures 4.90–4.92, and 4.96.

Root opening is "a separation at the joint root between the workpieces" forming the joint. The

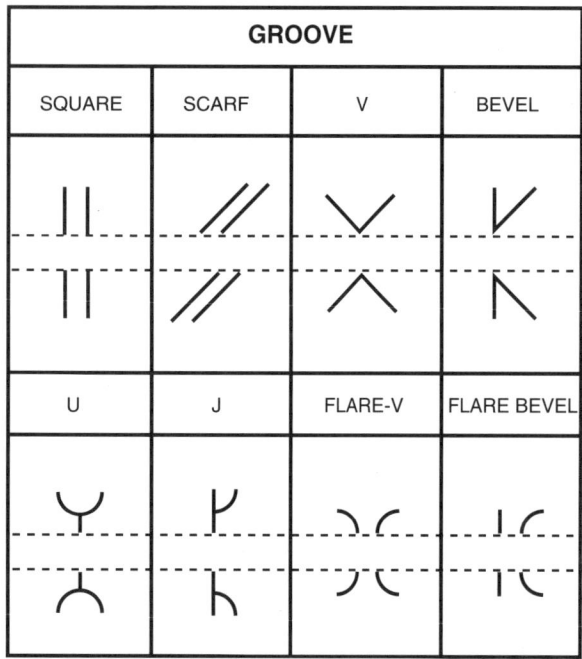

Note: The reference line is shown dashed for illustrative purposes.

Figure 4.84—Groove Weld Symbols

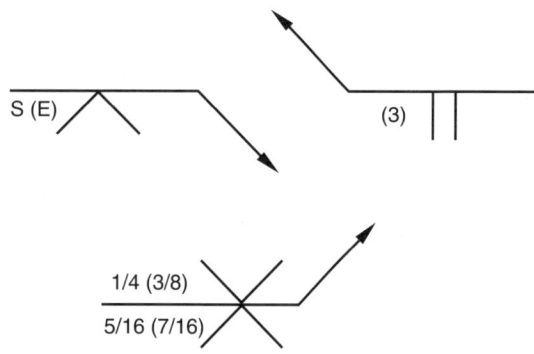

Figure 4.85—Depth of Bevel— Groove Weld Size

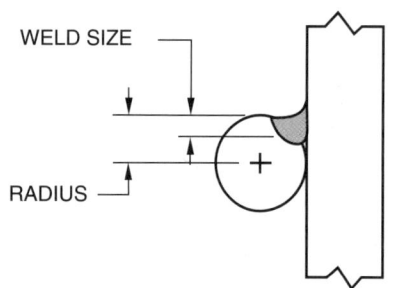

Figure 4.86—Flare Groove Size versus Radius

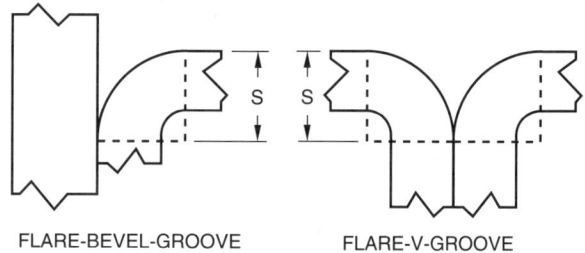

Figure 4.87—Flare Groove Depth of Bevel

root of a joint is either "open" or "closed." When a separation, the root opening, is indicated, the dimension appears inside the groove weld symbol. This dimension is specified only once for double groove welded joints, generally in the arrow side welding symbol (see Figure 4.97(D)).

Root openings affect member preparation when a drawing specifies design size (overall size of both members after fitup), rather than actual size (true size of a member after allowance or tolerances are applied). Allowances must be made when a root opening dimension is specified in the welding symbol and design size dimensions are specified for a particular object in the field of a drawing.

Groove angle is specified outside the weld symbol, positioned above or below the symbol dependent upon symbol placement on the reference line. Angle dimensioning is specified by degree, °, indicating the angle formed by the members to be welded. When the groove angle affects both members (such as a V or U), the bevel angle for each member equals half the given dimension. For example: a V-groove weld with a dimension of 60°, requires each member to be beveled at 30°. When combined, both members then form an included angle, the groove angle, of 60°. This is not the case when only one member is prepared. For example: a double-J groove weld specifying 15° on the arrow-side, and 20° on the other-side, is interpreted as 15° included angle arrow-side, 20° included angle other-side. In this case the arrow-side member is prepared on both sides (at different angles), but the other-side member remains square (see Figure 4.98(E)).

Radius and *root face* dimensions can apply for U- or J-groove joints. These dimensions do not appear in connection with the welding symbol. Radius and root faces are indicated by a reference to a particular drawing detail, cross section or other data in the tail of the welding symbol.

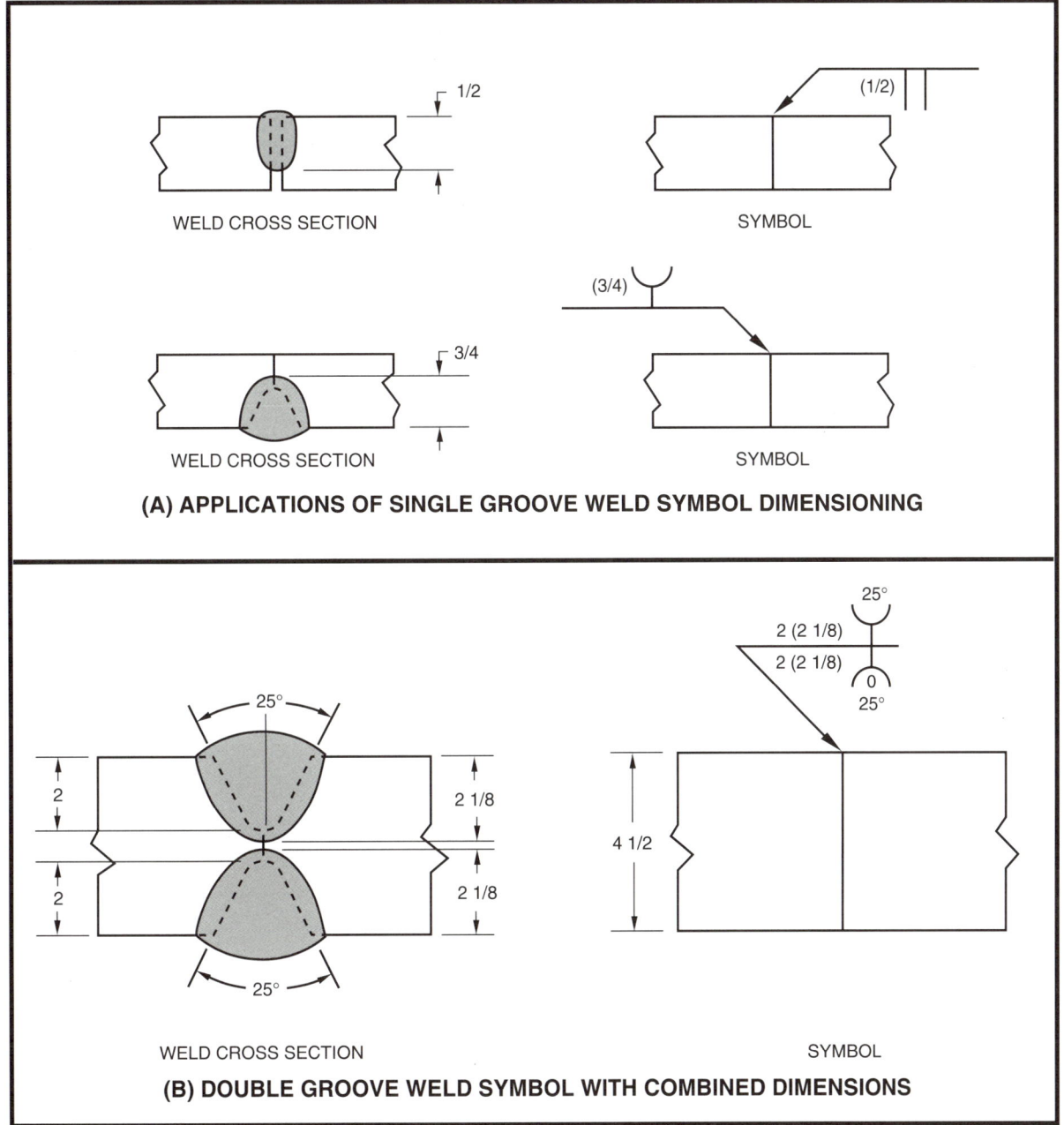

Figure 4.88—Applications of Dimensions to Groove Weld Symbols

Weld Size and Depth of Bevel Considerations

Groove weld size may be smaller than the dimension given for depth of bevel (partial joint penetration); at other times it will equal the depth of bevel (complete joint penetration). Where double grooves are specified, the groove weld size can be larger than the depth of bevel on either side of the joint, and the welds will overlap beyond depth of bevel (complete joint penetration) (see Figures 4.88, 4.97, and 4.98).

The inspector may encounter a groove welding symbol with no depth of bevel or weld size specified. When these dimensions are left out of the welding symbol, complete joint penetration is required. This rule holds true for all single-groove welds, and those double-groove welds having

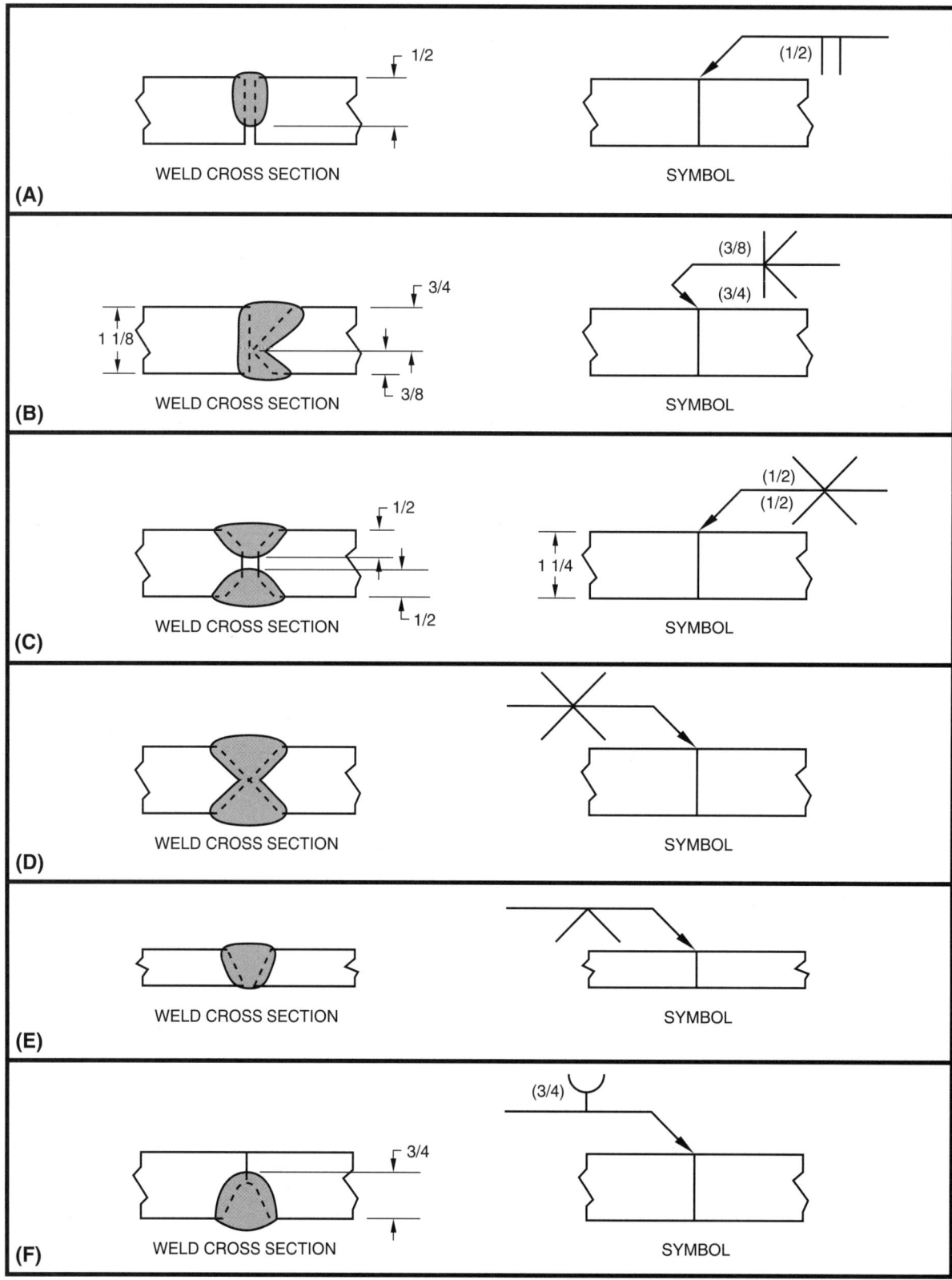

Figure 4.89—Groove Welds—Depth of Bevel Not Specified

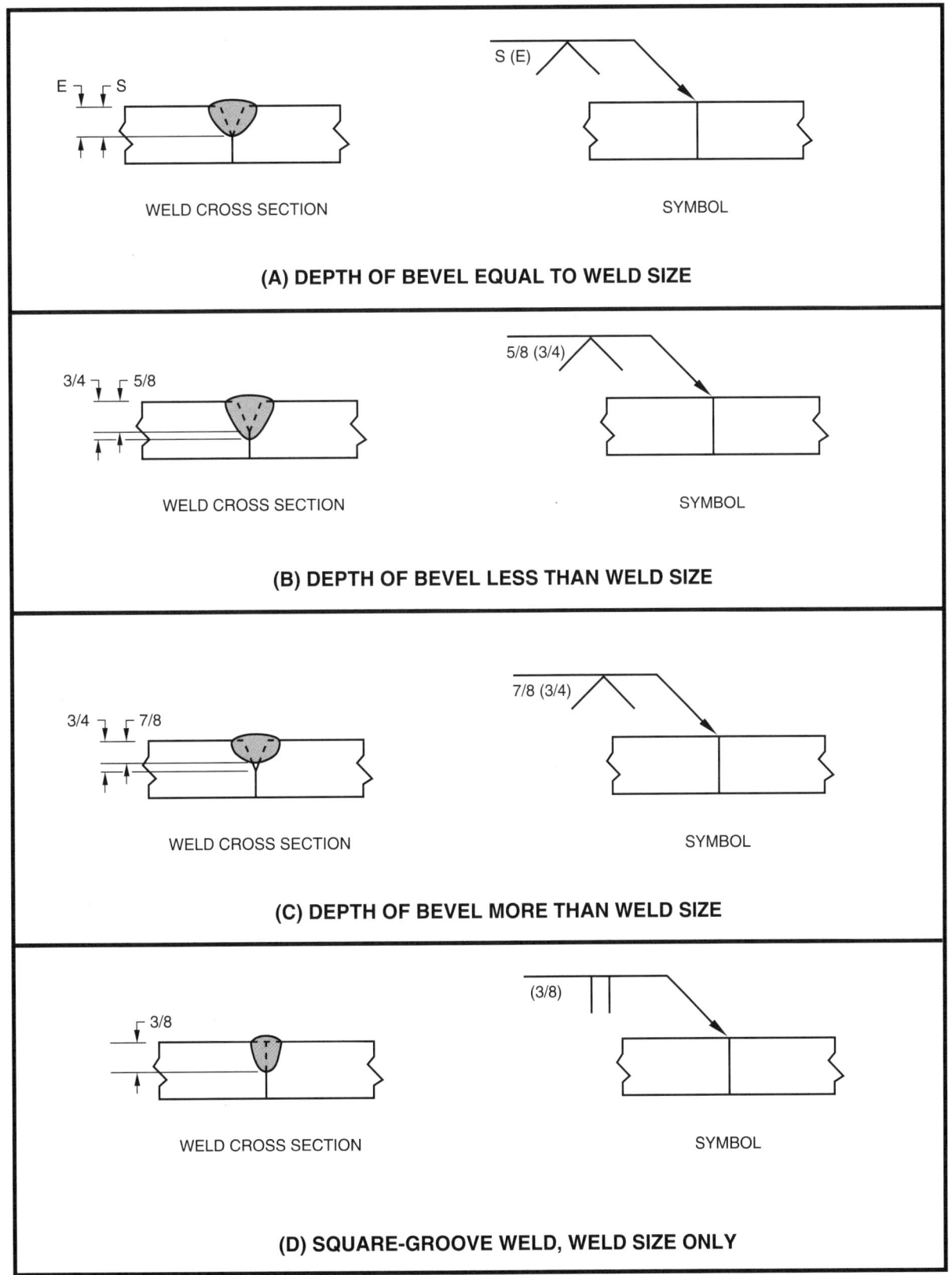

Figure 4.90—Groove Weld Size (E) Related to Depth of Bevel (S)

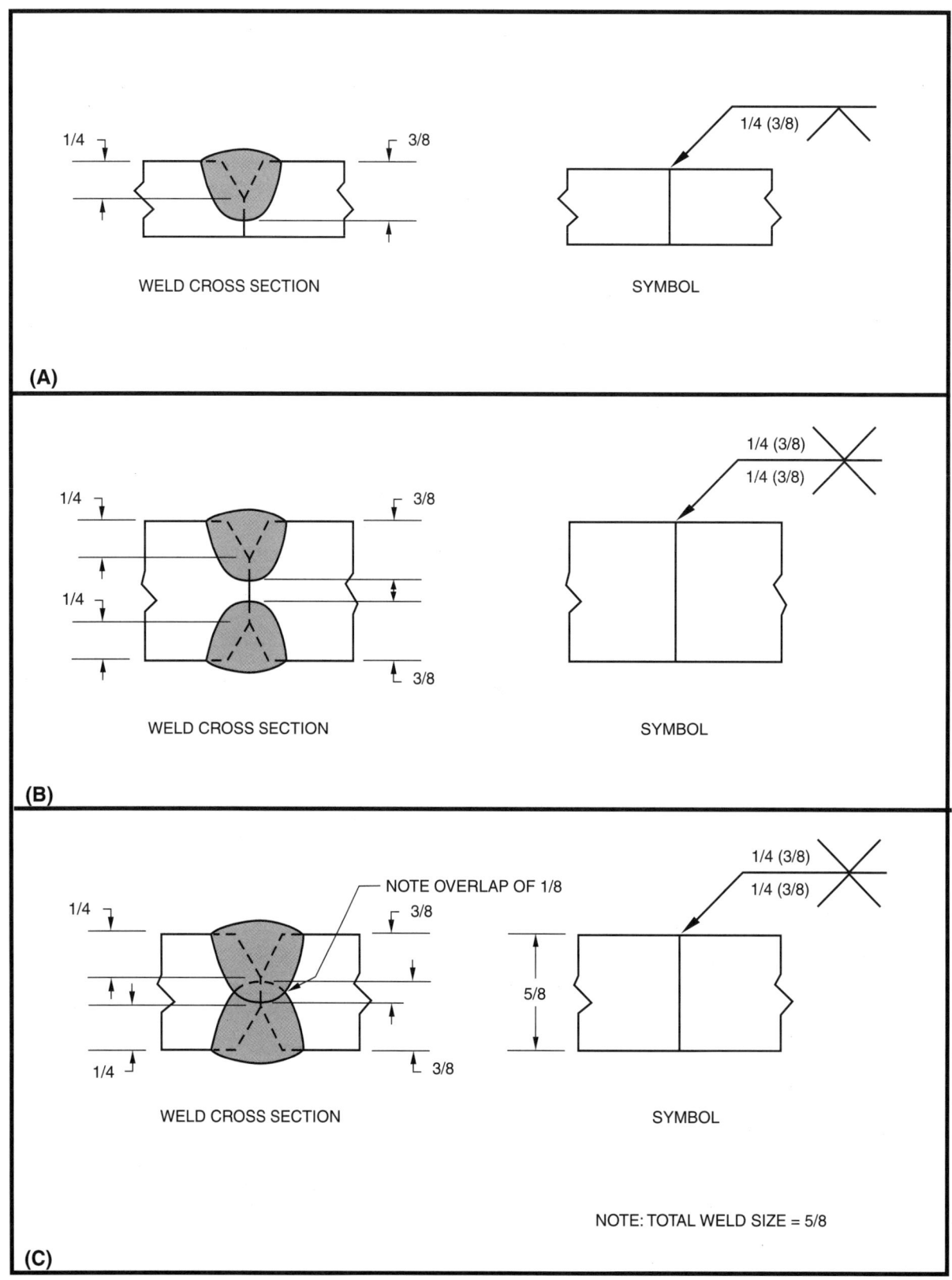

Figure 4.91—Specification of Groove Weld Size and Depth of Bevel

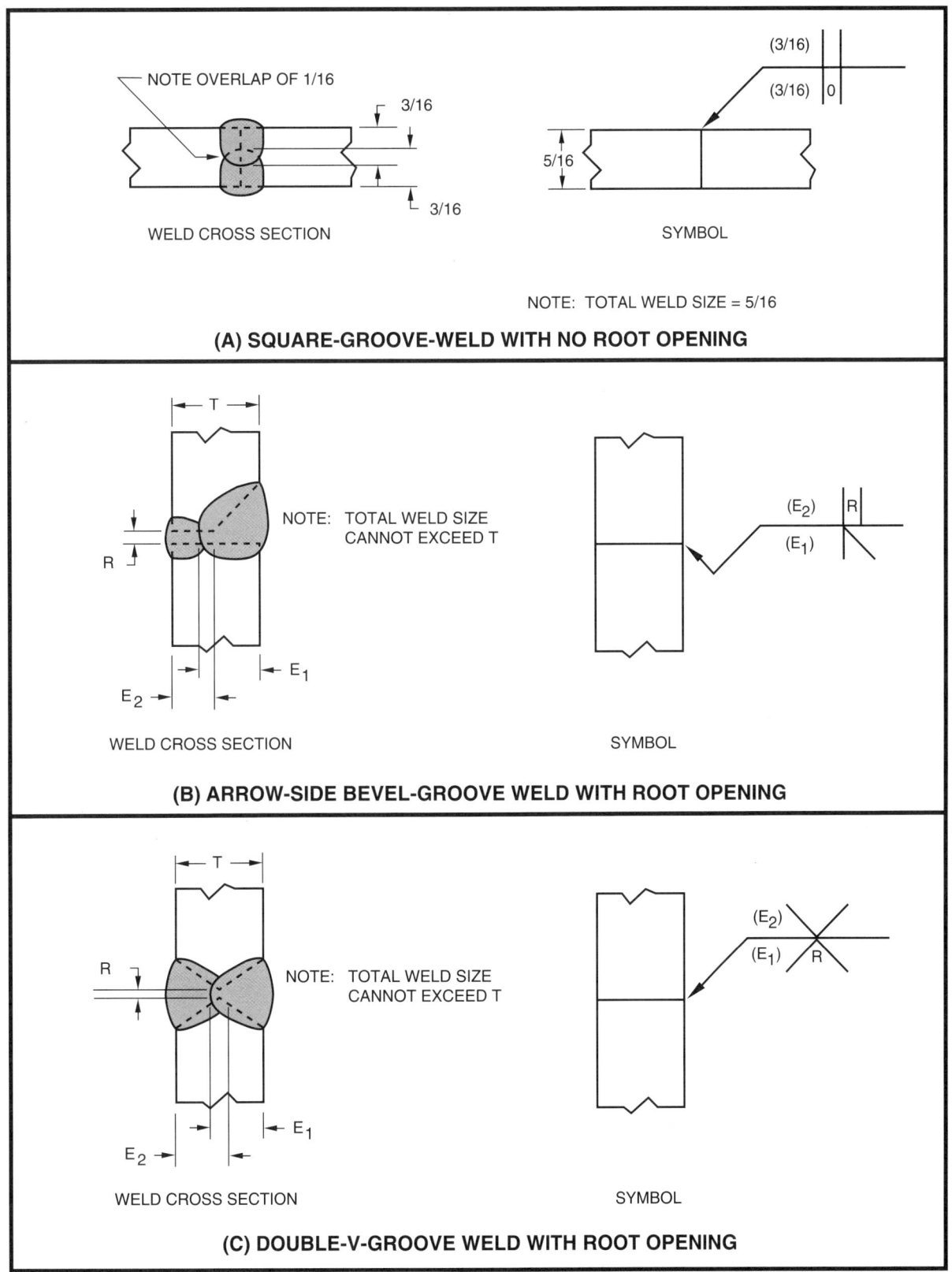

Figure 4.92—Specification of Groove Weld Size Only

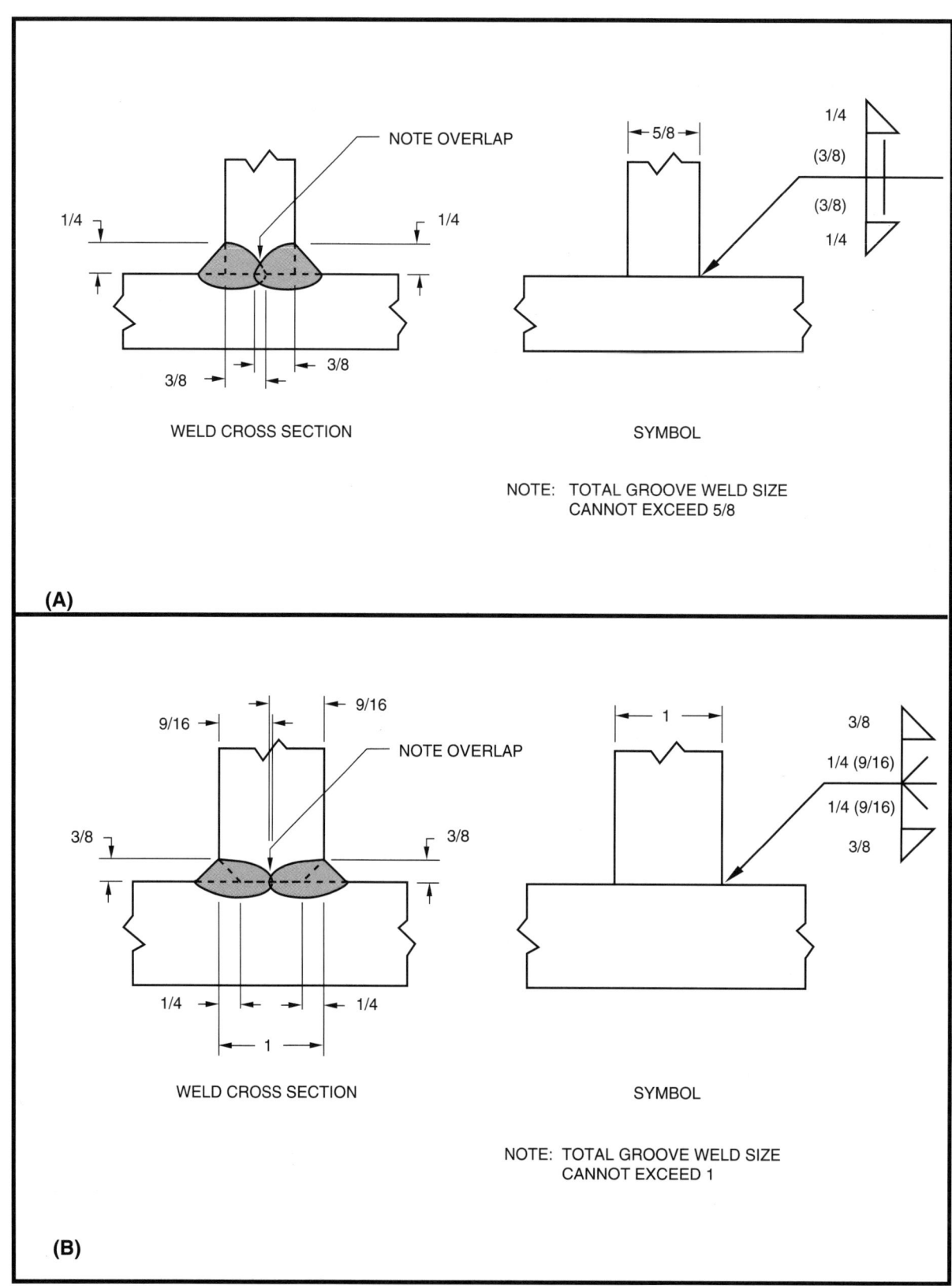

Figure 4.93—Combined Groove and Fillet Welds

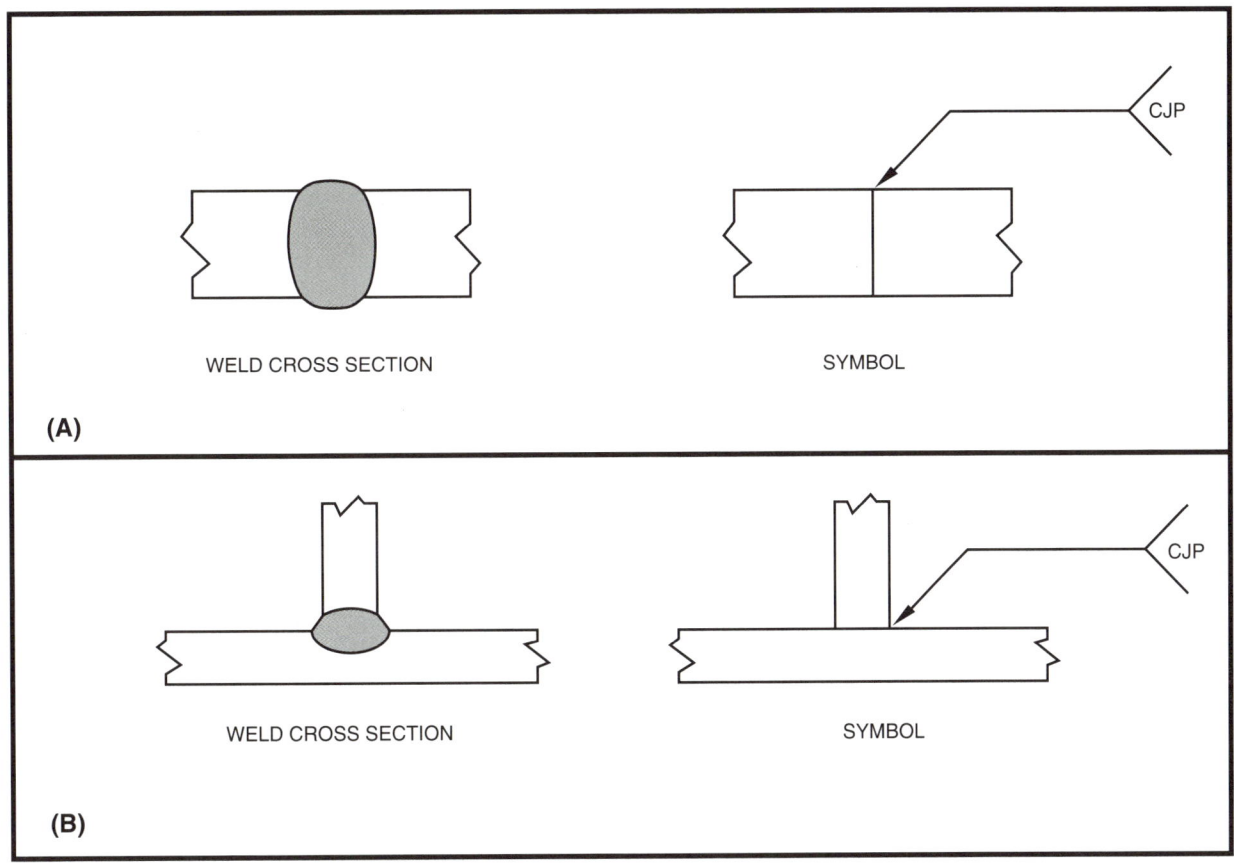

Figure 4.94—Complete Joint Penetration with Joint Geometry Optional

symmetrical joint geometries, with double-groove prepared members having the same edge shape on both sides of each member (see Figures 4.89(D) and (E), 4.98(A), (B), and (D), and 4.99). Asymmetrical groove welds that extend completely through the joint (complete joint penetration), require the use of groove weld size dimensions (see Figure 4.92(A) and (B)).

In some instances depth of bevel will not be specified by the welding symbol; only a groove weld size will appear. This groove weld dimensioning technique applies to groove welds that extend only partly through the joint (partial penetration) (see Figure 4.89(A), (C), and (F)).

At times, the type of groove welded joint, may not be specified on the drawing. Joint preparation becomes optional, and in some cases, determined by layout or fitting personnel. In these situations, the weld symbol is omitted. When no weld symbol is provided and the reference line and arrow point to the joint or weld area and the letters "CJP" are shown in the tail, this indicates complete joint penetration is required and joint geometry is optional (see Figure 4.94).

A second instance used to indicate optional joint geometry gives groove weld size only and also omits the weld symbol. Groove weld size is placed on either side of the reference line as required to show weld placement from the arrow or other side (see Figure 4.95).

For bevel-, V-, J- or U-grooves, depth of bevel may only be placed to the left of their respective weld symbols and groove weld size may appear elsewhere on the drawing. If this is the case, reference to the location of any required groove weld size will be made in the tail of the welding symbol.

Flare-Groove Welds

So far the discussion has addressed all groove weld symbols except flare grooves. In many

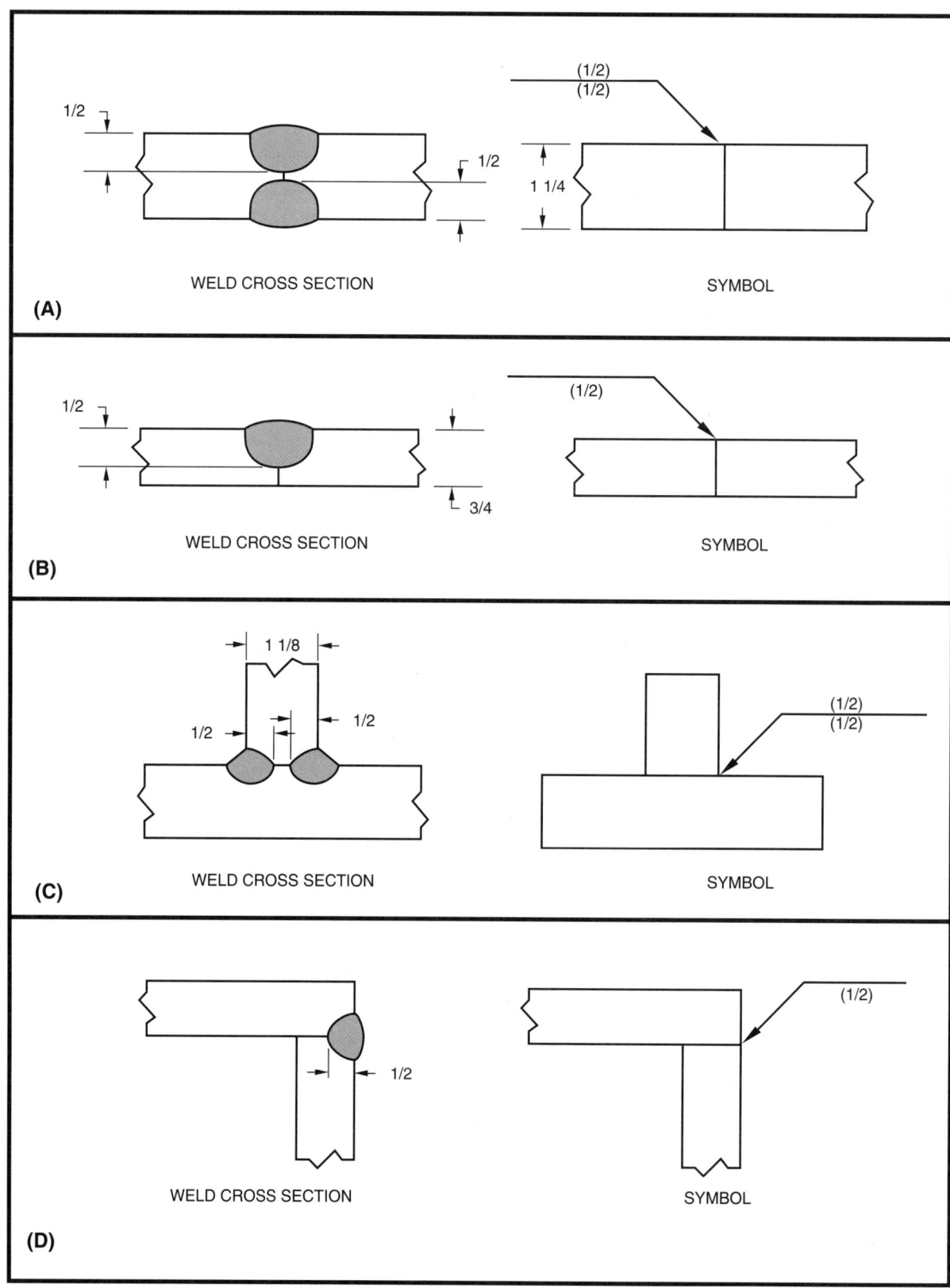

Figure 4.95—Partial Joint Penetration with Joint Geometry Optional

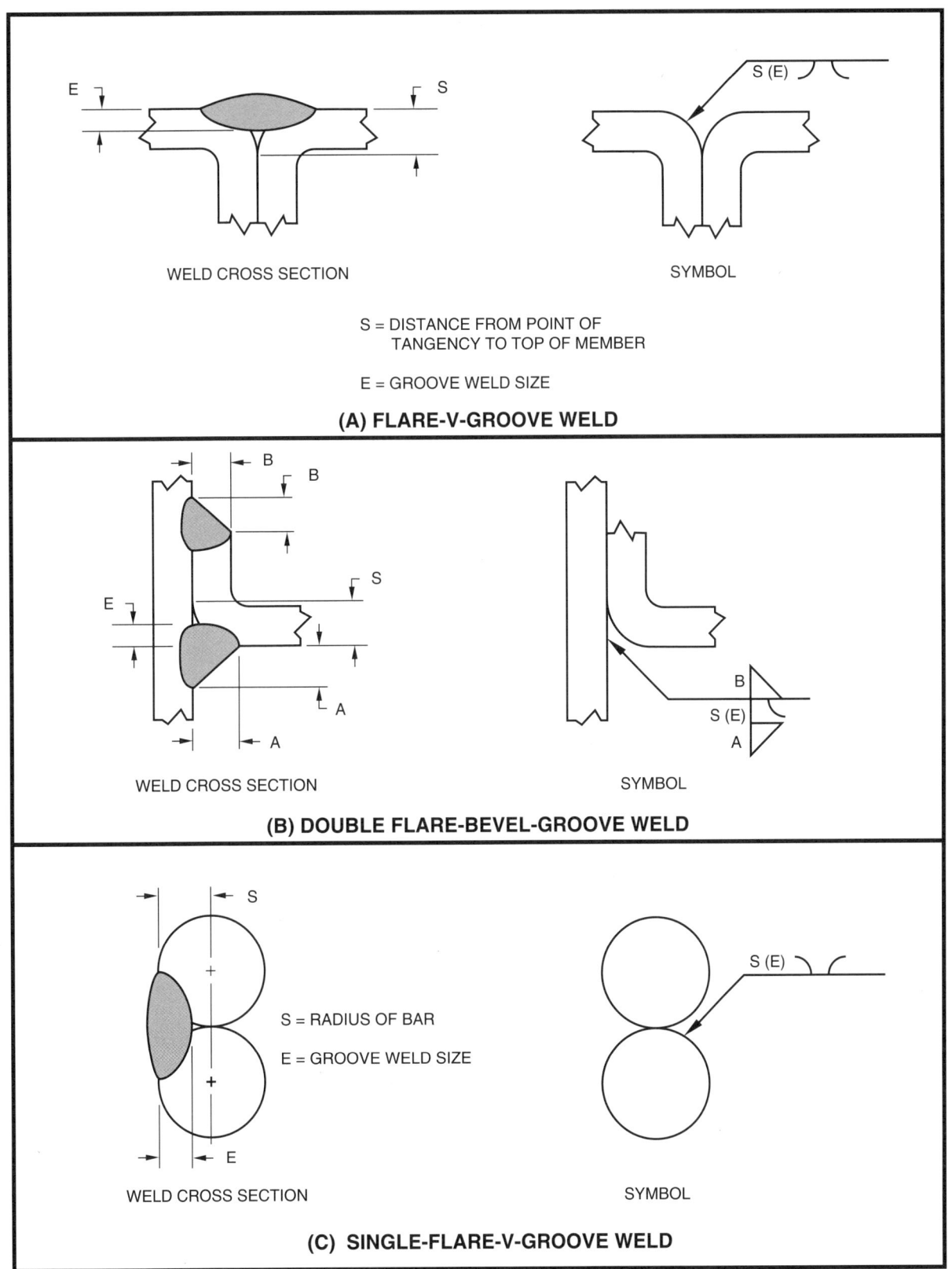

Figure 4.96—Applications of Flare-Bevel and Flare V-Groove Weld Symbols

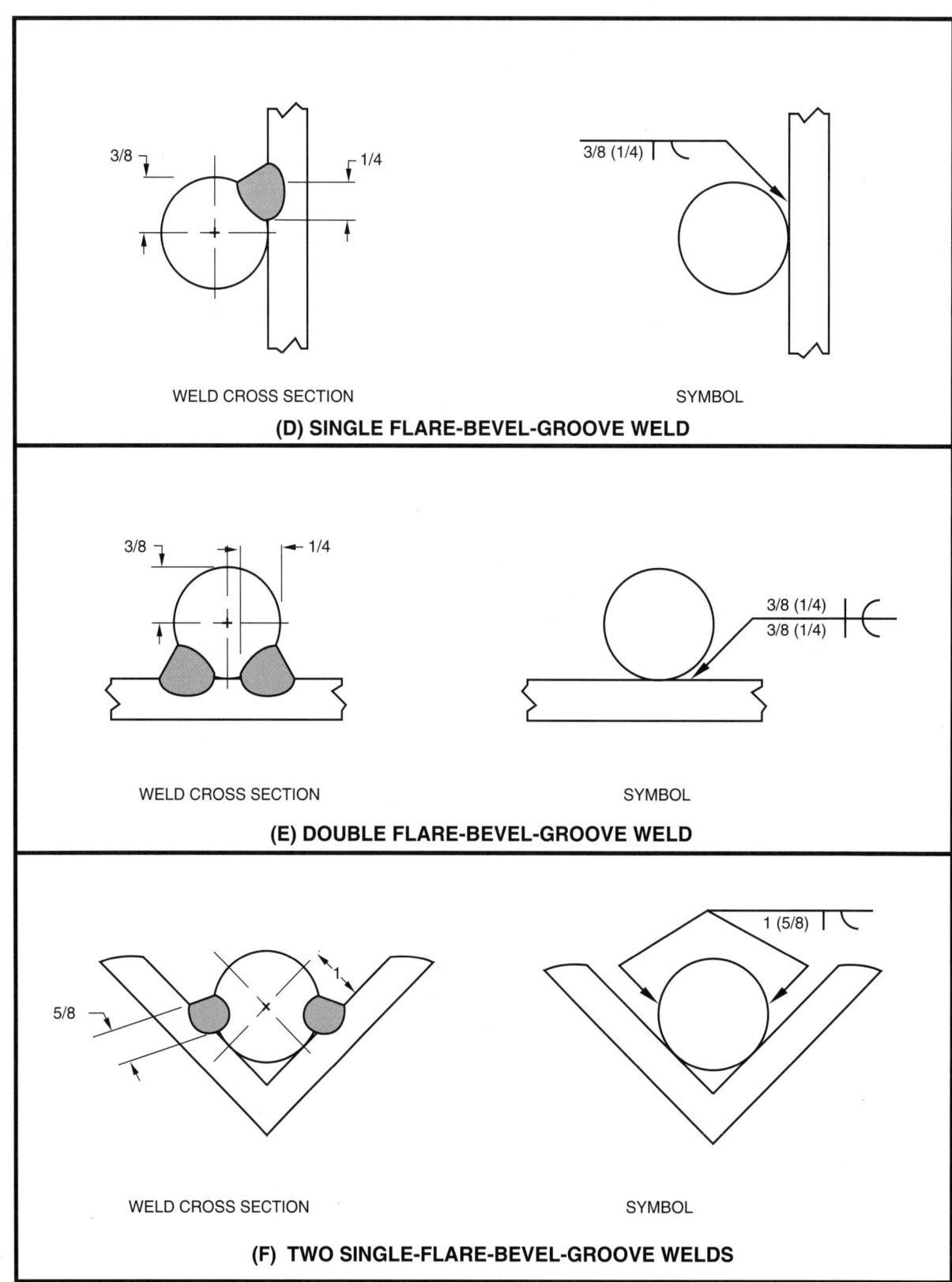

Figure 4.96 (Continued)—Applications of Flare-Bevel and Flare V-Groove Weld Symbols

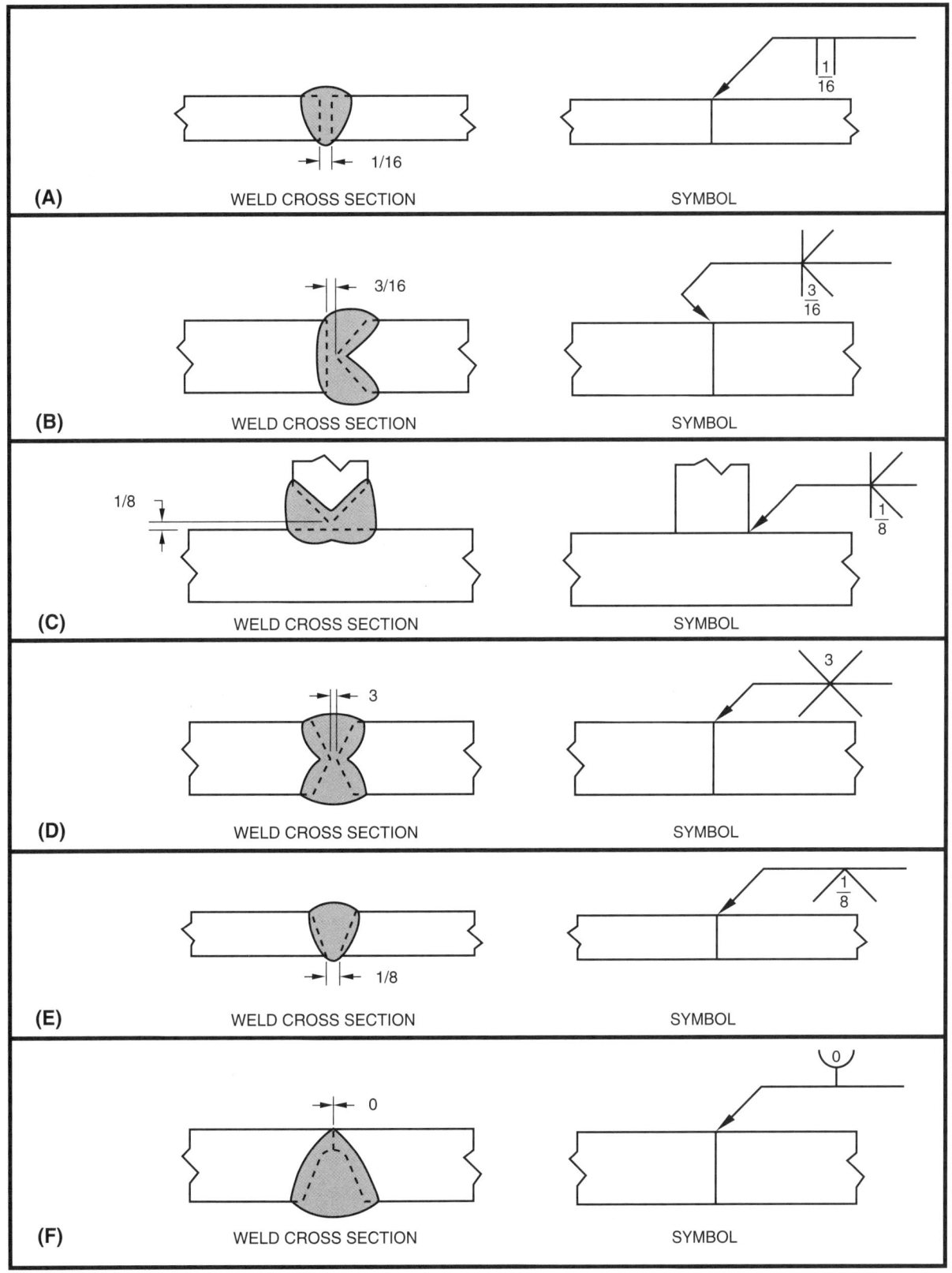

Figure 4.97—Specifications of Root Openings for Groove Welds

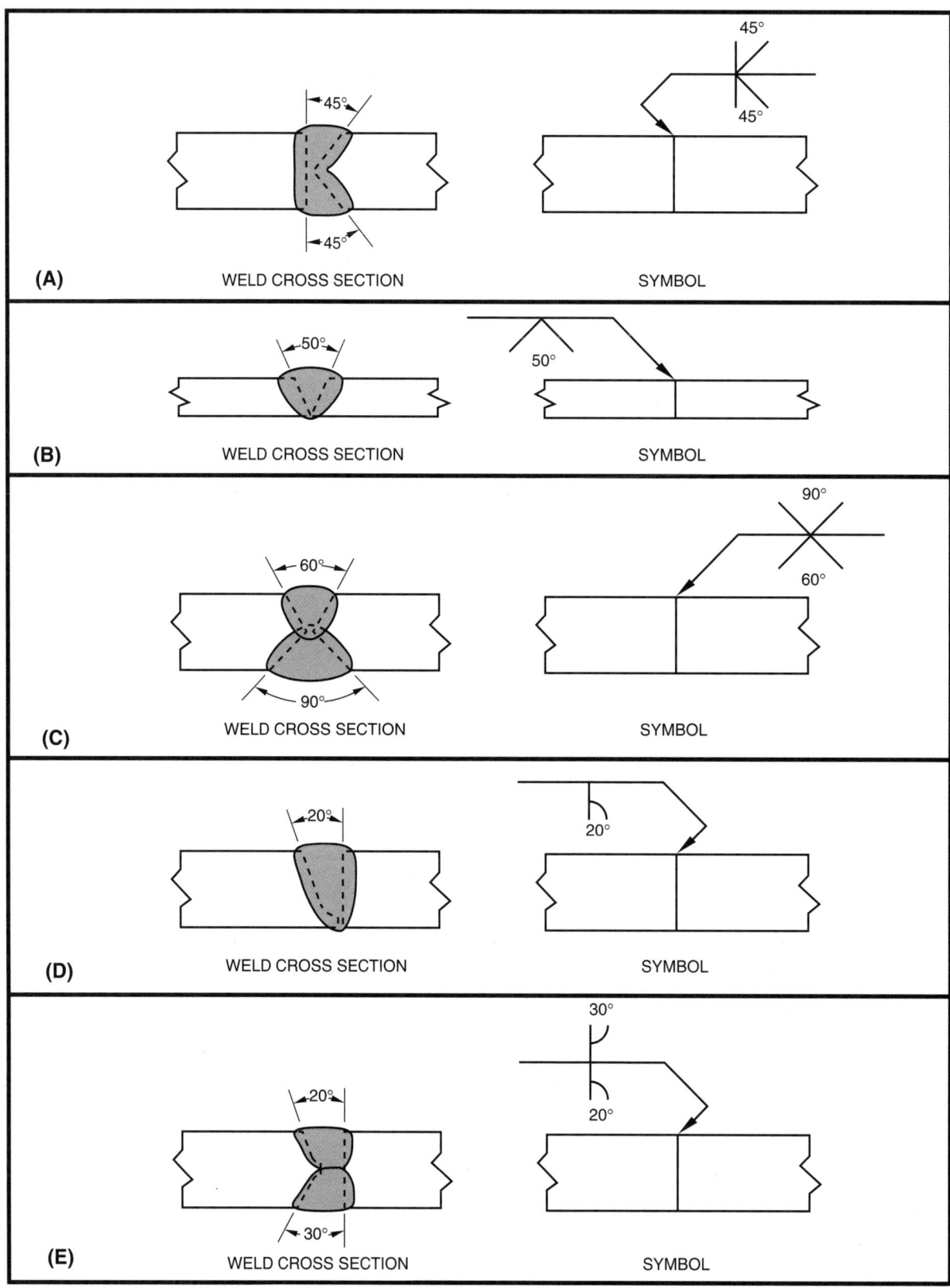

Figure 4.98—Specification of Groove Angle of Groove Welds

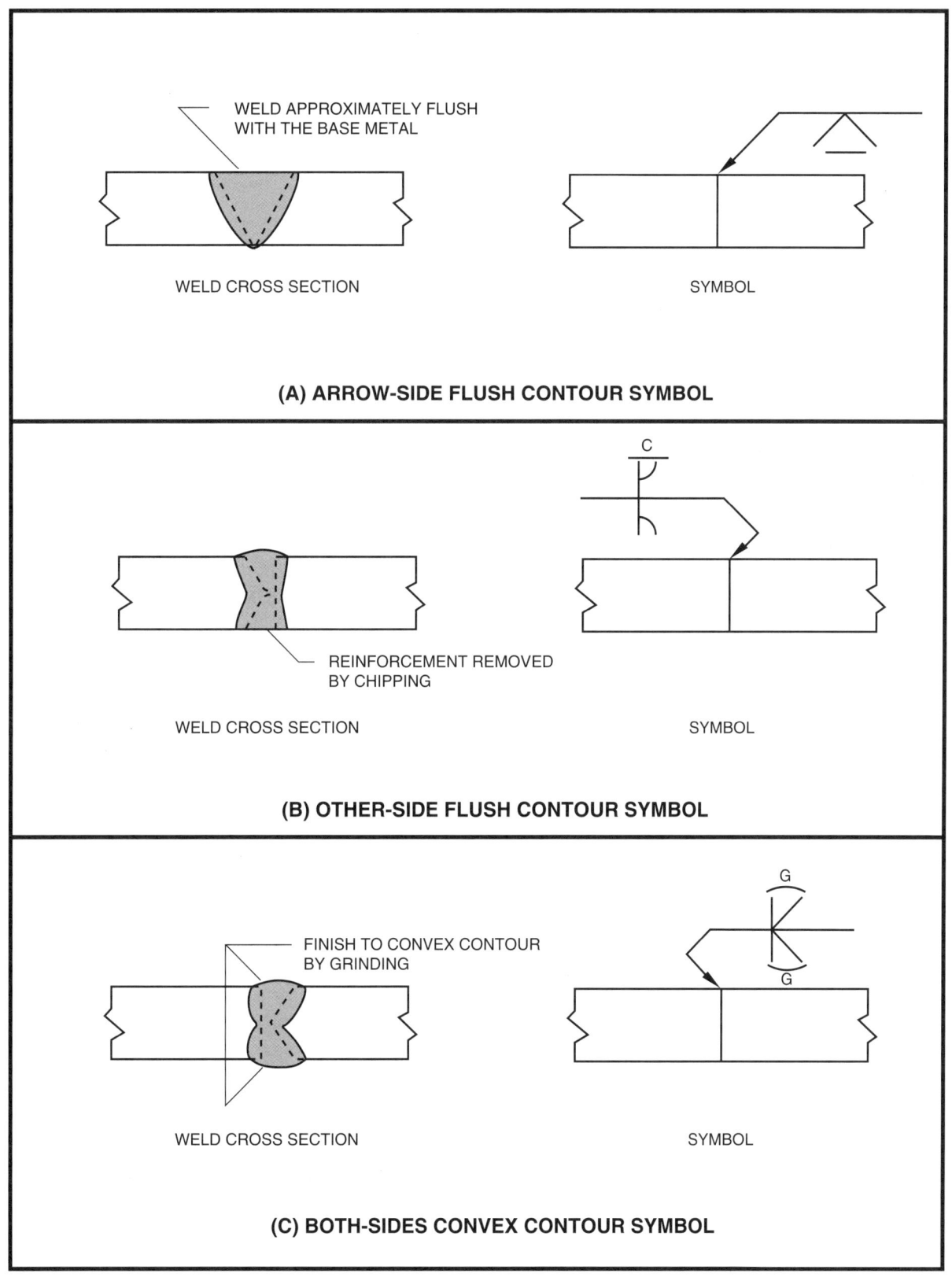

Figure 4.99—Applications of Flush and Convex Contour Symbols

respects flare grooves are special cases since they do not conform to all of the conventions associated with other types of groove welds. Dimensions related to depth of bevel and groove angle in a normal groove weld are functions of the edge shape. With flare groove welds these same dimensions are related to the curvature of the base metal and may be beyond the control of the welder. Complete penetration may not be obtainable in many flare groove welds since the fusion occurs along the surface of one or both members rather than through the thickness. Likewise, the amount of curvature in these type of grooves may result in a weld size that is only some fraction of the radius (see Figure 4.86).

A distinction also exists in the interpretation of the "S" dimension (depth of bevel) for flare-groove welds and its interpretation for other groove weld types. With flare groove joints, depth of bevel is defined as the radius, or the point of tangency, indicated by dimension lines shown in Figure 4.87.

Groove weld size "(E)," dimensioning also applies to flare-groove welds. Figure 4.96 illustrates the various dimensioning aspects of flare-groove welds.

Supplementary Symbols for Groove Welds

Groove weld *contours* that are obtained by welding will have face appearances which are approximately flush or convex. When post weld finishing is specified, the appropriate letter is applied above the contour symbol. This signifies the method used for obtaining the desired contour, but does not specify the degree of finish. The degree of finish is indicated by a drawing note or detail.

Groove welded joints that employ a backing material or device are specified by placing the backing symbol on the side of the reference line opposite the groove weld symbol. If the backing will be removed after welding, an "R" is placed in the backing symbol to indicate removal. The type of material or device used, and the dimensions of the backing are specified in the tail of the welding symbol or on the drawing. Use of a backing symbol should not be confused with the use of back or backing weld symbols. Although a backing weld is backing in the form of a weld, the backing symbol represents a material or device. Compare Figures 4.100 and 4.101.

If a particular groove welded joint requires use of a *spacer*, the specific groove weld symbol is modified to show a rectangle within it. When multiple reference lines are used in connection with groove welds and spacers, the spacer symbol will appear on the reference line closest to the arrow. Material and dimensions for spacers are indicated in the tail of the welding symbol or on the drawing (see Figure 4.101(B) and (C)).

Consumable inserts are also used with groove welded joints. When specified, the consumable insert symbol is placed on the side of the reference line opposite the groove weld symbol. *AWS Class of Consumable Insert* information is placed in the tail of the welding symbol (see Figure 4.102). Additional information regarding insert class can be obtained by consulting AWS A5.30, *Specification for Consumable Inserts*.

A common practice associated with groove welds and complete joint penetration involves backgouging. When backgouging is involved, the operation may be specified using either a single or multiple reference line welding symbol (see Figure 4.103).

Reference to backgouging is included in the tail of the welding symbol. When backgouging is used for asymmetrical double-groove welds, the welding symbol must show the depth of bevel on both sides, along with groove angle and root opening dimensions (see Figure 4.103(A)). When the operation involves single-groove welds or symmetrical double-groove welds, the only information required is weld symbols, with groove angles and root opening (see Figure 4.103(B) and (C)).

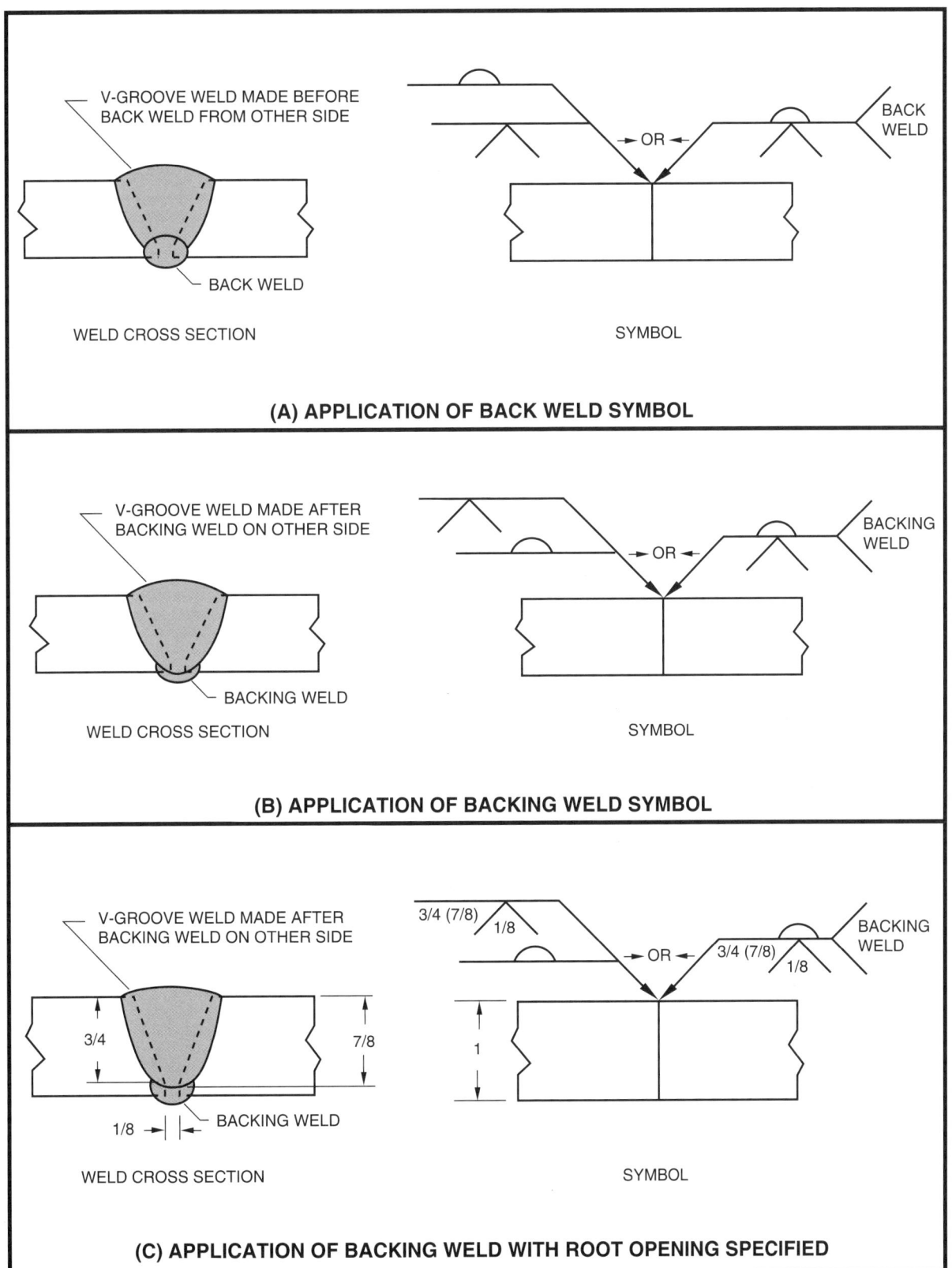

Figure 4.100 —Applications of Back or Backing Weld Symbol

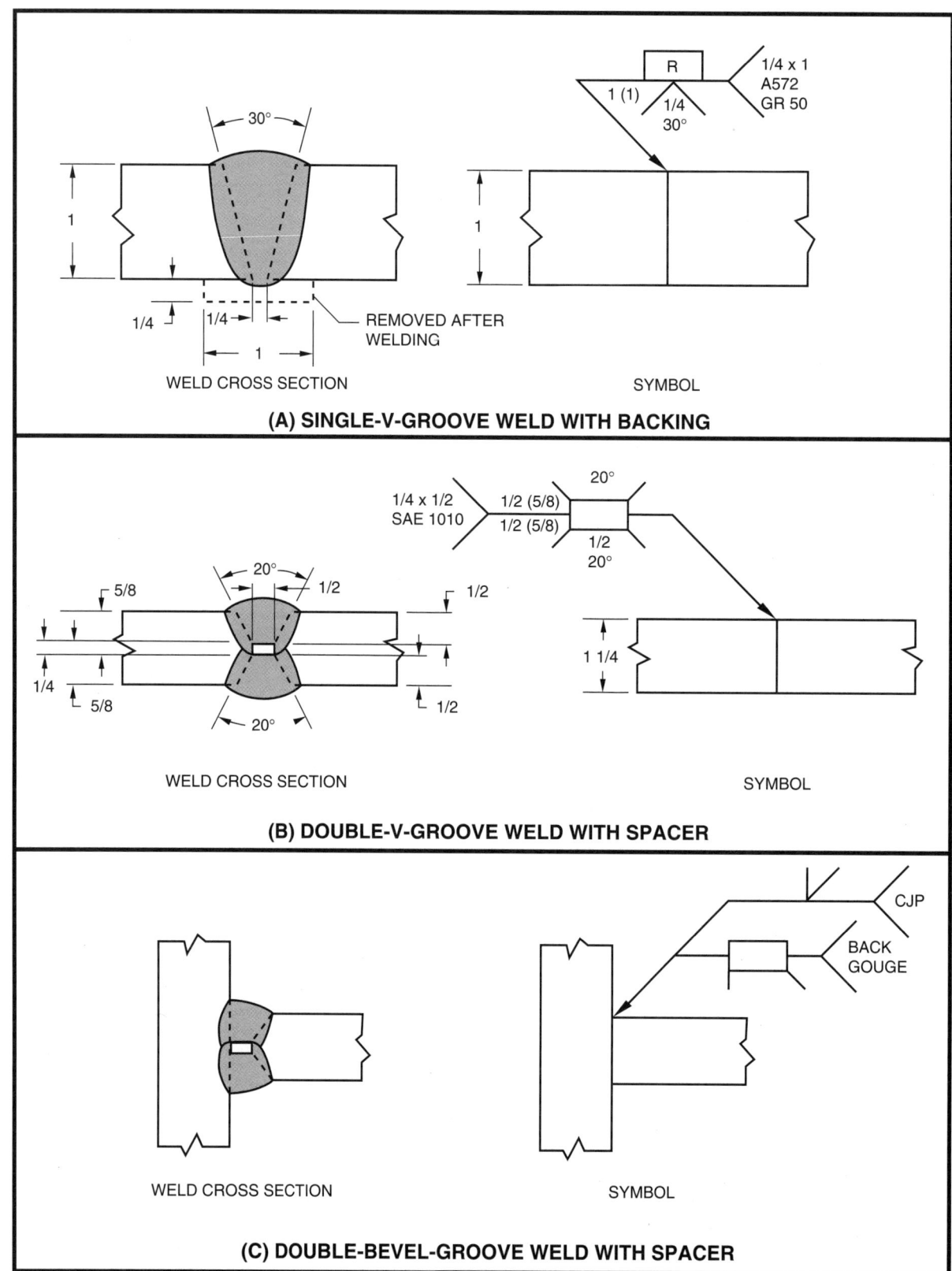

Figure 4.101—Joints with Backing or Spacers

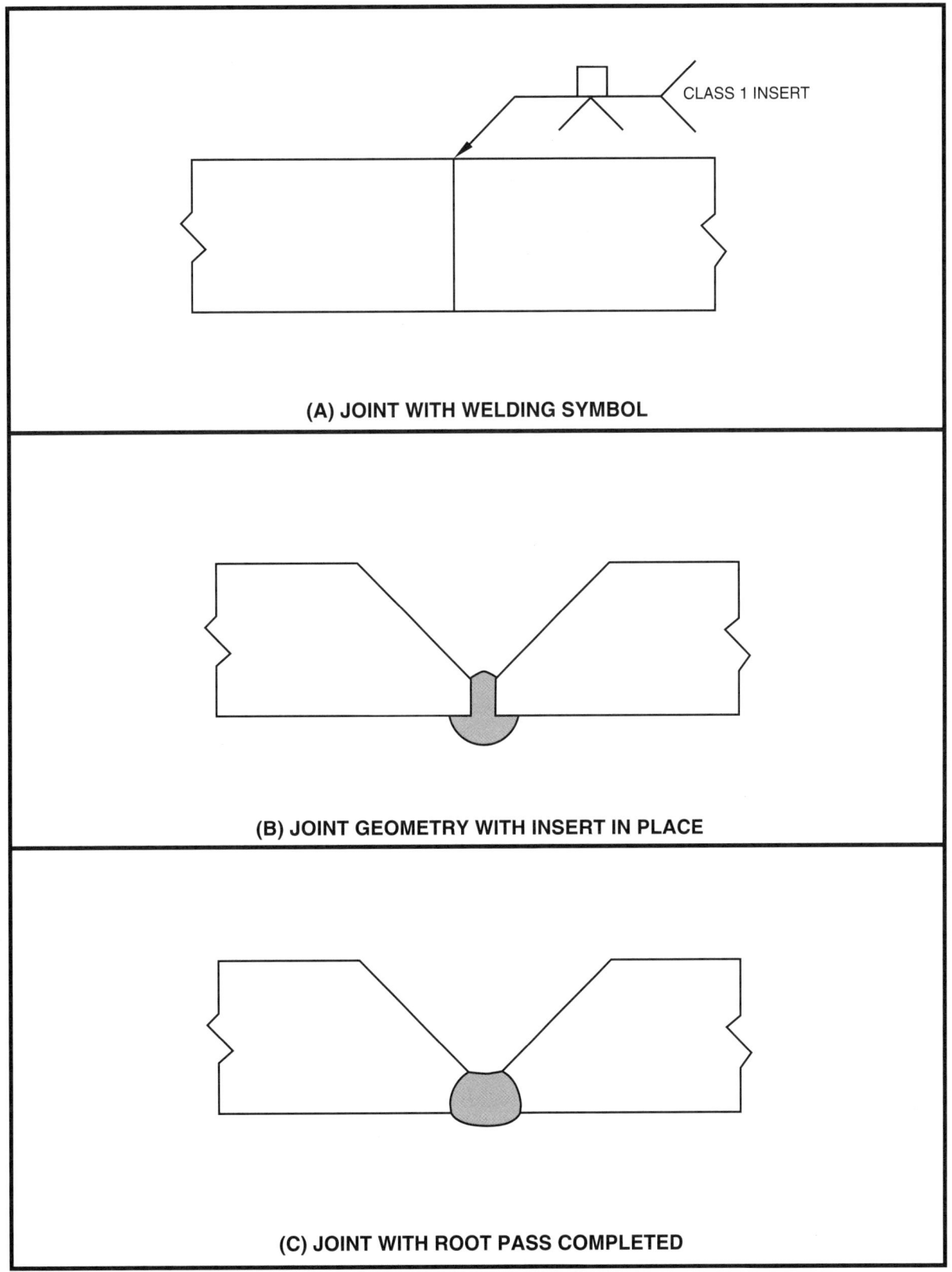

Figure 4.102—Application of Consumable Insert Symbol

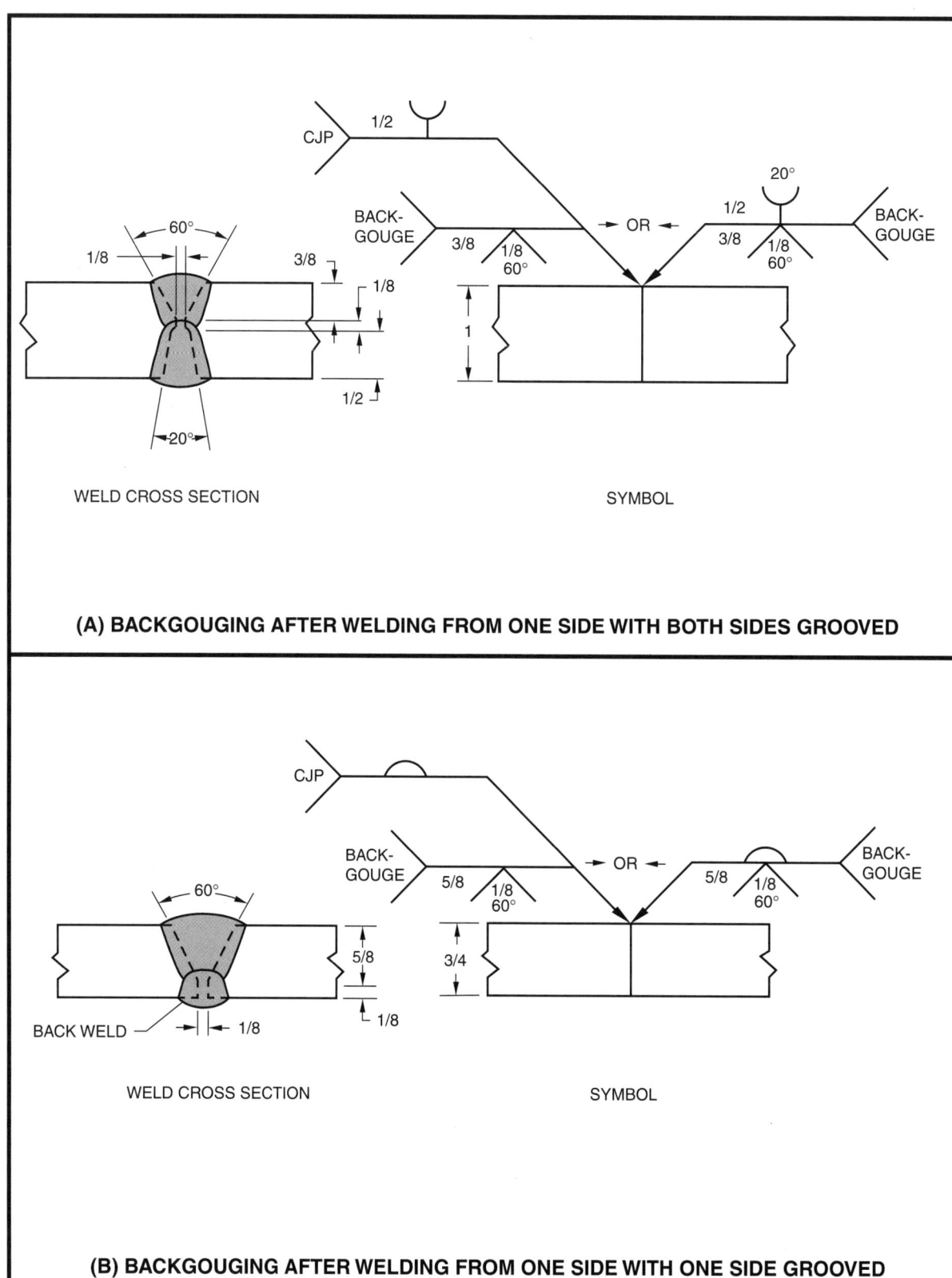

Figure 4.103—Groove Welds with Backgouging

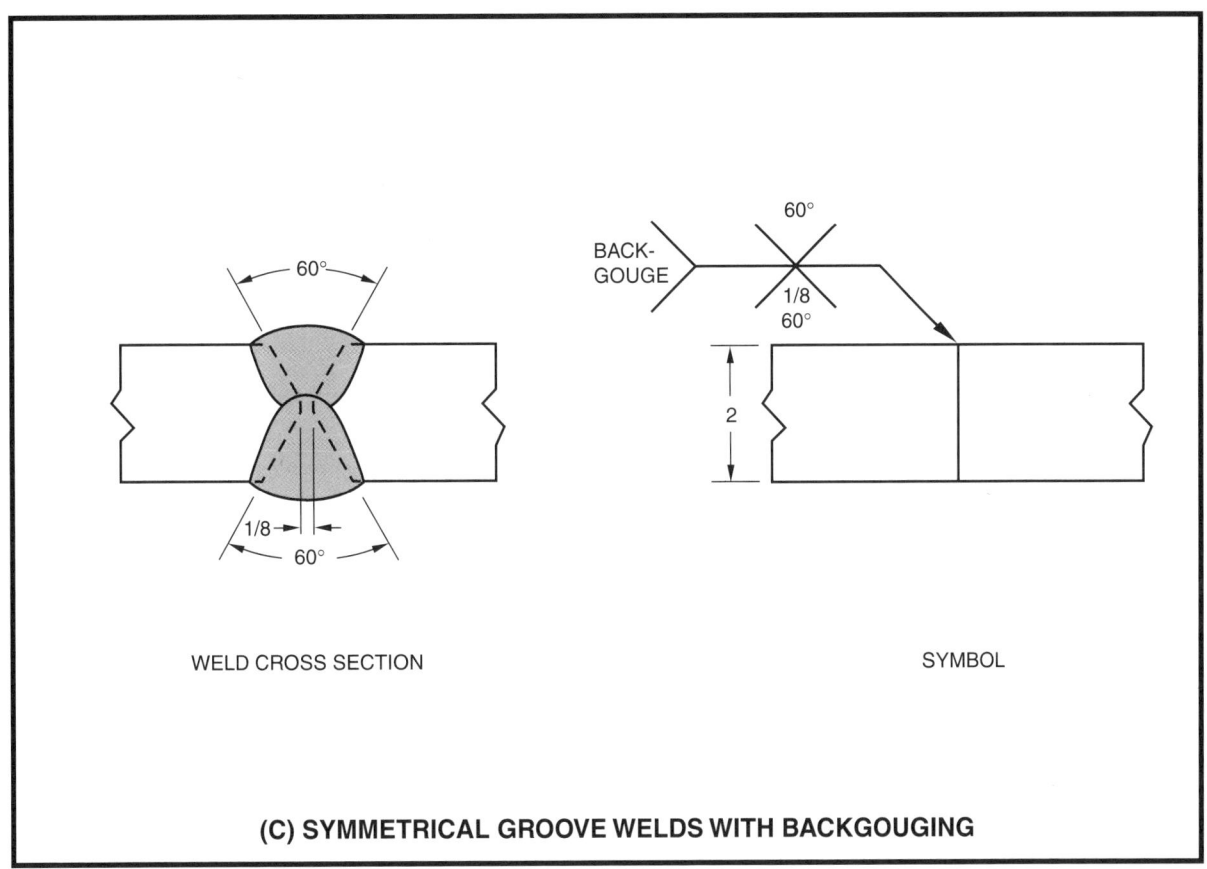

(C) SYMMETRICAL GROOVE WELDS WITH BACKGOUGING

Figure 4.103 (Continued)—Groove Welds with Backgouging

Key Terms and Definitions

actual throat—The shortest distance between the weld root and the face of a fillet weld (see Figure 4.27.)

back weld—A weld made at the back of a single groove welded joint after the completion of the groove weld (see Figure 4.19(E)).

backing—A material or device placed against the back side of the joint, or at both sides of a weld in electroslag and electrogas welding, to support and retain molten weld metal. The material may be partially fused or remain unfused during welding and may be either metal or nonmetal.

backing weld—Backing in the form of a weld, made before the primary weld (see Figure 4.19(F)).

backstep sequence—A longitudinal sequence in which weld passes are made in the direction opposite to the progression of welding (see Figure 4.35(A)).

bevel—An angular edge shape.

bevel angle—The angle between the bevel of a joint member and a plane perpendicular to the surface of the member (see Figure 4.14). This dimension equals one-half of the groove or included angle when the edges of both members are prepared at the same angle. When only one member is prepared at an angle, this dimension is still indicated in the same position on the welding symbol for groove angle, but equals the total degree of preparation for the groove.

bevel groove weld—A type of groove weld in which the mating members of the joint have one single-bevel or double-bevel edge and one square edge preparation (see Figure 4.15(D1, D2)).

block sequence—A combined longitudinal and cross sectional sequence for a continuous multiple-pass weld in which separated increments are completely or partially welded before intervening increments are welded (see Figure 4.35(B)).

boxing—The continuation of a fillet weld around a corner of a member as an extension of the principle weld (see Figure 4.34).

buildup—A surfacing variation in which surfacing material is deposited to achieve the required dimensions.

butt joint—A joint between two members aligned approximately in the same plane (see Figures 4.2(A), 4.3(A), 4.4, 4.5, upper left corner sketch, and 4.7).

buttering—A surfacing variation that deposits surfacing metal on one or more surfaces to provide metallurgically compatible weld metal for the subsequent completion of the weld.

butting member—A joint member that is prevented, by the other member, from movement in one direction perpendicular to its thickness dimension. For example, both members of a butt joint, or one member of a T-joint or corner joint (see Figure 4.5).

cascade sequence—A combined longitudinal and cross sectional sequence in which weld passes are made in overlapping layers (see Figure 4.35(C)).

chain intermittent fillet weld—An intermittent weld on both sides of a joint in which the weld increments (lengths) on one side are approximately opposite those on the other side (see Figures 4.16(F) and 4.33).

cladding—A surfacing variation that deposits or applies surfacing material usually to improve corrosion or heat resistance.

complete joint penetration—A joint root condition in a groove weld in which weld metal extends completely through the joint thickness (see Figure 4.26).

complete joint penetration weld—A groove weld in which weld metal extends completely through the joint thickness (see Figure 4.26).

concave fillet weld—A fillet weld having a concave face (see Figure 4.27).

concavity—The maximum distance from the face of a concave fillet weld perpendicular to a line joining the weld toes (see Figure 4.27).

convex fillet weld—A fillet weld having a convex face (see Figure 4.27).

convexity—The maximum distance from the face of a convex fillet weld perpendicular to a line joining the weld toes (see Figure 4.27).

corner joint—A joint between two members located approximately at right angles to each other in the form of an L (see Figures 4.2(B), 4.3(B), and 4.8).

depth of bevel—the perpendicular distance from the base metal surface to the root edge or the beginning of the root face (see Figure 4.14).

depth of fusion—The distance that fusion extends into the base metal or previous bead from the surface melted during welding (see Figure 4.23).

edge joint—A joint between the edges of two or more parallel or nearly parallel members (see Figures 4.2(E), 4.3(E), and 4.11).

edge preparation—The preparation of the edges of the joint members, by cutting, cleaning, plating, or other means.

edge shape—The shape of the edge of the joint member (see Figures 4.3–4.11).

edge weld—A weld in an edge joint, a flanged butt joint or a flanged corner joint in which the full thickness of the members are fused (see Figure 4.20(A) and (B)).

effective throat—The minimum distance, minus any convexity, between the weld root and the face of a fillet weld (see Figure 4.27).

face reinforcement—Weld reinforcement on the side of the joint from which welding was done (see Figure 4.21(A)).

faying surface—The mating surface of a member that is in contact with or in close proximity to another member to which it is to be joined.

fillet weld—A weld of approximately triangular cross section joining two surfaces approximately at right angles to each other in a lap-, T-, or corner-joint (see Figure 4.16).

fillet weld leg—The distance from the joint root to the toe of the fillet weld (see Figures 4.22 and 4.27).

flanged butt joint—A form of a butt joint in which at least one of the members has a flanged edge shape at the joint (see Figures 4.3(A) and 4.7).

flanged corner joint—A form of a corner joint in which the butting member has a flanged edge shape at the joint (see Figures 4.3(B) and 4.8).

flanged edge joint—A form of an edge joint in which at least one of the members has a flanged edge shape at the joint (see Figure 4.8).

flanged joint—A form of one of the five basic joint types in which at least one of the joint members has a flanged edge shape at the weld joint (see Figures 4.3 and 4.6–4.11).

flanged lap joint—A form of a lap joint in which at least one of the members has a flanged edge shape at the joint, and edge weld is not applicable (see Figures 4.3 and 4.10).

flanged T-joint—A form of a T-joint in which the butting member has a flanged edge shape at the joint, and an edge weld is not applicable (see Figures 4.3 and 4.9).

flare-bevel-groove weld—A type of groove weld in which the mating members of the joint can consist of one half-round, round or flanged edge shape, combined with one square edge shape (see Figure 4.15(G1, G2)).

flare-V-groove weld—A type of groove weld in which the mating members of the joint can consist of two half-round, two round or two flanged edge shapes (see Figure 4.15(H1, H2)).

fusion—The melting together of filler metal and base metal, or of base metal only, to produce a weld.

fusion face—A surface of the base metal that will be melted during welding (see Figure 4.23(A)) or the area of base metal melted as determined on the cross section of a weld.

groove angle—The total included angle of the groove between workpieces (see Figure 4.14). In joints where both edges of the workpieces are prepared at an angle this dimension is the total of both (shown as a degree dimension, placed directly above the weld symbol on other side designations, and directly below the symbol on arrow side designations).

groove face—The surface of a joint member included in the groove. The angular distance between the surface of the base metal to the root edge, including any root face (see Figure 4.13).

groove radius—The radius used to form the shape of a J or U groove weld (see Figure 4.14).

groove weld—A weld made in a groove between the workpieces (see Figure 4.15).

hardfacing—A surfacing variation in which surfacing material is deposited to reduce wear.

incomplete joint penetration—A joint root condition in a groove weld in which weld metal does not extend completely through the joint thickness (see Figures 4.25 and 4.26).

J-groove weld—A type of groove weld in which the mating members of the joint have one single-J or double-J and one square edge preparation (see Figure 4.15(F1, F2).

joint—The junction of members or the edges of members that are to be joined or have been joined.

joint design—The shape, dimensions and configuration of the joint.

joint filler—A metal plate inserted between the splice member and thinner joint member to accommodate joint members of dissimilar thickness in a spliced butt joint (see Figure 4.4).

joint geometry—The shape and dimensions of a joint in cross section prior to welding.

joint penetration—The distance the weld metal extends from the weld face into a joint, exclusive of weld reinforcement (see Figure 4.24).

joint root—That portion of a joint where the members approach closest to each other. When viewed in cross section the joint root may be either a point, a line, or an area (see Figure 4.12).

joint type—A weld joint classification based on five basic joint configurations such as a butt joint, corner joint, edge joint, lap joint and T-joint (see Figure 4.2).

lap joint—A joint between two overlapping members in parallel planes (see Figure 4.2(D), 4.3(D), 4.5, and 4.10).

nonbutting member—A joint member that is free to move in any direction perpendicular to its thickness dimension. For example, both members of a lap joint, or one member of a T-joint or corner joint (see Figure 4.5).

partial joint penetration weld—A joint root condition in a groove weld in which incomplete joint penetration exists as designed.

plug weld—A weld made in a circular hole in one member of a joint fusing that member to another member. A fillet-welded hole is *not* to be construed as conforming to this definition (see Figure 4.17(A)).

projection weld—A type of weld associated with a resistance welding process that produces a weld by the heat obtained from the resistance to the flow of the welding current. The resulting welds are localized at predetermined points by projections, embossments, or intersections (see Figure 4.18(C)).

root edge—is a root face of zero width (see Figure 4.13).

root face—that portion of the groove face within the joint root (see Figure 4.13) (also known as "land"). Although not shown by dimension on the weld symbol, when the depth of preparation for groove welds is subtracted from the thickness of the workpiece, the difference equals the root face of the joint.

root opening—a separation at the joint root between the workpieces (see Figure 4.14(A) and (E)).

root penetration—The distance the weld metal extends into the joint root (see Figure 4.24.

root reinforcement—Weld reinforcement opposite the side from which welding was done (see Figure 4.21(C)).

root surface—The exposed surface of a weld opposite the side from which welding was done (see Figure 4.21(C)).

scarf weld—A type of groove weld associated with brazing in which the mating members of the joint have single-bevel edge shapes. The groove faces of the joint are parallel (face the same way or same hand) (see Figure 4.15(A)).

seam weld—A continuous weld made between or upon overlapping members, in which coalescence may start and occur on the faying surfaces, or may have proceeded from the outer surface of one member. The continuous weld may consist of a single weld bead or a series of overlapping spot welds (see Figure 4.19(A)–(D)).

slot weld—A weld made in an elongated hole in one member of a joint fusing that member to another member. The hole may be open at one end. A fillet welded slot is not to be construed as conforming to this definition (see Figure 4.17(B)).

spliced joint—A joint in which an additional workpiece spans the joint and is welded to each member (see Figure 4.4).

splice member—The workpiece that spans the joint in a spliced joint (see Figure 4.4).

spot weld—A weld made between or upon overlapping members in which coalescence may start and occur on the faying surfaces or may proceed from the outer surface of one member. The weld cross section (plan view) is approximately circular (see Figure 4.18(A) and (B)).

square groove weld—A type of groove weld in which the mating members of the joint have square edge shapes (see Figure 4.15(B1) and (B2)).

staggered intermittent fillet weld—An intermittent weld on both sides of a joint in which the weld increments (lengths) on one side are alternated with respect to those on the other side (see Figures 4.16(E) and 4.33).

stringer bead—A type of weld bead made without appreciable weaving motion (see Figure 4.32).

stud weld—A type of weld associated with a general term for joining a metal stud or similar part to a workpiece. The weld can be made using arc, resistance, friction or other welding processes with or without an external gas shielding (see Figure 4.17(C)).

surface preparation—The operations necessary to produce a desired or specified surface condition. For example the holes or slots cut into one member of a lap joint to accommodate a spot or slot weld.

surfacing weld—A weld applied to a surface, as opposed to making a joint, to obtain desired properties or dimensions (see Figure 4.19(G)).

T-joint—A joint between two members located approximately at right angles to each other in the form of a T (see Figures 4.2(C), 4.3(C), 4.5, and 4.9).

theoretical throat—The distance from the beginning of the joint root perpendicular to the hypotenuse of the largest right triangle that can be inscribed within the cross section of a fillet weld. The dimension is based on the assumption that the root opening is equal to zero (see Figure 4.27).

U-groove weld—A type of groove weld in which the mating members of the joint both have single-j or double-j edge shapes (see Figure 4.15(E1) and (E2)).

V-groove weld—A type of groove weld in which the mating members of the joint have single-bevel or double-bevel edge shapes. The groove faces of the joint are opposed to each other (face the opposite way or opposite hand) (see Figure 4.15(C1) and (C2)).

weave bead—A type of weld bead made with transverse oscillation (see Figure 4.32).

weld bead—A weld resulting from a pass (see Figure 4.31).

weld face—The exposed surface of a weld on the side from which welding was done (see Figure 4.21(A).

weld groove—the channel in the surface of a workpiece or an opening between two joint members that provides space to contain a weld.

weld interface—The interface between weld metal and base metal in a fusion weld, between base metals in a solid-state weld without filler metal, or between filler metal and base metal in a solid-state weld with filler metal (see Figure 4.23).

weld layer—A single level of weld within a multiple-pass weld. A weld layer may consist of a single bead or multiple beads (see Figure 4.31).

weld pass—A single progression of welding along a joint. The result of a pass is a weld bead or layer (see Figure 4.31).

weld reinforcement—Weld metal in excess of the quantity required to fill a joint (see Figure 4.21(A)).

weld root—The points, shown in cross section, at which the root surface intersects the base metal surfaces (see Figure 4.7(C)).

welding sequence—The order of making welds in a weldment (see Figures 4.33, 4.34, and 4.35).

weld throat—See actual throat, effective throat, and theoretical throat.

weld toe—The junction of the weld face and the base metal (see Figure 4.21(B)).

Module 5
Documents Governing Welding Inspection and Qualification

Contents

Introduction	5-2
Drawings	5-2
Codes	5-6
Standards	5-7
Specifications	5-8
Control of Materials	5-14
Alloy Identification	5-17
Qualification of Procedures and Welders	5-20
Summary	5-39
Key Terms and Definitions	5-40

Module 5—Documents Governing Welding Inspection and Qualification

Introduction

The job of welding inspection requires that the inspector possess, or have access to, a great deal of information and guidance. Although welding inspection for different industries can be similar in many respects, each particular job can have requirements that make it unique. The simple statement, "The welds must be good," is not sufficient information for judging the weld quality. Many times inspectors are asked to evaluate other fabrication aspects besides weld quality. The condition of various materials used for welded structures will affect the overall quality. Welding inspectors cannot evaluate a welded structure without information from the designer or the welding engineer regarding weld quality. The inspector also needs to know when and how to evaluate the welding.

To satisfy this need, there are numerous documents available to the designer, welding engineer, and welding inspector that state what, when, where, and how the inspection is to be performed. Many of these documents also include acceptance criteria. They exist in various forms depending upon the specific application. Some of the documents the welding inspector may use include drawings, codes, standards, and specifications. Contract documents or purchase orders may also convey information such as which of the above documents will be used for that job. In the case where more than one of the above are specified, they are intended to be used in conjunction with each other. Job specifications may include supplemental requirements altering portions of the governing code or standard.

It is essential for the welding inspector to have an opportunity to study all applicable documents before the start of the job. This pre-welding effort provides the welding inspector with information about the upcoming inspection. Some of the information that can be gained from this document review includes the following:

- Part size and geometry
- Base and filler metals to be used
- Requirements for hold points
- Processing details
- Processes to be used
- Specification for nondestructive inspection
- Extent of inspection
- Acceptance/rejection criteria
- Qualification requirements for personnel
- Procedure and welder qualification
- Materials control requirements

Drawings

Drawings describe the part or structure in graphic detail. Drawing dimensions, tolerances, notes, weld and welding details, and accompanying documents should be reviewed by the inspector. This gives the welding inspector some idea of the part size and configuration. Drawings also help the inspector understand how a component is assembled. And, they can assist in the identification of problems that could arise during fabrication.

Dimensions provided on a blueprint have two basic functions:

- To provide the sizes needed to fabricate the parts.
- To indicate locations where the individual components of each part should be placed.

Dimensions are shown in several different ways on drawings. Shape and size of an object determine which method is chosen for each dimension. The location is then indicated by the use of a leader and an arrow. Placement of the dimension itself is dependent on the amount of space available. In the workplace of today, you will often encounter drawing dimensioning expressed in inches and decimals of an inch. As the workforce moves toward the future and the world becomes a marketplace for the goods and services of all nations, increasingly the inspector will be called upon to work with dimensioning expressed in the metric system. Presently, many companies doing business internationally use dual dimensioning systems on their drawings. This permits the manufactured parts to be fabricated in either U.S. units (Customary) or SI units (International System of Units). One dual-dimensioning practice puts the U.S. customary measurement first, and the SI measurement in parenthesis directly behind or slightly below it:

> 1-1/2 in. (38.1 mm)
>
> or
>
> 1.50 in.
> (38.1 mm)

Some companies also place a chart in the upper left-hand corner of the drawing which shows the SI equivalents. This is done to help workers begin to "think metric."

Another important piece of information shown on drawings is tolerances. Tolerances are the total amount of variation permitted from the design size of a part. Tolerances (see Figure 5.1), may be expressed three ways:

> (1) as a variation between limits,
> (2) as the design size followed by the tolerance,
> (3) when only one value is given, the other value is assumed to be zero.

Tolerances are also applied to location dimensions for other features such as holes, slots, notches, surfaces, welds, etc. Generally tolerances should always be as large as practical, all other factors considered, to reduce manufacturing costs. Tolerances may be very specific and given with a particular dimension value. They may also be more general and given as a note or included in the title block of the drawing. General tolerances will apply to all dimensions in the blueprint unless otherwise noted.

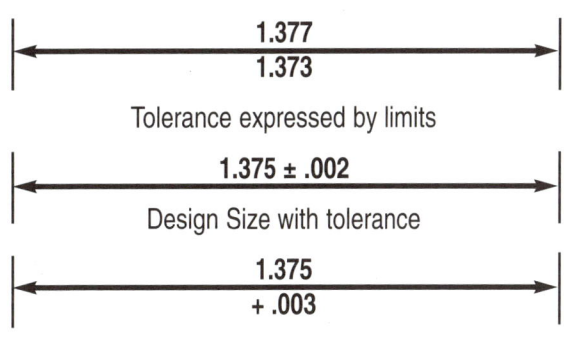

Figure 5.1—Examples of Tolerances

Tolerances give the CWI some latitude in terms of acceptance/rejection during size inspections of welds and weldments.

Drawing notes provide both instructions and information which are additions to the illustrations, as well as the information contained in the Title Block, or List of Materials (see Figure 5.2). Notes eliminate the need for repetition on the face of the drawing, such as the size of holes to be drilled, fasteners (hardware) used, machining operations, inspection requirements and so forth. When notes become too extensive, which is often the case in large, structural fabrication and architectural drawings, they are typed or printed on separate sheets and included along with the set of drawings or in the contractual documents. This is one reason why the inspector must review the contractual documents (sometimes referred to as a "specs package").

Notes can be classified as General, Local or Specifications depending on their application on the blueprint. General Notes apply to the entire drawing and are usually placed above or to the left of the title block in a horizontal position. These types of notes are not referenced in the list of materials and are not from specific areas of the drawings. If there are exceptions to general notes on the drawing field, the note will usually be followed by the phrase "*Except As Shown*," "*Except As Noted*," or "*Unless Otherwise Specified*." These exceptions are shown by a local note or data in the field of the drawing.

Local Notes or specific notes apply only to certain features or areas and are located near, and directed to, the feature or area by a "leader" (bold arrow with reference line). Local notes may also be referenced from the field of the drawing or the list

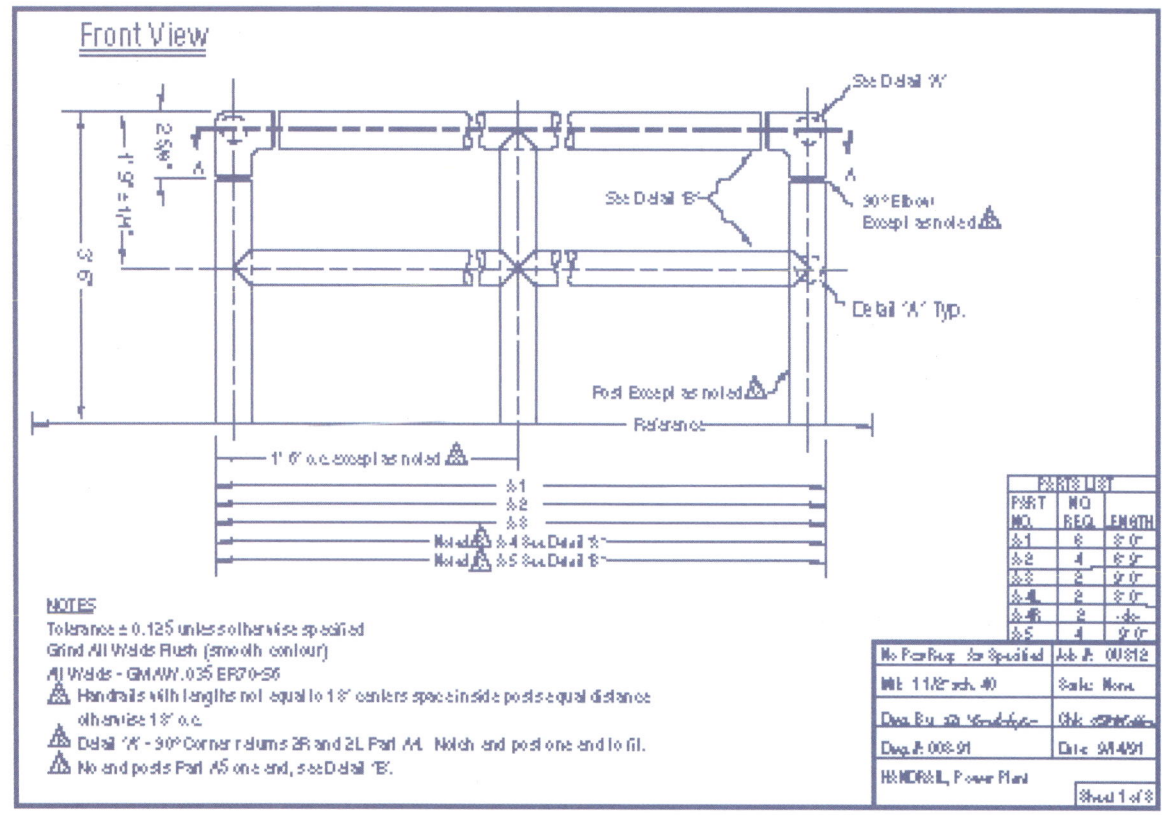

Figure 5.2—Example of Fabrication Drawing

of materials by the note number enclosed in an equilateral triangle (commonly referred to as a "flag").

Specifications presented as local notes will denote materials required, welding processes to be used, type and size of electrodes, and the kind and size of the welding rod. Specifications are located near a view when it refers specifically to that view. When the specifications are general and apply to all or several different features, they may be placed within a ruled space provided on the print for this purpose, included in the specifications package, or contractual documents.

The term "specification" will often be included in front of the information package to clarify that it is a specification for that particular drawing or job assignment. In the case of materials, however, the word specification is not necessarily used, but is implied. Further discussion will be presented about specifications later in this module.

Welding details shown on drawings or other documents include locations, lengths and sizes of welds, joint configurations, material call-outs, specification of nondestructive examination, and special processing requirements. Some materials require special techniques such as preheating. The welding inspector should be aware of this before the start of any welding.

Some of the applicable documents might also dictate "hold points" during the fabrication process. Hold points are specific, prearranged steps in the fabrication process to permit interim inspections. The inspector must be present to make an inspection or perform some specific operation during these steps. Only after the completed work has been inspected and approved can the fabrication continue. The welding inspector may be required to choose when those hold points are to occur or they may be outlined in the job specifications.

Processing details should also be noted in the information package. Such things as the specification of a certain amount of camber in a girder, or the use of a paint that requires special weld finishing are examples of features that necessitate additional attention by the inspector. The welding inspector should be aware of such details so that he or she can monitor the operation and review the results. The specification of the welding process for a particular job is an example of a processing detail that the inspector should be aware of. If the inspector knows the welding process and material to be used, suggestions can be made about what problems may occur and what methods could be applied to prevent their occurrence.

Before welding, the inspector should review welding procedures to ensure that all combinations of material, thickness, process, and position are adequately covered. These procedures will also indicate what important aspects of the welding operation should be monitored to aid in the achievement of a satisfactory weld.

Another example of a processing detail that might be included in the job specifications is the requirement for nondestructive examination of a finished weld. Nondestructive examination specifications must be accompanied by additional information, including method to be used, test procedure, location and extent of testing, and applicable acceptance/rejection criteria.

The specification documents should provide the inspector with a detailed description of the necessary visual inspection requirements as well. They should state the extent of visual inspection, indicating whether that inspection is to be continual or on a spot check basis. Accompanying this information should be statements regarding the quality requirements, including the specific acceptance or rejection criteria. The welding inspector cannot perform adequately if not provided with the appropriate acceptance/rejection criteria.

One final aspect of this information relates to the qualifications of the personnel who will perform the specified work. There may be specific requirements for qualified people in the areas of welding, visual inspection and nondestructive examination. The welding inspector may be actively involved in welder certification review or qualification testing. Inspectors should be aware of the requirements for these certifications and qualifications. Some contracts require certain qualification levels for persons performing visual weld inspection and nondestructive examination. If such a requirement exists, documentation must show evidence of the proper levels of qualification for each individual performing the inspections.

From the discussion above, it is evident that there is a wealth of information provided in the various types of documents. The documents must be made available to the welding inspector in time to perform an effective inspection. Experienced inspectors can identify possible trouble spots and locate details that might make assembly difficult. If found early enough in the fabrication process, allowances can be made so potential problems can be avoided. This preliminary step of the fabrication process is too often taken lightly. Many costly mistakes can be avoided if this preliminary review is performed by experienced individuals.

Our discussion to this point has been limited to a general treatment of the types of information contained in various documents governing weld quality. At this point it is appropriate to describe each document type in detail. The welding inspector will probably be required to refer to them during the course of the job. Three basic types (codes, standards and specifications) will be reviewed. This does not imply that these are the only documents with which the welding inspector will be concerned. Each inspector is responsible for the review of all documents that are involved in the projects inspected. Further, the inspector must become familiar with various requirements and methods that are described therein.

If you will recall from "Module 1—Welding Inspection and Certification," one important welding inspector attribute is knowledge of drawings, codes and standards. This does not mean that the welding inspector must memorize the contents of these documents. However, inspectors should be sufficiently familiar with a document to locate appropriate information promptly. All documents should be available for ready reference as questions arise. The welding inspector must be familiar with all specific documents relating to a particular job. Basic understanding of other documents and their areas of concern are also beneficial. This could prove to be helpful in explaining certain conditions. So, mention will be made of several of these standards, codes and specifications which could be consulted for answers to questions in various general areas. The following discussion deals specifically with three general document categories: codes, standards and specifications. A number of organizations are responsible for the production and revision of various documents. They include, but, are not limited to:

- American Welding Society (AWS)
- American Society of Mechanical Engineers (ASME)
- American National Standards Institute (ANSI)
- American Petroleum Institute (API)
- American Bureau of Shipping (ABS)
- Department of Transportation (DOT)
- Military Branches (Army, Navy, etc.)
- Other Government Agencies

Codes

The first category of document to be discussed is a code. By definition, a *code* is "a body of laws, as of a nation, city, etc., arranged systematically for easy reference." When a structure is built within the jurisdiction of a city or state, it must often comply with certain "building codes." Since a code consists of laws having legal status, it will always be considered mandatory. Therefore, we will see text containing words such as "shall" and "will." A specific code includes certain conditions and requirements for the item in question. Quite often it will also include descriptions of methods to determine if those conditions and requirements have been achieved.

The welding inspector will often inspect work according to some code. Several organizations including AWS and ASME have developed codes for various areas of concern. AWS has published six codes, each of which covers different types of industrial welding applications:

AWS CODES

- AWS D1.1 *Structural Welding Code—Steel*
- AWS D1.2 *Structural Welding Code—Aluminum*
- AWS D1.3 *Structural Welding Code—Sheet Steel*
- AWS D1.4 *Structural Welding Code—Reinforcing Steel*
- AWS D1.5 *Bridge Welding Code*
- AWS D9.1 *Sheet Metal Welding Code*

So, depending on the type of welding being performed, one or more of the above codes would be selected to detail the weld quality requirements.

ASME has also developed several codes that apply to pressure-containing piping and vessels. Two of these, ASME B31.1, *Power Piping,* and B31.3, *Process Piping,* detail those requirements for two types of pressure piping. While they carry an ANSI designation, they were developed by ASME. ASME has also developed a series of codes applicable to the design and construction of pressure vessels. Due to the variety of applications of these vessels, the ASME codes exist as a set of eleven separate sections. The eleven sections are:

ASME CODE SECTIONS

- Section I   Rules for Construction of Power Boilers
- Section II   Materials
- Section III   Subsection NCA—General Requirements for Division 1 and Division 2
- Section IV   Rules for Construction of Heating Boilers
- Section V   Nondestructive Examination
- Section VI   Recommended Rules for the Care and Operation of Heating Boilers
- Section VII   Recommended Guidelines for the Care of Power Boilers,
- Section VIII   Rules for Construction of Pressure Vessels, Divisions 1, 2, and 3
- Section IX   Welding and Brazing Qualifications
- Section X   Fiber-Reinforced Plastic Pressure Vessels
- Section XI   Rules for Inservice Inspection of Nuclear Power Plant Components

In addition to the eleven ASME Code Sections listed, some sections have further subdivisions. Welding inspectors who inspect according to ASME criteria may be required to refer to several individual sections of the code. For example, in the fabrication sequence of a carbon steel, unfired pressure vessel, the sections used might include:

- Section II   Part A—Ferrous Material Specifications
- Section II   Part B—Nonferrous Material Specifications
- Section II   Part C—Specifications for Welding Rods, Electrodes, and Filler Metals
- Section II   Part D—Properties
- Section V   Nondestructive Examination
- Section VIII   Rules for Construction of Pressure Vessels, Divisions 1, 2, or 3
- Section IX   Welding and Brazing Qualifications

With so many different sections involved, it is imperative that the welding inspector understand where each specific type of information can be found. It should be noted that Section II, Part C, is essentially identical to the AWS Filler Metal Specifications; ASME adopted the AWS specifications almost in their entirety. If the inspector specializes in a certain area, then only the section covering the topic concerned needs to be reviewed.

Standards

The next document type to be covered will be the standard. The dictionary describes a *standard* as, "something established for use as a rule or basis of comparison in measuring or judging capacity, quantity, content, extent, value, quality, etc." A standard is treated as a separate document classification; however, the term standard also applies to numerous types of documents, including codes and specifications. Other types of documents considered to be standards are procedures, recommended practices, groups of graphic symbols, classifications, and definitions of terms.

Some standards are considered to be mandatory. This means the information is an absolute requirement. A mandatory standard is precise, clearly defined and suitable for adoption as part of a law or regulation. Therefore, the welding inspector must make judgments based on the content of these standards. These mandatory standards use such words as "shall" and "will" because their requirements are not a matter of choice. Codes are examples of mandatory standards because they have legal status.

There are numerous standards that provide important information, but are considered to be nonmandatory. An example of a nonmandatory standard would be a recommended practice. They are nonmandatory because they may provide other ways in which objectives can be accomplished. Nonmandatory standards include words such as "should" and "could" in place of "shall" and "will." The implication here is that the information has been put forth to serve as a guideline for the performance of a particular task. However, it doesn't mean that something is rejectable just because it fails to comply with that guideline.

Even though a standard may be considered nonmandatory, it still provides important information that should not be ignored by the inspector. Nonmandatory standards can provide a basis for mandatory document development. Such is the case for ASNT's *Recommended Practice No. SNT-TC-1A.* ASNT prepared this document, "to establish guidelines for the qualification and certification of NDT personnel" (see Figure 5.3).

National standards are the result of elaborate voting and review procedures. They are developed according to rules established by the American National Standards Institute (ANSI). Standards produced by various technical organizations such as AWS and the American Society of Mechanical Engineers (ASME) are reviewed by ANSI. When adopted, they carry the identification of both organizations. Examples include: ANSI/ASME *Boiler & Pressure Vessel Code*, Section IX, and AWS D1.1, *Structural Welding Code—Steel* (see Figures 5.4 and 5.5).

Another common standard used by certain welding inspectors is the American Petroleum Institute's API 1104, *Standard for Welding of Pipelines and Related Facilities.* As the name implies, this standard applies to the welding of cross-country pipelines and other equipment used in the trans-

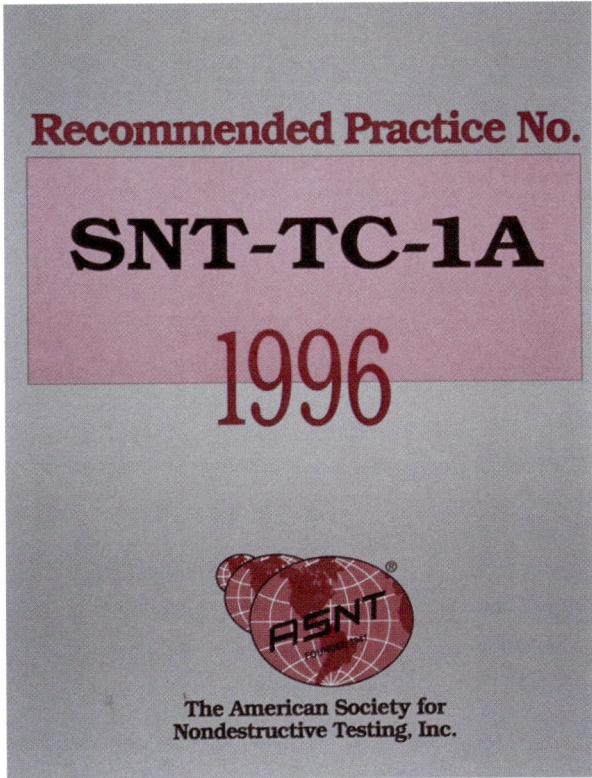

Figure 5.3—ASNT SNT-TC-1A

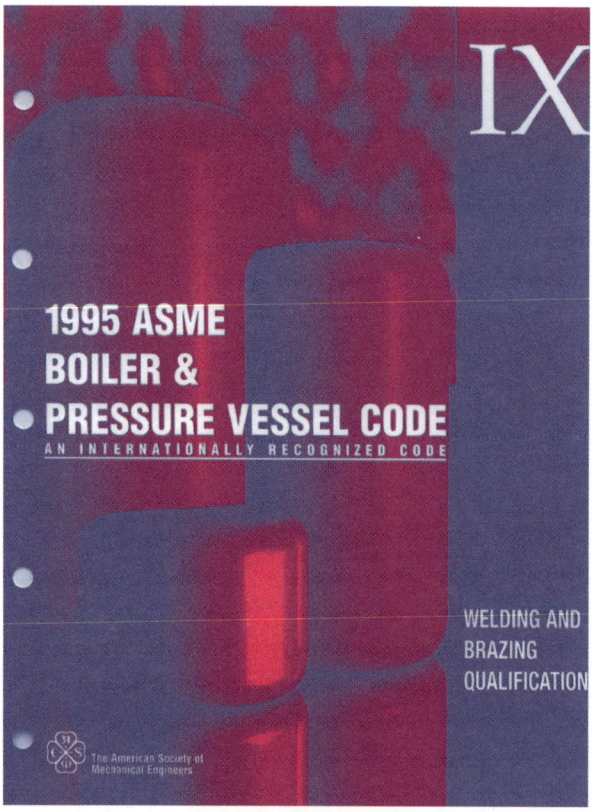

Figure 5.4—ASME *Boiler & Pressure Vessel Code*, Section IX

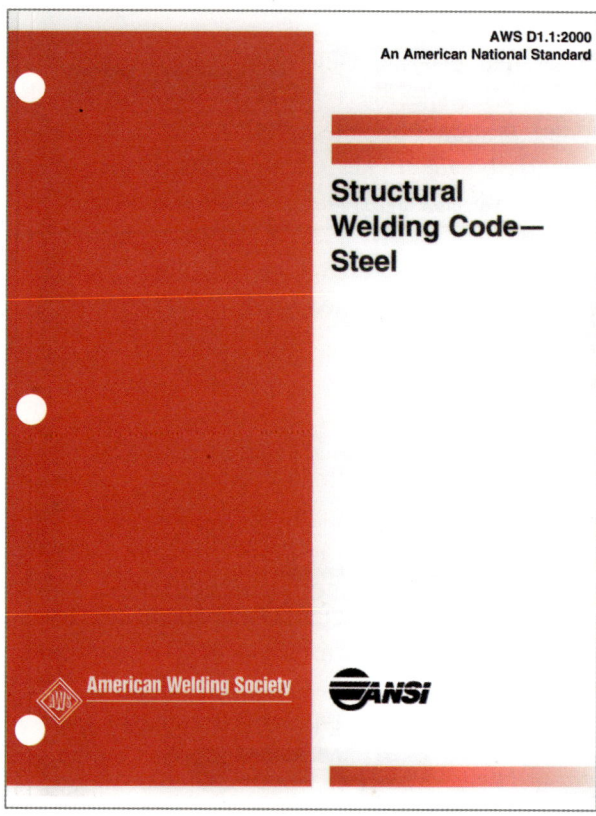

Figure 5.5—AWS D1.1

portation and storage of petroleum products. This standard covers the requirements for qualification of welding procedures, welders and welding operators. It applies to gas and arc welding of butt and T-joints in pipe used in the compression, pumping, and transmission of crude petroleum, petroleum products, and fuel gases. API 1104 also includes requirements for the visual and radiographic inspection of these welds (see Figure 5.6).

The American Society for Testing and Materials (ASTM) produces many volumes of specifications covering numerous materials and test methods. These standards include both metal and nonmetal products for many industries. As their name implies, they are also involved in the details of methods for evaluating these materials. These specifications are widely recognized by both buyers and suppliers. The result is a better understanding of the requirements for particular materials and test methods. When a specific material or test is required, it is easier to communicate the necessary information if the specification exists and is readily available (see Figure 5.7).

Specifications

The final document classification to be discussed is the specification. This type is described as, "a detailed description of the parts of a whole; statement or enumeration of particulars, as to actual or required size, quality, performance, terms, etc." A specification is a detailed description or listing of required attributes of some item or operation. Not only are those requirements listed, but there may also be some description of how they will be measured.

Depending upon a specific need, specifications can exist in different forms. Companies often develop in-house specifications describing the necessary attributes of a material or a process used in their manufacturing operation. The specification may be used entirely within the confines of that company, or it may be sent to suppliers to detail exactly what the company wants to purchase. When these requirements are put into writing, there is more assurance that the item or service that is supplied

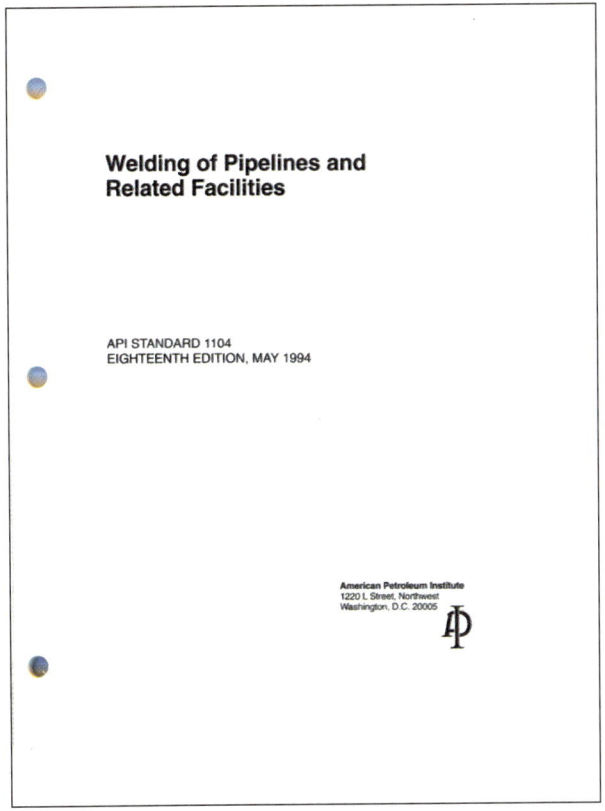

Figure 5.6—API Standard 1104

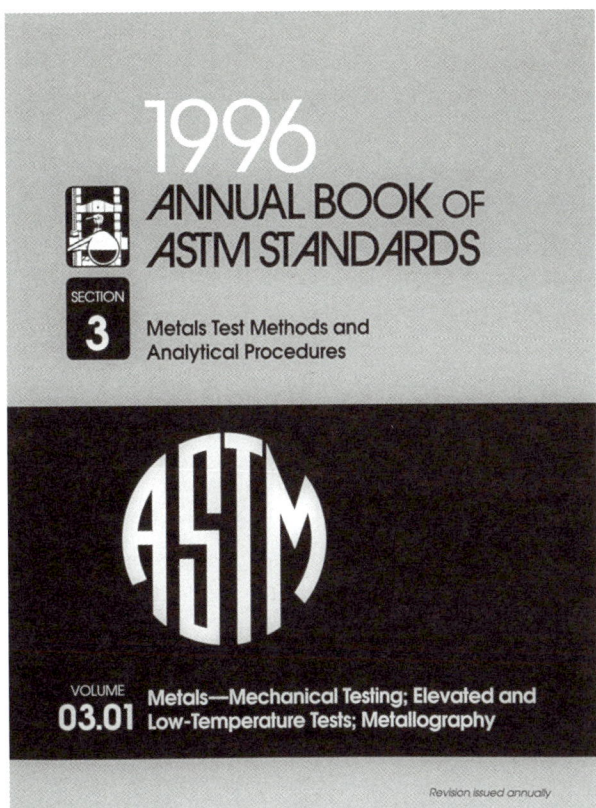

Figure 5.7—ASTM Standard

will meet the customer's needs. Both the engineering and purchasing departments rely heavily on specifications to describe their requirements.

Besides in-house or company specifications, various organizations publish specifications and standards that are available industry-wide. A partial listing of these organizations is shown in Table 5.1.

The interest of many of these groups overlap with regard to welding, and some agreements have been reached to reduce duplication of effort. Specifications that apply to a particular product are usually prepared by the group that has overall responsibility. Each organization that prepares consensus standards or specifications has volunteer committees or task groups to perform this function. Members of these committees or task groups are specialists in their fields. They prepare drafts of specifications or standards that are reviewed and approved by a larger group. Each main committee is selected to include persons with diverse interests including producers, users, and government representatives. To avoid control or undue influence by one interest group, consensus must be achieved by a high percentage of all members.

The federal government develops or adopts specifications and standards for items and services that are in the public domain rather than the private. Specifications or standards writing committees usually exist within a federal department or agency that has responsibility for a particular item or service.

Other organizations that have developed specifications for their particular industries are API and AWS. API specifications govern the requirements for materials and equipment used by the petroleum industry. AWS has developed a number of specifications that describe requirements for welding filler metals and specialized types of fabrication. The A5.XX series of specifications, AWS A5.1 through A5.33, cover the requirements for various types of welding consumables and electrodes.

For example, A5.1 details those requirements for carbon steel covered electrodes for shielded metal arc welding. Information provided includes

Table 5.1—Standards Producing Organizations

American Welding Society (AWS)[1]
550 N.W. LeJeune Rd.
Miami, FL 33126
tel. (305) 443-9353
fax (305) 443-7559

Aluminum Association (AA)[1]
900 19th Street, N.W.
Suite 300
Washington, DC 20006
tel. (202) 862-5100
fax (202) 862-5164

American Petroleum Institute (API)
1220 L Street, N.W.
Washington, DC 20005-8029
tel. (202) 682-8000
fax (202) 682-8115

American Association of State Highway and Transportation Officials (AASHTO)
444 N. Capital Street, N.W.
Suite 249
Washington, DC 20001
tel. (202) 624-5800
fax (202) 624-5806

American Bureau of Shipping (ABS)
Two World Trade Center
106th Floor
New York, NY 10048
tel. (212) 839-5000

Abrasives Engineering Society (AES)
108 Elliott Drive
Butler, PA 16001-1118
tel. (412) 282-6210
fax (412) 282-6210

American Gas Association (AGA)
1515 Wilson Boulevard
Arlington, VA 22209
tel. (703) 841-8400
fax (703) 841-8406

American Institute of Mining, Metallurgical, and Petroleum Engineers (AIME)
345 East 47th Street
14th Floor
New York, NY 10017
tel. (212) 705-7676
fax (212) 371-9622

Association for Facilities Engineering
8180 Corporate Park Drive
Suite 305
Cincinnati, OH 45242
tel. (513) 489-2473
fax (513) 247-7422

American Institute of Steel Construction (AISC)
One E. Wacker Drive
Suite 3100
Chicago, IL 60601-2001
tel. (312) 670-2400
fax (312) 670-5403

American Iron and Steel Institute (AISI)[1]
1101 17th Street, N.W.
Washington, DC 20036-4700
tel. (202) 452-7100
fax (202) 463-6573

American Nuclear Society (ANS)
555 North Kensington Avenue
La Grange Park, IL 60526
tel. (708) 352-6611
fax (708) 352-0499

American National Standards Institute (ANSI)
11 West 42nd Street
13th Floor
New York, NY 10036-8002
tel. (212) 642-4900
fax (202) 398-0023

American Railway Engineering Association (AREA)
50 F Street, N.W.
Suite 7702
Washington, DC 20001-2183
tel. (202) 639-2190
fax (202) 639-2183

American Society of Civil Engineers (ASCE)
345 East 47th Street
New York, NY 10017
tel. (212) 705-7496
fax (212) 355-0608

American Society of Mechanical Engineers (ASME)
345 East 47th Street
New York, NY 10017-2392
tel. (212) 705-7722
fax (212) 705-7674

Table 5.1 (Continued)—Standards Producing Organizations

American Society for Nondestructive Testing (ASNT)
1711 Arlingate Lane
P.O. Box 28518
Columbus, OH 43228-0518
tel. (614) 274-6003
fax (614) 274-6899

American Society for Quality Control (ASQC)
P.O. Box 3005
611 East Wisconsin Avenue
Milwaukee, WI 53201-3005
tel. (800) 248-1946
fax (414) 272-1734

American Society of Safety Engineers (ASSE)
1800 East Oakton
Des Plaines, IL 60018-2187
tel. (847) 699-2929
fax (847) 296-3769

American Society for Testing and Materials (ASTM)[1]
1916 Race Street
Philadelphia, PA 19103-1187
tel. (610) 832-9500
fax (610) 832-9555

American Welding Institute (AWI)
10628 Dutchtown Road
Knoxville, TN 37932
tel. (423) 675-2150
fax (423) 675-6081

American Water Works Association (AWWA)
6666 W. Quincy Avenue
Denver, CO 80235
tel. (303) 794-7711
fax (303) 795-1440

ASM International (ASM)
9539 Kimsmin Road
Materials Park, OH 44073
tel. (216) 338-5151

Association of American Railroads (AAR)
50 F Street, N.W.
Washington, DC 20001-1564
tel. (202) 639-2100
fax (202) 639-5546

Association of Iron and Steel Engineers (AISE)
Three Gateway Center
Suite 1900
Pittsburgh, PA 15222
tel. (412) 281-6323
fax (412) 281-4657

Canadian Standards Association (CSA)
178 Rexdale Boulevard
Rexdale, Ontario M9W 1R3
Canada
tel. (416) 744-4000

Canadian Welding Bureau (CWB)
254 Merton Street
Toronto, Ontario M4S 1A9
Canada
tel. (905) 487-5415

Canadian Welding Development Institute (CWDI)
391 Burnhamthorpe Road
East Oakville, Ontario L6J 4Z2
Canada
tel. (905) 845-9881

Copper Development Association[1]
260 Madison Avenue
New York, NY 10016-2401
tel. (212) 251-7200 or 1-800-232-3282
fax (212) 251-7234

Compressed Gas Association (CGA)
1725 Jefferson Davis Highway
Suite 1004
Arlington, VA 22202-4104
tel. (703) 412-0900
fax (703) 412-0128

Cryogenic Society of America (CSA)
c/o Huget Advertising
1033 South Boulevard
Oak Park, IL 60302
tel. (708) 383-6220
fax (708) 383-9337

Edison Welding Institute (EWI)
1250 Arthur E. Adams Drive
Columbus, OH 43221
tel. (614) 688-5000
fax (614) 688-5001

Table 5.1 (Continued)—Standards Producing Organizations

Fabricators & Manufacturers' Association International (FMA)
833 Featherstone Road
Rockford, IL 61107-6302
tel. (815) 399-8700
fax (815) 399-7279

Grinding Wheel Institute (GWI)
30200 Detroit Road
Cleveland, OH 44115-1967
tel. (216) 899-0010
fax (216) 892-1404

Industrial Accident Prevention Association (IAPA)
100 Front Street West
Royal York Hotel Arcade
Toronto, Ontario M5J1R3
Canada
tel. (416) 366-3711

Industrial Safety Equipment Association (ISE)
1901 N. Moore Street
Suite 808
Arlington, VA 22209
tel. (703) 525-1695
fax (703) 528-2148

International Institute of Welding (IIW)
550 N.W. LeJeune Road
Miami, FL 33126
tel. (305) 443-9353
fax (305) 443-7559

International Organization for Standardization (ISO)
(See American National Standards Institute)

International Oxygen Manufacturers' Association (IOMA)
P.O. Box 16248
Cleveland, OH 44116-0248
tel. (216) 228-2166
fax (216) 228-5810

International Titanium Association
1871 Folsom Street
Suite 200
Boulder, CO 80302-5714
tel. (303) 443-7515
fax (303) 443-4406

Material Handling Industry Association
8720 Red Oak Boulevard
Suite 201
Charlotte, NC 28217-3957
tel. (704) 522-8644
fax (704) 522-7826

National Association of Corrosion Engineers (NACE)
Box 218340
Houston, TX 77218-8340
tel. (713) 492-0535
fax (713) 492-8254

National Board of Boiler and Pressure Vessel Inspectors (NBBPVI)
1055 Crupper Avenue
Columbus, OH 43229
tel. (614) 888-8320
fax (614) 888-0750

National Electrical Manufacturers' Association (NEMA)
2101 L. Street, N.W.
Washington, DC 20037
tel. (202) 457-8400
fax (202) 457-8411

National Fire Protection Association (NFPA)
P.O. Box 91011
Batterymarch Park
Quincy, MA 02269-9101
tel. (617) 770-3000
fax (617) 770-0700

Naval Publication and Forms Center[2]
700 Andrews Robbins Avenue
Bldg. 4, Sect. D
Philadelphia, PA 19111
(215) 697-2179

Steel Tank Institute (STI)
570 Oakwood Road
Lake Zurich, IL 60047-1559
tel. (847) 438-8265
fax (847) 438-8766

Superintendent of Documents[3]
U.S. Government Printing Office
Washington, DC 20402
tel. (202) 783-3238

Table 5.1 (Continued)—Standards Producing Organizations

Ultrasonic Industry Association (UIA)
3738 Hilliard Cemetery Road
P.O. Box 628
Hilliard, OH 43026
tel. (614) 771-1972
fax (614) 771-1984

Underwriters Laboratories, Inc. (UL)
333 Pfingsten Road
Northbrook, IL 60062
tel. (312) 272-8800

Uniform Boiler and Pressure Vessel Laws Society (UBPVLS)
308 N. Evergreen Road
Suite 240
Louisville, KY 40243-1010
tel. (502) 244-6029
fax (502) 244-6030

Welding Research Council (WRC)
3 Park Avenue
27th Floor
New York, NY 10016
tel. (212) 591-7956
fax (212) 371-9622

Welded Steel Tube Institute (WSTI)
522 Westgate Tower
Cleveland, OH 44116
tel. (216) 333-4550

Note: For a detailed explanation of the agencies listed herein refer to AWS Welding Handbook, Vol. 1, 8th Ed., Chapter 13, "Codes and Standards."

[1] Number Assigner for "UNS"
[2] Military Specifications
[3] Federal Specifications

---

electrode classifications, chemical and mechanical properties of the weld deposit, required testing, details of tests, dimensional requirements, and packaging information. The AWS Specification A5.01, *Filler Metal Procurement Guidelines,* outlines procedures for ordering filler metals.

Another series of specifications has been developed by AWS to describe various fabrication requirements for individual types of apparatus. These are denoted by the numbers D14.1 through D14.6. Included in this group of documents are:

While each of the above refers to the general requirements of AWS D1.1, there are details provided which meet the specific needs of that particular structure or component.

- D14.1   Specification for Welding Industrial and Mill Cranes
- D14.2   Specification for Metal Cutting Machine Tool Weldments
- D14.3   Specification for Welding Earthmoving and Construction Equipment
- D14.4   Specification for Welded Joints in Machinery and Equipment
- D14.5   Specification for Welding Presses and Press Components
- D14.6   Specification for Rotating Elements of Equipment

The American National Standards Institute (ANSI) is a private organization responsible for coordinating national standards for use within the United States. ANSI does not actually prepare standards. Instead, it forms national interest review groups to determine whether proposed standards are in the public's interest. Each group is composed of persons from various organizations concerned with the scope and provisions of a particular document. If the consensus is reached for the general value of a particular standard, then it may be adopted as an American National Standard. However, adoption of a standard by ANSI does not, of itself, give it mandatory status.

Other industrial countries also develop and issue standards on the subject of welding. There is an International Organization for Standardization (ISO). Its goal is the establishment of uniform standards for use in international trade and exchange of services. ISO is made up of the standards-writing bodies of more than 80 countries and has adopted or developed over 4,000 standards. ANSI is the designated U.S. representative to ISO. ISO standards and publications are available from ANSI.

The American Welding Society (AWS) publishes numerous documents covering the use and quality control of welding. These documents include codes, specifications, recommended practices, classifications, methods and guides. AWS publications cover the following subject areas: Definitions and symbols; filler metals; qualification and testing; welding processes; welding applications; and safety (see Figure 5.8).

Control of Materials

In many industries, an important aspect of fabrication is the identification and traceability of materials. This certainly holds true in pressure vessel and nuclear work. Some inspectors may be required to assist in this material control program as part of their regular duties. If this is the case, the individual must be capable of properly identifying material and comparing that information with related documentation.

Materials for welded fabrication are often ordered with the stipulation that they meet a particular standard or specification. To demonstrate this compliance, the supplier can furnish documentation that describes the important characteristics of the material. This documentation for metals is sometimes referred to as an "MTR," which is the abbreviation for Material (or Mill) Test Report, or "MTC," which is the abbreviation for Material (or Mill) Test Certificate. These documents are usually notarized statements from the manufacturer tabulating the chemical and physical properties for the material. The attributes are usually listed as either "typical" or "actual," or both. Typical properties are simply those limits described by the particular specification. Actual attributes are material properties that have been physically measured and specifically listed on the MTR. Both indicate that the material complies with some specification. The actual limits describe the measured chemical and mechanical properties of tests representing that particular plate, pipe, bar, shape, welding filler metal, etc. See Figure 5.9.

When material ordered to some specification arrives at the fabrication site, the inspector may be responsible for reviewing the accompanying MTRs. This review can aid in determining whether or not the material meets all the applicable requirements of that specification. Normally, the material will be physically identified as to its type, grade, heat number, etc. This identification may be painted, stenciled, or otherwise noted in some conspicuous location on the material's surface. The inspector should compare that identification with the information contained on the MTR to ensure that the proper documentation has been provided and that the material is actually that which was ordered.

For a material control program to be successful, there must be some system whereby the received material can then be traced through the various fabrication steps. The goal is to be able to trace each piece of material used in some fabricated component all the way back to the MTR, and therefore, its manufacturer. While this is not an absolute requirement for many types of fabrication, there are indus-

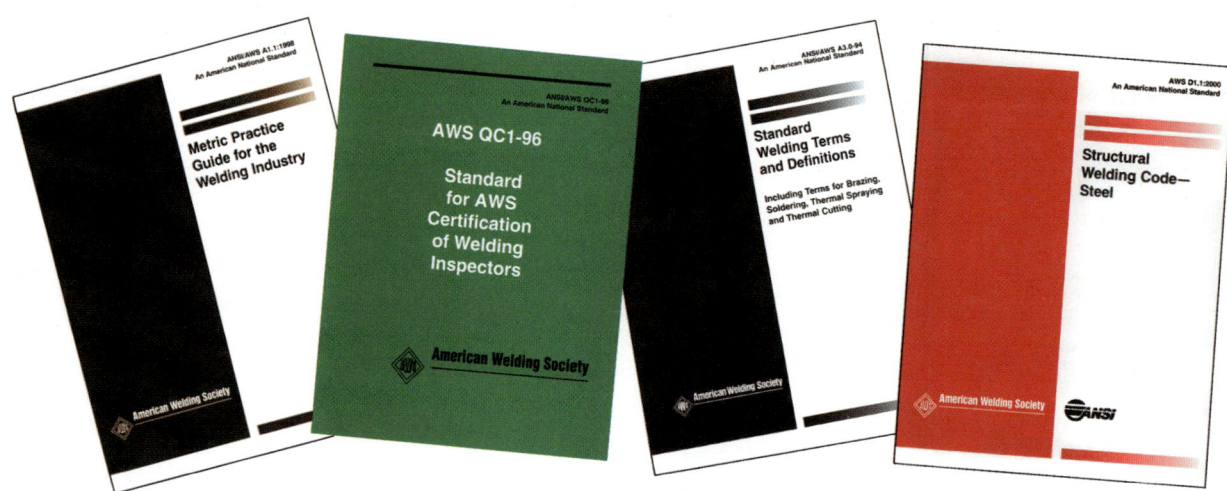

Figure 5.8—AWS Standards

```
                                        TRENT TUBE DIVISION
       Trent Tube                       2015 ENERGY DRIVE
                                        P.O. BOX 77
       A Division of Crucible Materials Corp
                        TEST REPORT     EAST TROY, WI 53120-0077

CUSTOMER ORDER NO.  CFO DATE RATING  ENTRY DATE  SHIP  MARKS           MILL ORDER NO.
T/CSM 331903                         110196                  A         8T7-48144-6        W
      TRIKENA                        TEKCOUR                 P

TRI-CLOVER INC.                      TEK SUPPLY
9201 WILMOT RD                       385 S PIERCE AVE
                                     COLORADO TECH CENTER
KENOSHA,              WI 53142-762   LOUISVILLE                   CO 80027-301
SPECIAL INSTRUCTIONS                    9F      4611          SHIPPING NOTICE
Area P
                                                F.O.B. SHP
                                                TERMS 30-15-10    FEET       300.000
  2000  065  E  200000  IV  321V SS  1.356UMT SOURCE              NET        407
  2.000 X  .065    20FT              CUT       300    FT          PCS         15
FULL FINISHED BRIGHT ANNEALED                         LBS QUANTITY GROSS      442
TRENTWELD STAINLESS SANITARY TUBING                   PCS ORDERED  Boxes       1
A270 90 EX OD TOL A269 94&SA249ID/OD POL 11/01/96     WANTED  Complete Shipment
TP304                                    11/04/96     PROMISED   B/L #E74678
INV: 030081-19                                                  DATE    04-NOV-96
CUST PO 2000628                                                 CF         PPD
Source: P7014706

HEAT # 871248
                         REPORT OF TESTS
                         ================
                         CHEMICAL ANALYSIS
HEAT #    C     MN    P     S     SI    NI    CR    MO    CU    N     CO
871248   .025  1.94 .0280 .0120  .38  8.24 18.59  .29         .090   .30

                      Mechanical and NDE Tests
                      ========================
HEAT #   FLG FLT FLR RFLT RBND ROTB FBD BND ECT  UT  RT LPT  HYDRO        AIR
871248   OK  OK      OK   OK            OK

                                                      RMS       C O R R O S I O N
HEAT #  LOT #   TENSILE  YIELD  ELON  HARDNS.  ID    OD GRAIN   TEST      VALUE
871248          98,282   51,335  49 X B 88/89

                              REMARKS
                              =======
DUAL CERTIFY TYPE 304/304L
20 RA MAX. ID
Strip/Plate Supplier - 871248        ALLEGHENY


     Mercury or mercury compounds are not used in the manufacture of tubing/pipe.

          We Hereby Certify This Report To Be True and Correct
          According To The Records in Possession of this Corporation.
               TRENT TUBE DIVISION - CRUCIBLE MATERIALS CORPORATION
Signed: _____Ronald Leach_____        Dated: 11/4/96

                              TEST REPORT
```

Figure 5.9—Example of a Mill Test Report

tries, such as the nuclear industry or aerospace industry, which are extremely concerned about this aspect of fabrication. Company policy will dictate what part, if any, the welding inspector will play in a material control system. The inspector may actually review documents for compliance or simply check to ensure that someone else has already performed this task. The inspector may be involved with the total material control system or just a particular aspect, such as the identification of materials for procedure qualification.

A successful system for material control has several important attributes. First, it should be as simple as possible. If a system is too complex, it may not be followed, resulting in loss of control. Simple systems that are understood by everyone have the best chance of providing satisfactory results. Another important feature is that they contain adequate checks and balances to ensure that the system will not break down and result in the loss of traceability.

There are several effective ways to maintain the necessary traceability of materials. Depending on the degree of control required, and the number of different types of material expected, a company can develop a system which meets their particular needs. If only two or three different types of material will be encountered, a simple system of segregation, or separation, might be sufficient. This method simply requires that individual types of material are stored separately. This separation could be achieved by using specially marked racks or by using different types of materials in separate areas of the fabrication facility.

Another effective way of maintaining control is accomplished with a color coding system. Individual types or grades of material are assigned a particular color marking with this approach. Upon material receipt, someone is responsible for marking each piece with the proper color. Color coding aids material identification during later fabrication steps. A note of caution with color coding: the color "fastness," or longevity, must be considered since many colored marking materials may change color when exposed to sunlight or weather conditions. A color change due to sunlight exposure can lead to serious errors in materials control. Another concern is that the marking materials must not be harmful to the materials; examples of this are high chloride marking materials causing damage to austenitic stainless steels, or high sulfur contents harming high nickel alloys.

Another method of material control is the use of an alphanumeric code. It is certainly possible to maintain a material's traceability by transferring its entire identification information to the piece. However, this information can be quite extensive and require a considerable amount of time and effort. The use of alphanumeric codes can eliminate the need to transfer all the information such as type, grade, size, heat number, lot number, etc., on each piece.

A short, specific alphanumeric code can be assigned to a specific group of material to simplify the operation while still maintaining traceability. When material of a given type, grade, heat, etc., is received, it is assigned some code such as A1, A2, A3, …, D1, D2, etc. The material information is then listed on a log sheet and associated with its proper alphanumeric code. Once this relationship is established, the specific code is all that is needed to trace that material through the fabrication steps. Abbreviated codes are more likely to be properly transferred and maintained than a long, complex identification.

A final method to be discussed is the "bar code" system which can be automated and is very effective for both material control and inventory control. This system uses a group of short, vertical lines of varying widths as the marker on the material. These bar codes can be applied manually in the field, or automatically in the manufacturing system. Bar code readers are available which scan the bar codes and translate the line information to the actual type, grade, chemical composition, etc. These systems are very effective and are becoming the choice for material control for many industries.

With any of these marking systems, the identification should be apparent. It is a good practice to ensure that the marking is placed in several locations on each part if that part is large. As a minimum, the marking should be placed on diagonally opposite corners of plate and both ends of pipes, shapes and bars. If the piece of material is then cut in half, the marking remains on both pieces. If the pieces are further cut, the marking must be transferred to every piece, including the "drop" that is placed back in storage. Many piping manufacturers are imprinting the pipe size, piping specification and heat number along the length of piping every three to six feet for ease of identification.

As mentioned in the preceding discussion, the welding inspector will be involved with a material control system only when dictated by his or her job description. That involvement may be the review and marking, or a simple check to ensure that the marking is present on the materials to be welded.

Alloy Identification

Alloy identifications are usually developed by industry associations such as the Society of Automotive Engineers (SAE), American Iron and Steel Institute (AISI), and the Copper Development Association (CDA). Alloy identification systems were created to assist those working within a particular industry, and often with little regard to industries outside their sphere of influence. Thus, the alloy specifications developed by these different associations often overlapped or even used identical alloy designations for completely different alloys, leading to confusion or even mistakes in alloy usage.

The Unified Number System (UNS) was developed in 1974 to help interconnect many nationally used numbering systems that are currently supported by societies, trade associations, and individual users and producers of metals and alloys (see Figure 5.10). The UNS is a means to avoid confusion caused by the use of more than one identification number for the same material, or the same identification numbers appearing for two or more entirely different materials.

The standard practice initiated by the Unified Numbering System aids the efficient indexing, record keeping, data storage, retrieval and cross referencing of metals and alloys. The system is not, however, a specification regarding form, condition, quality, etc., of the materials covered. It is for basic identification purposes only.

The UNS was devised to assign alphanumeric designations for each family of metals and alloys, considered as having a "commercial standing," or "production usage." This means metals and alloys which have an active industrial use, or are produced regularly. The UNS establishes 18 series of primary

Figure 5.10—UNS Metals & Alloys

numbers for metals and alloys. Each number consists of a single letter prefix, followed by five digits. In most instances the letter is suggestive of the family of metals identified, (such as A for aluminum or S for stainless steel). Figure 5.11 identifies the primary series of numbers and the metal or alloy classification for each. This information is found in the SAE HS-1086/ASTM DS-56 E, *Metals & Alloys in the Unified Numbering System* (a joint publication by both organizations).

To illustrate how existing metal and alloy numbers can be cross referenced, the following examples are provided:

```
AISI—1020 = CARBON STEEL
UNS—G10200 = CARBON STEEL
CDA—C36000 = FREE CUTTING BRASS
UNS—C36000 = FREE CUTTING BRASS
```

Within each series of "primary UNS numbers," a "secondary division" is created to classify metals and alloys covered in the primary designation. As shown by Figure 5.12, welding filler metals have

"UNS"—PRIMARY SERIES OF NUMBERS	
A00001–A99999	aluminum and aluminum alloys
C00001–C99999	copper and copper alloys
E00001–E99999	rare earth and similar metals and alloys
F00001–F99999	cast irons
G00001–G99999	AISI and SAE carbon and alloy steels
H00001–H99999	AISI and SAE H-steels
J00001–J99999	cast steels (except tool steels)
K00001–K99999	miscellaneous steels and ferrous alloys
L00001–L99999	low melting metals and alloys
M00001–M99999	miscellaneous nonferrous metals and alloys
N00001–N99999	nickel and nickel alloys
P00001–P99999	precious metals and alloys
R00001–R99999	reactive and refractory metals and alloys
S00001–S99999	heat and corrosion resistant steels (including stainless), valve steels, and iron-base "superalloys"
T00001–T99999	tool steels, wrought and cast
W00001–W99999	welding filler metals
Z00001-Z99999	zinc and zinc alloys

Figure 5.11—Primary UNS Numbers

"UNS"—SECONDARY SERIES OF NUMBERS	
W00000–W09999	weld, filler—carbon steels
W10000–W19999	weld, filler—manganese-molybdenum alloys
W20000–W29999	weld, filler—Ni steels
W30000–W39999	weld, filler—austenitic stainless steels
W40000–W49999	weld, filler—ferritic stainless steels
W50000–W59999	weld, filler—chromium low alloy steels
W60000–W69999	weld, filler—copper alloys
W70000–W79999	weld, filler—surfacing alloys
W80000–W89999	weld, filler—Ni alloys

Figure 5.12—Secondary UNS Numbers

"UNS"—NUMBER ASSIGNING ORGANIZATIONS	
AA	Aluminum Association
ASTM	American Society for Testing and Materials
AWS	American Welding Society
CDA	Copper Development Association
SAE	Society of Automotive Engineers
ZI	Zinc Institute, Inc.

Figure 5.13—Specific Alloy Type Organizations

been divided into a secondary series of numbers within the primary UNS classification. The reader should note, however, that this list is for filler metals as defined by chemical composition and the list should not be confused with the AWS designation "E" for electrode in its classification of welding electrodes based on weld deposit.

The information provided within this section of the module is for illustrative purposes, to provide you with an understanding of material specification numbering systems. For a more detailed explanation, or to obtain additional listings included in the Unified Numbering System for Metals and Alloys, the reader can contact one of the organizations listed in Figure 5.13 (addresses and phone numbers of these organizations were listed earlier).

Typical Steel Specification

The welding inspector is sometimes required to compare actual material properties with the requirements of the specified material specification. ASTM has developed numerous material specifications; those referring to metals contain much the same types of information. To become familiar with what type of information is provided as well as how it is presented, a typical steel specification will be discussed.

For this example, the ASTM specification A514, *Standard Specification for High Yield Strength, Quenched and Tempered Alloy Steel Plate, Suitable for Welding,* will be used to illustrate some of the details which may be included in a typical steel specification.

Some of the important sections and features of this specification are described to acquaint the welding inspector with the basic outline format of these specifications.

Scope. This statement explains exactly what is to be described by the specification. That is, it defines the limits of the specification's coverage.

Applicable Documents. This is a listing of other documents which may be referred to within the text of the specification.

General Requirements for Delivery. Here, there is a statement regarding the required condition of the material if ordered to comply with this specification. Steel specifications will normally refer to ASTM A6 rather than including all of those requirements in each individual specification.

Process. The approved method(s) of producing this product are listed.

Heat Treatment. For alloys requiring heat treatment, the details of that treatment will be stated.

Chemical Requirements. This section refers you to a table which lists the actual chemical composition requirements. It is important to note that several grades will usually be listed, and each grade has a separate required chemical composition.

Tensile Requirements. This paragraph refers to a table which defines the required tensile values for the alloy. Required tensile values are usually different for various thickness ranges.

Brinell Hardness Requirements. For materials requiring Brinell hardness testing, the extent and requirements are stated.

Test Specimens. Any information relating to the location, preparation and treatment of test specimens is stated here.

Number of Tests. The number of test specimens required to show compliance is stated.

Retest. This paragraph describes what procedures will be followed if any of the test specimens fail.

Marking. A statement is made regarding how this material will be identified.

Supplemental Requirements. Any additional details which may be required by the purchaser are stated. These are not considered to be requirements unless so stated by the purchaser in the purchase order.

Typical Filler Metal Specification

The welding inspector may also be required to review welding filler metal properties to check for compliance with the applicable specification. One of these specifications, AWS A5.1, *Specification for Covered Carbon Steel Arc Welding Electrodes,* will serve as an example of the type of information provided as well as a description of the meaning of that information.

Some of the important features of this specification are described below.

Scope. This describes the coverage of the specification.

Section A—General Requirements

Classification. The basis for classification is stated. Reference is made to various tables which list these classifications, based on type of current, type of covering, welding position, chemical composition, and mechanical properties.

Acceptance. States that electrodes will be accepted if they comply with the requirements of AWS A5.01.

Certification. States that the manufacturer must certify that his product meets all of the requirements of this specification.

Retests. If any test fails, two retests must be conducted and each must pass.

Method of Manufacture. Any method of manufacture which produces a product in accordance with this specification is satisfactory.

Marking. States what minimum identification must be visible on the outside of each package.

Packaging. Describes suitable packaging, including standard sizes and configurations.

Rounding-off Procedures. Explains how tensile data will be rounded to the nearest 1,000 psi.

Section B—Required Tests and Test Methods. Describes the various chemical and mechanical tests which may be required to judge the acceptability of a filler metal with this specification. Tests include chemical composition, all-weld-metal tensile, impact, soundness, transverse tensile, longitudinal guided bend, and fillet weld tests.

Section C—Manufacture, Packaging, and Identification. Details the specification requirements for these features.

Section D—Details of Tests. Describes the actual details of performing the various tests used to measure the suitability of a filler metal to meet this specification. It also describes which of those tests are required for each classification.

Appendix. Contains additional descriptive information about certain requirements found in the main body of the specification. Includes information related to the actual care and use of electrodes complying with this specification.

Qualification of Procedures and Welders

Part of every major welding project, whether completed in the shop or field, is the qualification of welding procedures and welders, or welding operators. It is one of the most important preliminary steps in the fabrication sequence. Too often projects are begun without the benefit of proven welding procedures and personnel. This can result in excessive reject rates in production due to some unsuspected deficiency in the technique, materials or operator skill.

During the performance of this qualification testing, the welding inspector may become involved. Individual company structures will dictate the degree of involvement in this process. Some codes require that the welding inspector witness the actual qualification welding and testing. Consequently, the welding inspector should be aware of the various steps in the qualification of welding procedures and welding personnel.

Most codes place the burden of responsibility for qualification on the fabricator or contractor. Therefore, welding qualifications are statements by the company verifying that the welding procedures and personnel have been tested in accordance with the proper codes and specifications and found to be acceptable.

However, smart welding fabricators and contractors realize that the qualification of welding procedures and personnel will actually result in a cost-saving. When people and methods are tested and found to be acceptable, it is less likely that there will be excessive costs caused by rejected welds and job delays. It is much more economical to find out that there is some deficiency during a test than during actual production.

The welding inspector may also become involved with these qualifications from a document review standpoint. One of the responsibilities may be to review both welding procedure and welder qualification forms to determine if they are in accordance with the code and job specifications. Experienced welding inspectors realize that numerous problem spots can be detected and corrected prior to welding if this review is done carefully. Further, most codes give the welding inspector the authority to request that welders be requalified in the event they continue to produce substandard work.

During this discussion of qualification testing, the references to welding do not imply that only welds require these qualification measures. Brazing, for example, also requires qualified procedures and personnel for satisfactory results. The specific qualification testing techniques for brazing can be found in ASME Section IX, which describes the various steps involved in the qualification of welding procedures and personnel. Since each major code (e.g., AWS D1.1, ASME Section IX, and API 1104) handles welding procedures slightly differently, some of these differences will be noted. You are urged, however, to always refer to the appropriate code for specific information about this topic.

Procedure Qualification

The very first step in the qualification process is the development of the welding procedure, and its performance within the procedure qualification. This must precede both the welder qualification and the production welding because it will determine if the actual technique and materials are compatible. In general, the welding procedure qualification is performed to show the compatibility of:

(1) base metal(s),
(2) weld or braze filler metal(s),
(3) process(es), and
(4) techniques

You will note that there is no mention of the skill level of the welder who performs the qualification test. Although most codes will consider the welder who performs the welding to be automatically qualified, the procedure qualification is not meant to specifically judge the welder's ability. Even though each code handles the qualification of welding procedures slightly differently, the general intent is the same.

There are three general approaches to procedure qualification. These include prequalified procedures, actual procedure qualification testing, and mock-up tests for special applications. The mock-up tests may be used to supplement the other more standard methods of procedure qualification.

Let's first discuss the system used by the American Welding Society in the AWS D1.1, *Structural Welding Code—Steel*. This system is unique in the welding industry, because there are numerous procedures which are deemed prequalified. That is, there is no need to perform actual qualification testing as long as the welding parameters are within certain prescribed limits. The D1.1 Code lists various welding processes, base metals, thicknesses, joint configurations, and welding techniques which, when used in specific combinations, are considered prequalified.

AWS D1.1 recognizes four welding processes as being prequalified, including shielded metal arc (SMAW), submerged arc (SAW), flux cored arc (FCAW), and gas metal arc (GMAW) except short circuiting transfer. However, this does not mean that these are the only welding processes that can be used. It simply implies that actual qualification testing is required if other welding processes are used for production welding. There are also numerous base metals which are considered acceptable and do not require qualification when used.

The thickness of the base metal will also have an effect on the effectiveness of the welding procedure. Therefore, the various prequalified weld joints have limitations on the thickness ranges covered. AWS D1.1 is limited to the welding of steel 1/8 in. thick and greater. The specific thickness ranges for various processes, positions and joint configurations are tabulated for each prequalified weld joint. Again, just because a certain condition places a procedure outside of these limitations, it does not imply that the procedure cannot be used. It simply means that qualification tests must be conducted to deem it acceptable.

Figure 5.14 is an example of how AWS D1.1 lists the limitations of the various aspects of prequalified weld joints.

Looking at this sketch and tabular values, you can see that this particular prequalified weld joint is for a single-V-groove butt joint, welded from one side, with steel backing material at the root. The tabular data show different requirements for the exact weld joint configuration depending on the process, thickness and position of welding. Further, for a given process, the root opening may vary with relation to the groove angle. Considering the SMAW process, there are three different choices of root opening and groove angle combinations: 1/4 in. root opening—45° groove angle, 3/8 in. root opening—30° groove angle and 1/2 in. root opening—20° groove angle. It is also important to note that the tolerances for both the root opening and groove angle appear in a table inserted in the upper right hand corner. There are listings for the "As detailed" and "As fit up" tolerances on those measurements. The "As detailed" tolerances relate to the dimensional freedom of the designer when he specifies these features. The "As fit up" tolerances relate to the permissible variations from these detailed dimensions during the actual assembly of the parts to be joined. Therefore, the welding inspector would apply these "As fit up" tolerances when inspecting the actual fit up of this joint in production.

In the next column, there is a listing of the positions for which that joint is considered to be prequalified. Following this, there is a column which states whether or not gas shielding is required when FCAW is used. When referring to these prequalified weld joint figures, it is important to pay attention to the notes which are referenced in the last column of the table. These notes may place further restrictions on the use of that prequalified weld joint.

The final judgment as to whether a procedure is considered to be prequalified is made after reviewing the contents of Sections 3 and 4 of the Code, which refer to prequalification and qualification, respectively. Section 5 defines many of the acceptable quality requirements for the preparation and completion of welds. In Section 5, there is also

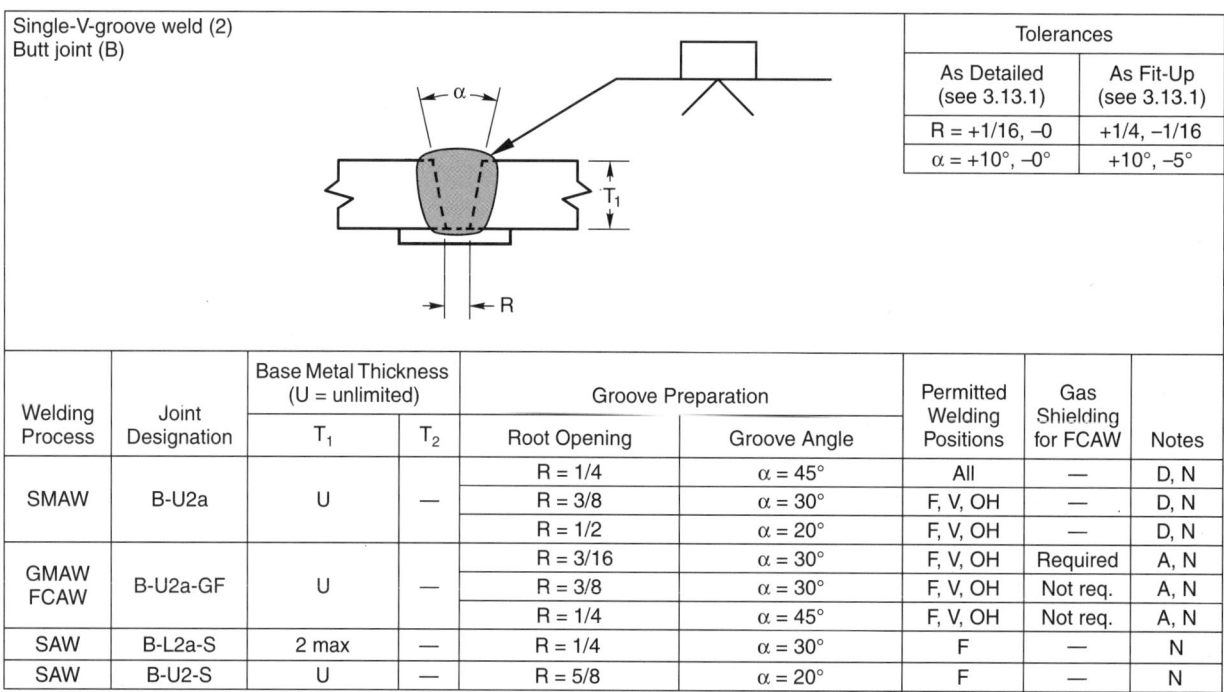

Figure 5.14—Example of AWS Prequalified Weld Joint Limitations

information about acceptable methods for the achievement of these quality levels. This section also details those specific requirements for the various welding processes, including their "essential variables."

Essential variables are those features of the welding process which, if changed beyond certain limits, require that a new welding procedure be established. That is, they are important enough that, if varied significantly, the resulting welds may be unsatisfactory. Essential variables are listed for each different welding process.

As previously mentioned, if any feature renders a welding procedure non-prequalified, it can be qualified by making a test weld and subjecting it to actual destructive tests. This is accomplished in much the same manner as qualification testing in accordance with ASME Section IX, which will be discussed next. One major difference, however, is that, for AWS, the procedure must be qualified for each position in which production welding will take place. The actual requirements for this testing appear in Section 4 of AWS D1.1. It details the test conditions, required test results, and the limitations of the various qualified procedures.

The next general method used for the qualification of procedures is the performance of actual qualification tests. This method is used, in somewhat modified versions, by both ASME and API for the welding procedure qualification. ASME Section IX, *Welding and Brazing Qualifications,* covers the qualification of brazing as well as welding for the fabrication of pressure vessels and piping. API 1104, *Standard for Welding of Pipelines and Related Facilities,* is primarily used by the cross-country pipeline industry for the qualification and inspection of welding procedures and personnel.

In both systems, there are certain essential variables which are defined. Like the AWS system, these essential variables dictate the extent of a given procedure qualification. That is, once these prescribed limitations have been exceeded, another procedure must be developed. Included in these essential variables are such items as welding processes, welding parameters, base metal types, base metal thicknesses, filler metal types and sizes, and specific welding techniques.

In the ASME system, these essential variables must be stated on a Welding Procedure Specification (WPS). It will list the range of each of the essential variables. Since these ranges may exceed the limits for various essential variables, numerous qualification tests may be required for full coverage. The actual test conditions are recorded on a second document, the Procedure Qualification Record (PQR). Consequently, there may be numerous PQRs referencing a single WPS.

Once these variables have been defined for a certain procedure so that it includes all conditions which will be encountered during production welding, an actual procedure test coupon is welded. It can be either plate or pipe for ASME to result in procedure qualification for both shapes. In API, the configuration is always tubular. Following welding, the required test specimens are removed and destructively tested to be judged as acceptable or rejectable based on the accompanying requirements. Figures 5.15 and 5.16 show some of the typical procedure qualification test coupons for qualification in accordance with ASME Section IX and API 1104, respectively.

For ASME, procedure qualification in the flat position qualifies that procedure in all positions. API requires that a procedure be qualified in the fixed or rolled position, or both, depending on job requirements. However, qualification in either one of these positions does not qualify that procedure in the other position.

The tests are designed to evaluate the effects of the welding techniques and the compatibility of the base and filler metals. Some of the more common tests which are used for procedure qualification testing are tensile, bend, nick-break, macro-etch, fillet break and nondestructive testing. Examples of some of these test specimen configurations are found in Module 6.

Figure 5.17 shows those test specimens required and the range of thickness qualified for various groove weld procedure qualifications according to ASME Section IX. Figure 5.18 lists those specimens required for procedure qualification of butt welds per API 1104.

Special service conditions may require additional tests to evaluate other weld properties. Some of these tests are impact, hardness, chemistry, and special service conditions (e.g., corrosion and abrasion resistance).

The specific code will dictate an appropriate test acceptance criteria. As a welding inspector you may be involved in this evaluation, as well as the actual testing operation. Perhaps the most important job that the welding inspector can perform during the qualification process is to carefully monitor the actual welding to assure that the procedure is being followed. If problems are encountered during procedure welding which are the result of inadequacies in the procedure itself, they can possibly be identified and corrected at this stage instead of during production welding.

The final method of qualifying a welding procedure is through the use of special test weld mockups. This technique is sometimes used for complex weldment configurations where there is a concern about how the overall shape or condition of that component might be affected by the welding operation. Such things as high levels of restraint and weld joint inaccessibility are possible causes of welding problems, but these are more difficult to evaluate using a standard qualification test. It is only through actual trial welds on joint mockups that these questions can be answered.

These mockup tests may be the exclusive test, or they could be used in conjunction with other more common qualification techniques. Regardless, these tests are helpful to the fabricator because he or she now has a feel for how a particular weld can be done and that a particular method can be applied successfully. This valuable experience can be gained through tests rather than having to debug the procedure during actual production.

To summarize this discussion of welding procedure qualification, let's look at a general sequence

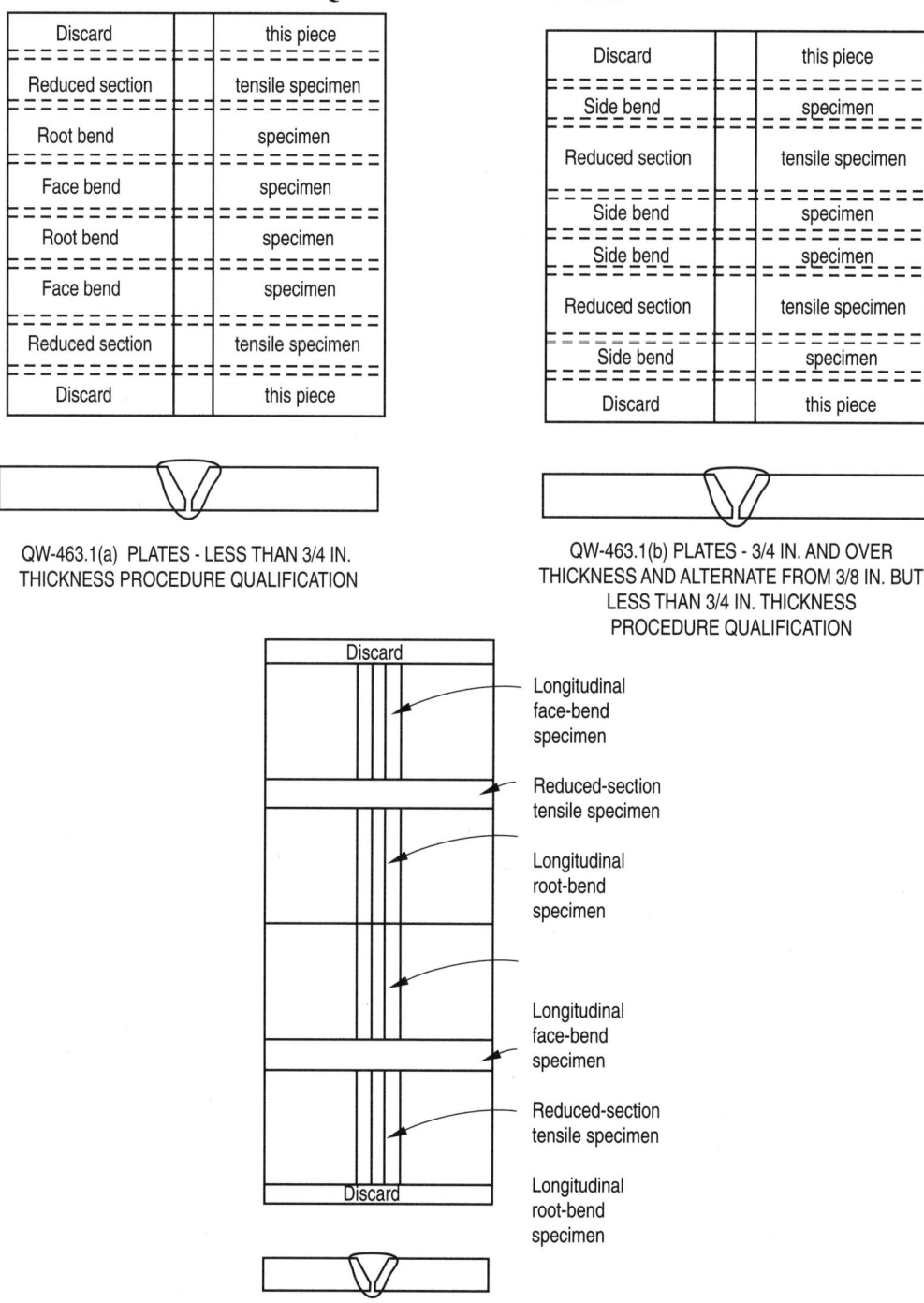

Figure 5.15—Typical ASME Procedure Qualification Coupons

QW-463 Order of Removal (Cont'd)

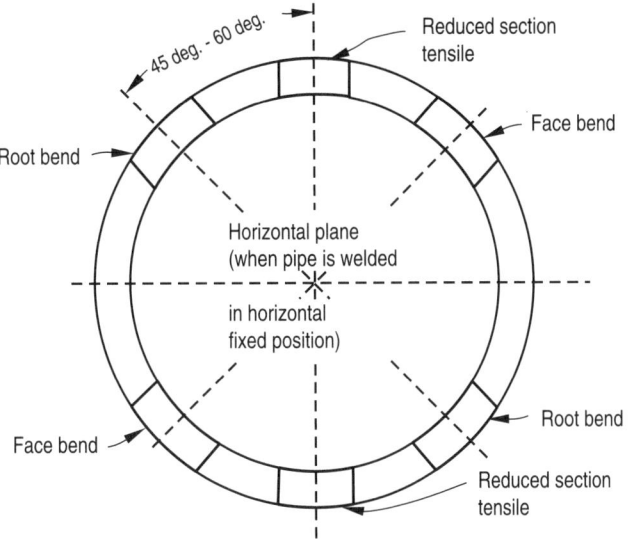

QW-463.1(d) PROCEDURE QUALIFICATION

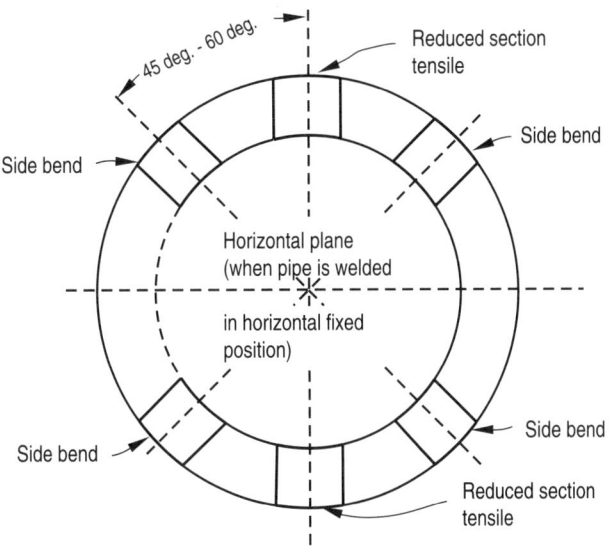

QW-463.1(e) PROCEDURE QUALIFICATION

Source: ASME Boiler & Pressure Vessel Code, Section IX

Figure 5.15 (Continued)—Typical ASME Procedure Qualification Coupons

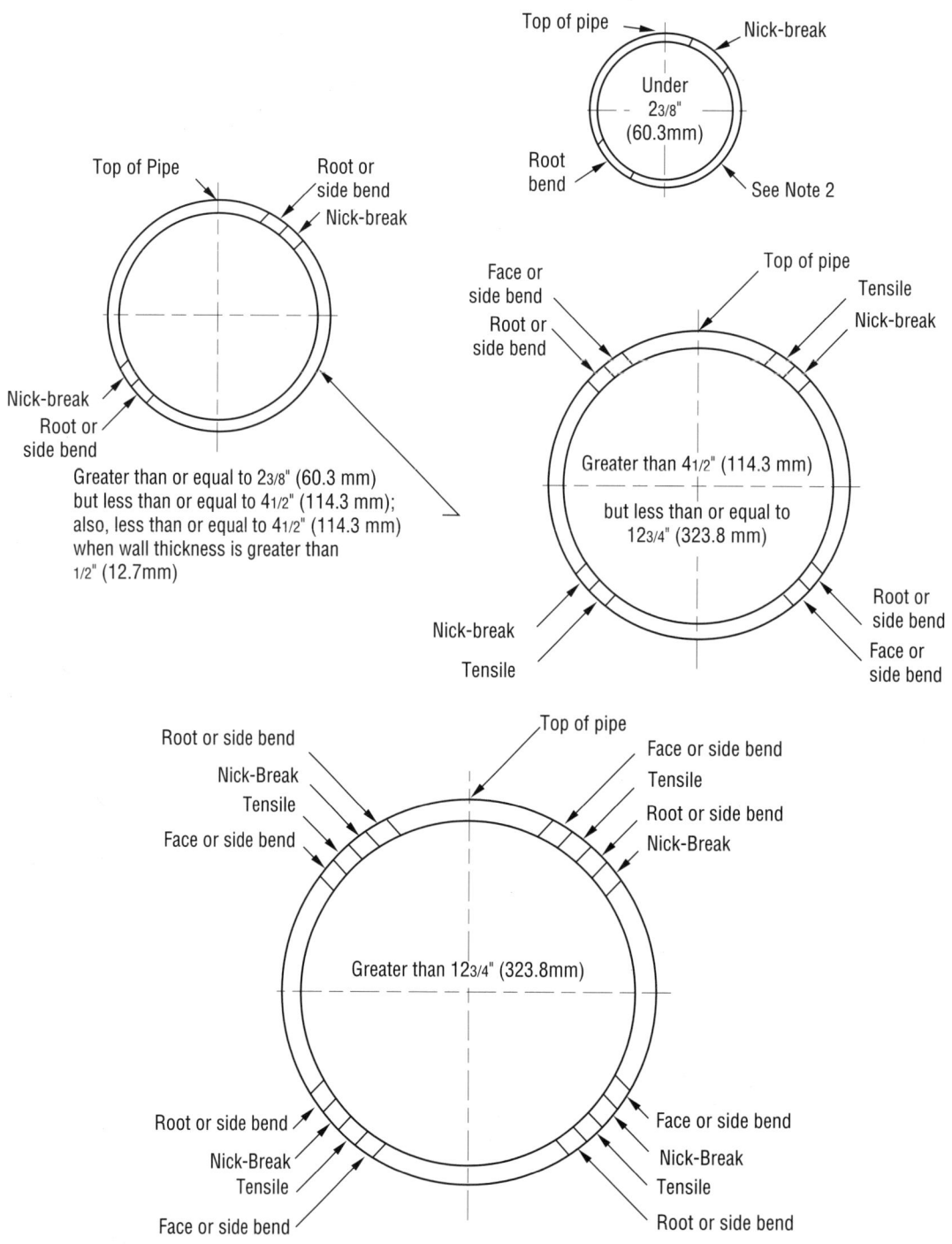

Figure 5.16—Typical API Procedure Qualification Coupons

QW-451.1
GROOVE-WELD TENSION TESTS AND TRANSVERSE-BEND TESTS

Thickness T of Test Coupon Welded, in.	Range of Thickness T of Base Metal Qualified, in. [Note (1)]		Thickness t of Deposited Weld Metal Qualified, in. [Note (1)]	Type and Number of Tests Required (Tension and Guided Bend Tests) [Note (4)]			
	Min.	Max.	Max.	Tension QW-150	Side Bend QW-160	Face Bend QW-160	Root Bend QW-160
Less than 1/16	T	2T	2t	2	...	2	2
1/16 to 3/8, incl.	1/16	2T	2t	2	Note (3)	2	2
Over 3/8, but less than 3/4	3/16	2T	2t	2	Note (3)	2	2
3/4 to less than 1-1/2	3/16	2T	2t when t < 3/4	2 (5)	4	...	...
3/4 to less than 1-1/2	3/16	2T	2T when t ≥ 3/4	2 (5)	4	...	...
1-1/2 and over	3/16	8 (2)	2t when t < 3/4	2 (5)	4	...	...
1-1/2 and over	3/16	8 (2)	8 (2) when t ≥ 3/4	2 (5)	4	...	...

NOTES:
(1) See QW-403 (.2, .3, .6, .9, .10), QW-404.32, and QW-407.4 for further limits on range of thickness qualified. Also see QW-202 (.2, .3, .4) for allowable exceptions.
(2) For the welding process of QW-403.7 only; otherwise per Note (1) or 2T, or 2t, whichever is applicable.
(3) Four side-bend tests may be substituted for the required face- and root-bend tests, when thickness T is 3/8 in. and over.
(4) For combination of welding procedures, see QW-200.4.
(5) See QW-151 (.1, .2, .3) for details on multiple specimens when coupon thicknesses are over 1 in.

Source: ASME Boiler & Pressure Vessel, Section IX

Figure 5.17—Test Specimens Required and Thicknesses Qualified for ASME Procedure Qualification

Outside Diameter of Pipe		Number of Specimens					
Inches	Millimeters	Tensile Strength	Nick-Break	Root Bend	Face Bend	Side Bend	Total
Wall Thickness ≤ 1/2 Inch (12.7 Millimeters)							
< 2-3/8	< 60.3	0[b]	2	2	0	0	4[a]
2-3/4 – 4-1/2	60.3 – 114.3	0[b]	2	2	0	0	4
> 4-1/2 – 12-3/4	> 114.3 – 323.8	2	2	2	2	0	8
> 12-3/4	> 323.8	4	4	4	4	0	16
Wall Thickness > 1/2 Inch (12.7 Millimeters)							
≤ 4-1/2	≤ 114.3	0[b]	2	2	0	2	4
> 4-1/2 – 12-3/4	> 114.3 – 323.8	2	2	2	0	2	4
> 12-3/4	> 323.8	4	4	4	0	8	16

[a] One nick-break and one root-bend specimen shall be taken from each of two test welds, or for pipe less than or equal to 1-5/16 inches (33.4 millimeters) in diameter, one full-section tensile-strength specimen shall be taken.
[b] For materials with specified minimum yield strengths more than 42 kips per square inch (ksi), a minimum of one tensile test shall be required.

Source: API 1104

Figure 5.18—Type and Number of Test Specimens Required for API Procedure Qualification of Butt Welds

for the qualification of a procedure through actual testing:

> **WELDING PROCEDURE QUALIFICATION SPECIFICATION**
>
> (1) Select welding variables.
> (2) Check equipment and materials for suitability.
> (3) Monitor weld joint fit up as well as actual welding, recording all important variables and observations.
> (4) Select, identify and remove required test specimens.
> (5) Test and evaluate specimens.
> (6) Review test results for compliance with applicable code requirements.
> (7) Release approved procedure for production.
> (8) Qualify individual welders in accordance with this specification.
> (9) Monitor the use of that procedure during production to assure that it continues to produce satisfactory results.

While this may vary slightly from company to company, most of these features are important enough to be considered. The welding inspector may be involved with each of these 9 steps or only a few, again depending on the structure of the particular company.

It must be understood that one of the most important parts of the procedure qualification process is the use of that procedure during actual production welding. Too often, companies perform qualification testing only to satisfy a customer's requirements. Once qualified, they are kept in a neat folder or binder buried on someone's shelf or in a file cabinet. This does not help the welder on the floor who needs to know the information stated on procedure qualification forms.

Procedures are welding instructions; therefore, they should be readily available to the welder during production. Due to the physical limitations of paper in a welding environment, some companies use plastic cards or paper enclosed in plastic which is durable enough to withstand being close to the welding operation. These contain all of the necessary information from the approved welding procedure, so the welder can easily reference these requirements should there be any questions. This also aids the welding inspector, because he or she can check procedure requirements and compare them to the actual parameters being used by the welder in production. Another purpose of the in-process monitoring of the welding is to spot any deficiencies in the procedure which might only show up during production welding. If noted, the welding inspector could report them to the supervisor or welding engineer so that corrective action can take place.

For each of the codes, standard forms have been developed to aid the summary of the procedure qualification information, and these are normally used for simplicity. Examples of the forms appear in each Code.

Welder Qualification

Once the procedure has been qualified, it is of no use until individual welders have been qualified to perform welding in accordance with that procedure. These are two separate operations because they serve different purposes, as will be explained in the following discussion. Let's now assume that the appropriate welding procedures have been established and approved through one method or another. It is now necessary to perform welder qualification tests to determine if individual welders possess sufficient skill to produce satisfactory welds using those procedures.

Before, the concern was with the compatibility of the materials and techniques. Once that has been proven, the individual welder qualifications are designed to judge the skill level of the production welders. Consequently, the welder qualification testing is done somewhat differently.

Although different in some respects, welder qualification also has certain similarities when compared to procedure qualification. Among these is the existence of essential variables. In the case of welder qualification, these may include welding position, joint configuration, electrode type and size, process, base metal type, base metal thickness, and specific welding technique. These features are all concerned with those aspects of the welding operation which are directly affected by the physical abilities of the welder.

Codes are generally specific as to the limitations of these essential variables. Figure 5.19 lists position limitations on certain weld types for welder qualification, according to AWS D1.1.

You can see that the range of qualification varies with the different weld configurations: plate

Figure 5.19—Type and Position Limitations for AWS Welder Qualification

		Production Plate Welding Qualified			Production Pipe Welding Qualified					Production Box Tube Welding Qualified				
		Groove	Groove	Fillet	Butt-Groove	Butt-Groove	T-, Y-, K-Groove	T-, Y-, K-Groove	Fillet	Butt-Groove	Butt-Groove	T-, Y-, K-Groove	T-, Y-, K-Groove	Fillet
Qualification Test		CJP	PJP		CJP	PJP	CJP	PJP		CJP	PJP	CJP	PJP	
Weld Type	Position[2]													
Groove[3]	1G	F	F	F, H	F	F			F,	F	F			F,
	2G	F, H	F, H	F, H	F, H	F, H			F, H	F, H	F, H			F, H
	3G	F, H, V	F, H, V	F, H, V	F, H, V	F, H, V			F, H, V	F, H, V	F, H, V			F, H, V
	4G	F, OH	F, OH	F, H, OH	F, OH	F, OH			F, H, OH	F, OH	F, OH			F, H, OH
	3G+4G	All	All	All	All	All			All	All	All			All
				Note 9	Note 4	Note 4		Note 4, 6	Note 9	Note 5			Note 6	Note 9
Fillet	1F			F,					F,					F,
	2F			F, H					F, H					F, H
	3F			F, H, V					F, H, V					F, H, V
	4F			F, H, OH					F, H, OH					F, H, OH
	3F+4F			All					All					All
				Note 9					Note 9					Note 9
Plug		Qualifies Plug and Slot Welding for Only the Positions Tested												
Groove[3] (Pipe or Box)	1G Rotated	F	F	F, H	F	F	F	F	F, H	F	F	F	F	F, H
	2G	F, H	F, H	F, H	F, H	F, H	F, H	F, H	F, H	F, H	F, H	F, H	F, H	F, H
	5G	F, V, OH	F, V, OH	F, V, OH	F, V, OH	F, V, OH	F, V, OH	F, V, OH	F, V, OH	F, V, OH	F, V, OH	F, V, OH	F, V, OH	F, V, OH
	6G	All	All	All	All	All	All	All	All	All	All	All	All	All
	2G+5G	All	All	All	All	All	All	All	All	All	All	All	All	All
	Note 10				Notes 5, 7	Note 7	Notes 6, 7	Notes 6, 7	Note 9			Note 6	Note 6	Note 9
	6GR (Fig. 4.27)	All	All	All	All	All	All	All	All	All	All	All	All	All
				Note 9	Notes 5, 7	Note 7	Notes 6, 7	Notes 6, 7	Note 9	Note 5		Note 6	Note 6	Note 9
	6GR (Fig. 4.27&4.28)	All	All	All								All Notes 6, 8	All Note 6	All Note 9
Pipe Fillet	1F Rotated			F,					F,					F,
	2F			F, H					F, H					F, H
	2F Rotated			F, H					F, H					F, H
	4F			F, H, OH					F, H, OH					F, H, OH
	5F			All					All					All
				Note 9					Note 9					Note 9

CJP — Complete Joint Penetration; PJP — Partial Joint Penetration

Notes: (Notes shown at the bottom of a column box apply to all entries.)
1. Not applicable for welding operator qualification (see Table 4.10).
2. See Figures 4.3, 4.4, 4.5 and 4.6.
3. Groove weld qualification also qualifies plug and slot welds for the test positions indicated.
4. Only qualified for pipe over 24 in. (610 mm) in diameter with backing, backgouging or both.
5. Not qualified for joints welded from one side without backing, or welded from two sides without backgouging.
6. Not qualified for welds having groove angles less than 30° (see 4.12.4.2).
7. Qualification using box tubing (Figure 4.27) also qualifies welding pipe over 24 in. (610 mm) in diameter.
8. Pipe or box tubing is required for the 6GR qualification (Figure 4.27). If box tubing is used per Figure 4.27, the macroetch test may be performed on the corners of the test specimen (similar to Figure 4.28).
9. See 4.25 and 4.28 for dihedral angle restrictions for plate joints and tubular T, Y, K connections.
10. Qualification for welding production joints without backing or backgouging requires using the Figure 4.24 joint detail. For welding production joints with backing or backgouging, either the Figure 4.24 or Figure 4.25 joint detail may be used for qualification.

Source: AWS D1.1-98

grooves, plate fillets and pipe grooves. It is apparent that qualification on a plate groove only provides limited coverage for welding on pipe. However, if the welder qualifies on pipe, he or she is automatically qualified for plate. It can be further noted that plate groove qualifications in the 3G and 4G positions will qualify that welder for all positions of plate. Also, qualification in either the 6G, or 2G and 5G pipe positions will qualify the welder for all positions of pipe except those in T-, K-, and Y- connections. The 6GR test position, however, will provide full coverage for all pipe positions and configurations.

These numeric designations for test positions are simply abbreviations and should be remembered by the welding inspector. Figures 5.20 through 5.23 are illustrations of the various test positions for plate grooves, plate fillets, pipe grooves, and pipe fillets, respectively.

Another important essential variable which determines what coverage is obtained from the completion of a specific qualification test is the thickness of the test plate or pipe. Figure 5.24 lists the thickness ranges qualified for various test plates and pipes, according to AWS D1.1. This table tells us that a 3/8 in. test plate will qualify a welder for welding production materials up to 3/4 in. thick. This is referred to as a limited thickness qualification. Further, successful qualification on a 1 in. test plate will qualify the welder for any thickness 1/8 in. or greater. This is termed an unlimited thickness qualification.

Another essential variable is the joint configuration itself. To determine this effect, standard test plates and pipes are used to approximate the necessary configurations. One of the more important aspects of the joint configuration is the presence or absence of weld backing. In D1.1, there are specific references to the direction of rolling of the plate materials when using backing. The ductility of the metal will vary depending on its rolling direction. If bend tests are performed on test specimens in which the plate rolling direction was in the transverse direction, the base metal may fail. It is therefore important to assure that the plates are properly oriented prior to qualification testing.

AWS has suggested an optional test plate configuration for weld tests in the horizontal position. It uses the same 45° groove angle as for the flat position, but only the upper plate is beveled. This provides a flat shelf upon which the welder can build up weld passes to more easily fill the groove.

Figure 5.25 shows the required tubular butt joint configurations for welds with and without backing.

The test plates for fillet weld qualifications are shown in Figures 5.26 and 5.27. Again, AWS D1.1 offers two methods for this type of qualification; the Fillet Weld-Break and Macroetch test (see Figure 5.26) and the Fillet Weld Root-Bend test (see Figure 5.27).

The final test joint configuration used in AWS D1.1 is referred to as the 6GR test, or test joint for T, K, and Y connections on pipe or square or rectangular tubing. It is shown in Figure 5.28. The initials T, K, and Y are simply a reference to the approximate shape of the joints. This test joint configuration is meant to simulate the access problems associated with welding T, K, and Y connections in tubular structures. This is accomplished by the addition of a restriction ring no more than 1/2 in. from the edge of the groove.

With some processes, requalification may be required if there is a change in the type of electrode specified. For example, Figure 5.29 shows the various types of SMAW electrodes which are grouped according to the skill level required for their operation.

The electrodes in Group F4 are considered to be the more difficult types to use; and similarly, the F1 Group includes those types which are considered to require the least manual ability. Normally, qualification with an electrode of a higher number group will automatically qualify that welder for welding with any electrode of a group bearing a lower number. Therefore, a qualification test performed with an E7018 electrode, which is in Group F4, will provide the welder with qualification coverage for all the carbon steel SMAW electrode types listed.

The specific welding technique used is also considered to be an essential variable for welder qualification. Changes in such details as the direction of welding for the vertical position (i.e., uphill or downhill) will require additional qualification testing. Other typical technique-related essential variables may include changes in the process, position, base metal type, base metal thickness, and tubing diameter.

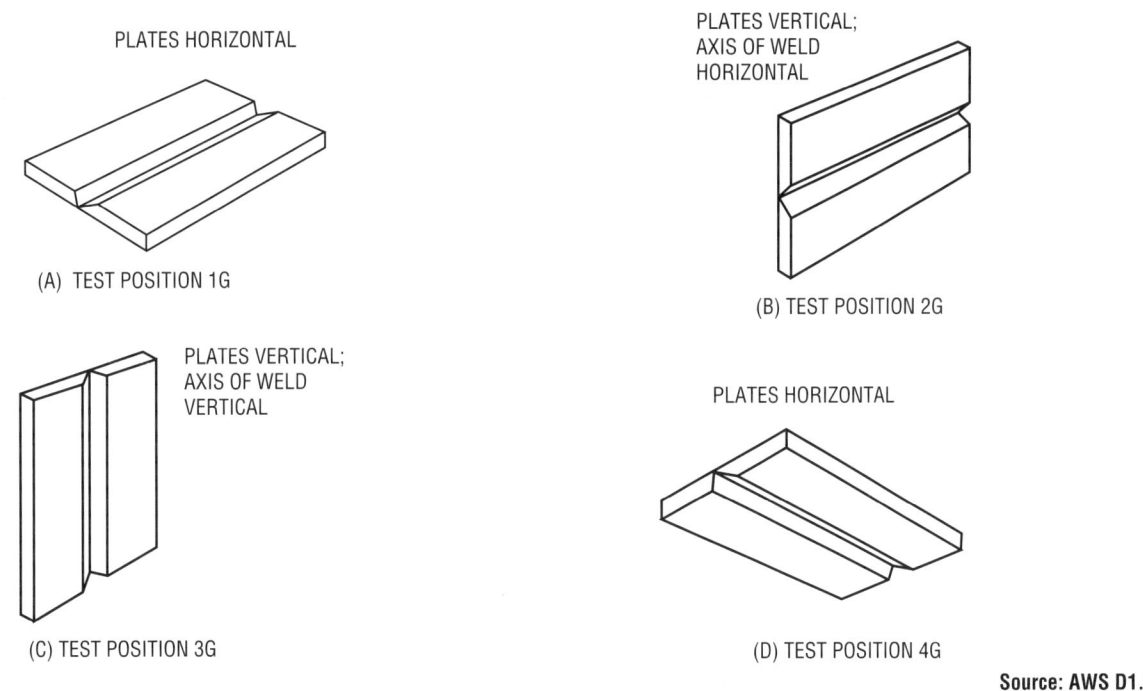

Figure 5.20—Positions of Test Plates for Groove Welds

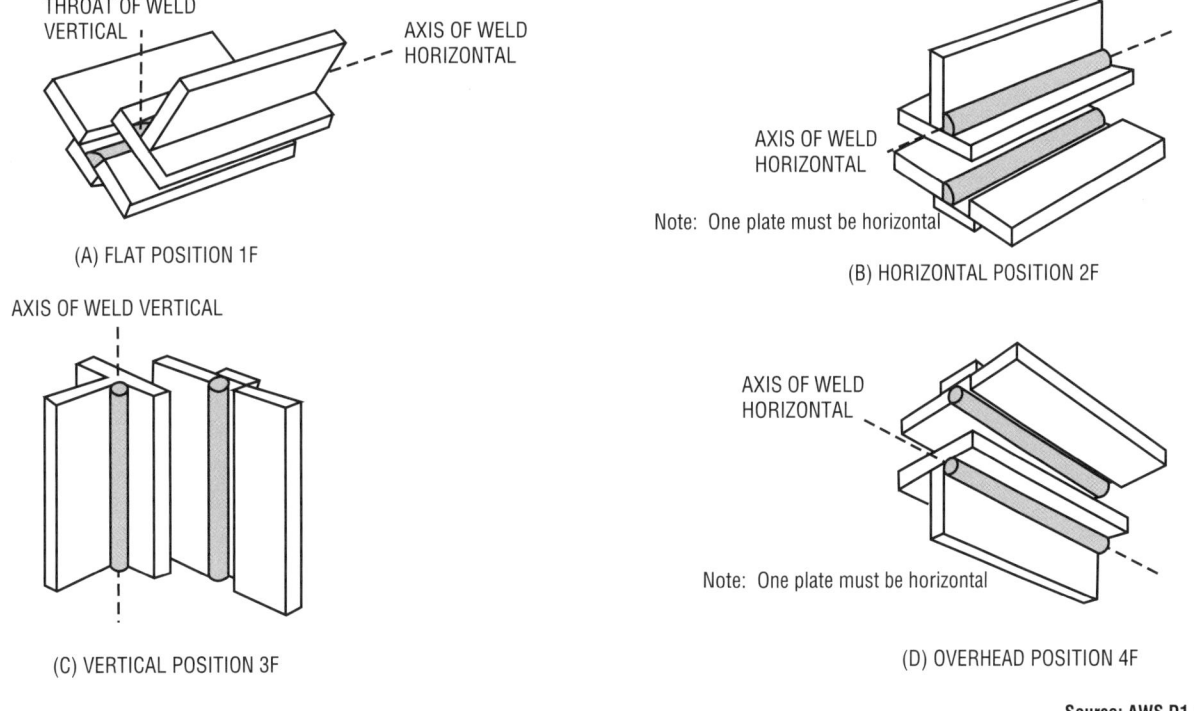

Figure 5.21—Positions of Test Plates for Fillet Welds

Source: AWS D1.1

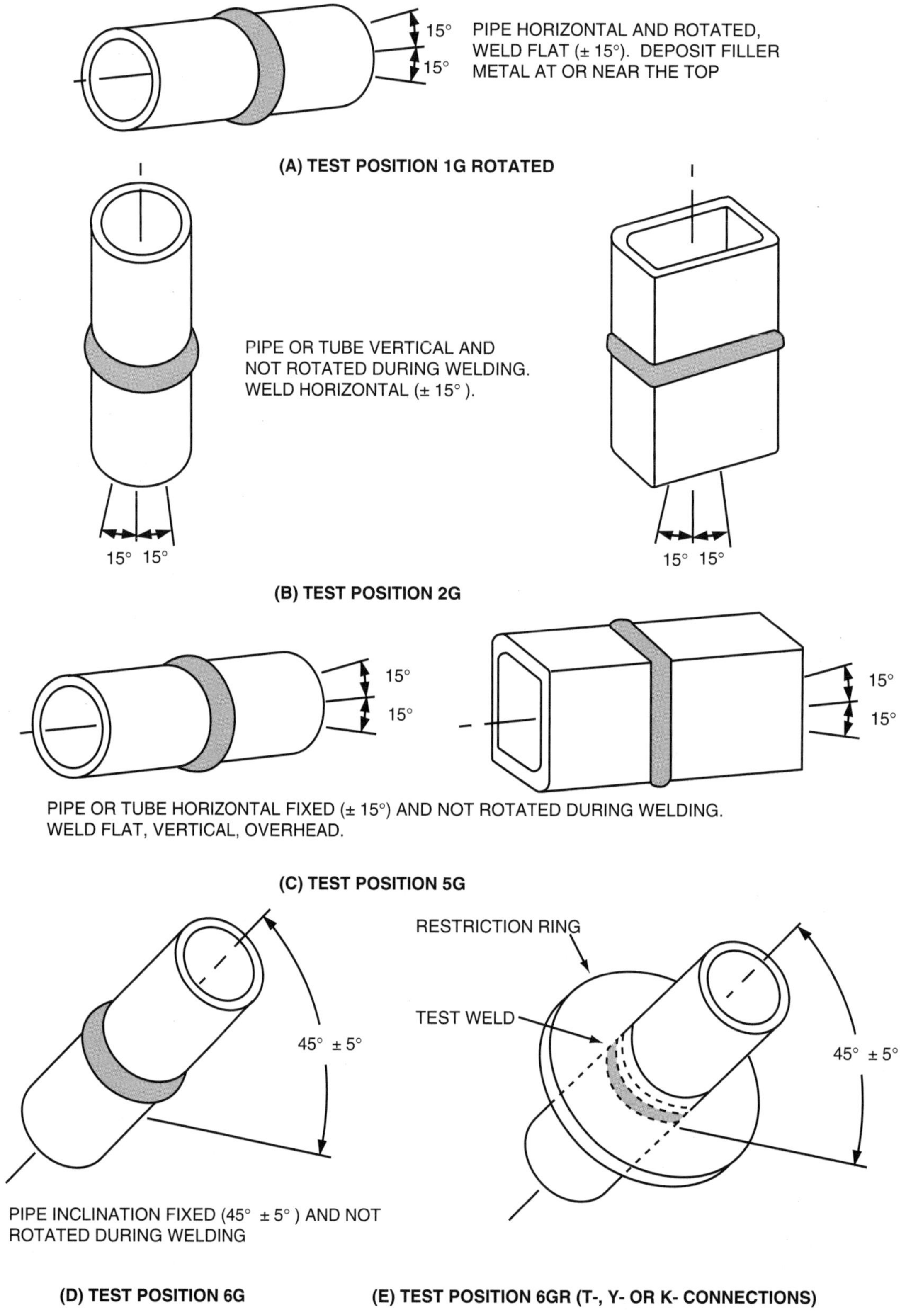

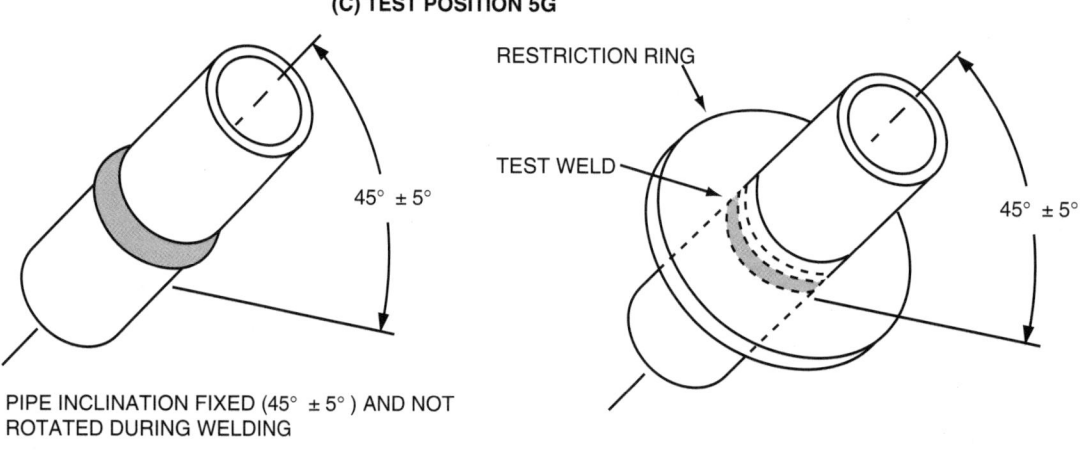

Figure 5.22—Positions of Test Pipes for Groove Welds

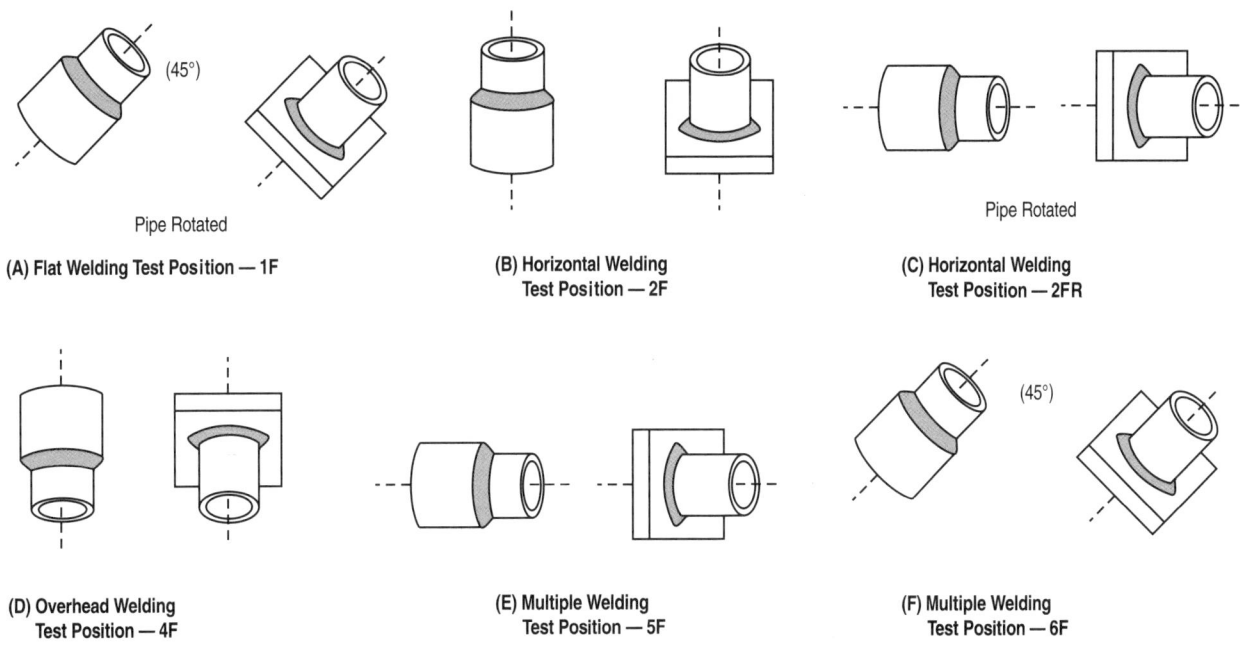

Source: ANSI/AWS A3.0-94

Figure 5.23—Positions of Test Pipes for Fillet Welds

(1) Test on Plate

Type of Weld	Thickness of Test Plate (T) as Welded, in.	Visual Inspection	Bend Tests* Face	Bend Tests* Root	Bend Tests* Side	T-Joint Break	Macro-etch Test	Plate Thickness Qualified, in.
Groove	3/8	Yes	1	1	—	—	—	1/8 to 3/4 max[3]
Groove	3/8 < T < 1	Yes	—	—	2	—	—	1/8 to 2T max[3]
Groove	1 or over	Yes	—	—	2	—	—	1/8 to Unlimited[3]
Fillet Option No. 1[1]	1/2	Yes	—	—	—	1	1	1/8 to Unlimited
Fillet Option No. 2[2]	3/8	Yes	—	2	—	—	—	1/8 to Unlimited
Plug	3/8	Yes	—	—	—	—	2	1/8 to Unlimited

Notes:
1. See Figure 4.36 as applicable.
2. See Figure 4.32 as applicable.
3. Also qualifies for welding fillet welds on material of unlimited thickness.
*Radiographic examination of the welder or welding operator test plate may be made in lieu of the bend test. (See 4.19.1.1)

Source: AWS D1.1-98

Figure 5.24—Number and Type of Specimens and Range of Thickness Qualified for AWS Welder Qualification on Plate

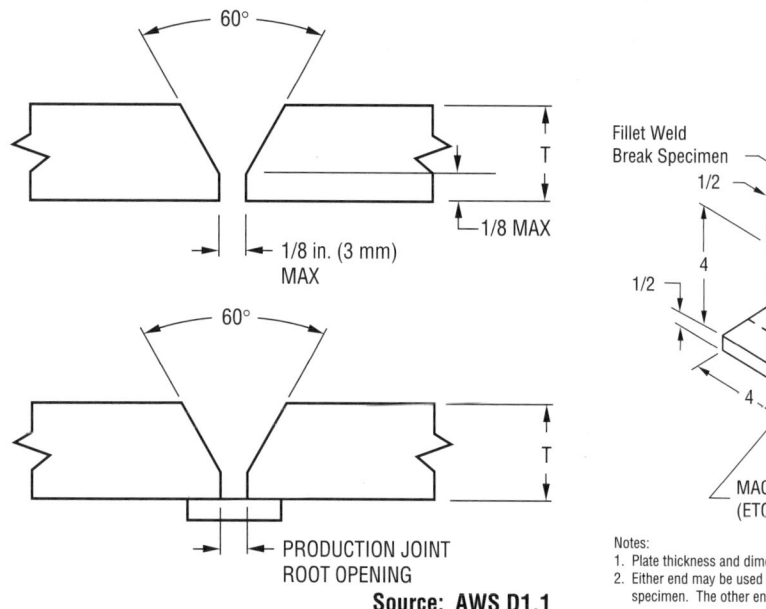

Figure 5.25—AWS Tubular Butt Joint Configurations for Welder Qualification (Without and With Backing)

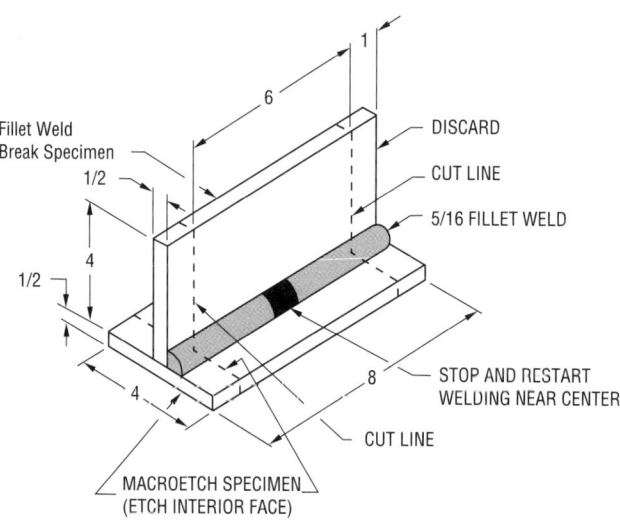

Figure 5.26—AWS Welder Qualification Test Plate for Fillet Welds— Option 1

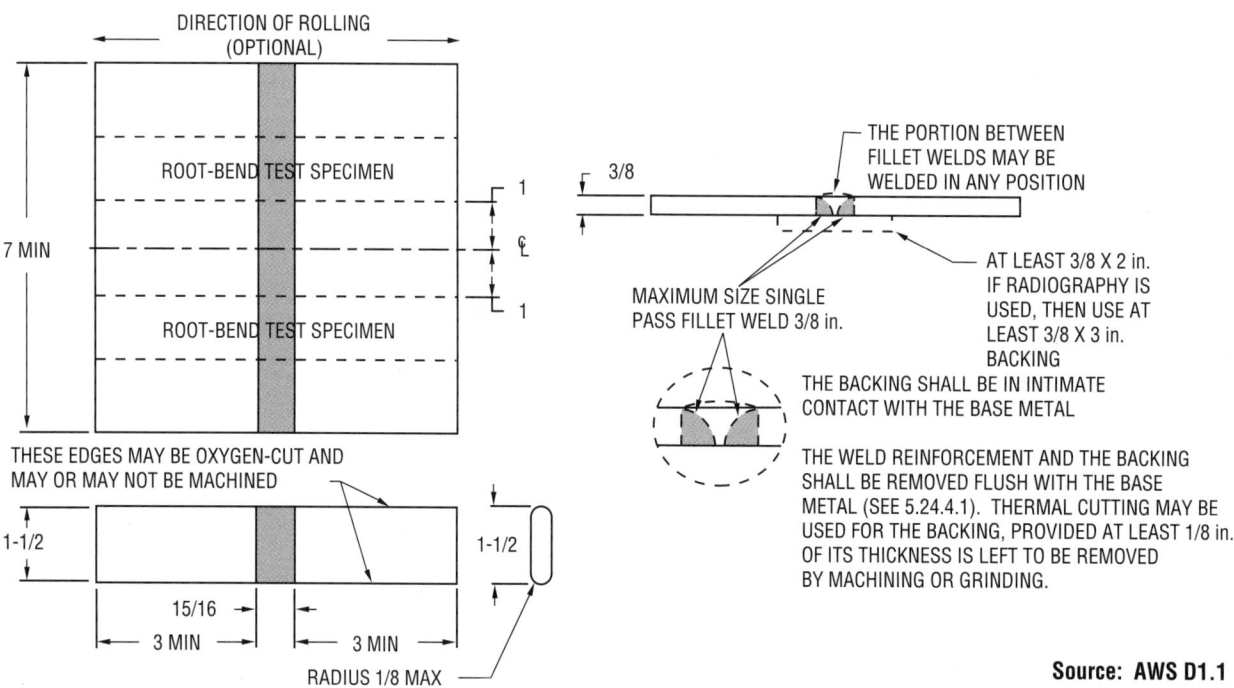

Figure 5.27—AWS Welder Qualification Test Plate for Fillet Welds—Option 2

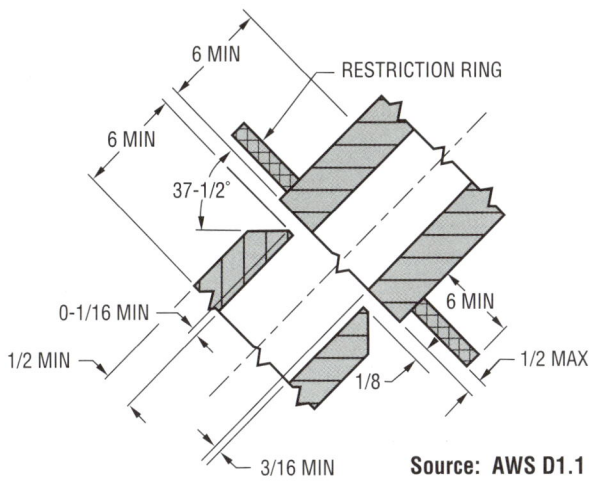

Figure 5.28—AWS D1.1, Welder Qualification Test Joint for T-, K-, and Y-Connections on Pipe or Box Tubing

Group Designation	AWS Electrode Classification*
F4	EXX15, EXX16, EXX18, EXX15-X, EXX16-X, EXX18-X
F3	EXX10, EXX11, EXX10-X, EXX11-X
F2	EXX12, EXX13, EXX14, EXX13-X
F1	EXX20, EXX24, EXX27, EXX28, EXX20-X, EXX27-X

*The letters "XX" used in the classification designation in this table stand for the various strength levels (60, 70, 80, 90, 100, 110, and 120) of electrodes.

Source AWS D1.1-98

Figure 5.29—Groupings of SMAW Electrode Classifications

Once the proper test coupon, position and technique have been selected to assure adequate coverage, the test weld is performed in accordance with the applicable procedure. The welding inspector may be asked to witness the actual welding to verify procedure compliance as well as note the ability of the welder. Careful attention to the techniques and abilities of the welder could reveal habits which might produce unsatisfactory welds.

The finished test coupon is then marked to identify it according to the welder's name, test position and the top of the coupon in the case of a pipe welded in the 5G, 6G, or 6GR positions. The code will then describe whether or not nondestructive examination is necessary, as well as both the type and number of test specimens required. In general, there are fewer specimens required for welder qualification than for procedure qualification. In fact, some codes permit the use of nondestructive examination only, such as radiography, in lieu of standard destructive tests for welder qualification.

Figure 5.24 identifies the type and number of test specimens required for welder qualification in accordance with AWS D1.1. You will note that only two bend tests are required for each welder qualification on plate. Face and root bends are used for thinner plates, while those over 3/8 in. require the use of side bends. This is due to the difficulty associated with the bending of extremely thick specimens.

Virtually all welder qualification test specimens are generally categorized as soundness tests, including bend tests, nick-break tests and fillet break tests. Their configurations and methods of testing are identical to those used for procedure qualification. For welder qualification on plate or pipe, specimens are removed according to the applicable code section. For pipe coupons welded in the 5G and 6G positions, specimens are removed in relation to the top of the pipe during actual welding.

Once properly identified, removed and tested, the specimens are then evaluated in accordance with the appropriate code requirements. If all test results are satisfactory, the welder is deemed qualified to perform welding within the limits of that qualification. The results of the tests, as well as a description of the welding procedure used, are listed on a report form to certify that the welder has satisfied the qualification requirements. A typical form used by AWS is shown in Figure 5.30. Welder qualification forms for use by ASME and API are shown in Figures 5.31 and 5.32.

At this point, it is appropriate to differentiate between the terms qualify and certify, as applied to these welder tests. If we say that a welder is qualified, we mean that he or she has demonstrated sufficient skill to perform a certain weld. Certification, however, applies to the document(s) which support this qualification. A welder who successfully passes a qualification test would then

WELDER, WELDING OPERATOR OR TACK WELDER QUALIFICATION TEST RECORD

Type of Welder _____
Name _____ Identification No. _____
Welding Procedure Specification No. _____ Rev _____ Date _____

Variables	Record Actual Values Used in Qualification	Qualification Range
Process/Type		
Electrode (single or multiple)		
Current/Polarity		
Position		
Weld Progression		
Backing (YES or NO)		
Material/Spec.	_____ to _____	
Base Metal		
Thickness: (Plate)		
Groove		
Fillet		
Thickness: (Pipe/tube)		
Groove		
Fillet		
Diameter: (Pipe)		
Groove		
Fillet		
Filler Metal		
Spec. No.		
Class		
F-No.		
Gas/Flux Type		
Other		

VISUAL INSPECTION
Acceptable YES or NO _____
Guided Bend Test Results

Type	Result	Type	Result

Fillet Test Results
Appearance _____ Fillet Size _____
Fracture Test Root Penetration _____ Macroetch _____
(Describe the location, nature, and size of any crack or tearing of the specimen.)

Inspected by _____ Test Number _____
Organization _____ Date _____

RADIOGRAPHIC TEST RESULTS

Film Identification Number	Result	Remarks	Film Identification Number	Result	Remarks

Interpreted by _____ Test Number _____
Organization _____ Date _____

We, the undersigned, certify that the statements in this record are correct and that the test welds were prepared, welded, and tested in accordance with the requirements of Section 4 of ANSI/AWS D1.1, (_____) Structural Welding Code—Steel.   year
Manufacturer or Contractor _____
Authorized by _____
Date _____
Form E-4   Source: AWS D1.1-98

Figure 5.30—AWS Welder and Welding Operator Qualification Test Record

QW-484 SUGGESTED FORMAT FOR MANUFACTURER'S RECORD OF WELDER OR WELDING OPERATOR QUALIFICATION TESTS (WPQ)
See QW-301, Section IX, ASME Boiler and Pressure Vessel Code

Welder's name _____ Clock number _____ Stamp no. _____
Welding process(es) used _____ Type _____
Identification of WPS followed by welder during welding of test coupon _____
Base material(s) welded _____ Thickness _____

Manual or Semiautomatic Variables for Each Process (QW-350)	**Actual Values**	**Range Qualified**
Backing (metal, weld metal, welded from both sides, flux, etc.) (QW-402)		
ASME P-No. _____ to ASME p-No. (QW-403)		
( ) Plate ( ) Pipe (enter diameter, if pipe)		
Filler metal specification (SFA): _____ Classification (QW-404)		
Filler Metal F-No.		
Consumable insert for GTAW or PAW		
Weld deposit thickness for each welding process		
Welding position (1G, 5G, etc.) (QW-405)		
Progression (uphill/downhill)		
Backing gas for GTAW, PAW, GMAW; fuel gas for OFW (QW-408)		
GMAW transfer mode (QW-409)		
GTAW welding current type/polarity		

Machine Welding Variables for the Process Used (QW-360)	**Actual Values**	**Range Qualified**
Direct/remote visual control		
Automatic voltage control (GTAW)		
Automatic joint tracking		
Welding position (1G, 5G, etc.)		
Consumable insert		
Backing (metal, weld metal, welded from both sides, flux, etc.)		

Guided Bend Test Results

Guided Bend Test Type ( ) QW-462.2 (Side) Results ( ) QW-462.3(a)(Trans. R & F) Type ( ) QW-462.3(b) (Long, R & F) Results

Visual examination results (QW-302.4) _____
Radiographic test results (QW-304 and Qw-305) _____
(For alternative qualification of groove welds by radiography)
Fillet Weld - Fracture test _____ Length and percent of defects _____ in.
Macro test fusion _____ Fillet leg size _____ in. x _____ in. Concavity/convexity _____ in.
Welding test conducted by _____
Mechanical tests conducted by _____

We certify that the statements in this record are correct and that the test coupons were prepared, welded, and tested in accordance with the requirements of Section IX of the ASME Code.

Organization _____

Date _____ By _____

This form (E00008) may be obtained from the Order Dept., ASME, 22 Law Drive, Box 2300, Fairfield NJ 07007-2300

Source ASME Boiler & Pressure Vessel Code, Section IX

Figure 5.31—ASME Welder or Welding Operator Qualification Test Record

COUPON TEST REPORT

Date _____ Test No. _____
Location _____
State _____ Weld Position: Roll ☐   Fixed ☐
Welder _____ Mark _____
Welding time _____ Time of day _____
Mean temperature _____ Wind break used _____
Weather conditions _____
Voltage _____ Amperage _____
Welding machine type _____ Welding machine size _____
Filler metal _____
Reinforcement size _____
Pipe type and grade _____
Wall thickness _____ Outside diameter _____

	1	2	3	4	5	6	7
Coupon stenciled							
Original specimen dimensions							
Original specimen area							
Maximum load							
Tensile strength per square inch of plate area							
Fracture location							

☐ Procedure   ☐ Qualifying test   ☐ Qualified
☐ Welder   ☐ Line test   ☐ Disqualified

Maximum tensile _____   Minimum tensile _____   Average tensile _____

Remarks on tensile-strength tests _____
1. _____
2. _____
3. _____
4. _____
Remarks on bend tests _____
1. _____
2. _____
3. _____
4. _____
Remarks on nick-break tests _____
1. _____
2. _____
3. _____
4. _____

Test made at _____ Date _____
Tested by _____ Supervised by _____
Note: Use back for additional remarks. This form can be used to report either a procedure qualification or a welder qualification test.

Source: API Standard 1104, 18th Ed., 1994

Figure 5.32—API Welder Qualification Test Record

be rightfully referred to as a qualified welder as opposed to a certified welder.

Once qualified, the welder will be permitted to weld in production as long as that welding does not involve positions, thicknesses, electrodes, etc., which are outside the limits of the qualification. Most codes allow qualification to remain in effect indefinitely as long as the welder continues to successfully use that process in production. However, if unsatisfactory performance is noted by the welding inspector or other supervisory personnel, the welder may be required to undergo another qualification test and/or further training. Certification (documentation of qualification) may be terminated when the welder leaves one employer and is hired by another. Since each manufacturer or contractor is responsible for the qualification of his own procedure and welders, codes generally require that a welder be qualified by each separate employer.

To summarize the above, the general sequence for the qualification of a welder is:

(1) Identify essential variables.
(2) Check equipment and materials for suitability.
(3) Check test coupon configuration and position.
(4) Monitor actual welding to assure that it complies with the applicable welding procedure.
(5) Select, identify and remove required test samples.
(6) Test and evaluate specimens.
(7) Complete necessary paperwork.
(8) Monitor production welding.

The qualification of individual welders provides the manufacturer or contractor with personnel to perform the production welding in accordance with qualified procedures. Once this production welding begins, the welding inspector will be required to monitor the welding to assure that welding is being performed in accordance with procedure requirements and that the finished welds are acceptable. Any deficiencies should be noted and corrected. If recurring problems are found, corrective measures may include either procedural and/or personnel changes.

While the existence of qualified procedures and personnel does not guarantee that all production welds will be satisfactory, it at least provides some assurance that the procedure and the personnel are capable of producing welds of adequate quality. It is important to remember, however, that qualification welds were most likely produced under more desirable conditions than those present in production. Consequently, variations in fitup, joint configuration, accessibility, etc., could introduce conditions which might increase the possibility of error. Therefore, the welding inspector should attempt to locate and identify these inconsistencies before they result in nonconforming welds.

Summary

Documents represent one side of the inspection equation. The other is, of course, the inspector, whose function is to establish product or part quality. Traditionally, inspection is viewed as a post production activity. Welding inspection is significantly different. Welding inspection embodies activities taking place before, during and after welding. Welding inspection is thus both predictive and reactive.

Quality is, by definition, "conformance to specification." As has been shown, the term "specification" may in fact refer to job or contract-invoked provisions embodied in:

- Drawings
- Codes
- Standards
- Specifications

Drawings give details of item size, form and configuration. Codes, Standards and Specifications give details of design, materials, methods and quality requirements to be satisfied. Included in the methods are the welding procedures and the skill of welding personnel; the qualification of which may well also involve the welding inspector.

Based on the concept of predictive action, welding inspection ideally covers all activities where problems may develop. As such, welding inspection and the documents setting out specific requirements are concerned with:

- Design of joints
- Materials, base metal, and filler metal
- Procedures of welding and workmanship
- Preparations, joint form, and dimensions
- Production before, during, and after welding

The welding inspectors' ability to read, interpret and fully understand the applicable documentation is basic to successful welding inspection.

Key Terms and Definitions

alphanumeric code—a short combination of letters and numbers used to identify a material type, grade, etc.

ANSI—American National Standards Institute.

API—American Petroleum Institute.

ASME—American Society of Mechanical Engineers.

ASNT—American Society for Nondestructive Testing.

ASTM—American Society for Testing and Materials.

AWS—American Welding Society.

bar code—a group of short, vertical lines representing a body of information.

camber—the permissible variation from straightness, as in girders or beams.

code—a body of laws, as of a nation, city, etc., arranged systematically for easy reference.

drawing—a graphic detail of a component, showing its geometry and size, with tolerances.

drop—in fabrication welding, the remaining piece of a material when a portion has been removed for use.

essential variables—those variables, if changed beyond certain limits, require a new welding procedure to be prepared and qualified.

heat number—a number assigned to each heat of steel by the manufacturing source.

hold point—a specific, prearranged step in the fabrication process where fabrication is stopped to permit an interim inspection. Fabrication can begin again only when the inspection shows the part meets the quality requirements.

in-house specification—a specification written by a company primarily for internal use.

ISO—International Organization for Standardization.

material call out—a list of materials required for fabrication of a component. The list will specify all required alloy types, grades, sizes, etc., for both base and filler metals.

MTC—Material (or Mill) Test Certificate.

MTR—Material (or Mill) Test Report.

NACE—National Association of Corrosion Engineers.

PQR—Procedure Qualification Record.

standard—something established for use as a rule or basis of comparison in measuring or judging capacity, quantity, content, extent, value, quality, etc.

specification—a detailed description of the parts of a whole; statement or enumeration of particulars, as to actual or required size, quality, performance, terms, etc.

tolerance—the amount of variation permitted from the design size of a part.

traceability—an attribute of a materials control system which permits tracing any part or material used in fabrication back to the source and certifying documents.

UNS—Unified Numbering System.

WPS—Welding Procedure Specification.

WPQ or WPQR—Welder or Welding Operator Performance Qualification Record.

Module 6
Metal Properties and Destructive Testing

Contents

Introduction 6-2

Mechanical Properties of Metals 6-2

Chemical Properties of Metals 6-9

Destructive Testing 6-13

Summary 6-31

Key Terms and Definitions 6-31

Module 6—Metal Properties and Destructive Testing

Introduction

In today's world, there are thousands of different metals available to serve as construction materials as both base metals and filler metals. From this selection, materials engineers and designers are able to choose the metals which best suit their particular needs. These metals may differ not only in their composition, but also in the manner in which they are manufactured. Within the United States, there are several organizations maintaining material standards, such as ASTM, ASME, and AWS. Additionally, there are material standards from many other countries and groups including Japan and the European countries.

It was noted in Module 1 that one of the responsibilities of the welding inspector may be to review documentation related to the actual properties of base and filler metals. The purpose of this module is to describe some of these mechanical and chemical properties to the extent that the welding inspector has some feeling for what the actual values mean. For the most part, the inspector must simply compare specification values with actual numbers to judge compliance. However, it will be helpful for the inspector to have additional information about these materials' properties. The additional information can help avoid problems which may occur during welding.

Another purpose of this module is to provide a base for the information which will be discussed in Module 8, "Welding Metallurgy for the Welding Inspector." Since the metallurgical makeup of a metal defines its properties, it will be shown how various metallurgical treatments may alter the properties of a metal.

Depending upon the mechanical and chemical properties of a metal, special fabrication techniques may be required to prevent the degradation of these properties. Preheating and postheating are examples of techniques that may be applied to maintain certain metal properties. For quenched and tempered steels, the welding inspector may be asked to monitor the welding heat input to prevent the degradation of base metal properties caused by overheating. In these examples, the welding inspector's involvement is not directly related to these material properties. However, effective monitoring can prevent problems caused by the alteration of expected properties from too much or too little heat.

Mechanical Properties of Metals

Some of the important mechanical properties of metals will now be reviewed; this discussion is limited to five categories of properties:

- Strength
- Ductility
- Hardness
- Toughness
- Fatigue Strength

Strength

Strength is defined as, "the ability of a material to withstand an applied load." There are numerous types of strength, each dependent upon how the load is applied to the material: tensile strength, shear strength, torsional strength, impact strength and fatigue strength.

The tensile strength of a metal is described as the ability of a metal to resist failure when subjected to a tensile, or pulling, load. Since metals are often used to carry tensile loads, this is one of the more important properties with which the designer is concerned. When a metal specification is examined, the tensile strength is usually expressed in two different ways. The terms used are ultimate tensile strength and yield strength. Both refer to different aspects of that material's behavior. Ultimate tensile strength, UTS, (sometimes referred to as simply tensile strength) relates to the maximum load-carrying capacity of that metal, or the strength of that metal at the exact point when failure occurs.

To define yield strength, it is necessary to understand what is meant when a metal behaves "elastically." Elastic behavior refers to the deformation of a metal under load which causes no permanent deformation when the load is removed. Elastic behavior can be illustrated with a familiar example; a rubber band is typical of an elastic material. It will stretch under a load, but returns to its original shape when the load is removed. When a metal is loaded within its elastic region, it responds with some amount of stretch, or elongation. In this elastic range, the amount of stretch is directly proportional to the applied load, so elastic behavior is referred to as being *linear*. When a metal behaves elastically, it can be stretched to some point, and returns to its original length when the applied load is removed. That is, it takes on no permanent deformation, or set. This is illustrated in Figure 6.1.

If a metal is stressed beyond its elastic limit, it no longer behaves elastically. Its behavior is now referred to as *plastic*, which means permanent deformation occurs. It also implies the stress-strain relationship is no longer linear. Once plastic deformation occurs, the material will not return to its original length upon removal of the applied load. It will now exhibit permanent deformation, or "set."

The point at which the material's behavior changes from elastic to plastic is referred to as its yield point. The yield strength, therefore, is that strength level at which the material's response to loading changes from elastic to plastic. This value is an extremely important value, since most designers will use it as a basis for the maximum load limit for some structure. This is necessary because a structure might be rendered useless if stressed beyond its yield point and becomes permanently deformed.

Both the ultimate tensile strength and yield strength are normally determined by a "tensile test." A specimen of known cross section is loaded so the stress, in pounds per square inch, can be determined. The specimen is loaded to failure and it is then possible to determine its load-carrying capacity on a pounds per square inch basis (psi). The examples which follow show how this relationship works for one material.

> Tensile Strength from tensile test is 60,000 psi. The maximum load which this metal can support is 60,000 psi x cross sectional area.
>
> For a 1 in. x 1 in. member (1 in.^{2} area):
> Maximum Load = 60,000 psi x 1 in.^{2}
> Maximum Load = 60,000 pounds
>
> For a 2 in. x 2 in. member (4 in.^{2} area):
> Maximum Load = 60,000 psi x 4 in.^{2}
> Maximum Load = 240,000 pounds

When a designer knows the tensile strength of a certain metal, he is able to determine how large the cross section of that material will be needed to carry a given load. The tensile test provides a direct measurement of the metal's strength; it is also possible to make an indirect measurement of strength using the hardness test. For carbon steels, there is a direct relationship between tensile strength and hardness.

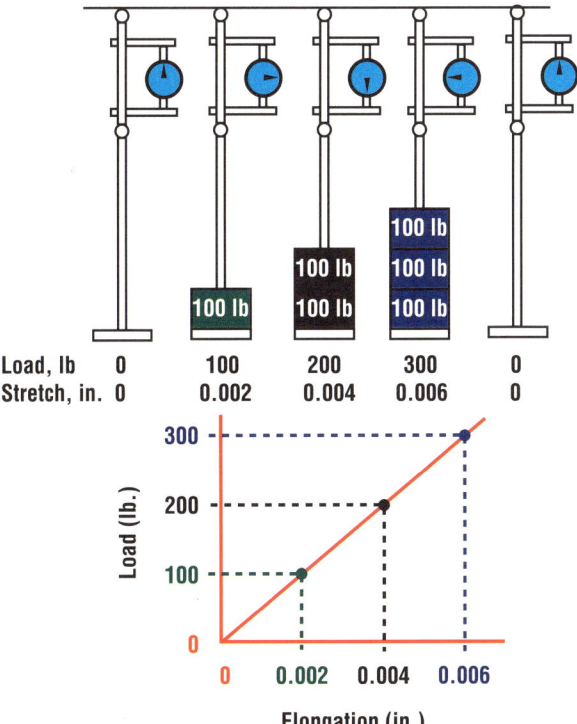

Figure 6.1—Illustration of Elastic Behavior of Metals

That is, if the hardness increases, the tensile strength increases also, and vice versa. The tensile test still provides the most accurate determination of tensile strength, but it is often convenient to perform a hardness test on carbon and low alloy steels to estimate their equivalent tensile strengths.

Figure 6.2 shows some typical values for tensile and yield strength, percent elongation and hardness for some common construction materials. It is interesting to note that recorded values may vary somewhat depending upon the materials' heat treatment, mechanical condition or mass. Such conditions, when changed, could alter the mechanical properties even though the material has the same chemical composition.

The temperature of the metal also has an effect on its strength. As temperature increases, the strength of a metal decreases. If a metal is to support some load at an elevated temperature, the designer must take into effect the reduction in material strength at that temperature. Temperature will also have an effect on the metal's ductility, the next subject to be discussed.

Ductility

Ductility is a term which relates to the ability of a material to deform, or stretch, under load without failing. The more ductile a metal is, the more it will stretch before it breaks. Ductility is an important property of a metal because it can affect whether the metal fails gradually or suddenly when loaded. If a metal exhibits high ductility, it will usually fail or break gradually. A ductile metal will bend before breaking, which is a good indicator that the metal's yield point is being exceeded. Metals having low ductility fail suddenly in a brittle manner, without any warning.

A metal's ductility is directly related to its temperature. As temperature increases, metal ductility increases, and as temperature decreases, ductility decreases. Metals behaving in a ductile manner at room temperature may fail in a brittle manner at sub-zero temperatures.

A metal having high ductility is referred to as being *ductile*, while a metal having low ductility is referred to as being *brittle*. Brittle materials exhibit little or no deformation before fracture. Glass is a good example of a brittle material. A common brittle metal is cast iron, especially white cast iron.

The difference in appearance of a ductile versus brittle failure can be readily seen. Figure 6.3 shows this difference in the halves of two separate Charpy specimens, one of which failed in a brittle manner while the other failed in a ductile fashion.

Ductility is the property which permits several members which may be of slightly different lengths to uniformly support some load without one of those members becoming overloaded to the point of failure. If one of the members is slightly shorter, but ductile, it is capable of deforming sufficiently to permit the load to be shared by the other members. A practical example of this phenomenon is the indi-

Base Metal or Alloy	Yield Strength			Tensile Strength			Elongation % in 2 in. (50 mm)	Hardness BHN
	lb/in.^{2}	MPa	kg/mm^2	lb/in.^{2}	MPa	kg/mm^2		
Aluminum	5,000	34.5	3.5	13,000	89.6	9.1	35	23
Copper (deoxidized)	10,000	68.9	7.0	33,000	227.5	23.2	40	30
Iron, cast	—	—	—	25,000	172.4	17.5	0.5	180
Iron, wrought	27,000	186.1	19.0	40,000	275.8	28.1	25	100
Steel, low alloy	50,000	344.7	35.1	75,000	517.1	52.7	28	170
Steel, high carbon	90,000	620.5	63.2	140,000	965.2	98.4	20	310
Steel, low carbon	36,000	218.2	25.3	60,000	413.6	42.2	35	120
Steel, manganese (14 Mn)	75,000	517.1	52.7	118,000	813.5	82.9	22	200
Steel, medium carbon	52,000	358.5	36.5	87,000	599.8	61.2	24	170
Steel, stainless (austenitic)	40,000	275.8	28.1	90,000	620.5	63.2	23	160
Titanium	40,000	275.8	28.1	60,000	413.6	42.2	28	—

Figure 6.2—Mechanical Properties of Some Metals

Figure 6.3—Brittle Versus Ductile Failure

rolling direction. In the through-thickness direction, the strength and ductility are even less. For some steels, the ductility in this direction is very low. Each of the three directions referred to above has been assigned a letter for identification. The rolling direction is referred to as the "X" direction, the transverse direction is the "Y" direction, and the through-thickness direction is the "Z" direction.

Perhaps you have witnessed a welder qualification bend test plate in which the coupon broke in the base metal. This is often the result of having the rolling direction of the plate oriented parallel to the weld axis. Even though a metal may exhibit excellent properties in the rolling direction, loading in either of the other two directions may result in premature failure.

The ductility of a metal is normally determined by the tensile test, at the same time as the metal's strength is being measured. The ductility is usually expressed in two ways, percent elongation and percent reduction of area.

Hardness

Hardness is one of the most commonly and easily measured mechanical properties. It is defined as the ability of a material to resist indentation, or penetration. It was previously noted that hardness and strength are directly related for carbon steels. Hardness increases with a material's strength and vice versa. Therefore, if a metal's hardness is known, it is possible to estimate its tensile strength, especially for the carbon and low alloy steels. This becomes extremely useful in estimating the strength of a metal without removing, preparing, and pulling a tensile specimen.

A metal's hardness can be determined in a number of different ways. However, most commonly used methods employ some type of indenter which is forced into the surface of the metal by an applied load. Various tests can be performed using this basic technique; they differ in the type and shape of indenter used as well as the magnitude of the applied load. The material's hardness is then determined as a function of either the depth or the size of the indentation. Figure 6.4 shows some of the commonly used hardness test indenters and the resulting shapes of the indentations.

From this wide variety of methods, it is possible to determine the hardness of a large area of the

vidual tightening of steel wires forming the cables for suspension bridges. Since this cannot be done with adequate precision, the wires are made of ductile metal. When the bridge is loaded, those wires momentarily carrying more than their share of the load can stretch to allow the other wires to assume their share.

Ductility becomes an even more important property for a metal which must undergo subsequent forming operations. For example, metals used for body components on automobiles must have sufficient ductility to permit forming into the desired shapes.

An important aspect relating to ductility, and strength, is the differences in magnitude versus the direction in which the load is applied relative to direction of rolling of material during its original manufacture. Rolled metals have very directional properties. Rolling causes the crystals, or grains, to be elongated in the direction of rolling much more than in the direction transverse to, or across, the rolling direction. The result is the strength and ductility of a rolled metal such as a sheet of steel is greatest in the direction of rolling. In the transverse direction of the material, the strength may be decreased as much as 30% and the ductility reduced as much as 50%, relative to the properties of the

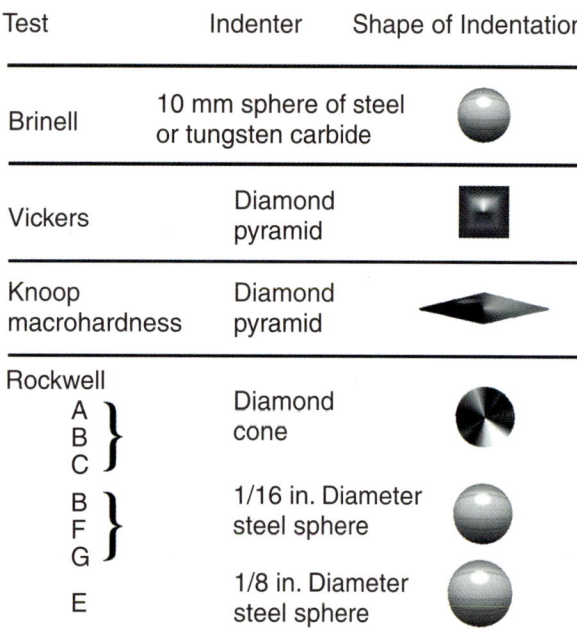

Figure 6.4—Hardness Tests, Indenters, and Shapes of Indentations

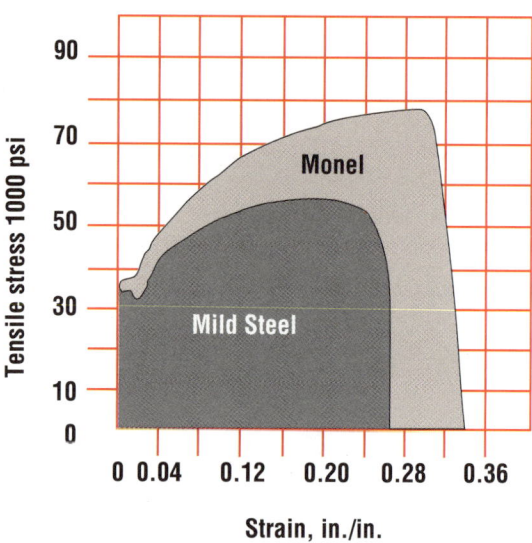

Figure 6.5—Toughness of Two Metals

metal surface or the hardness of an individual grain of the metal.

Toughness

The next mechanical property to be discussed is toughness. In general, toughness is the ability of a material to absorb energy. From a stress-strain diagram, produced during a tensile test, the toughness of the metal can be determined by calculating the area under the stress-strain curve, as shown in Figure 6.5. These curves show Monel® is a tougher material than mild steel because the area under its curve is greater.

Another common term is notch toughness. This differs from toughness in that it refers to the material's energy-absorbing ability when there are surface flaws present, whereas toughness refers to the energy absorption capacity of a smooth, unnotched sample. Notch toughness further differs from toughness in that toughness usually defines the material's behavior when loaded slowly, while notch toughness values reflect the energy absorption which occurs at a high rate of loading. For this reason, notch toughness is often referred to as impact strength.

The difference between these two terms is demonstrated by the analogy of breaking a string. If a steady load is applied, it takes more effort than if the string is pulled sharply to break it.

When discussing toughness or notch toughness, the area of interest is how much energy can be absorbed by a material before it fails. A metal which exhibits low toughness will fail at a low value, with little evidence of deformation. A tough metal, on the other hand, will fail at a considerably higher value with a significant amount of permanent deformation taking place.

Recalling the previous discussion of ductility, the difference between metals of low and high toughness is that low toughness values define brittle behavior while high values of toughness are related to ductile fracture. As is the case with ductility, the toughness of a metal will change as the temperature is changed. In general, as the temperature is reduced, the toughness of the metal decreases as well. Consequently, toughness properties of a metal are determined at a specific temperature. Without this additional test temperature information, the value for toughness has little meaning.

Since it is the presence of a notch or some other form of stress raiser making structural materials susceptible to brittle fracture under certain conditions, notch toughness becomes the primary concern. Many metals, especially high strength tool steels, are extremely sensitive to the presence of sharp surface irregularities. Figure 6.6 shows some typical features which create this notch effect.

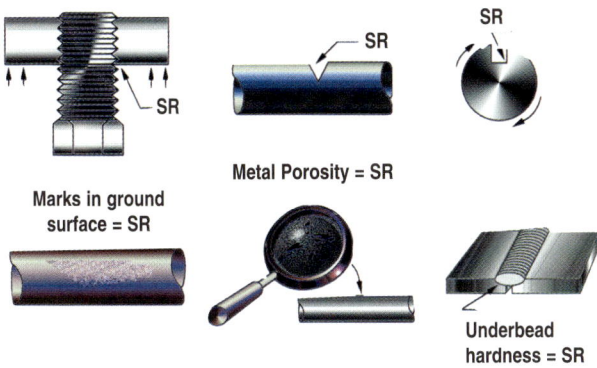

Figure 6.6—Various Conditions Which Produce a Stress Raiser

If a metal exhibits a high amount of notch toughness, this means it will perform well whether or not there is a notch present. However, if a metal is notch-sensitive, meaning that it exhibits low notch toughness, it could more easily fail during impact or repetitive loading. In general, a metal's notch toughness decreases as its hardness increases and its temperature is reduced.

In performing tests to determine the notch toughness of a metal, one usually tries to determine that temperature at which the fracture behavior changes from ductile to brittle. This temperature is referred to as the metal's *transition temperature*.

There are several types of tests used to determine the notch toughness of a metal. However, they differ primarily in the way in which a notch is introduced and the loading is applied. Most include some type of impact load which is applied when the metal has been brought to some temperature. Some of these more common notch toughness or impact tests include Charpy, drop-weight nil-ductility, explosion bulge, dynamic tear, and crack tip opening displacement (CTOD).

Fatigue Strength

The final mechanical property of metals for review is fatigue strength. To define fatigue strength, one must first understand what is meant by a fatigue failure of a metal. Metal fatigue is caused by a cyclic or repeating mechanical action on a member. That is, the load alternately changes between a high stress and some lower stress, or a stress reversal. This action can occur quickly, as in the case of a motor's rotation, or slowly where the cycles could be measured in days. An example of fatigue failure would be the repeated bending of a motor shaft to produce a break. This type of failure will usually occur below the tensile strength of the shaft.

The fatigue strength of a metal is defined as that strength necessary to resist failure under repeated load applications. Knowledge of fatigue strength is important because the vast majority of metal failures are the result of fatigue. Fatigue strength data are often reported in relation to a specific number of cycles required to cause a failure; typical cycles are one million or ten million.

The fatigue strength can be determined through fatigue testing. While this can be performed in a number of different ways, fatigue tests are commonly conducted by applying a stress in tension and then at the same level in compression. This type of testing is referred to as *reverse bending*. As the maximum stress applied is increased, the number of cycles required to produce failure decreases. If a number of these tests are conducted at various stress levels, an S-N curve can be produced, as shown in Figure 6.7. The S-N curve is a graphic description of how many fatigue cycles are necessary to produce a failure at various stress levels.

These curves show steel exhibits a well-defined endurance limit, but the curve for aluminum does

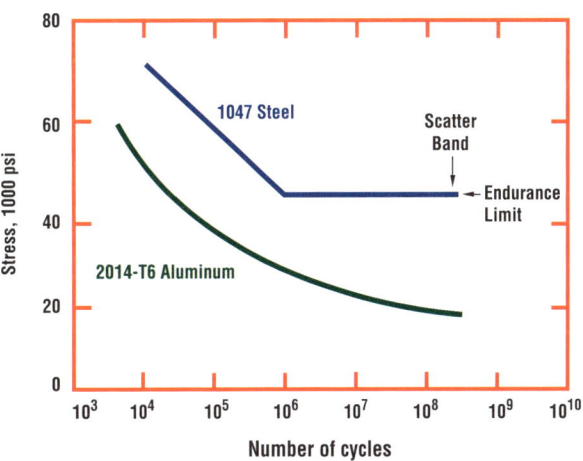

Figure 6.7—Typical S-N Curves for Steel and Aluminum

not. The endurance limit is the maximum stress at which no failure will occur, no matter how many cycles the load is applied. The curve shows aluminum will eventually fail, even at low stress levels. However, the steel will last indefinitely as long as the stress remains below this endurance limit. Often, the fatigue strength of carbon steels is roughly equal to half its tensile strength.

Fatigue strength, like impact strength, is extremely dependent upon the surface geometry of the member. The presence of any notch or stress risers can increase the stress at that point to above the metal's endurance limit. Upon the application of a sufficient number of cycles, fatigue failure will result. Figure 6.8 shows the effect of notch sharpness on the fatigue strength of a metal. The surface finish also has an effect on the fatigue strength, as shown in Figure 6.9.

A major concern in welding relates to a metal's fatigue strength. That concern is not for the metallurgical changes that may occur, however. The overriding factor, of which welding is a part, is the presence of some sharp surface irregularity. Unless ground smooth after welding, the weld itself creates a surface irregularity. Weld surface discontinuities such as undercut, overlap, excessive reinforcement or convexity, can have an effect on a member's fatigue strength. Such conditions create a sharp notch which can act as a fatigue crack initiation site. Examples of some of these surface irregularities are shown in Figure 6.10.

While fatigue failures can result from internal weld discontinuities, those on the surface represent a more significant concern. That is, a surface dis-

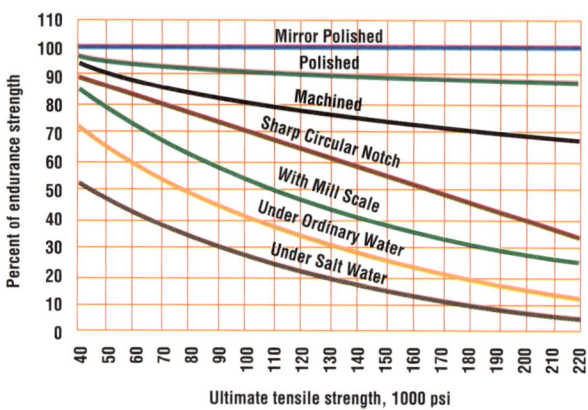

Figure 6.9—Effect of Surface Finish on Fatigue Strength

continuity will more quickly lead to fatigue failure than will a subsurface discontinuity. The reason for this is that surface stress levels are usually higher than the internal stress levels. For that reason, the welding inspector can play a large role in preventing fatigue failures by performing a careful visual examination of surfaces. Discovery and correction of sharp surface irregularities will greatly improve the fatigue properties of any structure. In many fatigue situations, a small weld with a smooth contour will perform better than a much larger weld having sharp surface irregularities.

Chemical Properties of Metals

The mechanical properties of a metal can be altered by the application of various mechanical and

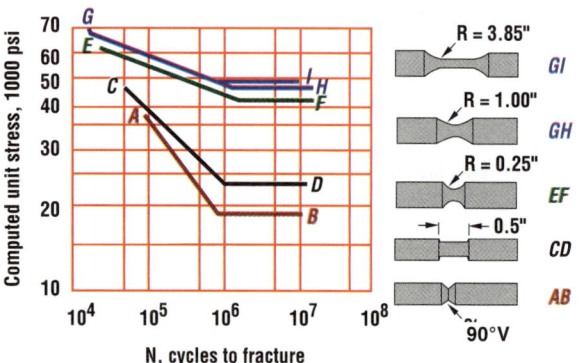

Figure 6.8—Effect of Notch Sharpness on Fatigue Performance

Figure 6.10—Examples of Weld Surface Discontinuities

thermal treatments. However, drastic changes can also occur if the chemical composition is changed. From a welding standpoint, of primary interest are alloys or mixtures of different elements, both metallic and nonmetallic. The most common example is steel, which is a mixture of iron and carbon, plus other elements occurring in various amounts.

In addition to mechanical properties, a metal's chemical composition will also have an effect on its corrosion resistance and weldability (the ease with which a metal can be successfully welded). Therefore, part of a welding inspector's duties might include the verification of a metal's chemical composition by comparing its actual chemistry with its specification requirements.

Alloy Groups

A welding inspector may be exposed to a number of different metal alloys. Metals can be grouped into many alloy categories; some common categories are steel, aluminum, nickel, and copper. This discussion is primarily concerned with steel alloys, further divided into three subcategories: plain carbon steels, low-alloy steels, and high alloy steels.

Based on tonnage, plain carbon steels are the most widely used. They contain primarily iron but also small amounts of carbon, manganese, phosphorus, sulfur and silicon. The amount of carbon present has the greatest effect on the metal's properties. Figure 6.11 shows the carbon content and some characteristics of plain carbon steels.

Low-alloy steels contain minor additions of other elements such as nickel, chromium, manganese, silicon, vanadium, columbium, aluminum, molybdenum, and boron. The presence of these elements in various amounts can result in remarkable differences in mechanical properties. These low-alloy steels can be generally classified as high strength low-alloy structural steels, automotive and machinery steels, steels for low temperature service, or steels for elevated temperature service. Many of these low-alloy steels have been classified according to their chemical composition, as shown in Figure 6.12. This classification was developed by the American Iron and Steel Institute (AISI) and the Society of Automotive Engineers (SAE) and is often used in steel fabrication.

The last group of steels is the "high alloy" types. Stainless and other corrosion resistant alloys are examples of this steel alloy group. Stainless steels contain at least 12% chromium, and many grades also contain significant amounts of nickel. Figure

Common Name	Carbon Content	Typical Use	Weldability
Ingot Iron	0.03% maximum	Enameling, galvanizing and deep drawing sheet and strip	Excellent
Low-Carbon Steel	0.15% maximum	Welding electrodes, special plate and shapes, sheet, strip	Excellent
Mild Steel	0.15%–0.30%	Structural shapes, plate and bar	Good
Medium Carbon Steel	0.30%–0.50%	Machinery parts	Fair (Preheat and frequently postheat required)
High Carbon Steel	0.50%–1.00%	Springs, dies, railroad rails	Poor (Difficult to weld without adequate preheat and postheat)

Figure 6.11—Types of Plain Carbon Steels

Series Designation	Types and Classes
10xx	Non-resulphurized carbon steel grades
11xx	Resulphurized carbon steel grades
13xx	Manganese 1.75%
23xx	Nickel 3.50%
25xx	Nickel 5.00%
31xx	Nickel 1.25%—chromium 0.65% or 0.80%
33xx	Nickel 3.50%—chromium 1.55%
40xx	Molybdenum 0.25%
41xx	Chromium 0.50—0.95%—molybdenum 0.12% or 0.20%
43xx	Nickel 1.80%—chromium 0.50% or 0.80%—molybdenum 0.25%
46xx	Nickel 1.55% or 1.80%—molybdenum 0.20% or 0.25%
47xx	Nickel 1.05%—chromium 0.45%—molybdenum 0.25%
48xx	Nickel 3.50%—molybdenum 0.25%
50xx	Chromium 0.28 or 0.40%
51xx	Chromium 0.80, 0.90%, 0.95%, 1.00%, or 1.45%
5xxxx	Carbon 1.00%—chromium 0.50%, 1.00%, or 1.05%
61xx	Chromium 0.80 or 0.95%—vanadium 0.10% or 0.15% minimum
86xx	Nickel 0.55%—chromium 0.50% or 0.65%—molybdenum 0.20%
87xx	Nickel 0.55%—chromium 0.50%—molybdenum 0.25%
92xx	Manganese 0.85%—silicon 2.00%
93xx	Nickel 3.25%—chromium 1.20%—molybdenum 0.12%
94xx	Manganese 1.00%—nickel 0.45%—chromium 0.40%—molybdenum 0.12%
97xx	Nickel 0.55%—chromium 0.17%—molybdenum 0.20%
98xx	Nickel 1.00%—chromium 0.80%—molybdenum 0.25%

Figure 6.12—AISI-SAE Designations of Carbon and Low-Alloy Steels

6.13 shows the compositions of some of these stainless steel types, which are divided into five groups: austenitic, martensitic, ferritic, precipitation hardening, and the duplex grades.

Effects of Chemical Elements in Steel

The following discussion describes the effects of various alloying elements on the properties of steel, including weldability.

Carbon—generally considered to be the most important alloying element in steel and can be present up to 2% (although most weldable steels have less than 0.5%). The carbon can exist either dissolved in the iron, or in a combined form such as iron carbide (Fe_3C). Increased amounts of carbon increase hardness and tensile strength, as well as response to heat treatment (hardenability). On the other hand, increased amounts of carbon reduce weldability.

Sulfur—usually an undesirable impurity in steel rather than an alloying element. Special effort is often made to eliminate it during steel making. In amounts exceeding 0.05% it tends to cause brittleness and reduce weldability. Alloying additions of sulfur in amounts from 0.10 to 0.30% will tend to improve the machinability of a steel. Such types may be referred to as "resulfurized" or "free-machining." The free-machining alloys are not intended for use where welding is required.

Phosphorus—generally considered to be an undesirable impurity in steels. It is normally found in amounts up to 0.04% in most carbon steels. In hardened steels, it may tend to cause embrittlement. In low-alloy high-strength steels, phosphorus may be added in amounts up to 0.10% to improve both strength and corrosion resistance.

Silicon—Usually only small amounts (0.20%) are present in rolled steel when it is used as a deoxidizer. However, in steel castings, 0.35 to 1.00% is com-

	Nominal Composition %					
AISI TYPE	17-7	C	0.09 max	Mn, max	—	Si, max
Austenitic Steels						
304	329	0.08 max	0.08 max	2.00	—	1.00
304L	3RE60	0.03 max	0.03 max	2.00	—	1.00
310	44LN	0.25 max	0.03 max	2.00	—	1.50
316		0.08 max		2.00		1.00
321		0.08 max		2.00		1.00
Martensitic Steels						
403		0.15 max		1.00		0.50
410		0.15 max		1.00		1.00
Ferritic Steels						
420		0.15 min		1.00		1.00
Precipitation Hardening						
430		0.12 max		1.00		1.00
446		0.20 max		1.50		1.00
Duplex						
15-5		0.07 max		—		—
17-4		0.07 max		—		—

Notes
(a) Other elements in addition to those shown above are as follows:
Phosphorus is 0.04% max in types 304, 304L, 310, 316, 321. Sulfur is 0.030% max in types 304, 304L, 310, 316, 321, 403, 410, 420, 430, and 446.

Figure 6.13—Compositions of Some Stainless Steels

monly present. Silicon dissolves in iron and tends to strengthen it. Weld metal usually contains approximately 0.50% silicon as a deoxidizer. Some filler metals may contain up to 1% to provide enhanced cleaning and deoxidation for welding on contaminated surfaces. When these filler metals are used for welding of clean surfaces, the resulting weld metal strength will be markedly increased. The resulting decrease in ductility could present cracking problems in some situations.

Manganese—Steels usually contain at least 0.30% manganese because it acts in a threefold manner: (1) assists in the deoxidation of the steel, (2) prevents the formation of iron sulfide inclusions, and (3) promotes greater strength by increasing the hardenability of the steel. Amounts up to 1.5% are found in carbon steels.

Chromium—a powerful alloying element in steel. It is added for two principle reasons; first, it strongly increases the hardenability of steel, and second, it markedly improves the corrosion resistance of alloys in oxidizing media. Its presence in some steels can cause excessive hardness and cracking in, and adjacent to, the weld. Stainless steels contain chromium in amounts exceeding 12%.

Molybdenum—This element is a strong carbide former and is usually present in alloy steels in amounts less than 1.0%. It is added to increase hardenability and elevated temperature strength. It is added to the austenitic stainless steels to improve pitting corrosion resistance.

Nickel—added to steels to increase their hardenability. It performs well in this function because it often improves the toughness and ductility of the steel, even with the increased strength and hardness it brings. Nickel is frequently used to improve a steel's toughness at low temperatures.

Aluminum—added to steel in very small amounts as a deoxidizer. It is also a grain refiner for improved toughness; steels with moderate aluminum additions are referred to as having been made to a *fine grain practice*.

Vanadium—The addition of vanadium will result in an increase in the hardenability of a steel. It is very effective in this role, so it is generally added in minute amounts. In amounts greater than 0.05%, there may be a tendency for the steel to become embrittled during thermal stress relief treatments.

Niobium (Columbium)—like vanadium, this is generally considered to increase the hardenability of a steel. However, due to its strong affinity for carbon, it may combine with carbon in the steel to result in an overall decrease in hardenability. It is added to austenitic stainless steels as a stabilizer to improve as-welded properties. Niobium is also known as columbium.

Dissolved Gases—Hydrogen (H_2), oxygen (O_2), and nitrogen (N_2) all dissolve in molten steel and can embrittle steel (and cause porosity) if not minimized. Steel refining processes are designed to eliminate as much of these gases as possible. Special fluxes or shielding gases are used to prevent their solution into the molten weld metal.

Alloy Groups

Aluminum Alloys—probably the largest group of nonferrous alloys used in the metalworking industry today. Available in both the wrought and cast forms, they are generally considered weldable. Aluminum is very desirable for applications requiring good strength, light weight, high thermal and electrical conductivity and good corrosion resistance. Commercially pure aluminum in the annealed or cast state has a tensile strength about 1/5 that of structural steel. Cold working increases the strength considerably, as does alloying the aluminum with other metals. Alloying with copper, silicon or zinc permits heat treating to increase strength. In some cases, the strength is increased to a point where it is comparable with steel.

There are two general categories in which the aluminum alloys may be placed: heat treatable and nonheat treatable. The heat treatable types get their hardness and strength from a process known as "precipitation hardening." The nonheat treatable grades are strengthened only by strain hardening (cold working) and by additions of alloying elements. Figure 6.14 lists the Aluminum Association designations for the various types of aluminum alloys, according to the major alloying elements present.

To indicate the condition of these various grades, a suffix can be added to the numeric designation. These standard temper designations are shown in Figure 6.15.

Nickel Alloys—Nickel is a tough, silvery metal of about the same density of copper. It has excellent resistance to corrosion and oxidation even at high temperatures. Nickel readily alloys with many materials and is a basis for a number of alloys in combination with iron, chromium and copper. Many of the high temperature alloys and corrosion resistant alloys have nickel percentages in the 60 to 75% range. These would include several proprietary alloys such as Monel 400®, Inconel 600®, and Hastelloy C-276®. Welding procedures similar to those used on steel are used on nickel and nickel alloys, and all of the common welding methods may be used.

Copper Alloys—Copper is probably best known for its high electrical conductivity, explaining why it is used extensively for electrical applications. It is approximately 3 times more dense than aluminum and has thermal and electrical conductivities which

Major Alloying Element	Aluminum Association Number
Pure Aluminum*	1xxx
Copper	2xxx
Manganese	3xxx
Silicon	4xxx
Magnesium	5xxx
Magnesium and Silicon	6xxx
Zinc	7xxx

*(99% minimum)

Figure 6.14—Aluminum Association Alloy Groups

Designation	Condition
F	As-fabricated
O	Annealed
H1	Strain hardened only
H2	Strain hardened and partially annealed
H3	Strain hardened and thermally stabilized
W	Solution heat-treated
T1	Cooled from an elevated temperature shaping process and naturally aged
T2	Cooled from an elevated temperature shaping process, cold worked, and naturally aged
T3	Solution heat-treated, cold worked, and naturally aged
T4	Solution heat-treated and naturally aged
T5	Cooled from an elevated temperature shaping process and then artificially aged
T6	Solution heat-treated and then artificially aged
T7	Solution heat-treated and stabilized
T8	Solution heat-treated, cold worked, and then artificially aged
T9	Solution heat-treated, artificially aged, and then cold worked
T10	Cooled from an elevated temperature shaping process, cold worked, and then artificially aged

Figure 6.15—Basic Temper Designation for Aluminum Alloys

are nearly 1-1/2 times greater. Copper is resistant to oxidation below temperatures of about 400°F in fresh and salt water, ammonia-free alkaline solutions, and many organic chemicals. However, copper reacts readily with sulfur and its compounds to produce copper sulfide. Copper and copper alloys are widely used for water tubing, valves and fittings, heat exchangers, and chemical equipment.

The alloys of copper can be divided into eight major groups, including:

- Coppers
- High copper alloys
- Brasses (Cu-Zn)
- Bronzes (Cu-Sn)
- Copper-nickels (Cu-Ni)
- Copper-nickel-zinc alloys (nickel silvers)
- Leaded coppers
- Special alloys

Although most of the copper alloys are weldable and/or brazable to some degree, their high thermal conductivity does present some problems. This factor tends to draw the welding or brazing heat away from the joint quite rapidly. Tenacious surface oxides which are present can also cause difficulties, so cleaning is critical. However, these alloys can be joined quite effectively using a wide variety of welding and brazing processes.

Destructive Testing

Once it is recognized that metal properties are important to the suitability of a metal or a weld, it becomes necessary to determine the actual values. That is, now the designer would like to put a number on each of these important properties so he or she can effectively design a structure using materials having the desired characteristics.

There are numerous tests used to determine the various mechanical and chemical properties of metals. While some of these tests provide values for more than one property, most are designed to determine the value for a specific characteristic of the metal. Therefore, it may be necessary to perform several different tests to determine all the desired information.

It is important for the welding inspector to understand each of these tests. The inspector should know when the test is applicable, what results it will provide, and how to determine if the test results are in compliance with the specification. It may also be helpful if the welding inspector understands some of the methods used in the testing, even though not directly involved with the actual testing.

Test methods are usually grouped into two classes, destructive or nondestructive. Destructive tests render the material or part useless for service once the test has been performed. These tests often determine how materials behave when loaded to failure. Nondestructive testing does not affect the component regarding its further usage, and will be discussed in Module 10.

Throughout this discussion, no mention will be made regarding the specific destructive test being used to determine a base metal or weld metal property. For the most part, this does not represent a significant change in the manner in which the test is performed. There will be occasions when a test is performed to specifically test the base metal or weld metal, but the mechanics of the testing operation will vary little, if any.

Tensile Testing

The first material property previously reviewed was strength, so the first destructive test method will be the tensile test. This one test provides us with a great amount of information about a metal. Some of the properties that can be determined as the result of the tensile test include:

- Ultimate Tensile Strength
- Yield Strength
- Ductility
- Percent Elongation
- Percent Reduction of Area
- Modulus of Elasticity
- Proportional Limit
- Elastic Limit
- Toughness

Some tensile test values can be determined through direct reading of a gauge. Others can be quantified only after analysis of the stress-strain diagram which can be produced during the test. The values for ductility can be found by making comparative measurements of the tensile specimen before and after testing. The percentage of that difference then describes the amount of ductility present.

When performing a tensile test, one of the most important aspects of that test involves the preparation of the tensile specimen. If this part of the test operation is conducted carelessly, the validity of the test results will be severely reduced. Small imperfections in the surface finish, for example, can result in significant reductions in the apparent strength and ductility of the tensile specimen.

Sometimes, the sole purpose of the tensile test of a welded sample is to simply show if the weld zone will perform as well as the base metal. For such an evaluation, all that is necessary is to remove a specimen (sometimes referred to as a strap) transverse to the longitudinal axis of the weld having the weld roughly centered in the specimen. The two cut sides should be parallel using a saw or cutting flame, but no further surface treatment is essential, including the removal of any existing weld reinforcement. However, often the weld reinforcement is ground flush.

This approach is used for procedure and welder qualification testing in accordance with API 1104. A successful tensile test done to this specification is described as a specimen which fails in the base metal, or in the weld metal if above the specified base metal strength.

For most cases in which the tensile test is required, however, there is a need to determine the actual amounts of strength and other properties exhibited by that metal, not just if the weld is as strong as the base metal. When the determination of these values is necessary, the specimen must be prepared in a configuration providing a reduced section somewhere near the center of the length of the specimen, as shown in Figure 6.16.

This reduced section is intended to localize the failure. Otherwise the failure might tend to occur preferentially near the grips, making subsequent measurements difficult. Also, this reduced section results in the increased uniformity of the stresses throughout the cross section of the specimen. This reduced section must exhibit the following three features in order that valid results can be obtained:

(1) The entire length of the reduced section must be a uniform cross section.

(2) The cross section should be a configuration which can be easily measured so a cross sectional area can be calculated.

(3) The surfaces of the reduced section should be free of surface irregularities, especially if perpendicular to the longitudinal axis of the specimen.

For these reasons, as well as the actual mechanics of preparing a specimen, the two most common cross sectional configurations for tensile specimens are circular and rectangular. Both are readily prepared and measured. If required to actually perform a tensile test, the welding inspector may have to be

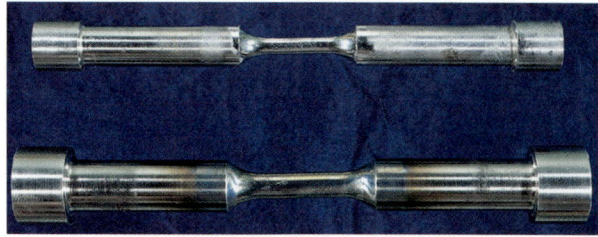

Figure 6.16—Tensile Specimen with Reduced Section

(handwritten at top: Load/Area = Tensile Strength)

able to calculate the actual cross sectional area of the reduced section of the tensile specimen.

Examples 1 and 2 show how these calculations are made for both common cross sections.

> **Example 1: Area of a Circular Cross Section**
>
> Area (circle) = π × r² or, = π × $\frac{d^2}{4}$
>
> Sample diameter, d = 0.505 in. (measured)
> Sample radius, r = d/2 = 0.2525 in.
> Area = 3.1416 × 0.2525²
> Area = 0.2 in.²
>
> or, calculated using the diameter directly,
>
> Area = 3.1416 × $\frac{0.505^2}{4}$
> A = 0.2 in.²

> **Example 2: Area of Rectangular Cross Section**
>
> Measured width, w = 1.5 in.
> Measured thickness t = 0.5 in.
> Area = w × t
> Area = 1.5 × 0.5
> Area = 0.75 in.²

The determination of this area prior to testing is critical because that value will be used to finally determine the strength of the metal. This strength will be calculated by dividing the applied load by the original cross sectional area. Example 3 shows this calculation for the standard circular cross section specimen used in Example 1.

> **Example 3: Calculation of Tensile Strength**
>
> Load = 12,500 lb. to break sample
> Area = 0.2 in.² (see Example 1)
> Tensile Strength = Load/Area
> Tensile Strength = 12,500/0.2
> Tensile Strength = 62,500 psi (lb/in.²)

The previous example shows a typical tensile strength calculation for a standard circular specimen. This is a standard specimen because it yields an area of exactly 0.2 in.². This is convenient since dividing a number by 0.2 is the same as multiplying that same number by 5. Therefore, if this standard tensile specimen is used, the calculation for tensile strength can be performed very simply as shown in Example 4.

> **Example 4: Alternate Tensile Strength Calculation**
>
> Load = 12,500 lb
> Area = 0.2 in.²
> Tensile Strength = 12,500 × 5
> Tensile Strength = 62,500 psi (lb/in.²)

The result of this calculation is identical to that of Example 3. The use of this standard size tensile specimen was very popular years ago before the advent of the modern calculators. At that time, it was easier to accurately machine a tensile specimen to this exact size than it was to arithmetically determine the strength by dividing the load by some more complicated number. However, today we can easily calculate the exact tensile strength no matter what the actual area happens to be.

Another operation that must be performed prior to testing is to accurately mark a gauge length on the reduced section. This gauge length is normally marked using a pair of center punches held at some prescribed distance apart. The most common gauge lengths are 2 and 8 in. After testing, the new distance between these marks is measured and compared to the original distance to determine the amount of elongation, or stretch, exhibited by that specimen when loaded to failure.

Percent elongation refers to the amount that the specimen has stretched between two "gauge marks" during the tensile test. It is calculated by dividing the difference between the final and original length between the marks by the original length, and multiplying the result by 100 to determine a percentage. An example of calculating the percent elongation follows:

> Original gauge length of 2.0 in.
>
> Final length between marks of 2.5 in.
>
> Percent Elongation = $\frac{2.5 - 2.0}{2.0}$ × 100 = 25%

When a ductile specimen is subjected to a tensile test, a portion of it will exhibit "necking," resulting from the application of the longitudinal tensile load. If we remeasure and calculate the final area of this smaller "necked-down" region, subtract it from the original cross sectional area, divide the remainder by the original area, and multiply the result by 100, this will provide a value for the percent reduction of area. An example of the percent reduction of area (RA) follows:

Original Cross Section Area of 0.20 in.²
Final Cross Section Area of 0.10 in.²

$$\text{Percent RA} = \frac{0.20 - 0.10}{0.20} \times 100 = 50\%$$

Once properly measured and marked, the specimen is then placed firmly in the appropriate grips of the stationary and moving heads of the tensile machine, such as the one shown in Figure 6.17.

Once in place, the tensile load is applied at a steady rate. Differences in this rate of loading could result in testing inconsistencies. Before load application, a device known as an extensometer is connected to the specimen at the gauge length marks. During the application of the load, the extensometer will measure the amount of elongation which results from a certain load. Both the load and elongation data are fed into a strip chart recorder to result in a plot of the variation in the elongation as a function of the applied load. This is referred to as a load versus deflection curve. However, we normally see tensile test results expressed in terms of stress and strain.

Stress is proportional to strength, since it is the applied load at any time divided by the cross sectional area. The strain is the amount of stretch apparent in a given length. Stress is expressed in terms of psi (lb/in.²) while strain is a dimensionless value expressed as in/in. When these values are plotted for a typical mild steel tensile test, the result would appear as in Figure 6.18.

The stress-strain diagram exhibits several important features that will be discussed. The test begins with the stress and strain both equal to zero. As the load is applied, the amount of strain increases linearly with stress. This zone shows what was previously referred to as elastic behavior, where the stress and strain are proportional. For any given material, the slope of this line is a constant value. This slope is the modulus of elasticity.

For steel, the modulus of elasticity (or Young's Modulus) at room temperature is approximately equal to 30,000,000 psi as compared to 10,500,000 psi for aluminum. What this number actually defines is the stiffness of the metal. That is, the higher the modulus of elasticity, the stiffer the metal.

Eventually, the strain will begin to increase faster than the stress, meaning that the metal is stretching more for a given amount of applied stress. This change marks the end of elastic behavior and the onset of plastic, or permanent deformation. The point on the curve showing the extent of the linear behavior is referred to as the elastic, or proportional, limit. If the load were removed at any time up to this point, the specimen would return to its original length.

Figure 6.17—Tensile Testing Apparatus

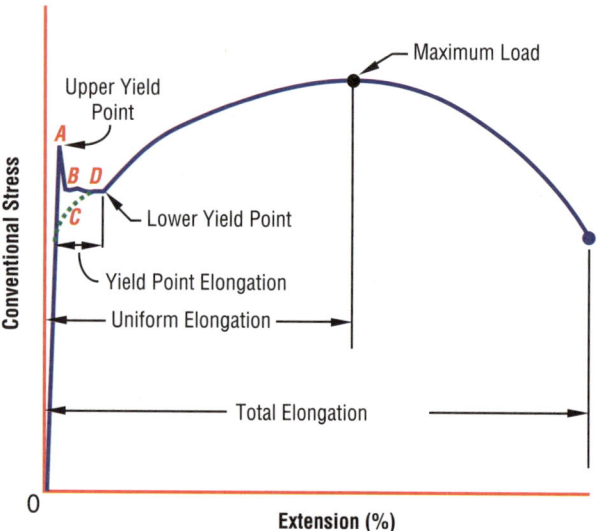

Figure 6.18—Typical Stress-Strain Curve for a Mild Steel

Many metals tend to exhibit a drastic departure from the initial elastic behavior. As can be seen in Figure 6.18, not only are stress and strain no longer proportional, but the stress may actually drop or remain steady while the strain increases. This phenomenon is characteristic of yielding in ductile steel. The stress increases to some maximum limit and then drops to some lower limit. These limits are referred to as the upper and lower yield points, respectively. The upper yield point is that stress at which there is a noticeable increase in strain, or plastic flow, without an increase in stress. The stress then drops and remains relatively constant at the lower yield point while the strain continues to increase during what is known as the yield point elongation.

For a metal exhibiting this behavior, the yield strength is the stress corresponding to the upper yield point, or some point midway between the upper and lower yield points. During the tensile test, the yield point can be seen as a drop in the gauge or recording device. Yield strength can be determined by observing and noting this load reduction. When this method is used, we refer to it as the "drop-beam" technique.

During this yielding phenomenon, the plastic flow of the metal is increasing at such a rate that stresses are being relieved faster than they are formed. When this plastic flow occurs at room temperature, it is referred to as *cold working*. This action causes the metal to become stronger and harder and it is said to be work hardened. The yielding will therefore continue until the metal becomes work hardened to the extent that it now requires additional stress to produce any further elongation. Corresponding to this, the curve begins to climb in a nonlinear fashion.

The stress and strain continue to increase at varying rates until some maximum stress is reached. This point is referred to as the maximum stress, or *ultimate tensile strength*. Figure 6.19 shows when this maximum stress is reached, it is followed by an apparent decrease in stress even though the strain continues to increase (engineer's curve). This phenomenon is due to the specimen beginning to "neck down" so that the actual cross section resisting the applied stress is less than the original area. Since the stress is calculated based on the original area, this gives the appearance that the

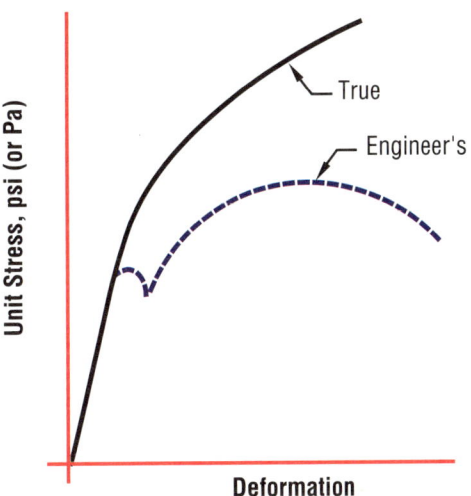

Figure 6.19—Comparison of True and Engineer's Stress-Strain Diagrams

load is dropping when in actuality it continues to increase on a psi basis.

If a tensile test is conducted where the stress is continually calculated based on the actual area resisting the applied load, the true stress-strain diagram can be plotted. A comparison between this true curve and the previously discussed engineer's curve is shown in Figure 6.19. It shows the specimen's strain continues to increase with increasing stress. This true curve shows failure occurs at both the maximum stress and strain.

For less ductile metals, there may not be a pronounced change in behavior between elastic and plastic deformation. Therefore the drop-beam method cannot be used to determine their yield strength. An alternate method is referred to as the offset technique. Figure 6.20 shows the typical stress-strain behavior for a less ductile metal.

When the offset method is used, a line is drawn parallel to the modulus of elasticity at some prescribed amount of strain. The amount of strain is usually described in terms of some percentage. A common offset is 0.2% (0.002) of the strain; however, other amounts may also be specified. Figure 6.21 shows how this offset line is drawn to provide a 0.2% offset.

The stress corresponding to the intersection of this offset line with the stress-strain curve is defined as the yield stress. It should be reported as a 0.2% offset yield stress so that others know how it was determined.

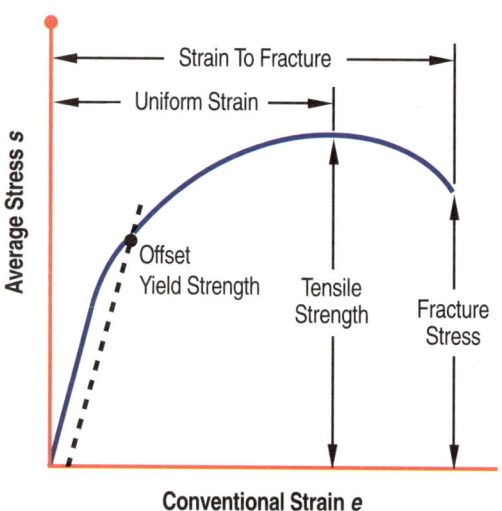

Figure 6.20—Typical Stress-Strain Diagram for Less Ductile Steel

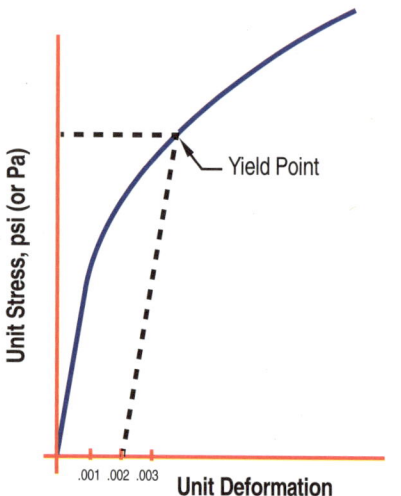

Figure 6.21—Determination of Yield Point by Offset Method

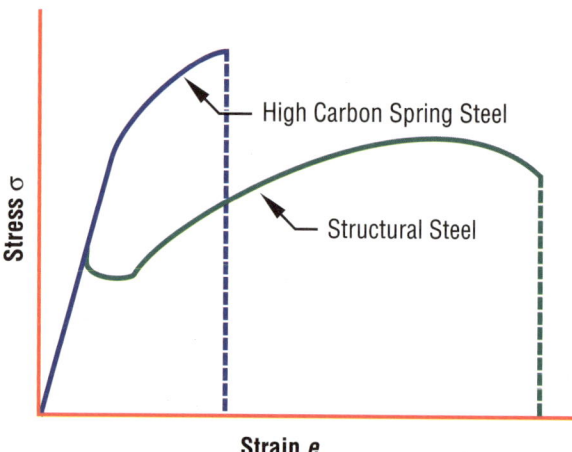

Figure 6.22—Stress-Strain Diagrams for High and Low Toughness Steels

tural steel is larger primarily due to the greater overall elongation even though the spring steel exhibited a higher tensile strength. The structural steel is therefore a tougher metal.

Following the actual tensile testing, it is now necessary to make a determination of the metal's ductility. This is expressed in two ways; percent elongation and percent reduction of area. Both methods involve making measurements before and after testing.

To determine the percent elongation, it is necessary to have placed gauge marks on the specimen before stressing. After the specimen has failed, the two pieces are placed together and the new distance between these gauge marks is measured. With data for the original and final length between the gauge marks, it is possible to calculate the percent elongation as shown in Example 5.

The final piece of information which can be gained from the stress-strain diagram is the amount of toughness exhibited by the metal. You recall that toughness is a measure of a metal's ability to absorb energy. You also learned that for slow, steady applications of load, toughness can be determined by the area under the stress-strain diagram. So, a metal having higher values for stress and strain is considered to be tougher than a metal with low values. Figure 6.22 shows a comparison between the stress-strain diagrams for a high carbon spring steel and a structural steel. If the areas under the two curves are compared, it is evident that the area under the struc-

---

Example 5: Determination of Percent Elongation

Original Gauge Length = 2.0 in.

Final Gauge Length = 2.6 in.

$$\% \text{ Elongation} = \frac{\text{Final Length} - \text{Original Length}}{\text{Original Length}} \times 100$$

$$\% \text{ Elongation} = \frac{2.6 - 2.0}{2.0} \times 100$$

$$\% \text{ Elongation} = \frac{0.6}{2.0} \times 100$$

% Elongation = 30%

Ductility can also be expressed in terms of how much it necks down during the tensile test. This is referred to as percent reduction of area, where the original and final areas of the tensile specimen are measured and calculated for comparison. Example 6 shows this calculation.

Example 6: Determination of Percent Reduction of Area (%RA)

Original Area = 0.2 in.²

Final Area = 0.1 in.²

% Reduction of Area =

$$\frac{\text{Original Area} - \text{Final Area}}{\text{Original Area}} \times 100$$

$$\% \text{ RA} = \frac{0.2 - 0.1}{0.2} \times 100$$

$$\% \text{ RA} = \frac{0.1}{0.2} \times 100$$

% RA = 50%

While both percent elongation and percent reduction of area represent expressions for the amount of ductility exhibited by a tensile specimen, their values will seldom, if ever, be equal. Typically, percent reduction of area will be approximately twice the value for percent elongation. Percent reduction of area is thought to be a representative expression for determining the ductility of a metal in the presence of some notch. However, we most often see percent elongation specified if only a single method is used.

Hardness Testing

Hardness is the ability of a metal to resist penetration or indentation. A metal's hardness permits an approximation of its tensile strength. Consequently, hardness tests are performed using some type of penetrator which is forced against the surface of the test object. Depending upon the type of hardness test being used, either the diameter or depth of the resulting indentation is measured. Hardness can also be measured using a variety of electronic or ultrasonic units, but this discussion is limited to the indentation methods.

Hardness of a metal is easily determined, primarily due to the vast variety of methods which can be used to determine hardness. Three basic groups of indentation hardness tests shall be discussed: Brinell, Rockwell and microhardness. In general, the three types differ from one another in the size of indentation produced. The Brinell is the largest and microhardness is the smallest.

The Brinell method is commonly used for determining the hardness of metal stock. It is well suited for this purpose because the indentation covers a relatively large area, eliminating problems associated with localized hard or soft spots in the metal. The characteristically higher loads used for Brinell tests also assist in reducing errors produced by surface irregularities.

Prior to Brinell testing, it is necessary to properly prepare the surface; this includes grinding or sanding the surface to achieve a relatively flat test area. The surface should also be of sufficient smoothness so that the size of the indentation can be accurately measured.

To perform a Brinell test, a penetrator is forced into the surface of the test object at some prescribed load. Once this load is removed, the diameter of the indentation is then measured using a graduated magnifier. Based on the size and type of the indenter, the applied load, and the resulting diameter of the impression, a Brinell Hardness Number (BHN) can be determined. Since this is a mathematical relationship, a BHN can be determined with a variety of indenter types and loads. Also, this BHN can be related to the actual tensile strength of carbon steels. That is, the BHN multiplied by 500 is approximately equal to the metal's tensile strength. This relationship does not apply to all alloys, but only carbon and low alloy steels.

A common Brinell test uses a 10 millimeter (mm) hardened steel ball and a 3000 kilogram (kg) load. However, as test conditions may change, such as specimen hardness and thickness, variations in the type and diameter of ball and the amount of the applied load may also be required. Other types of balls which can be used include the 5 mm hardened steel ball and the 10 mm tungsten carbide ball. For soft metals, loads as low as 500 kg may be used. Other loads between 500 and 3000 kg can also be used with equivalent results. Field testing with the Brinell method often uses a hammer blow to create indentations in both the test piece and a calibration block of known hardness. The hardness of the test piece is then determined by comparing its indentation diameter with the diameter created in the calibration test block.

BHN is normally determined by measuring the diameter of the impression and reading the hardness value from a table (see Figure 6.23). The usual steps for Brinell testing are:

1. Prepare test surface.
2. Apply test load.
3. Hold at load for prescribed time.
4. Measure impression diameter.
5. Determine BHN from table.

One important feature to note in the above procedure is a requirement to hold at the test load for some specified time. For iron and steel, this will be 10 to 15 seconds. Softer metals require holding times of approximately 30 seconds. When using some portable models, this hold time is simulated by maintaining the hydraulic load once the test load has been reached. Other test units may require an impact, and there is no hold time.

It is evident from this simple procedure how easily the Brinell test can be applied. Even with its simplicity, the test results will be quite accurate, provided sufficient care has been taken during each step of the operation. For additional information regarding Brinell testing, refer to ASTM E10, *Standard Test Method for Brinell Hardness of Metallic Materials*.

Quite often, there is a need for the testing of objects too large to be placed in a stationary Brinell test machine. In such cases, a portable test machine can be used. These come in a variety of types and configurations, but the basic test principals are identical.

The next type of hardness test to be discussed is the Rockwell method. This type encompasses numerous variations of the same basic principal but uses indenters of several diameters. The indenters used are the diamond Brale, shown in Figure 6.24, and 1/16, 1/8, 1/4, and 1/2 in. diameter hardened steel balls. The Rockwell tests result in smaller indentations than Brinell testing. This permits localized testing of a relatively small metal area.

Using one of these indenters, various loads can be applied to test most materials. The loads applied are much lower than those used for Brinell testing, ranging from 60 to 150 kg. There is also a group of Rockwell tests designated as "superficial." These are primarily used to determine the hardness of thin metal samples and wires; therefore the loads are significantly lower than those for the other types of Rockwell tests.

Just as with the Brinell test, the test surface must be properly prepared prior to applying a Rockwell test. Good technique is imperative for accurate hardness testing. Once the sample is prepared, the correct scale should be selected based on the approximate range of hardness expected. The "B" and "C" scales are by far the most commonly used scales for steel, with the "B" scale chosen for softer alloys and the "C" scale for the harder types. When in doubt as to which scale might be chosen for an unknown alloy, the "A" scale could be used because it includes a range of hardnesses covering both the "B" and "C" scales. Conversion tables have been prepared for converting hardness data from one scale to another.

Once the proper scale has been selected, and the test unit calibrated, the test object is placed on the anvil in the Rockwell bench testing machine. The anvil can be of various shapes depending upon the configuration of the test piece. The object must be adequately supported or test errors will result. The Rockwell method relies on the extremely accurate measurement of the depth of penetration of the indenter. So, if the test object is not properly supported, this measurement could be inaccurate. A variation in this depth measurement of only 0.00008 in. will result in a change of one Rockwell number. The bench model does the depth measurement automatically.

No matter which Rockwell scale is being used, the basic test steps are essentially the same. They are listed below.

1. Prepare test surface.
2. Place test object in Rockwell tester.
3. Apply minor load using elevating screw.
4. Apply major load.
5. Release major load.
6. Read dial.
7. Release minor load and remove part.

The minor load is used to take any looseness or slack out of the system, improving the accuracy of the test. Figure 6.26 shows graphically each of the test steps.

Indentation diameter mm	Brinell Hardness Load, kg			Indentation Diameter mm	Brinell Hardness Load, kg		
	500	1500	3000		500	1500	3000
2.45	104	314	627	4.45	30.5	91.5	183
2.50	100	301	601	4.50	29.8	89.5	179
2.55	96.3	289	578	4.55	29.1	87.0	174
2.60	92.6	278	555	4.60	28.4	85.0	170
2.65	89.0	267	534	4.65	27.8	83.5	167
2.70	85.7	257	514	4.70	27.1	81.5	163
2.75	82.6	248	495	4.75	26.5	79.5	159
2.80	79.6	239	477	4.80	25.9	78.0	156
2.85	76.8	231	461	4.85	25.4	76.0	152
2.90	74.1	222	444	4.90	24.8	74.5	149
2.95	71.5	215	429	4.95	24.3	73.0	146
3.00	69.1	208	415	5.00	23.8	71.5	143
3.05	66.8	201	401	5.05	23.3	70.0	140
3.10	64.6	194	388	5.10	22.8	68.5	137
3.15	62.5	188	375	5.15	22.3	67.0	134
3.20	60.5	182	363	5.20	21.8	65.5	131
3.25	58.6	176	353	5.25	21.4	64.0	128
3.30	56.8	171	341	5.30	20.9	63.0	126
3.35	55.1	166	331	5.35	20.5	61.5	123
3.40	53.4	161	321	5.40	20.1	60.5	121
3.45	51.8	156	311	5.45	19.7	59.0	118
3.50	50.3	151	302	5.50	19.3	58.0	116
3.55	48.9	147	293	5.55	18.9	57.0	114
3.60	47.5	143	285	5.60	18.6	55.5	111
3.65	46.1	139	277	5.65	18.2	54.5	109
3.70	44.9	135	269	5.70	17.8	53.5	107
3.75	43.6	131	262	5.75	17.5	52.5	105
3.80	42.4	128	255	5.80	17.2	51.5	103
3.85	41.3	124	248	5.85	16.8	50.5	101
3.90	40.2	121	241	5.90	16.5	49.6	99.2
3.95	39.1	118	235	5.95	16.2	48.7	97.3
4.00	38.1	115	229	6.00	15.9	47.8	95.5
4.05	37.1	112	223	6.05	15.6	46.9	93.7
4.10	36.2	109	217	6.10	15.3	46.0	92.0
4.15	35.3	106	212	6.15	15.1	45.2	90.3
4.20	34.4	104	207	6.20	14.8	44.4	88.7
4.25	33.6	101	201	6.25	14.5	43.6	87.1
4.30	32.8	98.5	197	6.30	14.2	42.8	85.5
4.35	32.0	96.0	192	6.35	14.0	42.0	84.0
4.40	31.2	93.5	187	6.40	13.7	41.3	82.5
				6.45	13.5	40.5	81.0

Figure 6.23—BHNs for Various Indentation Diameters and Loads

The results gained from the Rockwell test can then be related to Brinell values and therefore the tensile strength of the metal. Figure 6.27 shows how Brinell, Rockwell and tensile strength values are related.

For further information regarding the Rockwell tests, refer to ASTM E18, *Standard Test Methods for Rockwell Hardness and Rockwell Superficial Hardness of Metallic Materials.*

Like Brinell testing, there are also portable units which can be used to determine the Rockwell hardness of a metal. Although their operation may vary slightly from that of the bench models, the results will be equivalent.

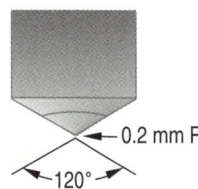

Figure 6.24—Rockwell Diamond Brale Indenter

The next type of hardness tests to be discussed is referred to as microhardness. They bear this name because their impressions are so small that high magnification is required to facilitate measurement. Microhardness testing is very beneficial in the investigation of metal microstructures, because they can be performed on single grains of a metal to determine the hardness in that microscopic region. Therefore, the metallurgist is primarily interested in this type of hardness testing.

There are two major types of microhardness tests, Vickers and Knoop. Both use diamond indenters, but their configurations are slightly different. Diagrams of the two types of resulting indentations are shown in Figure 6.25.

The square-based Vickers indenter provides an indentation in which the two diagonals are approximately equal. The Knoop indenter, however, makes an indentation having a long and a short dimension. As with the other test methods, you have a selection of test loads as well as indenter types. The term microhardness implies that the applied loads will range from 1 to 1000 grams (g). However, the majority of microhardness tests use loads in the range of 100 to 500 g.

To perform either Vickers or Knoop microhardness testing, the preparation of the surface is of utmost importance. Even the smallest surface irregularity can cause inaccuracies. Normally, for microhardness testing, the sample surface is prepared just as it would be for other metallographic investigations. The importance of this surface finish increases as the applied test load is reduced.

Once prepared, the specimen is securely clamped in a test fixture or holder so that the indentations can be accurately placed. Many microhardness machines employ a moving stage which facilitates accurate movement of the specimen without the need for its removal and readjustment. Such a device is required when taking a number of readings across a region of the metal. An example of this type of application would be the determination of the hardness variation across the weld heat-affected zone (HAZ). The result would be referred to as a microhardness traverse.

The steps used in the microhardness testing of a sample would be as follows:

1. Prepare test surface.
2. Place specimen in holding fixture.
3. Locate area of interest, using microscope.
4. Make indentation.
5. Measure indentation using microscope.
6. Determine hardness using tables or calculation.

The use of hardness testing will provide a great deal of useful information about a metal. However, the hardness method must be specified for a given application.

Toughness Testing

Another metal property of interest is toughness. You have learned that this property describes the metal's ability to absorb energy. When tensile testing was discussed, you learned that the toughness of a metal could be described as the area under the stress-strain curve. This is a value for the amount of energy that can be absorbed by a metal when the load is applied gradually.

However, in the discussion of toughness, you remember that when the load is applied rapidly, the concern is with notch toughness, or impact strength. The discussion which follows is concerned with tests

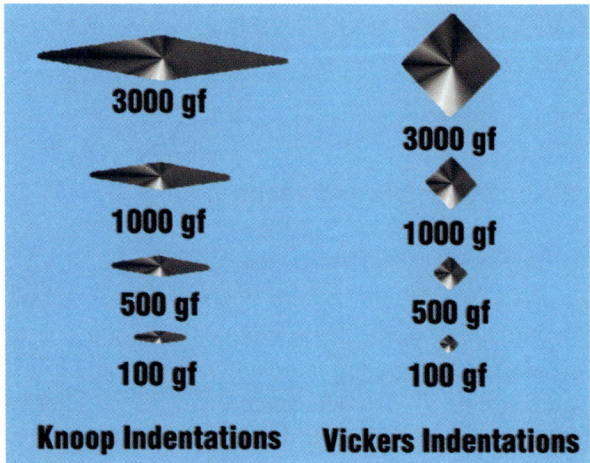

Figure 6.25—Microhardness Indentations

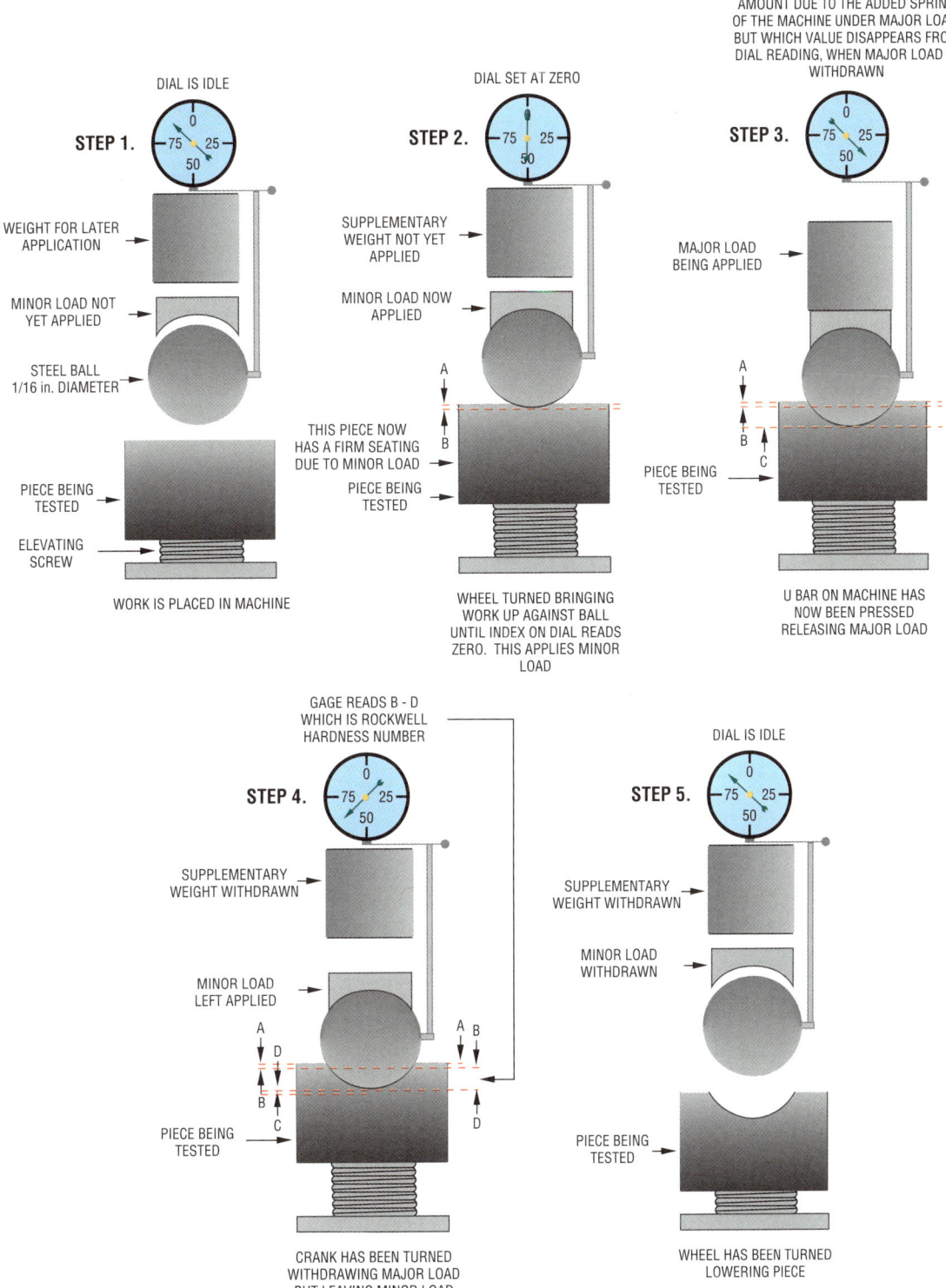

Figure 6.26—Steps Involved in a Rockwell Hardness Test

BRINELL		VICKER HARDNESS #	ROCKWELL		SHORE SCLEROSCOPE #	TENSILE STRENGTH 1000 psi
Diameter in. mm 2000 kg Load 10 mm Ball	Hardness No.		C 150 kg Load 120° Diamond Cone	B 100 kg Load 1/16 in. Diameter Ball		
2.05	898	.....	.....	.....	.....	440
2.10	857	.....	.....	.....	.....	420
2.15	817	.....	.....	.....	.....	401
2.20	780	1150	70	.....	106	384
2.25	745	1050	68	.....	100	368
2.30	712	960	66	.....	95	352
2.35	682	885	64	.....	91	337
2.40	653	820	60	.....	87	324
2.45	627	765	62	.....	84	311
2.50	601	717	58	.....	81	298
2.55	578	675	57	.....	78	287
2.60	555	633	55	120	75	276
2.65	534	598	53	119	72	266
2.70	514	567	52	119	70	256
2.75	495	540	50	117	67	247
2.80	477	515	49	117	65	238
2.85	461	494	47	116	63	229
2.90	444	472	46	115	61	220
2.95	429	454	45	115	59	212
3.00	415	437	44	114	57	204
3.05	401	420	42	113	55	196
3.10	388	404	41	112	54	189
3.15	375	389	40	112	52	182
3.20	363	375	38	110	51	176
3.25	352	363	37	110	49	170
3.30	341	350	36	109	48	165
3.35	331	339	35	109	46	160
3.40	321	327	34	108	45	155
3.45	311	316	33	108	44	150
3.50	302	305	32	107	43	146
3.55	293	296	31	106	42	142
3.60	285	287	30	105	40	138
3.65	277	279	29	104	39	134
3.70	269	270	28	104	38	131
3.75	262	263	26	103	37	128
3.80	255	256	25	102	37	125
3.85	248	248	24	102	36	122
3.90	241	241	23	100	35	119
3.95	235	235	22	99	34	116
4.00	229	229	21	98	33	113
4.05	223	223	20	97	32	110
4.10	217	217	18	96	31	107
4.15	212	212	17	96	31	104
4.20	207	207	16	95	30	101
4.25	202	202	15	94	30	99
4.30	197	197	13	93	29	97
4.35	192	192	12	92	28	95
4.40	187	187	10	91	28	93
4.45	183	183	9	90	27	91
4.50	179	179	8	89	27	89
4.55	174	174	7	88	26	87
4.60	170	170	6	87	26	85
4.65	166	166	4	86	25	83
4.70	163	163	3	85	25	82
4.75	159	159	2	84	24	80
4.80	156	156	1	83	24	78
4.85	153	153	.....	82	23	76
4.90	149	149	.....	81	23	75
4.95	146	146	.....	80	22	74
5.00	143	143	.....	79	22	72

Note: Figures in *italic* type are an approximation and are to be used only as a guide.

Figure 6.27—Hardness Conversion Table—Steel

which can be used to determine this particular metal property. Therefore, the various tests used to determine the notch toughness of a metal will use a specimen containing some type of machined notch and the load will be applied in a very rapid manner. You further recall that the temperature of the specimen has a significant effect on the test results, so the testing must be performed at a prescribed temperature.

Since the advent of interest in the notch toughness of metals, numerous different tests have been developed to measure this important property. When the energy absorption capabilities of a metal are discussed, it must be understood that the metal absorbs energy in steps. First, there is a certain amount of energy required to initiate a crack. Then, additional energy is required to cause that crack to grow, or propagate.

Some of the notch toughness tests can measure the energy of propagation separately from the energy of initiation while other methods simply provide us with a measure of the combined energy of initiation and propagation. It will be up to the engineer to specify the test method which will provide the desired information.

Although numerous types of notch toughness tests exist, probably the most common one used in this country is the Charpy V notch test. The standard specimen used for this test is a bar 55 mm long and 10 mm by 10 mm square. One of the long sides of the specimen has a carefully machined V-shaped notch 2 mm deep. At the base of this notch, there is a radius of precisely 0.25 mm. The machining of this radius is extremely critical, since small inconsistencies will result in large variations in test results. Standard Charpy specimens are shown in Figure 6.28.

Reduced cross section sizes commonly used when the metal sample is too small for the standard size sample include the three quarter, half, and one quarter samples. The square cross sections for these are 7.5 mm, 5.0 mm, and 2.5 mm, respectively. A caution when using these subsize samples; the toughness data generated from subsize samples is usually higher than the data from standard sized samples for the same material due to the mass effect. Thus, subsized Charpy sample data should not be compared with the standard sized sample data unless correction factors have been determined for the specific material. ASTM standard E23 covers impact testing in detail, and should be consulted for questions pertaining to size.

Once the specimen has been carefully machined, it is then cooled to the prescribed test temperature, if it is a temperature below room temperature. This can be accomplished using a variety of liquid or gaseous mediums; ice and water are often used for moderately cold temperatures, and dry ice and acetone are used for very cold temperatures. After the specimen is stabilized at the required temperature, it is then removed from the low temperature bath and quickly placed in the anvil of the testing machine. The arrangement of the anvil and the placement of the specimen is shown in Figure 6.29.

The machine used to perform Charpy impact testing is shown in Figure 6.30. The Charpy impact

Figure 6.28—Standard Charpy Specimens

Figure 6.29—Placement of Charpy Specimen in Anvil

Figure 6.30—Typical Charpy Testing Machine

tester consists of a pendulum with a striker head, an anvil, a release lever, a pointer, and a scale. Since we are intending to measure the amount of energy absorbed during the fracturing of a specimen, a given amount of energy is supplied by raising the pendulum to a specified height. Upon release, the pendulum will fall and continue through its stroke until it reaches a maximum height on the opposite side of its travel path. If it encounters no resistance, it will rise to a height which is designated as zero energy absorption. When it contacts the Charpy specimen, there is a certain amount of energy required to initiate and propagate a fracture in the specimen. This causes the pendulum to rise to a level below that for zero energy absorption. The maximum height of this swing is indicated by the pointer on the scale. Since this scale is calibrated, we can read the amount of energy required to break the specimen directly from the scale.

This value, referred to as breaking energy, is the primary information gained from the Charpy impact test. This energy is expressed in terms of foot-pounds of energy. While most Charpy results are expressed in terms of foot-pounds of energy absorption, there are other means of describing the notch toughness of a metal. They are determined by measuring various features of the failed Charpy specimen. These values are lateral expansion and percent shear. Lateral expansion is a measure of the amount of lateral deformation produced during the fracturing of the specimen. It is measured in terms of mils, or thousandths of an inch. Percent shear is an expression for the amount of the fracture surface which failed in a ductile, or shearing, fashion.

No matter which of these methods of measurement is used, we are usually concerned with the results from a whole series of tests. Once we have tested a number of specimens at various temperatures, we can determine how the values change with temperature. If we plot these values versus temperature, we will get curves which have upper and lower horizontal shelves with a near-vertical zone between. For each measurement category, there is some temperature at which the values drop rather abruptly. These temperatures are referred to as transition temperatures, which means that the behavior of the metal changes from relatively ductile to brittle at that temperature. The designer then knows that the metal should behave satisfactorily above that temperature. Examples of some of these transition curves are shown in Figure 6.31.

In addition to the Charpy test, there are others which can be applied for various applications. Other tests used to measure a metal's notch toughness include the drop weight nil-ductility, explosion bulge, dynamic tear, and crack tip opening displacement (CTOD). These tests employ different types of specimens as well as various ways of applying the load to the specimen.

Soundness Testing

This group of tests is designed to aid in the determination of the metal's soundness, or its freedom from imperfections. Soundness tests are routinely used for the qualification of welding procedures and welders. After a test plate has been welded, specimens are removed and then subjected to a soundness test to determine if the weld metal contained any imperfections or defects.

There are three general types of destructive soundness tests: bend, nick-break, and fillet break. (Soundness can also be determined nondestructively—RT and UT are often used.) The first type, bend testing, can be performed in a number of different ways. This is probably the most common test used to judge the adequacy of a welder's qualification test coupon.

The different types of bend tests are usually named based on the orientation of the weld with respect to the bending action. There are three types

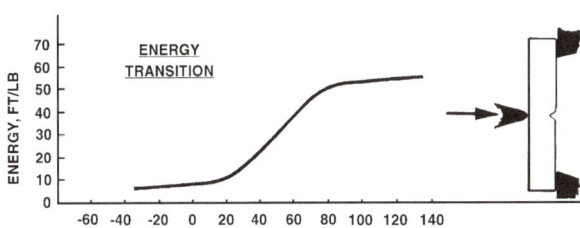

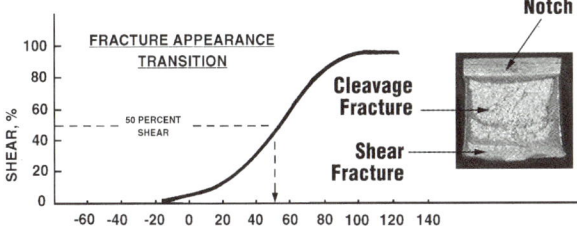

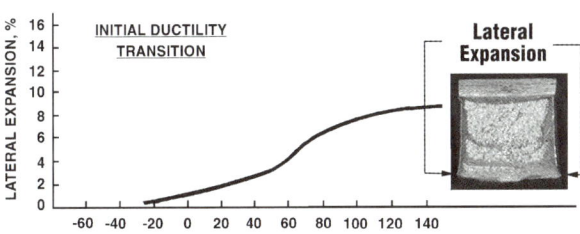

Figure 6.31—Transition Temperature Determination

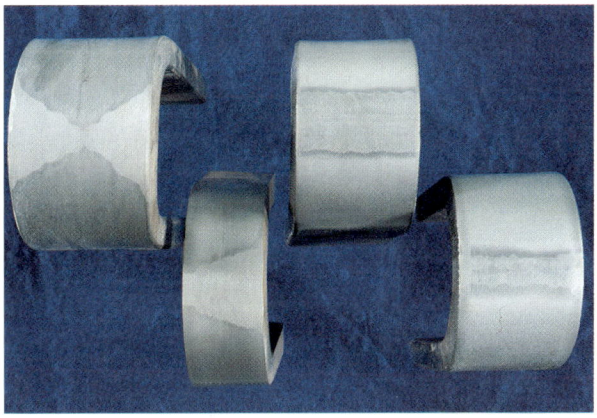

Figure 6.32—Typical Transverse Weld Bend Specimens

Figure 6.33—Guided Bend Test Jig

of transverse weld bend specimens: face, root, and side bends. With these three types, the weld lies across the longitudinal axis of the specimen and its type refers to the side of the weld which is placed in tension during the test. That is, the face of the weld is stretched in a face bend, the root of the weld is stretched in a root bend, and the side of a cross section of the weld is stretched in a side bend. Figure 6.32 shows sketches of these three types of bend specimens.

Bend tests are normally performed using some type of bend jig. There are three basic types: guided bend, roller equipped guided bend, and wrap-around guided bend. The standard guided bend test jig, shown in Figure 6.33, consists of a plunger (also referred to as mandrel or ram) and a matching die which forms the previously straight bend specimen into a U-shape.

To perform a bend test, the specimen is placed across the shoulders of the die with the side to be placed in tension facing down toward the inside of the die. The plunger is then situated over the area of interest and forced toward the die causing the specimen to be bent 180° and become U-shaped. The specimen is then removed and evaluated.

The second type of guided bend test jig is similar to the standard guided bend jig except it is equipped with rollers instead of hardened shoulders on the die portion. This reduces the friction against the specimen allowing for lower loads to achieve the bending. The last common type of guided bend jig is referred to as the wrap-around jig. It bears this name because the specimen is bent by being wrapped around a stationary pin, as shown in Figure 6.34.

Some qualification tests for mild steel require that the specimen be bent around a mandrel having a diameter four times the thickness of the specimen.

Figure 6.34—Wrap-around Bend Test Jig

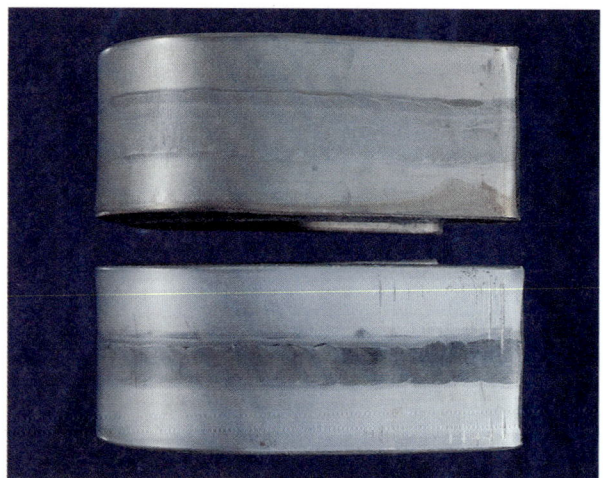

Figure 6.35—Longitudinal Bend Specimen

Therefore, a 3/8 in. specimen would be bent around a 1-1/2 in. diameter mandrel. This results in about 20% elongation of the outer surface of the bend specimen. If a smaller bend mandrel is used, the amount of elongation would increase.

In qualification test coupons where the weld metal is much stronger than the base metal, there may be a tendency for the specimen to kink in the base metal next to the weld rather than forming smoothly around the mandrel. If a wrap-around jig is not available, it may be beneficial to select longitudinal bend specimens rather than the standard transverse types. The weld lies in line with the longitudinal axis of the longitudinal bend specimen. This is shown in Figure 6.35.

With any of these bend tests, the specimens must be carefully prepared to prevent test inaccuracies. Any grinding or sanding marks on the tension surface should be oriented in the same direction as bending so they don't provide transverse notches (stress risers) which could cause the specimen to fail prematurely. The corners of the specimen should be radiused or chamferred to relieve the corner stress concentration as well. For specimens removed from pipe coupons, the side of the bend specimen against the ram can be ground flat to eliminate the bending in the direction transverse to the bending direction.

The acceptability of bend test specimens is normally judged based on the size and/or number of discontinuities which appear on the tension surface. The governing code or specification will dictate the exact acceptance/rejection criteria.

The next type of soundness test to be discussed is the nick-break test. This test is used almost exclusively by the pipeline industry as described in API 1104. This method judges the soundness of the weld by fracturing the specimen through the weld so the fracture surface can be examined for the presence of discontinuities. The fracture is localized in the weld zone by the use of saw cuts along two or three surfaces. A typical nick-break specimen is shown in Figure 6.36.

Once the specimen has been saw cut, it is then broken by pulling in a tensile machine, hitting the center with a hammer while supporting the ends or hitting one end with a hammer while the other end is held in a vise. The method of breaking is not significant because the interest is not in how much effort is required to break the specimen. The goal is to fracture the specimen through the weld zone so it can be determined if any imperfections are present. The fracture surface is then examined for any areas of slag inclusions, porosity or incomplete fusion. If present, they are measured and accepted or rejected

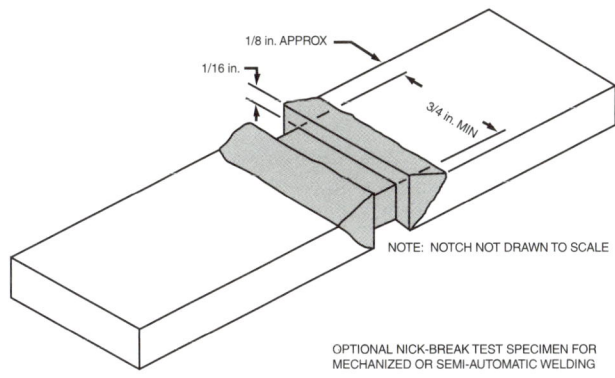

Figure 6.36—Nick Break Specimen

based on the code limitations. The requirements for API 1104 are shown graphically in Figure 6.37.

The last soundness test to be mentioned is the fillet weld break test. Like the other two types, this soundness test is used primarily in the qualification of welders. This is the only test required for the qualification of tackers in accordance with AWS D1.1. A fillet weld break sample is shown in Figure 6.38. Once welded, the specimen is broken by loading it as shown in Figure 6.39.

With this test, the inspector is looking for a weld having satisfactory surface appearance. Further, the fracture surface is examined to assure that the weld has evidence of fusion to the joint root and that there are no areas of incomplete fusion to the base metal or porosity larger than 3/32 in. in their greatest dimension.

These soundness tests are used routinely in many different industries. Their application and evaluation appear to be quite straight-forward.

However, the welding inspector should be aware that the evaluation of these tests may not be as simple as the various codes and specifications might imply. For this reason, it is desirable for the welding inspector to actually spend some time performing these tests to become familiar with their performance and interpretation.

Fatigue Testing

The last mechanical test method to be discussed is fatigue testing. This is a type of test which enables the determination of the fatigue strength of a metal. Fatigue loading is the cyclic loading of a member. Fatigue tests help designers determine how well a metal will resist failure when cyclically loaded in fatigue. Normally a series of fatigue tests are completed to arrive at the endurance limit for a metal. Tests are conducted at various stress levels until some maximum stress is found below which the metal should exhibit infinite fatigue life.

Figure 6.38—Fillet Weld Break Specimen

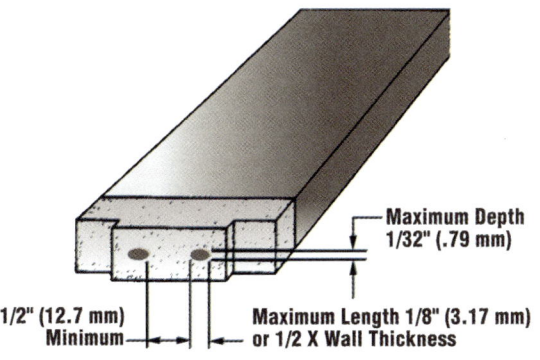

Figure 6.37—Evaluation of Nick Break Specimen

Figure 6.39—Method of Rupturing Fillet Break Specimen

Since fatigue is so strongly influenced by surface finish and configuration, the preparation of fatigue specimens is extremely critical. Only minor blemishes are necessary to cause significant changes in the results. So, if sufficient care is not taken at this stage, all final results could be invalid.

Fatigue tests can be performed in a number of different ways. The specific test to be used depends upon the expected loading of the metal in service. That loading could be planar bending, rotational bending, torsion, axial tension, axial compression, or combinations of these. When loading in the axial, or longitudinal direction, the cycles could be such that the specimen is loaded in alternating tension and compression. This is usually the most severe case. The welding inspector must be aware of the fatigue aspects of metals, but is seldom involved in fatigue testing of metals.

Destructive Tests for Chemical Properties

The tests which have been previously discussed are used to determine the mechanical properties of a metal. There are also important chemical properties of metals. In fact, the chemical makeup of a metal and its heat treatment determine to a great degree the mechanical properties of that metal. There is often a need to determine the chemical composition of a metal. Three common methods are spectrographic, combustion, and wet chemical analysis.

The welding inspector will rarely be required to actually perform chemical analysis. However, he or she may have to assist in the taking of samples for analysis, or the review of the analyses to determine if a metal complies with a particular specification. For more information regarding chemical analysis of metals, refer to the ASTM specifications which cover this subject. The particular methods used for steel are listed in ASTM A 751, *Standard Methods, Practices, and Definitions for Chemical Analysis of Steel Products.*

Metal analysis can be done in the field using X-ray fluorescence techniques. While this technique has limitations in elemental analysis, it can be very helpful in avoiding material mixups and the sorting of alloy types. When only sorting is needed, there are test kits based on magnetic properties and qualitative color changes from reagents which are very helpful. There are also portable spectrographic units available when more accurate field analyses are needed.

Another group of tests which can generally be classified as chemical tests are corrosion tests. These are specific tests designed to determine the corrosion resistance of a metal or combination of metals. Losses from corrosion of metals cost industry billions of dollars annually. One estimate places this annual loss at 120 billion dollars. Designers are very concerned about how a metal will behave in a particular corrosive environment. The tests used to determine the degree of corrosion resistance are designed to simulate as closely as possible the actual conditions which the metal will encounter during its service. Some of the considerations which must be addressed when setting up a corrosion test are chemical composition, corrosive environment, temperature, presence of moisture, presence of oxygen or other metals, and amount of stress. If any of these features are ignored, the corrosion test may yield invalid results.

Metallographic Testing

Another way to learn about the characteristics of a metal or a weld is through the use of various metallographic tests. These tests basically consist of removing a section of a metal or a weld and polishing it to a high degree. Once prepared, the specimen can then be evaluated with the unaided eye or with the use of magnification.

Metallographic testing is generally classified as either macroscopic or microscopic. They differ in the amount of magnification that is used. Macro tests are generally performed at magnifications of 10X or lower. Micro tests, on the other hand, use magnifications greater than 10X, usually 100X or higher.

A number of different features can be observed on a typical macro specimen. A weld cross section can provide a macro specimen for determining such things as depth of fusion, depth of penetration, effective throat, weld soundness, degree of fusion, presence of weld discontinuities, weld configuration, number of weld passes, etc. A picture of a macro specimen is referred to as a photomacrograph. Typical photomacrographs are shown in Figure 6.40.

Micrographic samples can be used to determine various features as well. Included are microstruc-

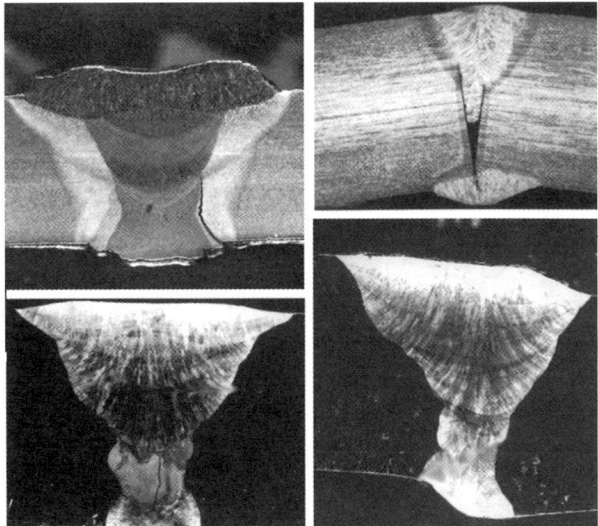

Figure 6.40—Weld Photomacrographs

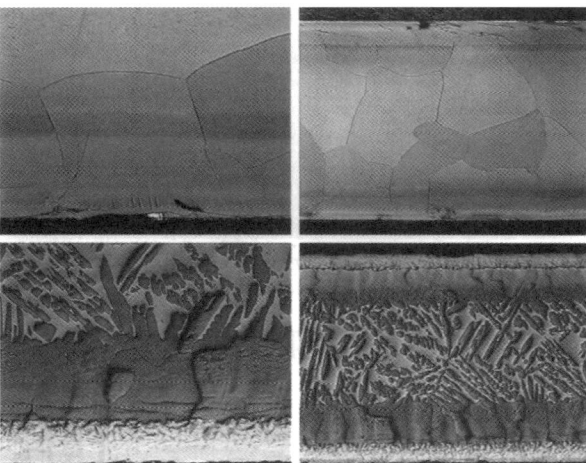

Figure 6.41—Typical Photomicrographs

tural constituents, presence of inclusions, presence of microscopic defects, nature of cracking, etc. Similarly, pictures of micrographic specimens are called photomicrographs. Figure 6.41 shows some typical photomicrographs. Both metallographic tests can be very helpful in such matters as failure analysis, weld procedure and welder qualification, and process control testing.

The two types of specimens also differ in the amount of preparation required. Some macro specimens need only be rough ground with an 80 grit finish, while micro specimens require very fine grinding at 600 grit and polishing to produce a mirror finish. Both usually require etching to reveal the structure. Etching is done by applying etching reagents which remove the surface layer, exposing the grain structure below. Considerable information can be gained about the properties of metals by making simple macro and micro evaluations. Metallographic testing is an important tool for both the welding inspector and the engineer.

Summary

There have been numerous details provided in this Module regarding some of the more important destructive test methods available to the welding inspector to determine the various material properties. While the welding inspector may never be responsible for the actual performance of the testing, it is important that he or she understand what information these tests can provide so they can be used as examination tools. While many of these tests appear to be straightforward, they often involve more than meets the eye. Therefore, the inspector should work with an experienced person before trying to perform any of these operations.

Key Terms and Definitions

alloy—a mixture of elements creating a metal. Steel is an alloy of iron and carbon.

annealed—the heat treat condition of metals exhibiting minimum strength and hardness.

API—American Petroleum Institute.

ASTM—American Society for Testing and Materials.

austenite—a metal phase found in steels at elevated temperature and stainless steels at room temperature.

austenitic—a term applied to the group of stainless steels whose room temperature stable phase is austenite.

Brinell—a type of macrohardness test.

brittle—the behavior of metals that break without deformation; materials with little or no ductility.

carbide former—an element which promotes the formation of its metallic or nonmetallic carbides.

Charpy—a type of impact test.

cold work—permanent deformation of a metal below its transformation temperature.

crystals—in metals, the very small, individual zones which form on solidification from the liquid. Also referred to as grains.

density—the ratio of mass per unit volume. For metals, density is usually noted as grams/cubic centimeter.

directional properties—the differences in a metal's mechanical properties depending on rolling direction during manufacture.

discontinuity—any interruption of the normal pattern of a metal; examples are porosity, incomplete fusion and slag inclusions. A rejectable discontinuity is also referred to as a defect.

ductile—the behavior of metals that exhibit ductility under load to failure.

ductility—the ability of a metal to deform under load without breaking.

duplex—a term referring to a group of stainless steels having two stable phases at room temperature.

elastic behavior—deformation of a metal under load without permanent deformation. As the load is removed, the metal returns to its original shape.

elastic limit—see Proportional Limit.

elongation—the stretching of a material, either elastic or plastic. Percent elongation is a measure of a metal's ductility.

endurance limit—in metals, the applied stress at which the metal will not fail, regardless of the number of fatigue cycles.

fatigue—in reference to design, an applied cyclic stress; a mode of failure when metals are subject to cyclic loading.

fatigue strength—the relative ability of a metal to withstand cyclic loading, as in stress reversal.

ferrite—a phase found in steels; also referred to as *alpha*.

ferritic—a term referring to a group of stainless steels whose room temperature stable phase is ferrite.

gauge length—in tensile testing, the distance between two small marks placed on the sample prior to applying the load. Usually a distance of 2 or 8 in.

grains—see Crystals.

hardenability—the relative ability of a metal to be hardened, usually by rapid quenching.

hardness—the ability to resist indentation or penetration.

HAZ—heat-affected zone; the zone of the base metal, not melted during welding but undergoing metallurgical changes from heat.

impact strength—the relative ability of a metal to absorb an impact load.

impact testing—a group of tests that rapidly apply a load, an impact, to a metal sample. Examples are Charpy, explosion bulge, and drop-weight nil-ductility tests.

kg—abbreviation for kilogram. One kg is approximately equal to 2.2 pounds.

Knoop—a type of microhardness test.

ksi—strength or pressure in thousands of pounds per square inch. A 70,000 psi tensile strength can also be written as 70 ksi.

lateral expansion—a measurement of the deformation of a Charpy sample on breaking.

martensite—a phase found in steels which forms on rapid cooling or quenching.

martensitic—a term applied to a group of stainless steels whose room temperature stable phase is martensite.

mil—linear measure; one mil equals 0.001 in.

mm—abbreviation for millimeter. One mm is approximately equal to 39.37 mils.

modulus of elasticity—the ratio between the stress applied and the resulting elastic strain; the slope of a metals' elastic limit curve; a relative measure of a materials' stiffness. Also called Young's Modulus.

neck-down—a reduction of cross sectional area of a ductile metal at the fracture point when a tensile load causes failure.

notch sensitive—referring to a metal which has low notch toughness.

notch toughness—the ability of a metal to absorb energy without failure when surface notches are present.

pi—a constant number, derived by dividing the diameter of a circle into its circumference. It is 3.14159 (5 places). The symbol is π.

plastic behavior—permanent deformation of a metal under applied load. The metal does not return to its original shape after the load is removed.

postheating—heating of completed weld and base metal after welding.

precipitation hardening—a term applied to alloys which are hardened by the formation of a hardening precipitate on heat treatment. A group of stainless steels.

preheating—heating of base metal and/or filler metal prior to welding.

proportional limit—the elastic limit of a metal, beyond which yield or plastic deformation occurs.

psi—pounds per square inch. The units of measurement used for strength and pressure.

quenching—cooling very rapidly from an elevated temperature. A method of increasing the hardness of heat treatable steels.

Rockwell—a type of macrohardness test.

rolling direction—in metal manufacture, the longitudinal direction of rolling; in the same direction as the rolling.

shear—in metal fractures, a ductile mode of failure.

slag inclusion—a discontinuity in metals, usually non-metallic as an oxide or sulfide.

S-N curve—a curve generated from data relating the number of cycles and the applied stress levels to cause failures of metal samples.

soundness—in metal testing, reference to freedom from imperfections. Soundness tests include bend, nick-break and fillet break.

spectrographic testing—a testing technique for determining a metal's chemistry.

stainless steel—by definition, an alloy containing 12% chromium or greater.

strain hardening—an increase in metal hardness and strength due to the application of a strain (permanent deformation or cold work) to a metal.

stress raiser—any surface blemish or geometry that increases the applied stress at a particular point on a component. Examples are weld ripples, shaft keyways, surface scratches.

stress relief—a controlled heat treatment which relieves residual stress in metals.

tempering—a heat treatment, usually of a quenched steel, that reduces hardness and restores ductility and toughness.

tensile strength—see Ultimate Tensile Strength.

transition temperature—in impact testing, the temperature at which the metal fracture changes from ductile to brittle.

torsion—a twisting or rotational force.

toughness—ability of a metal to absorb slowly applied energy. See Notch Toughness and Impact Strength.

transverse—lying across, as side to side when compared to the rolling direction of metal.

ultimate tensile strength—maximum load carrying capacity of a material. Abbreviated as UTS.

Vickers—a type of microhardness test.

weldability—the capacity of a material to be welded under the imposed fabrication conditions into a specific, suitably designed structure, and to perform satisfactorily in the intended service.

X-ray fluorescence—a nondestructive testing technique for determination of a metal's chemistry.

yield strength—the load at which a material will begin to yield, or permanently deform. Also referred to as the Yield Point.

Young's Modulus—see Modulus of Elasticity.

Module 7
Metric Practice for Welding Inspection

Contents

Introduction 7-2

Positive and Negative Number Usage 7-4

Scientific Notation 7-6

Round Off Convention 7-8

Conversion Factors 7-8

Additional SI Style and Usage Conventions 7-12

Summary 7-13

Key Terms and Definitions 7-16

Module 7—Metric Practice for Welding Inspection

Introduction

For many years, there have been efforts to convert the official United States system of measurement to the far more widely used international system. To date, the conversion remains voluntary, and is still not the *law of the land*. However, the Federal Government has initiated the requirement that all new Federal scientific and engineering documents are to be published using the international system. This international system is used by every other major nation in the world, and is referred to as "Le Systeme Internationale d'Unites," abbreviated as SI. The common name for this system in the United States is the metric system. The system currently in use in the United States is referred to as U.S. Customary and is abbreviated as US.

The international system offers many advantages over our existing system, but it is resisted by industry for many reasons. A major reason is economic; conversion to a new system requires expenditures for retooling, workforce retraining, manufacture of new drawings, and even design changes in many cases. However, many industries are making the measurement conversion voluntarily to improve their global marketing position, and within the United States the current system used is a mixture of the old and the new. One example of metric use in the United States is the soft drink and distilled spirit industries which now market their product in liters, or milliliters, rather than the US system of pints, quarts, and gallons. Another example is metric-sized fasteners found in abundance on new U.S. manufactured cars.

Thus, it is fast becoming a requirement for the United States workforce to have a knowledge of both systems to work more accurately and effectively. For those industries choosing to compete in global markets, use of the metric system, or SI, is an economic imperative. Fabrication drawings, product dimensions, shipping cartons, weights, etc., must all be converted to the international system to conform to world-wide requirements. The welding industry is no exception, and this module will discuss the common terms and systems necessary for operating in both the conventional US system and the SI system. Converting from one system to the other does require learning some new rules, especially for calculations; these rules will be covered and examples given for clarity.

The American Welding Society developed a standard, AWS A1.1, *Metric Practice Guide for the Welding Industry* (see Figure 7.1) to assist the welding industry in its transition to the use of the SI system. The Foreword from that document states:

"(This Foreword is not a part of AWS A1.1:1998, Metric Practice Guide for the Welding Industry, but is included for information purposes only.)

The present AWS Policy on Metrication states, in part, that 'The AWS supports a timely transition to the use of SI units. The AWS recognizes that the U.S. customary system of units will eventually be replaced by the SI units. To delay the transition to SI units and to lengthen unnecessarily the transition period results in greater costs and confusion and increases the loss of compatibility with the international market.'

At the present time, the U.S. stands alone as the only industrial country that still predominantly uses the inch-pound system of measurement. Since the signing of the Metric Act of 1975 by President Ford and an initial flurry of transition, the voluntary feature of the Act allowed the impetus to stagnate. We now find ourselves at odds not only with other industrial countries, but also, in many cases, with each other.

Figure 7.1—AWS A1.1, Metric Practice Guide for the Welding Industry

Many major companies—including General Motors Corporation, Ford Motor Company, Chrysler Corporation and an estimated 70 percent of the Fortune 500—have made the switch in some aspect of their businesses. But smaller firms—typically those with fewer international interactions—have been slower to change.

More recently, the Omnibus Trade and Competitiveness Act, which was signed by President Reagan in August 1988, designated the metric system as the preferred measurement method in trade and commerce. Specifically, this act requires each federal agency to use the metric system in its procurements, grants and other business-related activities by the end of fiscal year 1992.

This standard is intended to facilitate this transition.

Comments and suggestions for the improvement of this standard are welcome. They should be sent to the Secretary, A1 Committee on Metric Practice, American Welding Society, 550 N.W. LeJeune Road, Miami, Florida 33126."

It is evident from the preceding statement from the standard AWS A1.1 that AWS supports the conversion to the SI system, but has not made it mandatory as of this date for its documents. AWS A1.1 is a review of the SI system, noting specifically the standard conventions for its usage, and it also lists those common terms related to the welding industry. Excerpts from AWS A1.1 will be used throughout this module to point out the proper usage of the SI system, but one must keep in mind, the usage is voluntary, not mandatory. The information is being presented to increase your general knowledge of the SI system and to improve your effectiveness in dealing with today's global markets.

To begin the review for SI conversion, a discussion of our existing system is valuable to see just how complicated it really is. Since most are familiar with its complexity, it is often thought of as being simple, but in fact, it is very complex. For starters, think how many terms, or unit values, there are for the simple measurement of length. Commonly, units of inches, feet, yards, and miles are used for measuring length, as well as some others such as furlongs, leagues, fathoms, and many, many more. All these different terms to measure just one dimension, that of length. And while one can convert each of these units to any other, the conversion factors are very inconvenient and seldom in multiples of 10. Most have had to learn there are 12 inches in a foot, 36 inches or 3 feet in a yard, and 5,280 feet or 1,760 yards in a mile.

And there is a similar problem for measuring liquid volumes in our US system: liquid ounces, pints, quarts, gallons, cubic feet, etc. To make it even more confusing, sometimes the same word is used for two different cases. An example is the base unit, ounce, used for both volume and mass (weight). Ounce can signify a volume, as in 128 per gallon, or a weight, as in 16 per pound. But the US system is preferred in the United States because of our familiarity with it and people are resistant to change.

The metric system, when compared to our US system, is very simple. Since most Americans are unfamiliar with it, it seems difficult to many, especially those with many years of past usage of the US system. However, the metric system is quickly learned, and offers many advantages over our present system primarily because it has only one primary base unit for each needed measurement, and it operates consistently with multiples of 10 on the base units for larger values. Using a multiplier base of 10 also permits the use of a simple decimal system for values less than one. Examples of several common base units are shown in Table 7.1.

Note that length is always expressed in the base unit of meters; mass, or weight as commonly used, is always expressed in kilograms. *(Editor's Note: The SI system uses the gram as the basic mass unit, but AWS standard A1.1 has chosen the kilogram as its preferred base mass unit. Thus, the AWS base mass unit includes the prefix "kilo" and it is incorrect to apply additional prefixes to the kilogram.)* Liquid volume is measured in liters. Larger or smaller values for these measurements require a prefix, or multiplier, placed in front of the base unit; Table 7.2 lists several common prefixes (the kilogram is the only exception to this last note; mass is always denoted by AWS in kg). Thus, distances between cities are measured in kilometers (one kilometer is equal to 1000 meters), while small dimensions may be measured in millimeters (one millimeter is 1/1000 of a meter). In addition to the measurement units shown in Table 7.1, there are several terms relating to welding and these are shown in Table 7.3.

A common challenge for the welding inspector is the conversion of SI units to the US units, or vice versa. Since these conversions usually require multiplying the given unit by a multiplier, and these multipliers are frequently given in the shorthand "Scientific Notation" form, a review of some basic arithmetic principles is very helpful. The following discussion will enable the inspector to confidently handle conversion multipliers in the various forms such as 7.89×10^7 or 3.45×10^{-3}.

Property	SI Unit	Symbol
Length	meter	m
Mass	kilogram	kg
Volume	liter	L
Temperature	Celsius	C
Time	second	s
Pressure, Stress	pascal	Pa
Energy	joule	J
Electric current	ampere	A
Frequency	hertz	Hz

Table 7.1—Common SI Units of Measurement

Exponential Expression	Multiplication Factor	Prefix	Symbol
10^6	1,000,000	mega	M
10^3	1,000	kilo	k
10^{-1}	0.1	deci	d
10^{-2}	0.01	centi	c
10^{-3}	0.001	milli	m
10^{-6}	0.000001	micro	μ

Table 7.2—Common SI Prefixes and Symbols

Positive and Negative Number Usage

The first step is to review the use of positive and negative numbers and common operations using these numbers. The second step is to review the use of scientific notation, which is used in conversion of SI units to US Customary and vice versa. Positive numbers are numbers greater than zero, and negative, or minus numbers, are numbers less than zero. Everyone should be very familiar with operations using positive numbers, such as adding 5 and 7 to get an answer of 12, or subtracting 6 from 11 to get 5. There is no plus sign written by the numbers in these two examples because the standard convention is **if a number does not have a sign in front of it, it is considered to be a positive number.** Or, stated another way, **positive numbers do not require the + sign to be written.** However, 5 and 7 in the first example can be written as a (+5) and a (+7) for clarity, using parentheses, and this can be helpful when first learning or reviewing the usage rules.

Plus and minus signs are also used as *operators*; a positive sign between two numbers means they are to be added together, while a minus sign between two numbers means the second is to be subtracted from the first. The operation sign must be considered separately from the sign of the individual number.

The first addition operation above could then be shown as:

$$(+5) + (+7) = (+12)$$

Read as: "A plus 5 added to a plus 7 equals a plus 12."

Property	Unit	Symbol
area dimensions	square millimeter	mm²
current density	ampere per square millimeter	A/mm²
deposition rate	kilogram per hour	kg/h
electrical resistivity	ohm meter	Ω•m
electrode force	newton	N
flow rate (gas and liquid)	liter per minute	L/min
fracture toughness	meganewton meter$^{-3/2}$	MN•m$^{-3/2}$
impact strength	joule	J = N•m
linear dimensions	millimeter	mm
power density	watt per square meter	W/m²
pressure (gas and liquid)	kilopascal	kPa = 1000 N/m²
pressure (vacuum)	pascal	Pa = N/m²
tensile strength	megapascal	MPa = 1 000 000 N/m²
thermal conductivity	watt per meter kelvin	W/ (m•K)
travel speed	millimeter per second	mm/s
volume dimensions	cubic millimeter	mm³
wire feed speed	millimeter per second	mm/s

Table 7.3—Common SI Units Pertaining to Welding

When we introduce the use of negative, or minus, numbers, **the negative sign must be shown in front of a negative number to distinguish it from a positive number**. The use of parentheses is recommended initially to aid in the understanding and use of combining positive and negative numbers, and to separate it from the operation signs. As experience is gained, the parentheses can be eliminated to save time and effort, but the use of parentheses is always very helpful in separating the number sign from the operation sign. Adding a plus 5 to a minus 7 results in an answer of a minus 2. This operation can be written, using signs and parentheses as:

$(+5) + (-7) = (-2)$

Read as: "A plus 5 added to a minus 7 equals a minus 2."

Additional examples of adding positive and negative numbers are shown:

$(-9) + (+3) = (-6)$

"A minus 9 added to a plus 3 equals a minus 6."

$(+11) + (-6) = (+5)$

"A plus 11 added to a minus 6 equals a plus 5."

$(-3) + (-4) = (-7)$

"A minus 3 added to a minus 4 equals a minus 7."

$(-6) + (-9) = (-15)$

"A minus 6 added to a minus 9 equals a minus 15."

These examples show the usage of adding positive and negative numbers. The next step is to review how these positive and negative numbers are subtracted. Examples of subtraction are:

$(+5) - (-8) = (+13)$

"A plus 5 minus a minus 8 equals a plus 13."

In this last example the numerical value of 8 was added to the value of 5 because **the subtraction of a negative number is the same as the addition of that number's numerical or absolute value**. Another way of viewing this rule is "**two negatives make a positive**." The minus sign of the 8, combined with the minus sign of the operation, results in two negatives, leading to a positive, and the answer of +13.

It can be helpful to write the last example in two steps to clarify the operation:

$(+5) - (-8) = ?$

The two minus signs above make a positive, shown below:

$5 + 8 = 13$

The value of 8 is added to the value of 5.

Other examples of subtraction, showing the operation in two steps, are shown:

$(+9) - (-5) = (+14)$
$9 + 5 = 14$

"A plus 9 minus a minus 5 equals a plus 14."

$(-8) - (-4) = (-4)$
$-8 + 4 = -4$

"A minus 8 minus a minus 4 equals a minus 4."

$(-9) - (+6) = (-15)$
$-9 - 6 = -15$

"A minus 9 minus a plus 6 equals a minus 15."

For **multiplication** of positive and negative numbers, two additional rules apply:

1. **A negative number multiplied by a negative number equals a positive answer.**
2. **A negative number multiplied by a positive number equals a negative answer.**

Examples of these are shown as follows:

$(-6) \times (-5) = (+30)$

"A minus 6 times a minus 5 equals a plus 30."

$(-4) \times (-6) = (+24)$

"A minus 4 times a minus 6 equals a plus 24."

$(-5) \times (+3) = (-15)$

"A minus 5 times a plus 3 equals a minus 15."

$(+7) \times (-2) = (-14)$

"A plus 7 times a minus 2 equals a minus 14."

For **division**, similar rules apply:

1. **Dividing a negative number by a positive number results in a negative answer.**
2. **Dividing a positive number by a negative number results in a negative answer.**
3. **Dividing a negative number by a negative number results in a positive answer.**

Examples of each of these 3 rules for division are shown, using the forward slash mark "/" as the division operation symbol:

$(-6) / (+2) = (-3)$

"A minus 6 divided by a plus 2 equals a minus 3."

$(+8) / (-4) = (-2)$

"A plus 8 divided by a minus 4 equals a minus 2."

$(-9) / (-3) = (+3)$

"A minus 9 divided by a minus 3 equals a plus 3."

Scientific Notation

The next subject to review is **Scientific Notation (SN)**, which expresses numbers as powers of a base number, using a **root**, a **base number**, and an **exponent**. A common base number is 10, and the number 100 can be written as 1×10^2, (read as, "one times ten squared") where 1 is the root, 10 is the base, and 2 is the exponent of the base. An exponent is equal to the number of times the base number is multiplied by itself; 10^2 equals 10×10 or 100, and 10^3 equals $10 \times 10 \times 10$ or 1,000. 100 can also be written as 10^2 where it is not necessary to write the root number and the multiplier sign when the root is exactly equal to 1. For this case, only the base number and exponent are used. The exponent also indicates the number of spaces which the decimal point is moved, either to the left or right. Further examples of SN, with base 10 exponents, for some common whole numbers greater than one are shown below. For numbers greater than one the exponent is positive.

$10^0 = 1$ (by definition, any number raised to a zero power = 1)
$10^1 = 10$
$10^2 = 100$
$10^3 = 1,000$
$10^4 = 10,000$
$10^5 = 100,000$
$10^6 = 1,000,000$

For numbers less than one, but greater than zero, the exponent is negative. The following examples have eliminated the "1 ×" preceding the base number.

$10^{-1} = 0.1$
$10^{-2} = 0.01$
$10^{-3} = 0.001$
$10^{-4} = 0.0001$
$10^{-5} = 0.00001$
$10^{-6} = 0.000001$

As stated earlier, it is not necessary to write the number 1 and the multiplier sign when a value is exactly equal to 1 times the exponent; the 1 is understood to be present in such cases. Additional examples of numbers written as both standard and SN form are shown below. In SN, the style rule is **the decimal is always moved to the immediate right of the first digit other than zero**. Examples of this convention are shown below:

Standard Form	Scientific Notation
2,345	= 2.345×10^3
1,450,000	= 1.45×10^6
0.348	= 3.48×10^{-1}
0.0078	= 7.8×10^{-3}

The prefixes in Table 7.2 are necessary to aid in handling very large or very small values normally found in our everyday work. For example, a common fabrication material, plain carbon steel, has an approximate tensile strength of about 70,000 pounds per square inch (psi) in our current US system. Converting 70,000 psi to the SI unit of pascal (Pa) for tensile strength results in a very large numerical value because there are 6,895 pascals in each psi. This conversion is shown below:

Example 1
70,000 psi = ?? Pa
= 70,000 psi × 6,895 Pa/Psi
= 482,650,000 Pa

The magnitude of the above answer is a bit awkward to use because of its size, so we can apply the prefix of "mega" from Table 7.2 to simplify it. The prefix mega has a value of 10^6 or 1,000,000, and we apply it to the answer and move the decimal point accordingly. This results in a simpler answer without all the zeros by moving the decimal point 6 places to the left after adding the prefix.

Example 2
70,000 psi = 482.65 MPa

There will be several more examples of converting from one system to the other, but first some simple arithmetical conventions required for adding, subtracting, multiplying or dividing must be addressed. To begin, the number line will be reviewed to make sure of the nomenclature used for referring to a particular number's position on the number line. An example noting the positions of all digits in a very large number containing many digits after the decimal point follows:

Example 3
For the number, 1,234,567.987654

The numbers to the left of the decimal are greater than one and are referred to as:

The 7 is in the units position
The 6 is in the tens position
The 5 is in the hundreds position
The 4 is in the thousands position
The 3 is in the ten thousands position
The 2 is in the hundred thousands position
The 1 is in the millions position

Addressing the same number again, and looking at the numbers to the right of the decimal we can refer to each of their positions:

1,234,567.987654

The numbers to the right of the decimal point, which are less than one, are referred to as:

9 is in the tenths position
8 is in the hundredths position
7 is in the thousandths position
6 is in the ten thousandths position
5 is in the hundred thousandths position
4 is in the millionths position

Keeping these various positions in mind will aid in handling the calculations of conversions.

From the above examples, it is evident a movement of the decimal point one space to the left is equivalent to dividing by ten, and moving the decimal point one space to the right is equivalent to multiplying by ten. The negative exponent in scientific notation means the number is less than one.

An advantage of scientific notation is the ease of calculations with very large or small numbers. When multiplying two numbers, both written in scientific notation, it is only necessary to multiply the two root numbers together, add the exponents, or powers of 10, of each number, and reconfigure the answer in scientific notation. Division of two numbers consists of dividing the two roots as normally done, subtracting the denominator number exponent from the numerator number exponent, and then reconfiguring the answer back into scientific notation. Several examples are shown:

Example 4
Multiplication (add exponents)

$(2.0 \times 10^3) \times (1.5 \times 10^5) = 3.0 \times 10^8$
$(1.0 \times 10^8) \times (4.5 \times 10^7) = 4.5 \times 10^{15}$
$(3.5 \times 10^{-3}) \times (2.0 \times 10^6) = 7.0 \times 10^3$
$(5 \times 10^2) \times (12 \times 10^{-6}) = 60 \times 10^{-4}$
$\qquad\qquad\qquad\qquad\text{or} = 6.0 \times 10^{-3}$

Example 5
Division (subtract exponents)

$(3.0 \times 10^4) \div (1.5 \times 10^2) = 2.0 \times 10^2$
$(6.0 \times 10^{-7}) \div (3.0 \times 10^3) = 2.0 \times 10^{-10}$
$(4.5 \times 10^4) \div (1.5 \times 10^{-5}) = 3.0 \times 10^9$
$(8.0 \times 10^{-6}) \div (2.0 \times 10^{-9}) = 4.0 \times 10^3$

For addition or subtraction of numbers in scientific notation, the first step is to reconfigure them so that all have the same exponent; then do the normal operation of addition or subtraction. For ease of operation, the smaller number should be reconfigured.

Example 6
Addition

$(2.3 \times 10^4) + (3.54 \times 10^5) =$
$(0.23 \times 10^5) + (3.54 \times 10^5) = 3.77 \times 10^5$

$(3.78 \times 10^{-6}) + (7.45 \times 10^{-4}) =$
$(0.0378 \times 10^{-4}) + (7.45 \times 10^{-4}) = 7.4878 \times 10^{-4}$

Example 7
Subtraction

$(7.8 \times 10^6) - (9.4 \times 10^4) =$
$(7.8 \times 10^6) - (0.094 \times 10^6) = 7.706 \times 10^6$

$(3.9 \times 10^{-4}) - (6.1 \times 10^{-5}) =$
$(3.9 \times 10^{-4}) - (0.61 \times 10^{-4}) = 3.29 \times 10^{-4}$

Note that the standard rules apply for adding and subtracting positive and negative numbers. The final answer should always be configured into scientific notation, having only one digit greater than one to the left of the decimal point by adjusting the exponent.

Round Off Convention

The next subject to review is the *round off convention*; most everyone is familiar with some sort of rules for rounding, but the AWS convention used is:

> **Rule 1**— Increase the last retained digit by one if the next digit to its right is 5 or larger.
>
> **Rule 2**— Retain the last digit unchanged if the next digit to its right is less than 5.

Example 8
8,937 = 9,000 rounded to nearest thousand
8,937 = 8,900 rounded to nearest hundred
8,937 = 8,940 rounded to nearest ten

Another example shows the rounding convention for a number containing decimals rounded to different positions:

Example 9
4.4638 = 4   rounded to nearest units
4.4638 = 4.5   rounded to nearest tenth
4.4638 = 4.46   rounded to nearest hundredth
4.4638 = 4.464   rounded to nearest thousandth

Rounding should always be a single operation only; that is, do not round off the very last digit, then the next one forward, and so on to get to the desired digit. The single operation approach helps avoid rounding errors in calculations; rounding should always begin at the proper position on the number line for the desired value, and then rounded in a single step. These rounding conventions are also noted in AWS standard A1.1 with additional examples for illustration.

Conversion Factors

It was shown earlier in Example 1 how a steel with a tensile strength of 70,000 psi can be converted to pascals. Then, to make the number more manageable, a prefix of "mega" was applied to eliminate several zeros.

These prefixes are very convenient, and are abbreviations of the multiplier of the number. An example of a common term found in the newspapers daily is the prefix, "kilo." It means 1,000, so if it is applied to the metric unit of length, a kilometer is 1,000 meters. Similarly, "milli" means one thousandth, so a millimeter is one thousandth of a meter; there are 1,000 millimeters in one meter. Examples of the use of prefixes are shown:

Example 10
456,000,000 Pa = 456 MPa
56 km = 56,000 m
234,000 mm = 234 m
456 g = 0.456 kg

Because conversions of SI units to US units, or vice versa, are commonly needed, conversion factor charts have been developed to aid these conversions. Table 7.4 on the next page shows many of these conversion factors used in welding. The use of the chart is simple; find the property to convert from, and multiply the number to be converted by the conversion factor given. The welding inspector should make no effort to memorize any of the conversion factors shown in Table 7.4; these will be provided when they are needed for conversion of data. The CWI must be capable of computing the numbers to arrive at a solution.

Looking further at the conversion factor table, there are a few important features. You see that the table is arranged in four columns entitled, *Property, To Convert From, To,* and *Multiply By*. You will use these columns in the same order as they are listed.

For any conversion exercise, the first step is to decide what particular property is described by the given units that are to be converted. Once the proper category has been chosen from the "Property" column, look at the second column, "To Convert From," and locate the line that contains the unit given. This is the unit used for conversion. Moving straight across this same line to the right, look for the unit that matches the unit used for conversion in the "To" column. Next, locate the line that contains both the known and desired units; the value found in the last column, "Multiply By," is the appropriate conversion factor. At this point, multiply the number of the known units by this conversion factor. The result is the number in the desired units. Several examples appear below to show how this table is used to perform typical conversions:

Example 11
An oxygen gauge shows a pressure of 40 psi. What is this pressure in kilopascals?

1. Property = pressure (gas or liquid)
2. Known unit = 40 psi
3. Desired unit = kilopascals (kPa)
4. Conversion factor = 6.895

   40 psi × 6.895 = 275.8 kPa

And the computed answer can then be rounded off and placed into scientific notation.

$$275.8 \text{ kPa} = 276 \text{ kPa}$$
$$= 2.76 \times 10^2 \text{ kPa}$$

Example 12
A tensile specimen was pulled and displayed an ultimate tensile strength of 625 MPa. This corresponds to how many psi?

1. Property = tensile strength
2. Known unit = 625 MPa
3. Desired unit = psi
4. Conversion factor = 1.450×10^2

   625 MPa × 1.450×10^2 = 906.25×10^2
   $= 9.06 \times 10^4$ psi

The calculator will give an answer of 906.25×10^2 but this can be rounded to 9.06×10^4 psi in scientific notation.

Example 13
What is the diameter in millimeters of a 5/32 inch (0.156 in.) electrode?

1. Property = linear measurements
2. Known unit = 5/32 in. (0.156 in.)
3. Desired unit = mm
4. Conversion factor = 25.4

   0.156 × 25.4 = 3.9624 mm
   = 3.96 mm

Here, the answer has been rounded off to three places.

Property	To Convert From:	To:	Multiply By:
area dimensions	in.^{2}	mm^2	6.452×10^2
	mm^2	in.^{2}	1.550×10^{-3}
current density	A/in.^{2}	A/mm^2	1.550×10^{-3}
	A/mm^2	A/in.^{2}	6.452×10^2
deposition rate	lb/hr	kg/hr	0.454
	kg/hr	lb/hr	2.205
flow rate	ft^3/h	l/min	4.719×10^{-1}
	l/min	ft^3/h	2.119
heat input	J/in.	J/m	39.37
	J/m	J/in.	2.54×10^{-2}
linear measure	in.	mm	25.4
	mm	in.	3.937×10^{-2}
	ft	mm	3.048×10^2
	mm	ft	3.281×10^{-3}
mass	lb	kg	0.454
	kg	lb	2.205
pressure	psi	kPa	6.895
	psi	MPa	6.895×10^{-3}
	kPa	psi	0.145
	MPa	psi	1.450×10^2
	bar	psi	14.50
	psi	bar	6.9×10^{-2}
temperature	°F	°C	(°F − 32) / 1.8
	°C	°F	(°C × 1.8) + 32
tensile strength	psi	MPa	6.895×10^{-3}
	MPa	psi	1.450×10^2
travel speed	in./min	mm/s	4.233×10^{-1}
	mm/s	in./min	2.362
vacuum	Pa	torr	7.501×10^{-3}
wire feed speed	in./min	mm/s	0.423
	mm/s	in./min	2.362

Table 7.4—Welding Usage Conversion Chart—U.S. Customary and SI

Area of Square or Rectangle

Area = length x width   or: Area = width x thickness

Area or Circle

Area = $\pi \times \text{radius}^2$ or: Area = $\dfrac{\pi \times \text{diameter}^2}{4}$ or: Area = $0.7854 \times \text{diameter}^2$

Percent Elongation

% Elongation = $\dfrac{\text{Final Gauge Length} - \text{Original Gauge Length}}{\text{Original Gauge Length}} \times 100$

Percent Reduction of Area

% Reduction of Area = $\dfrac{\text{Original Area} - \text{Final Area}}{\text{Original Area}} \times 100$

Tensile Strength

General

UTS = $\dfrac{P\ max}{\text{Area}}$ where: P max = load to break specimen
Area = specimen's original cross-sectional area

Pipe

UTS for full section pipe = $\dfrac{P\ max}{0.7854\,(OD^2 - ID^2)}$

Yield Strength

YS = $\dfrac{\text{Load at specified offset}}{\text{Original cross-sectional area}}$

Welding Heat Input

J/in. = $\dfrac{V \times A \times 60}{\text{Travel Speed (ipm)}}$ where: J = Joules (energy)
V = welding voltage
A = welding amperage
ipm = inches per minute

Carbon Equivalent

CE = %C + $\dfrac{\%Mn}{6}$ + $\dfrac{\%Ni}{15}$ + $\dfrac{\%Cu}{13}$ + $\dfrac{\%Mo}{14}$

Table 7.5—WIT—Useful Formulae

Example 14

Welding parameters were adjusted to produce a weld metal deposition rate of 7.3 kg/h. What is that deposition rate in terms of lb/h?

1. Property = deposition rate
2. Known unit = 7.3 kg/h
3. Desired unit = lb/h
4. Conversion factor = 2.205

$$7.3 \times 2.205 = 16.0965 \text{ lb/h}$$
$$= 16 \text{ lb/h}$$

The calculator gives an answer of 16.0965, and it can be rounded to two places, resulting in an answer of 16 lb/h.

Temperature conversions for Fahrenheit to Celsius and vice versa are found in Tables 7.4 and 7.6.

Additional SI Style and Usage Conventions

The following are some additional excerpts from AWS A1.1, including the paragraph numbers for easy cross referencing, to show additional style and usage conventions used in the SI system. It must be remembered, AWS A1.1 is a guide, not a mandated system, and should be used accordingly.

6. Style and Usage
6.1 Application and Usage of Prefixes

6.1.1 Prefixes should be used with SI units to indicate orders of magnitude. Prefixes provide convenient substitutes for using powers of ten and they eliminate insignificant digits.

Preferred	Nonpreferred
12.3 km	12 300 m, 12.3 × 10³ m

6.1.2 Prefixes in steps of 1000 are recommended. The use of the prefixes hecto, deka, deci, and centi should be avoided.

Preferred	Nonpreferred
mm, m, km	hm, dam, dm, cm

6.1.3 Prefixes should be chosen so that the numerical value lies between 0.1 and 1000.

6.1.3.1 For special situations, such as tabular presentations, the same unit, multiple, or submultiple may be used even though the numerical value exceeds the range of 0.1 to 1000.

6.1.4 Multiple and hyphenated prefixes should not be used.

Correct	Incorrect
pF, GF, GW	µ µF, Mkg, kMW, G-W

6.1.5 It is generally desirable to use only base and derived units in the denominator. Prefixes are used with the numerator unit to give numbers of appropriate size (see 6.1.3).

Preferred	Nonpreferred
200 J/kg	0.2 J/g
1 Mg/M	1 kg/mm
5 Mg/m³	5g/mm³

6.1.6 Prefixes are to be attached to the base SI units with the exception of the base unit for mass, the kilogram, which contains a prefix. In this case, the required prefix is attached to gram.

6.1.7 Prefixes should not be mixed unless magnitudes warrant a difference.

Correct
5 mm long × 10 mm high

Incorrect
5 mm long × 0.01 m high

Exception
4 mm diam × 50 m long

6.1.8 The pronunciation of the prefixes is always the same, regardless of the accompanying base unit. For example, the accepted pronunciation for kilo is "kill-oh." The slang expression "keelo" for kilogram should never be used.

6.2 Use of Nonpreferred Units

6.2.1 The mixing of units from different systems should be avoided.

Preferred	Nonpreferred
kg/m³	kg/gal

6.8 Capitalization. SI unit names are capitalized only at the beginning of a sentence (examples: newton, pascal, meter, kelvin, and hertz). In "degree Celsius," degree is lower case and Celsius is always capitalized.

SI unit symbols are not capitalized except for those derived from a proper name.

A (ampere), K (kelvin), W (watt),
N (newton), J (joule), etc.
m (meter), kg (kilogram), etc.

Only five prefix symbols are capitalized, namely, E (exa), P (peta), T (tera), G (giga), and M (mega).

6.9 Plurals. Unit symbols are the same for singular and plural. Unit names form their plurals in the usual manner.

Correct	Incorrect
50 newtons (50N)	50 newton's (50Ns)
25 grams (25g)	25 grams' (25 gs)

6.10 Punctuation. Periods are not to be used after SI unit symbols except at the end of a sentence.

Periods (not commas) are used as decimal markers.

Periods are not used in unit symbols or in conjunction with prefixes.

Correct	Incorrect
5.7 mm	5.7 m.m.

6.11 Number Grouping

6.11.1 Numbers made up of five or more digits should be written with a space separating each group of three digits counting both to the left and right of the decimal point. With four digit numbers, the spacing is optional.

6.11.2 Spaces (not commas) should be used between the groups of three digits.

Correct	Incorrect
1 420 462.1	1,420,462.1
0.045 62	0.04562
1452 or 1 452	1,452

6.12 Miscellaneous Styling

6.12.1 A space is to be used between the numerical value and the unit symbol.

Correct	Incorrect
4 mm	4mm

6.12.2 Unit symbols and names are never used together in a single expression:

Correct	Incorrect
meter per second (m/s)	meter/s

6.12.3 Numbers are expressed as decimals, not as fractions. The decimal should be preceded by a zero when the number is less than unity.

Correct	Incorrect
0.5 kg, 1.75 m	1/2 kg, .5 kg., 1 3/4m

6.12.4 SI unit symbols should be printed in Roman (upright) type rather than *italic* (slanted or script).

6.12.5 Typed rather than hand-drawn prefixes should be used when possible. The spelled word may be used in preference to the use of hand-drawn symbols.

6.12.6 When it is necessary or desirable to use U.S. inch-pound units in an equation or table, SI units should be restated in a separate equation, table, or column in a table. As an alternative, a note may be added to an equation or table giving the factors to be used in converting the calculated result in U.S. inch-pound to preferred SI units. The SI equivalents may follow and be inserted in parentheses or brackets.

Summary

The above examples are part of the job math which the welding inspector may be required to perform. As a minimum, he or she will be asked to perform some of these conversions on the AWS CWI examination. The examples above are typical of problems which will appear on the AWS CWI examination. No matter how large or small the numbers may be, the problems are all dealt with in the same manner. Simply go through the various steps and use the conversion factor table to provide a multiplier. Then, all that is left to do is perform the arithmetic according to the rules and conventions noted previously.

Conversions for Fahrenheit–Celsius Temperature Scales

Find the number to be converted in the center (boldface) column. If converting Fahrenheit degrees, read the Celsius equivalent in the column headed "°C." If converting Celsius degrees, read the Fahrenheit equivalent in the column headed "°F."

°C		°F	°C		°F	°C		°F	°C		°F
–273	**–459**		–40	**–40**	–40	24.4	**76**	168.8	199	**390**	734
–268	**–450**		–34	**–30**	–22	25.6	**78**	172.4	204	**400**	752
–262	**–440**		–29	**–20**	–4	26.7	**80**	176.0	210	**410**	770
–257	**–430**		–23	**–10**	14	27.8	**82**	179.6	216	**420**	788
–251	**–420**		–17.8	**0**	32	28.9	**84**	183.2	221	**430**	806
–246	**–410**		–16.7	**2**	35.6	30.0	**86**	186.8	227	**440**	824
–240	**–400**		–15.6	**4**	39.2	31.1	**88**	190.4	232	**450**	842
–234	**–390**		–14.4	**6**	42.8	32.2	**90**	194.0	238	**460**	860
–229	**–380**		–13.3	**8**	46.4	33.3	**92**	197.6	243	**470**	878
–223	**–370**		–12.2	**10**	50.0	34.4	**94**	201.2	249	**480**	896
–218	**–360**		–11.1	**12**	53.6	35.6	**96**	204.8	254	**490**	914
–212	**–350**		–10.0	**14**	57.2	36.7	**98**	208.4	260	**500**	932
–207	**–340**		–8.9	**16**	60.8	37.8	**100**	212.0	266	**510**	950
–201	**–330**		–7.8	**18**	64.4	43	**110**	230	271	**520**	968
–196	**–320**		–6.7	**20**	68.0	49	**120**	248	277	**530**	986
–190	**–310**		–5.6	**22**	71.6	54	**130**	266	282	**540**	1004
–184	**–300**		–4.4	**24**	75.2	60	**140**	284	288	**550**	1022
–179	**–290**		–3.3	**26**	78.8	66	**150**	302	293	**560**	1040
–173	**–280**		–2.2	**28**	82.4	71	**160**	320	299	**570**	1058
–168	**–270**	–454	–1.1	**30**	86.0	77	**170**	338	304	**580**	1076
–162	**–260**	–436	0.0	**32**	89.6	82	**180**	356	310	**590**	1094
–157	**–250**	–418	1.1	**34**	93.2	88	**190**	374	316	**600**	1112
–151	**–240**	–400	2.2	**36**	96.8	93	**200**	392	321	**610**	1130
–146	**–230**	–382	3.3	**38**	100.4	99	**210**	410	327	**620**	1148
–140	**–220**	–364	4.4	**40**	104.0	100	**212**	414	332	**630**	1166
–134	**–210**	–346	5.6	**42**	107.6	104	**220**	428	338	**640**	1184
–129	**–200**	–328	6.7	**44**	111.2	110	**230**	446	343	**650**	1202
–123	**–190**	–310	7.8	**46**	114.8	116	**240**	464	349	**660**	1220
–118	**–180**	–292	8.9	**48**	118.4	121	**250**	482	354	**670**	1238
–112	**–170**	–274	10.0	**50**	122.0	127	**260**	500	360	**680**	1256
–107	**–160**	–256	11.1	**52**	125.6	132	**270**	518	366	**690**	1274
–101	**–150**	–238	12.2	**54**	129.2	138	**280**	536	371	**700**	1292
–96	**–140**	–220	13.3	**56**	132.8	143	**290**	554	377	**710**	1310
–90	**–130**	–202	14.4	**58**	136.4	149	**300**	572	382	**720**	1328
–84	**–120**	–184	15.6	**60**	140.0	154	**310**	590	388	**730**	1346
–79	**–110**	–166	16.7	**62**	143.6	160	**320**	608	393	**740**	1364
–73	**–100**	–148	17.8	**64**	147.2	166	**330**	626	399	**750**	1382
–68	**–90**	–130	18.9	**66**	150.8	171	**340**	644	404	**760**	1400
–62	**–80**	–112	20.0	**68**	154.4	177	**350**	662	410	**770**	1418
–57	**–70**	–94	21.1	**70**	158.0	182	**360**	680	416	**780**	1436
–51	**–60**	–76	22.2	**72**	161.6	188	**370**	698	421	**790**	1454
–46	**–50**	–58	23.3	**74**	165.2	193	**380**	716	427	**800**	1472

Table 7.6—Conversions for Fahrenheit-Celsius Temperature Scales

°C	°F		°C	°F		°C	°F		°C	°F	
432	**810**	1490	682	**1260**	2300	932	**1710**	3110	1182	**2160**	3920
438	**820**	1508	688	**1270**	2318	938	**1720**	3128	1188	**2170**	3938
443	**830**	1526	693	**1280**	2336	943	**1730**	3146	1193	**2180**	3956
449	**840**	1544	699	**1290**	2354	949	**1740**	3164	1199	**2190**	3974
454	**850**	1562	704	**1300**	2372	954	**1750**	3182	1204	**2200**	3992
460	**860**	1580	710	**1310**	2390	960	**1760**	3200	1210	**2210**	4010
466	**870**	1598	716	**1320**	2408	966	**1770**	3218	1216	**2220**	4028
471	**880**	1616	721	**1330**	2426	971	**1780**	3236	1221	**2230**	4046
477	**890**	1634	727	**1340**	2444	977	**1790**	3254	1227	**2240**	4064
482	**900**	1652	732	**1350**	2462	982	**1800**	3272	1232	**2250**	4082
488	**910**	1670	738	**1360**	2480	988	**1810**	3290	1238	**2260**	4100
493	**920**	1688	743	**1370**	2498	993	**1820**	3308	1243	**2270**	4118
499	**930**	1706	749	**1380**	2516	999	**1830**	3326	1249	**2280**	4136
504	**940**	1724	754	**1390**	2534	1004	**1840**	3344	1254	**2290**	4154
510	**950**	1742	760	**1400**	2552	1010	**1850**	3362	1260	**2300**	4172
516	**960**	1760	766	**1410**	2570	1016	**1860**	3380	1266	**2310**	4190
521	**970**	1778	771	**1420**	2588	1021	**1870**	3398	1271	**2320**	4208
527	**980**	1796	777	**1430**	2606	1027	**1880**	3416	1277	**2330**	4226
532	**990**	1814	782	**1440**	2624	1032	**1890**	3434	1282	**2340**	4244
538	**1000**	1832	788	**1450**	2642	1038	**1900**	3452	1288	**2350**	4262
543	**1010**	1850	793	**1460**	2660	1043	**1910**	3470	1293	**2360**	4280
549	**1020**	1868	799	**1470**	2678	1049	**1920**	3488	1299	**2370**	4298
554	**1030**	1886	804	**1480**	2696	1054	**1930**	3506	1304	**2380**	4316
560	**1040**	1904	810	**1490**	2714	1060	**1940**	3524	1310	**2390**	4334
566	**1050**	1922	816	**1500**	2732	1066	**1950**	3542	1316	**2400**	4352
571	**1060**	1940	821	**1510**	2750	1071	**1960**	3560	1321	**2410**	4370
577	**1070**	1958	827	**1520**	2768	1077	**1970**	3578	1327	**2420**	4388
582	**1080**	1976	832	**1530**	2786	1082	**1980**	3596	1332	**2430**	4406
588	**1090**	1994	838	**1540**	2804	1088	**1990**	3614	1338	**2440**	4424
593	**1100**	2012	843	**1550**	2822	1093	**2000**	3632	1343	**2450**	4442
599	**1110**	2030	849	**1560**	2840	1099	**2010**	3650	1349	**2460**	4460
604	**1120**	2048	854	**1570**	2858	1104	**2020**	3668	1354	**2470**	4478
610	**1130**	2066	860	**1580**	2876	1110	**2030**	3686	1360	**2480**	4496
616	**1140**	2084	866	**1590**	2894	1116	**2040**	3704	1366	**2490**	4514
621	**1150**	2102	871	**1600**	2912	1121	**2050**	3722	1371	**2500**	4532
627	**1160**	2120	877	**1610**	2930	1127	**2060**	3740	1377	**2510**	4550
632	**1170**	2138	882	**1620**	2948	1132	**2070**	3758	1382	**2520**	4568
638	**1180**	2156	888	**1630**	2966	1138	**2080**	3776	1388	**2530**	4586
643	**1190**	2174	893	**1640**	2984	1143	**2090**	3794	1393	**2540**	4604
649	**1200**	2192	899	**1650**	3002	1149	**2100**	3812	1399	**2550**	4622
654	**1210**	2210	904	**1660**	3020	1154	**2110**	3830	1404	**2560**	4640
660	**1220**	2228	910	**1670**	3038	1160	**2120**	3848	1410	**2570**	4658
666	**1230**	2246	916	**1680**	3056	1166	**2130**	3866	1416	**2580**	4676
671	**1240**	2264	921	**1690**	3074	1171	**2140**	3884	1421	**2590**	4694
677	**1250**	2282	927	**1700**	3092	1177	**2150**	3902	1427	**2600**	4712

Table 7.6 (Continued)—Conversions for Fahrenheit-Celsius Temperature Scales

°C	°F	°C	°F	°C	°F	°C	°F
1432	2610	4730	1488	2710	4910	1543	2810	5090	1599	2910	5270
1438	2620	4748	1493	2720	4928	1549	2820	5108	1604	2920	5288
1443	2630	4766	1499	2730	4946	1554	2839	5126	1610	2930	5306
1449	2640	4784	1504	2740	4964	1560	2840	5144	1616	2940	5324
1454	2650	4802	1510	2750	4982	1566	2850	5162	1621	2950	5342
1460	2660	4820	1516	2760	5000	1571	2860	5180	1627	2960	5360
1466	2670	4838	1521	2770	5018	1577	2870	5198	1632	2970	5278
1471	2680	4856	1527	2780	5036	1582	2880	5216	1638	2980	5396
1477	2690	4874	1532	2790	5054	1588	2890	5234	1643	2990	5414
1482	2700	4892	1538	2800	5072	1593	2900	5252	1649	3000	5432

°C = 5/9 (°F − 32)   °F = (9/5 °C) + 32

Table 7.6 (Continued)—Conversions for Fahrenheit-Celsius Temperature Scales

Key Terms and Definitions

AWS A1.1—the *Metric Practice Guide for the Welding Industry*, a standard published by AWS.

conversion factor—a number established to aid in converting from one unit to another.

exponent—the number used as the power of ten; 2 is the exponent in the expression 10^2.

number line—the set of numbers, both greater and less than one, that make up the number system used to assign values.

prefix—a word placed in front of another word that changes its meaning or value.

rounding—in mathematics, the practice of adjusting the size of the last digit retained in a number based on the next digit's size relationship to 5.

scientific notation—the numbering system which uses the powers of ten, the exponential system, to simplify the handling of very large or very small numbers.

SI—Le Systeme Internationale d'Unites (the abbreviation used to denote the metric system).

SN—the abbreviation used for scientific notation.

US—the abbreviation for the current measurement system used in the United States (for the U.S. customary system).

Module 8
Welding Metallurgy for the Welding Inspector

Contents

Introduction	8-2
Basic Metal Structures	8-2
Metallurgical Considerations for Welding	8-11
Welding Metallurgy of Commonly Used Materials	8-15
Summary	8-18
Key Terms and Definitions	8-18

Module 8—Welding Metallurgy for the Welding Inspector

Introduction

Metallurgy is the science which deals with the internal structure of metals and the relationship between those structures and the properties exhibited by metals. When referring to welding metallurgy, the concerns are about the various changes that occur in metals when joined by welding, especially those affecting the mechanical properties.

It is certainly appropriate for the welding inspector to be knowledgeable in the basics of welding metallurgy. Granted, it is unlikely that the inspector will be responsible for the specification of base or weld metal alloys or their treatment. However, an understanding of the basics of welding metallurgy is not only helpful to the welding inspector, but often a requirement for many inspection functions. One reason for this is that the mechanical properties of metals, such as strength, hardness, ductility, toughness, fatigue strength, and abrasion resistance are all affected by the metallurgical transformations as a result of welding.

These properties are affected by various metallurgical factors, including alloy additions, thermal treatments and mechanical treatments. The welding inspector who has an understanding of these properties will have a better feel for why certain fabrication operations are necessary. Certain fabrication requirements, such as preheat, postheat, interpass temperature control, heat input control, peening, thermal stress relief, and other heat treatments can all result in some type of metallurgical change which, in turn, will affect the metal's mechanical properties. Therefore, this section will primarily describe certain aspects of ferrous (iron based) welding metallurgy with emphasis on the need for fabrication methods to control the changes which can occur.

Since the topic of welding metallurgy includes numerous facets, it would be unreasonable to think that this discussion could cover them all. So, we will limit the coverage to those more important changes which may occur during the welding operation. These changes can be summarized and divided into two categories.

The first category includes those changes which occur in a metal as it is heated from room temperature to a higher temperature and those changes occurring when a metal is cooled from a high temperature to a lower temperature. The second category is the effect on the metal's properties versus the rate at which these temperature changes occur. More specifically, we are concerned with how quickly a hot metal cools to room temperature; that is, the metal's cooling rate.

Our discussion will begin with specific references to the changes which occur in metals as they are heated and cooled uniformly. However, it must be noted that welding presents some very different problems since the welding operation tends to heat very localized areas of the metal. Consequently, this nonuniform heating/cooling creates the need for some additional considerations.

Basic Metal Structures

To gain an understanding of the metallurgical properties of metals, it is necessary to start the discussion by describing some of the properties of the particles which comprise all forms of matter. These basic particles, which combine to form solid, liquid and gaseous materials, are referred to as atoms. These atoms are so small that they cannot be seen, even with the most powerful microscopes. However, by starting the discussion at this level and explaining the properties of these atoms and their structures, we will be able to better understand some of the phenomena which we can observe macroscopically, or with the naked eye.

One of the important properties of these atoms is that, at certain temperature ranges, they tend to form substances having specific shapes. This is because there are definite forces acting between these individual atoms when they are placed within certain distances of each other. These forces tend to both pull, or attract, the atoms toward one another while at the same time the atoms are pushed away, or repelled, from one another. Therefore, the individual atoms are held in their particular "home" positions relative to all of the other atoms around them by these counteracting forces (see Figure 8.1). These atoms in their home positions are aligned row upon row, layer upon layer in a three dimensional, symmetrical, crystalline lattice structure or pattern.

They are not, however, stationary in these positions. In reality, they tend to vibrate about an equilibrium position to maintain a balanced spacing. At a given temperature, they will remain at an equilibrium spacing for that particular temperature. When there is a balance between the attractive and repulsive forces, we say that the internal energy of the metal is at a stable level.

Any attempt to force the atoms closer together will be counteracted by repulsive forces which increase as the atoms are pushed closer together. This behavior is evidenced by the fact that metals exhibit extremely high compressive strengths. Similarly, any attempt to pull the atoms further apart will result in a counteracting attractive force. These attractive forces, however, tend to decrease as the atoms are pulled further apart.

Evidence of this latter behavior can be observed in a tensile test. Below a metal's yield point, loading elongates the tensile specimen, and the spacing between individual atoms is increased. Upon release of the load, the specimen will behave elastically; that is, it will return to its original size on a macroscopic level, which means that the atoms return to their original equilibrium spacing.

If the load on the tensile specimen is increased beyond the metal's yield point, it will then behave plastically. Now, it will no longer return to its original size or interatomic spacing, because the atoms have been forced far enough away from each other that the attractive forces are no longer strong enough to hold them in their original position. When the interatomic spacing further increases to the point that the attractive forces are no longer strong enough to hold the atoms together, the metal will fail.

It was noted before that the metal atoms exhibit a very specific spacing at a given temperature, or internal energy. Since heat is a form of energy, the internal energy of a metal is increased when its temperature is raised. This additional energy tends to cause the atoms to vibrate more which increases their interatomic spacing. We can observe the result of this additional energy, visually, because the overall size of the piece of metal will increase as the individual atoms move apart. Conversely, any decrease in the temperature of the metal will result in the atoms moving closer together which, in turn, is observed as a contraction of the metal.

As additional heat is added to the metal, the vibration of the atoms continues to increase, causing the spacing to increase and, consequently, the metal to expand. This will continue to some point at which the interatomic spacing is so great that the atoms are no longer attracted enough to exhibit any specific structure. The solid metal then transforms into a liquid (see Figure 8.2). The temperature associated with this change is referred to as the melting point. Further heating would eventually transform the liquid into a gas; this last transformation occurs at a temperature known as the vapor point.

Solid metal has the lowest internal energy and the shortest interatomic spacing. Liquid metal has a higher internal energy with greater interatomic spacing, and is considered to be amorphous, which means that it is unstructured. Gaseous metal has the

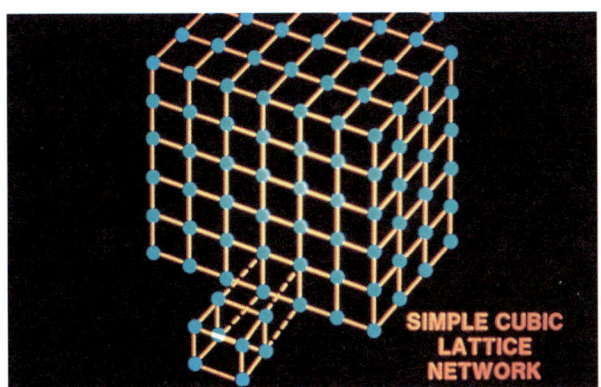

Figure 8.1—Atomic Structure—Showing "Home" Location of Atoms and Electrons

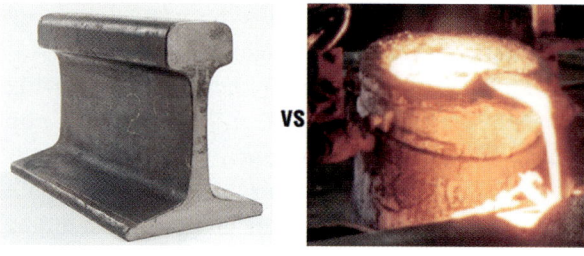

Figure 8.2—Solid versus Liquid

highest internal energy, the greatest interatomic spacing, and is also unstructured.

While all of this is rather intriguing, it is more significant to realize why it is important to you, as a welding inspector. It is obvious that welding and cutting introduce heat into a metal; this heating will result in an expansion of the metal. If we were considering the uniform heating of a metal, we would be able to measure the change in length, or size, of a piece of metal as it is heated. Each metal alloy has associated with it a specific coefficient of thermal expansion. That is, there is a certain numerical value which describes how much a metal will expand for a given increase in temperature.

With welding, however, the heat is not applied uniformly. That is, part of the metal is raised to some very high temperature while the metal adjacent to the weld zone remains at a lower temperature. This results in different amounts of expansion of the metal at different locations relative to the weld zone. The portion of the metal being directly heated will tend to expand, with this expansion being resisted by that metal which is at some lower temperature.

Figure 8.3 illustrates the dimensional changes which occur in a straight bar (see Figure 8.3A) that is heated on one side by a welding arc. In Figure 8.3B, the arc is struck and the plate begins to heat under the influence of the arc. The heated portion expands (see Figure 8.3C) and, because it is partially restrained by the portion of the bar that is not heated, the bar tends to bend in an arc at each end away from the heat source. Since the hot portion is weaker (part of it is actually liquid and is very weak) it does not succeed in forcing the bar to bend very much. The hot part is less restrained in the sideways direction, so it tends to get wider on the side where heat is applied.

When the arc is extinguished (see Figure 8.3D), the hot, molten portion begins to cool and shrink.

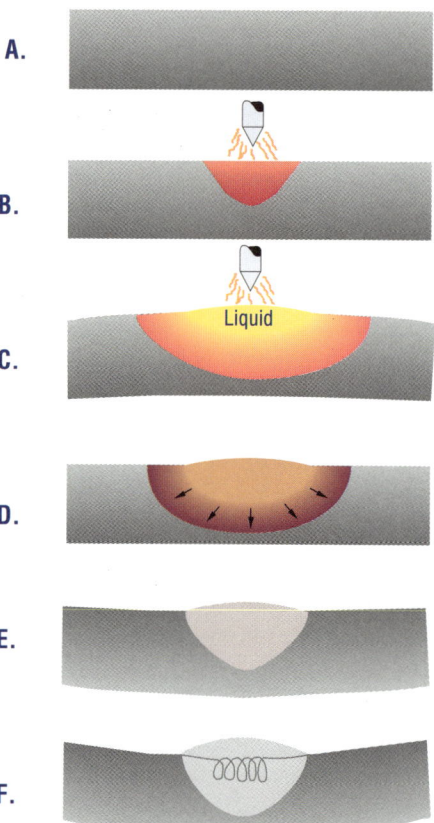

Figure 8.3—Shrinkage in a Weld Caused by Expansion and Contraction

Heat always flows from the hot area to the cold, so during cooling, the heat flows into the previously cool region to warm it. Now, as the hot, expanded portion begins to cool, it contracts, reversing the direction of the deforming forces which ultimately causes the length along the top of the bar to shorten and the ends of the bar to lift upward giving the bar a concave shape as it cools, shown in Figure 8.3E. Therefore, as we apply heat to a part in a nonuniform manner, as is the case for welding, the result is a dimensional change from the thermal stresses developed causing the part to be distorted or warped when it cools. Figure 8.3F represents the resolidified bar with a residual stress level remaining in the bar, denoted by the "coiled spring" representation.

Whenever a metal is melted in a small, localized zone, as in welding, shrinkage stresses are created. Even if the bar had been externally restrained during this heating and cooling cycle, the cooled part still contains stresses caused by this differential heating and cooling. We refer to these stresses as residual stresses. The residual stresses tend to keep the bar in its bent shape. However, the bar will not bend any more because it has cooled to room temperature and is now stronger than the forces exerted by the residual stress. The residual stresses will remain in the bar unless something is done to relax the stress.

There are several ways of reducing or eliminating residual stresses. It can be done thermally where the entire part or a large band containing the weld zone is heated uniformly and held at some temperature for a prescribed time period. The result of this method is that the uniform heating allows the residual stress to relax because the metal's strength is now reduced. Slow uniform cooling to room temperature will then produce a part with much lower residual stress. There are also methods of providing this stress relief treatment by the application of a vibratory, or mechanical, treatment. Both methods have been shown to be effective in many applications.

A third method of reducing residual stresses which can be done in conjunction with the welding operation is known as peening (see Figure 8.4). This is also a mechanical treatment. Peening involves the use of a heavy pneumatic hammer (not a deslagging hammer) which is used to pound on the face of intermediate layers of a multipass weld. This hammering action tends to deform the surface causing the thickness of the layer to decrease. This deformation tends to spread out the face of the weld to make it longer and wider. Since the metal is spread out slightly, the residual stresses are reduced.

When heavy peening is used for stress relieving, care should be taken to prevent the cracking of the weld by this aggressive mechanical treatment. It is not advisable to peen the root layer which may be easily fractured by this hammering. Normally, the final layer is not peened either, but for a different reason; a heavily peened surface can mask the presence of discontinuities, making inspection more difficult. When properly applied, peening provides a very effective way of reducing residual stresses when welds are made in heavy sections, or in situations where the welds are rigidly restrained.

Crystal Structures

In a solid metal, the atoms tend to align themselves into orderly lines, rows, and layers to form three dimensional crystalline structures. Metals are, by definition, crystalline and any discussion of failures due to "crystallization" is of course incorrect. When a metal solidifies, it always does so in a crystalline pattern. The "crystalline" fracture surface appearance mistakenly referred to is usually typical of a fatigue or brittle fracture surface.

The smallest number of atoms that can completely describe their orderly arrangement is referred to as a "unit cell." It is important to realize that unit cells do not exist as independent units, but share atoms with adjoining unit cells in a three dimensional array.

The most common crystal structures, or phases, are body-centered cubic (BCC), face-centered cubic (FCC), body centered tetragonal (BCT), and hexagonal close-packed (HCP). These are shown in Figure 8.5. Some metals, such as iron, exist as one solid phase at room temperature and as another solid phase at elevated temperatures. This change with temperature from one phase to another in a solid metal is known as an allotropic, or solid state, phase transformation. A metal crystal possessing different structures but the same chemical composition is referred to as allotropic. This will be discussed in greater detail later.

The BCC unit cell can be described as a cube with an atom at each of the eight corners and a single atom at the center of the cell. Among the common BCC metals are iron, carbon steels, chromium, molybdenum, and tungsten.

The FCC unit cell can be envisioned as a cube with atoms at each of the eight corners and with one atom at the center of each of the six faces. Among the common FCC metals are aluminum, copper, nickel, and austenitic stainless steels.

Figure 8.4—Peening of Intermediate Weld Layers to Relieve Welding Residual Stresses

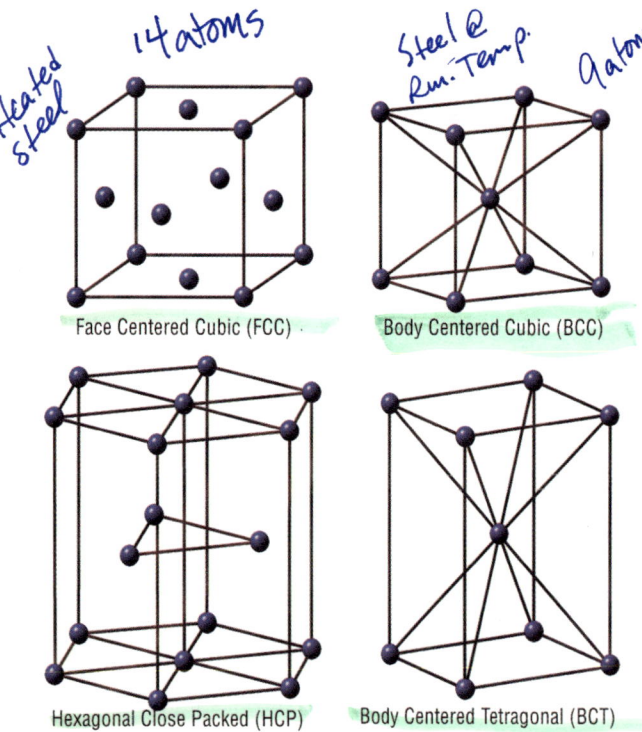

Figure 8.5—Common Crystal Structures of Metals and Alloys

The BCT unit cell can be described by taking the basic BCC unit cell, and elongating one axis to make it rectangular shaped, with an atom in the center. Martensite, a phase of steel formed by rapid quenching, is a BCT structure.

The HCP unit cell is a hexagonal prism. It can be envisioned as two hexagons (six-sided shapes) forming the top and bottom of the prism. An atom is located at the center and at each point of the hexagon. Three atoms, one at each point of a triangle, are located between the top and bottom hexagons. Among the common HCP metals are zinc, cadmium and magnesium.

Solidification of Metals

A metal solidifies into a crystalline structure by a process known as nucleation and growth. Upon cooling, clusters of atoms nucleate (solidify) at impurities or at locations on a liquid-solid boundary, such as at the interface between the molten weld metal and the cooler unmelted heat-affected zone. These clusters are called nuclei and they occur in great numbers. In the weld metal, the nuclei tend to attach themselves to existing grains of the heat-affected zone at the weld interface. Atoms continue to solidify and attach themselves to the nuclei. Each nucleus grows along a preferred direction, with the atoms aligning themselves in the arrangement described by the appropriate unit cell to form an irregularly shaped grain, or crystal.

Figure 8.6 shows how the weld metal grains form as the weld metal solidifies. In Figure 8.6A, the initial crystals begin to form at the weld interface. Figure 8.6B shows the solid grains formed as these initial nuclei grow. Since the nuclei are oriented differently, grain boundaries are formed when adjacent grains grow together. Figure 8.6C shows the completed solidification of the weld metal. Grain boundaries are considered to be discontinuities, because they represent interruptions in the uniform arrangement of the atoms. From our previous discussion, residual stress is present in the solidified metal.

Mechanical properties can be dependent upon the grain size of the metal. A metal exhibiting a small grain size will have improved room temperature tensile strength, because the grain boundaries tend to inhibit the deformation of individual grains when the material is stressed. However, at elevated temperatures, the atoms in the boundaries can move easily and slide past one another, thus reducing the material's strength at these higher temperatures. As a result, fine-grained materials are preferred for room and low temperature service while

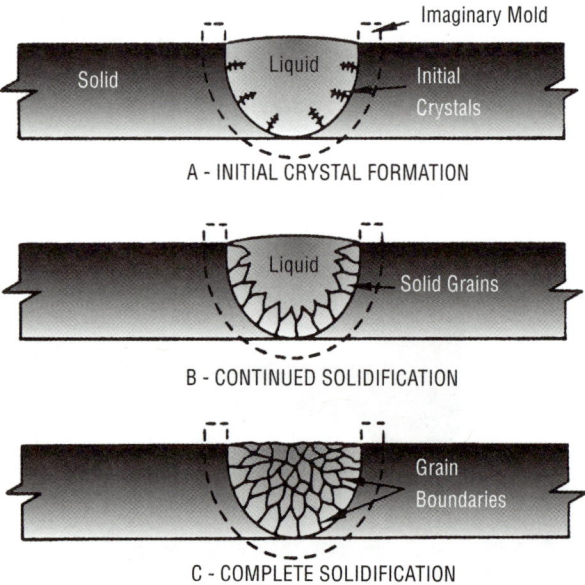

Figure 8.6—Nucleation and Solidification of Molten Weld Metal

coarse-grained materials are desirable for high temperature service. Fine-grained metals generally exhibit better ductility, notch toughness and fatigue properties.

As a quick review before continuing; metals are crystalline structures formed by atoms in ordered patterns. This ordered pattern, or arrangement, is known as a phase and is described by a unit cell. Metals solidify from many locations at once and grow in preferred directions to form grains or crystals. The junction between individual grains is referred to as a grain boundary. The grain size will dictate the amount of grain boundary area present in a metal which, in turn, determines to a certain degree the mechanical properties of the metal.

Alloying

The properties of metallic elements can be altered by the addition of other elements, which may or may not be metallic. Such a technique is known as alloying. The metal which results from this combination is referred to as an alloy. For example, the metallic element zinc is added to the metal copper to form the alloy brass. The nonmetal carbon is one of the alloying elements added to iron to form the alloy steel.

Alloying elements are included in the base metal lattice (the general arrangement of individual atoms) in various ways depending upon the relative sizes of the atoms. Smaller atoms, such as carbon, nitrogen and hydrogen, tend to occupy sites between the atoms that form the lattice structure of the base metal. This is known as interstitial alloying and is shown in 2-dimensions in Figure 8.7. For example, small amounts of carbon can occupy the interstitial sites between the iron atoms in steel.

Alloying elements with atoms close to the size of those of the base metal tend to occupy substitutional sites. That is, they replace one of the base metal atoms in the lattice structure. This is called substitutional alloying and is shown in Figure 8.8. Examples of this are both copper in nickel and nickel in copper.

Like the presence of grain boundaries, the addition of alloying elements produces irregularities in the lattice structure. As can be seen in both Figures 8.7 and 8.8, the presence of alloying elements exerts varying degrees of atomic attraction and repulsion to result in a lattice arrangement which is somewhat

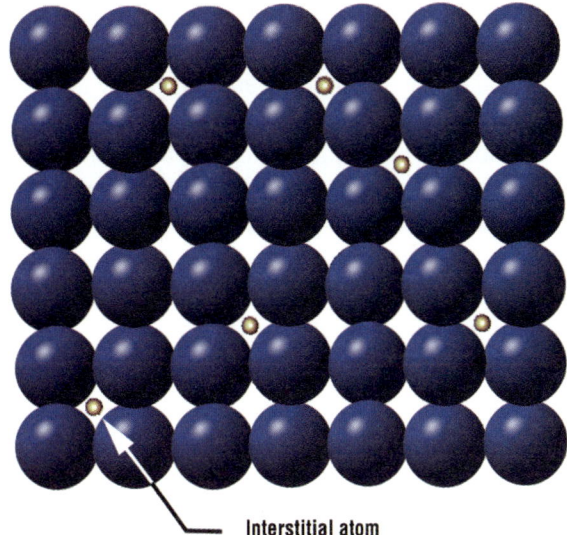

Figure 8.7—Interstitial Alloying

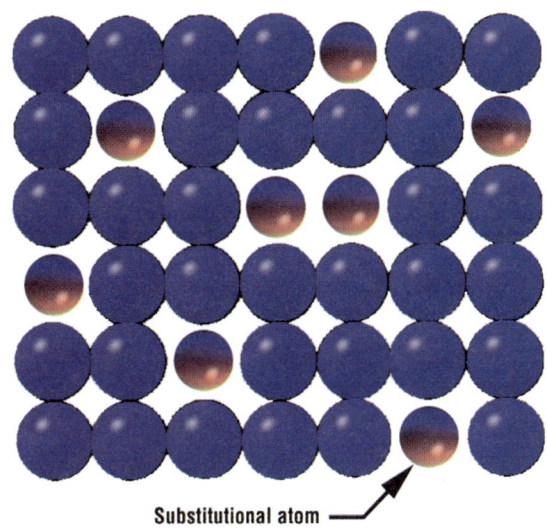

Figure 8.8—Substitutional Alloying

distorted, or strained. This tends to increase the internal energy of the metal and can result in improved mechanical properties.

Nearly all engineering metals are alloys consisting of one major element and variable amounts of one or more additional elements. Alloys usually consist of many randomly oriented grains with each grain arranged in a specific way and containing one or more of the characteristic phases that exist for the alloy. If more than one phase is present, each will have its own characteristic crystalline structure.

Microstructural Constituents of Carbon Steel

The overall arrangement of grains, grain boundaries, and phases present in a metallic alloy is called its microstructure. The microstructure is primarily responsible for the properties of the alloy. This microstructure is affected by the composition or alloy content, and by other factors such as forming and heat treating operations. The microstructure is greatly affected by the welding operation which, in turn, influences the properties of the alloy.

While all metals exhibit various microstructures, this discussion will deal exclusively with the microstructural changes that occur in plain carbon steel, which is an alloy consisting of iron in combination with carbon. Other alloying elements may be included as well, but their effects on the microstructure will not be as significant as those of carbon.

To introduce this topic, it is important to realize that iron and steels undergo changes in their crystallographic arrangement as a result of temperature changes. That is, as iron-carbon alloys are heated or cooled, phase changes occur. The occurrence of this phenomenon allows us to change the mechanical properties of a specific alloy through the application of various heat treatments. To understand the changes that occur, the metallurgist uses a chart, or phase diagram, which graphically displays the ranges of various microstructural components for the iron-carbon system. Referred to as the "Iron-Carbon Phase Diagram," it is shown in Figure 8.9.

This diagram describes the nature of the phases present in iron-carbon alloys under "near equilibrium" conditions, i.e., very slow heating and cooling. It should be noted that many of these microstructural constituents have multiple names which are used interchangeably. For example, pure iron at room temperature is referred to as alpha iron or ferrite. The iron carbide which is present at room temperature is called cementite or Fe3C. The face-centered cubic structure which occurs at intermediate temperatures is referred to as either austenite or gamma iron.

Looking at the diagram, notice that the vertical axis describes the temperature changes while the horizontal axis indicates the amount of carbon present. Consequently, for a given carbon content, a vertical line can be drawn through that horizontal axis intersect. Moving vertically up that line, it can be

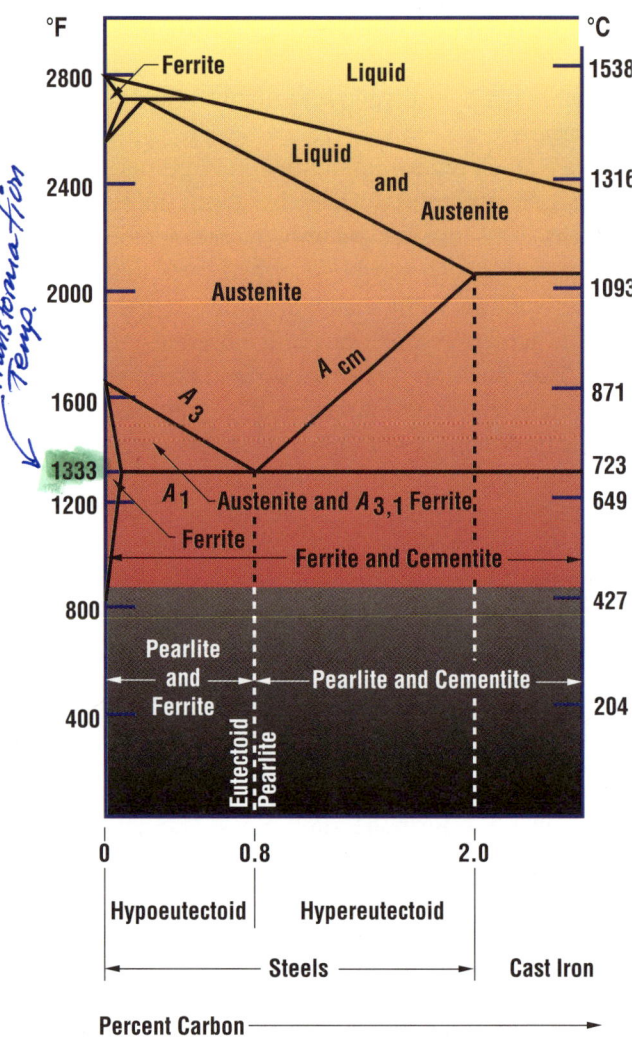

Figure 8.9—Iron-Carbon Phase Diagram

determined what microstructures will exist at various temperatures.

As shown in the notation below the horizontal axis, steels are considered to include those iron-carbon alloys having from 0.008% to 2% carbon. Within this range, the steels are further divided into hypoeutectoid, eutectoid and hypereutectoid types, with the eutectoid point (0.8% carbon) being the dividing line. Hypoeutectoid steels are those alloys with less than 0.8% C which exist at room temperature as combinations of pearlite and ferrite as opposed to hypereutectoid types which contain more than 0.8% C and exist as combinations of pearlite and cementite. The room temperature equilibrium microstructure for a eutectoid steel (exactly

0.8% carbon) is pure pearlite. Pearlite is a layered mixture of cementite and ferrite. The technique of polishing and using an acid etch reveals the microstructures which are shown in Figures 8.10–8.12.

Figure 8.10—Microstructure of Commercially Pure Iron; White Grains are Ferrite. The Grain Boundaries are shown, and the darker globules are nonmetallic inclusions (~100X Magnification)

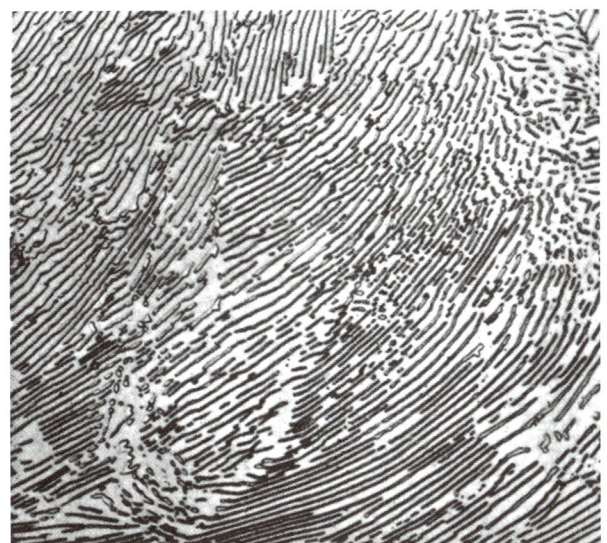

Figure 8.11—Lamellar Appearance of Pearlite (1500X Magnification)

Figure 8.12—Quenched Martensite—Showing Acicular Structure (500X)

Figure 8.10 shows the typical microstructure of commercially pure iron (ferrite) with almost no carbon content. Figure 8.11 shows the typical appearance of pearlite when polished, etched with acid, and viewed under a high power microscope (1500X). The light areas are ferrite and the dark areas are cementite.

One of the important transformations which occurs in steel is the transformation from the various room temperature constituents (ferrite, pearlite, cementite, and combinations of these) to austenite, which is the face-centered cubic structure of iron and carbon. Upon heating, this transformation will begin to occur at 1333°F; the horizontal line representing this transformation temperature is referred to as the A1. Except at a carbon content of 0.8%, the eutectoid percentage, this transformation will occur over a range of temperatures with complete transformation occurring only when the temperature is raised above the sloping line designated as the A_3. In pure iron, the transformation is complete at 1670°F, while a eutectoid steel will undergo this complete transformation at 1333°F.

Upon very slow cooling, these same changes will occur in reverse. It is the existence of this transformation that permits us to harden or soften steels through the use of various heat treatments. When a steel has been heated into the austenitic range and then allowed to cool slowly through this transformation range, the resulting microstructure will contain pearlite. This structure can only occur when sufficient time is allowed for the atoms to diffuse into that arrangement. Diffusion is simply the migration of the atoms within the solid metal structure. The higher the temperature, the more mobile the atoms become in the lattice structure. When cooling from austenite occurs slowly enough, pearlite will form. Steels that are heat treated to produce pearlite are generally very soft and ductile.

When the cooling from the austenitic range occurs more rapidly, there are significant changes in this transformation for a given steel alloy. First, the transformation will occur at a lower temperature. Secondly, the resulting microstructure is drastically changed and the hardness and tensile strength of the steel increase significantly, with a corresponding decrease in ductility. At faster cooling rates, the principal microstructures produced include fine pearlite, bainite and martensite.

With a slight increase in cooling rate, the transformation temperature is depressed somewhat, producing a finer pearlitic structure with the lamellae more closely spaced. This structure is slightly harder than the coarse pearlite and has somewhat lower ductility. At still faster cooling rates, and lower transformation temperatures, pearlite no longer forms. Instead, bainite is formed and this structure has a feathery arrangement of fine carbide needles in a ferrite matrix. Bainite has significantly higher strength and hardness and lower ductility, and is very difficult to see under the microscope.

Upon very rapid cooling, or quenching, insufficient time is available for diffusion to take place. Consequently, some of the carbon becomes trapped in the lattice. If the cooling rate is fast enough and the amount of carbon present is high enough, martensite will be formed. The formation of martensite is a diffusionless process (the cooling rate is so rapid the atoms do not have time to move around). The transformation from austenite to martensite results from a shear-type, or mechanical, action. The resulting crystalline structure is referred to as body-centered tetragonal, which is a distortion of a body-centered cubic structure into a rectangular one. Due to the presence of this distorted lattice arrangement, the martensitic structure exhibits a higher internal energy or strain which results in extremely high hardness and tensile strength. However, martensite has a characteristically low ductility and toughness. Figure 8.12 shows the appearance of martensite at a high magnification (500X).

To improve the ductility and toughness without significantly decreasing the hardness and tensile strength of the martensite, a process referred to as *tempering* is employed. This heat treatment consists of reheating the quenched martensitic structure to some temperature below the lower transformation temperature (1333°F). This permits the as-quenched, unstable martensite to change to tempered martensite by allowing the carbon to precipitate in the form of tiny carbide particles. The desired strength and ductility can be controlled by choosing the proper tempering times and temperatures. Higher tempering temperatures result in softer and more ductile properties. Quenching and tempering heat treatment is frequently used to enhance the properties of machinery steels, since it develops high yield and tensile strengths, high yield strength/tensile strength ratios, and improved notch toughness as compared to the rolled, annealed, or normalized properties. An example of the effects of various tempering temperatures for a particular steel alloy is illustrated in Figure 8.13.

To aid in the determination of what microstructural constituents will result from rapid cooling rates, the metallurgist uses another diagram which is referred to as a TTT diagram, or Time-Temperature-Transformation diagram. These are also called isothermal transformation diagrams (IT). As the name implies, it describes the microstructural products which occur after specific times at a particular temperature for a particular steel composition. A similar diagram, the CCT, or Continuous Cooling Transformation, diagram shows the changes that occur during continuous cooling from the austenitic range. These two types of diagrams agree quite well. Figure 8.14 depicts the continuous cooling transformation characteristics of type 8630 steel.

This diagram shows the microstructural products as a function of both temperature and time. Various cooling rates are shown to illustrate how the diagram is used. The resulting transformation products depend on the regions through which the

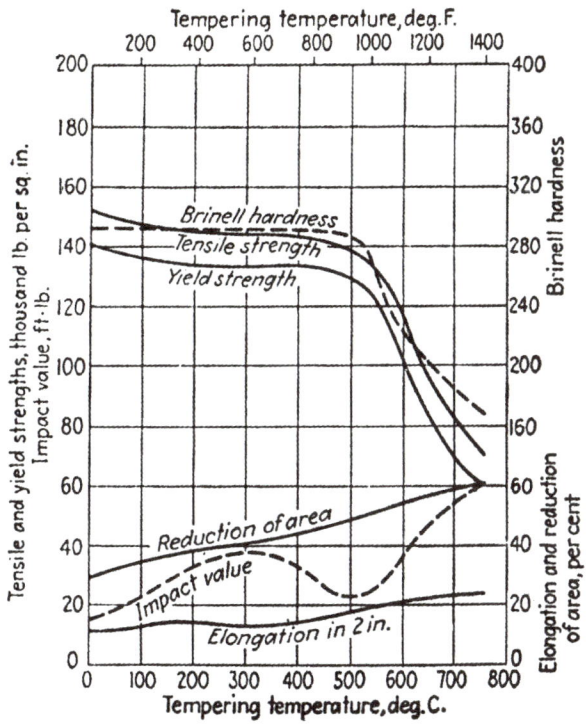

Figure 8.13—Effects of Tempering Temperature on Mechanical Properties of a 12.2% Cr Alloy

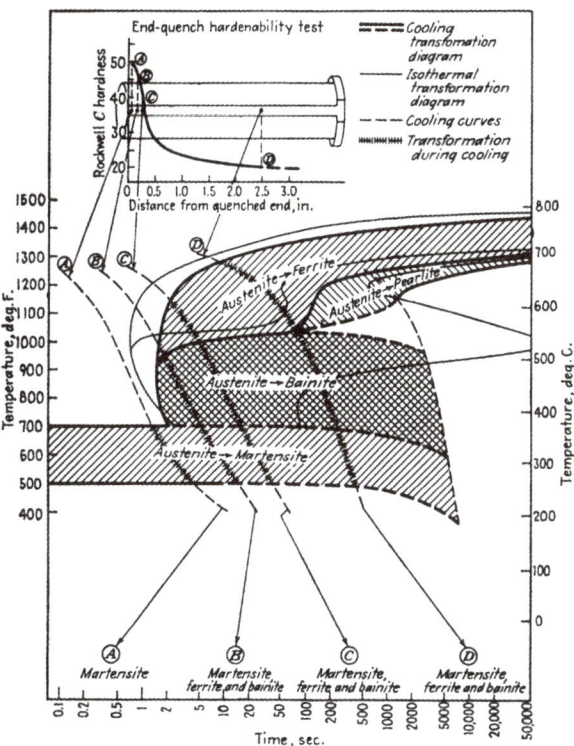

Figure 8.14—Continuous Cooling and Jominy Hardenability Diagrams for Type 8630 Steel

cooling curves pass and the amount of time it takes for the cooling curves to pass through those regions. As an example, curve "A" only passes through the austenite to martensite region, so the resulting microstructure is 100% martensite. The slower cooling rate characterized by curve "D" shows that the resulting microstructural components will be primarily ferrite with only minor amounts of bainite and martensite. Since martensite can only transform from austenite, any austenite which transforms to ferrite or bainite cannot transform to martensite.

Metallurgical Considerations for Welding

Since welding can result in significant changes in both the temperature of the metal as well as the rate of cooling from that elevated temperature, it is important to understand what metallurgical changes can result from the welding operation. Figure 8.15 illustrates the relationship between the peak temperatures exhibited in various regions of the weld zone and the iron–iron carbide equilibrium diagram.

As can be seen, depending on the location of a point within or adjacent to the weld, there can be various metallurgical structures produced. Within the weld, the region of highest temperatures, the metal can cool from the liquid through the various phase regions noted earlier. Adjacent to the weld, in the heat-affected zone (HAZ), no melting occurs but extremely high temperatures can be reached. The HAZ is that region of the base metal adjacent to the weld metal which has been raised to temperatures from just below the transformation temperature to just below the melting point of the steel. Cooling rates in this heat-affected zone are among the most rapid because of a phenomena known as contact quenching. Changes in the welding conditions can have a very significant effect on the formation of the various phases, because the welding conditions have a significant effect on the resulting cooling rate for the weld. Some of the welding conditions which could produce changes include the amount of heat input, use of preheat, carbon

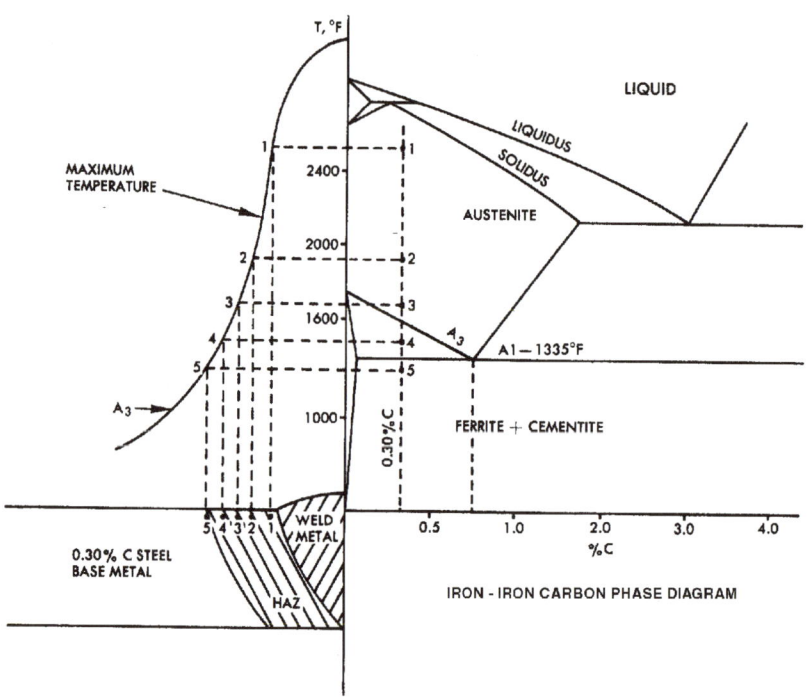

Figure 8.15—Relation Between Peak Temperatures of Various Regions of a Weld, and Correlation with the Iron-Carbon Phase Diagram

equivalent of the base metal, and the base metal thickness.

As the heat input increases, the cooling rate decreases. The use of smaller diameter welding electrodes, lower welding currents, and faster travel speeds will all tend to decrease the heat input and therefore increase the cooling rate. For any arc welding process, the heat input can be calculated. It is dependent only on the apparent welding current, arc voltage and travel speed, as measured along the longitudinal axis of the weld joint. The formula for welding heat input is shown below.

$$\text{Heat Input} = \frac{\text{Welding Current} \times \text{Welding Voltage} \times 60}{\text{Welding Travel Speed, (in./min.)}}$$

For this formula, heat input is expressed in terms of joules/inch, and travel speed in inches per minute. Joules can also be expressed as watt-seconds. Therefore, the 60 occurring above the line in the formula converts the minutes in the travel speed into seconds. The welding inspector may be asked to monitor the welding heat input to control the resulting microstructural properties occurring in the heat-affected zone.

Another item which has a significant effect on the resulting microstructure of the heat-affected zone is the use of preheat. In general, the use of preheat will tend to reduce the cooling rate in the weld and HAZ resulting in improved ductility. When no preheat is used, the heat-affected zone is relatively narrow and exhibits its highest hardness. In some cases, depending on alloy content, martensite may be formed. However, when preheat is included, the heat-affected zone is wider and the resulting hardness is significantly lower due to the slower cooling rate which allows for the formation of ferrite, pearlite, and possibly bainite instead of martensite. Therefore, the welding inspector may be asked to monitor the preheat required for a particular welding operation. This requirement is primarily concerned with slowing down the cooling rate in the heat-affected zone to produce microstructures having desirable properties.

Another important factor for the welding of a steel is its carbon equivalent. Since carbon has the most pronounced effect on the hardenability (the ease with which a metal hardens upon cooling from an austenitic temperature, or its ability to form martensite) of the steel, we are most concerned with how much of it is present in a particular alloy. The higher the carbon content, the more hardenable the steel.

Other alloy elements will also promote hardenability, to various degrees. An equivalent carbon content is therefore an empirical expression which is used to determine what the combined effects of various alloying elements are on the hardenability of that steel. An example of a typical carbon equivalent content (C.E.) is shown below.

$$C.E. = \%C + \frac{\%Mn}{6} + \frac{\%Ni}{15} + \frac{\%Cr}{5} + \frac{\%Cu}{13} + \frac{\%Mo}{4}$$

This formula is intended for carbon and alloy steels containing not more than 0.5% Carbon, 1.5% Manganese, 3.5% Nickel, 1% Chromium, 1% Copper, and 0.5% Molybdenum.

Once an equivalent carbon content has been determined, we can predict the approximate range of preheat that will be necessary for best results in welding. The table below summarizes some suggested preheat temperatures for various ranges of carbon equivalents.

Carbon Equivalent	Suggested Preheat Temperature
Up to 0.45	Optional
0.45 to 0.60	200 to 400°F
Above 0.60	400 to 700°F

Using these guidelines, the welding engineer can make a preliminary decision as to what preheat temperature would be satisfactory for a given application. Other factors will affect this decision, but this at least provides a starting point.

Base metal thickness also has an effect on the cooling rate; generally, welds on thicker base metal cool more rapidly than welds on thin sections. The larger heat capacity, or heat sink, associated with the thicker sections produces faster cooling of the weld bead. So, when thicker sections are being welded, various welding requirements, such as preheat, may be specified to reduce the cooling rate to improve the resulting mechanical properties of the heat-affected zone. Therefore, when welding heavy sections, preheat and interpass requirements are normally increased to aid in slowing down the resulting cooling rate.

Heat Treatments

There has been mention of some of the heat treatments that may be applied to metals. They can be applied to the base metal prior to welding or to the completed weldment to produce specific mechanical properties. As a welding inspector, one of your jobs may be to monitor these heat treatment operations to assure that the time and temperature requirements are being observed.

The basic heat treatments include annealing, normalizing, quenching, tempering, preheating, postheating, and thermal stress relieving.

Annealing is a softening treatment used to increase the metal's ductility at the expense of its strength. To accomplish annealing, the metal is raised just into the austenitic range, held for one hour per inch of thickness or a minimum of one hour, and then cooled very slowly. In a furnace heat treatment, this cooling is usually done by turning off the furnace power and allowing the part to cool to room temperature while remaining in the furnace.

Normalizing also softens the metal, but not as significantly as annealing. It is considered an "homogenizing" heat treatment by making the metal structure very uniform throughout its cross section. The normalizing treatment is initiated by raising the metal's temperature into the austenitic range, holding for a short time, and then allowing it to slow-cool in still air. This cooling occurs faster than a furnace cool, so the resulting properties include slightly higher hardness and strength and possibly lower ductility as compared to annealing. Normalized carbon and low alloy steels are usually easily weldable.

Quenching differs from annealing and normalizing in that the resulting mechanical properties show significantly increased hardness and strength and decreased ductility. This hardening treatment is accomplished by raising the metal's temperature into the austenite range, holding for a time, and rapidly cooling to room temperature by immersing the part in a quenching medium, such as water, oil, or brine (salt water). Quenching is performed to produce a primarily martensitic structure which has characteristically high hardness and strength and low ductility. To improve the ductility without significant degradation of the metal's strength characteristics, a tempering treatment is usually performed. To temper, the metal is reheated to a temperature below the lower transformation temperature, held for a short time to allow the highly stressed martensitic structure to relax somewhat, and then cooled.

Preheat treatments are used, as discussed previously, to slow down the cooling rate of the base metal adjacent to the weld to allow for the formation of microstructural constituents other than martensite. Preheat is applied prior to welding. Postheat treatments are used to reduce residual stresses and to temper hard, brittle phases formed during cooling or quenching. Postheat is applied after the welding has been completed. Generally, postheat temperatures are higher than those used for preheat.

The final heat treatment to be discussed is thermal stress relief, which falls under the category of postheat treatment. This was discussed earlier as a method to reduce the amount of residual stresses which are present following welding. Thermal stress relief is done at temperatures below the lower transformation temperature of 1333°F. By raising the temperature of the weld and base metal gradually and uniformly, the thermal stresses created by the localized heat of welding are allowed to relax. Stress relief occurs because the strength of a metal is reduced as its temperature is increased, allowing the residual stress to relax and the metal to recover. The component is cooled at a moderate rate after stress relief. This treatment will aid in the elimination of problems associated with distortion.

There are two other aspects of welding metallurgy to be discussed since they will also aid the inspector's understanding of the physical principals involved in the various metallurgical changes discussed earlier. These are diffusion and solid solubility.

Diffusion

We have previously noted that atoms in the liquid state can move about easily with respect to each other; however, under certain conditions, even atoms in the solid state can change positions. In fact, any atom may "wander" away, step by step, from its home position. These changes of atom position in the solid state are called diffusion.

An example of diffusion is shown if smooth, flat bars of lead and gold are clamped tightly together (see Figure 8.16). If they are left clamped together at room temperature for several days, the two sheets of metal remain attached when the clamps are removed. This attachment is due to the atoms of lead and gold each migrating, or diffusing, into the other metal, forming a very weak metallurgical bond. This bond is quite weak, and the two metals can be broken apart by a sharp blow at their joint line. If the two metals' temperatures are increased, the amount of diffusion increases, and at a temperature above the melting point of both, complete mixing occurs.

Another example of diffusion occurs when hydrogen, a gas, is allowed in the vicinity of molten metal, such as a weld. The most common source of hydrogen is moisture (H_2O), or organic material contamination on the surfaces of the parts to be welded. Many of the contaminants normally found on metals are organic compounds, like oil, grease, etc., and they contain hydrogen in their chemical makeup. The heat of welding will break down the water or organic contaminants into individual atoms, which includes the hydrogen atom (H^+).

The hydrogen atoms are quite small, and they can easily diffuse into the base metal structure. As they enter the base metal, the hydrogen atoms often re-combine into the hydrogen molecule (H_2), a combination of two atoms of hydrogen, which is much larger than a single atom of hydrogen. The larger molecules often become trapped in the metal at discontinuities such as grain boundaries or inclu-

Figure 8.16—Diffusion of Lead and Gold Atoms

sions. These hydrogen molecules, because of their larger size, can cause high stresses in the internal structure of the metal, and for metals of low ductility, can cause cracking. Hydrogen cracking is often referred to as *underbead* or *delayed* cracking.

The primary cure for hydrogen cracking problems is to eliminate the source of hydrogen; the first step is to thoroughly clean all surfaces to be welded. Another approach is to specify the "low hydrogen" electrodes for use on carbon and low alloy steels. These low hydrogen electrodes are specially formulated to keep their hydrogen content quite low, but they do require special handling to avoid moisture pickup after opening the sealed shipping containers. Preheating the base metals is also effective in eliminating hydrogen pickup because hydrogen will diffuse out of most metals at temperatures of 200°–450°F. The methods noted can all aid in reducing the possibility of hydrogen cracking in those metals which are susceptible.

Solid Solubility

Most of us are familiar with the normal solubility of solids into liquids. Adding a spoonful of salt to a glass of water and stirring will dissolve the salt. However, most of us are not familiar with one solid dissolving into another solid. In the example given above with the lead and gold, the two metals were diffusing through solid solution into each other. And, returning to our example of salt and water, if additional salt is added, we find that some of it will not dissolve regardless of how much we stir.

What has happened is that for that volume of liquid, and its temperature, we reach a "critical solubility limit." No amount of stirring will dissolve any more salt. In order to dissolve more salt, the liquid volume would either have to be increased, or its temperature raised. Thus we see that for a solid dissolving into a liquid, there is a critical solubility limit depending on liquid volume and temperature. Metals behave similarly, except through diffusion, and they "dissolve" into each other when both are solid.

But just like the salt and water, there are solubility limits of one solid dissolving into another and the critical solubility limit is dependent on temperature. The higher the temperature of a metal, the more dissolving of a second element will occur. Thus, we can get metals combining even when both are solid. Of course, as the metal temperatures are raised, the amount of diffusion and solubility increase.

An example of a solid dissolving into another solid is a method we use for increasing the surface hardness of a steel. If the steel is packed into a bed of carbon particles, and then heated to a temperature of about 1600°–1700° F, which is well below the melting point of both the carbon and the steel, some of the carbon will diffuse (dissolve) into the surfaces of the steel. This added carbon in the steel's surface makes the surface much harder, and is useful for resisting wear and abrasion. This technique is commonly called *pack carburizing*.

The surface of steels can also be made hard by exposing the steel to an ammonia environment at similar temperatures to carburizing. The ammonia (NH_3) breaks down into its individual components of nitrogen and hydrogen, and the nitrogen atoms enter the surface. This technique is called *nitriding*. Both of these surface hardening techniques demonstrate the diffusion and solid solubility of metals. Knowledge of diffusion and solid solubility will aid the welding inspector in understanding the importance of cleanliness in welding, and the need for proper shielding during all welding operations.

Welding Metallurgy of Commonly Used Materials

To this point, the primary alloy used for discussion has been carbon steel and low alloy steel. This final section will review the welding metallurgy of three commonly used materials: stainless steel, aluminum and copper and the various alloys of each.

Stainless Steels

The word "stainless" is a bit of a misnomer when applied to the classes of metals referred to as stainless steels, since it usually means they resist corrosion. However, in severe corrosive environments, many of the stainless steels corrode at very high rates. The stainless steels are defined as having at least 12% chromium. There are many types of stainless steels, and the welding inspector should recognize this when discussing them and use the proper designation for each type.

The five main classes of stainless steels are ferritic, martensitic, austenitic, precipitation hardening, and the duplex grades. The first three categories refer to the stable room temperature phase found in each class. The fourth one, often called "PH" stainless steels, refers to the method of hardening them by an "aging" heat treatment, a precipitation hardening mechanism as opposed to the quenching and tempering mechanism known as transformation hardening. The last one, the duplex grades, are approximately half ferrite and half austenite at room temperature with improved resistance to chloride stress corrosion cracking.

The stable room temperature phase found in stainless steels depends on the chemistry of the steel, and some stainless steels may contain a combination of the different phases. The more common stainless steels are the austenitic grades, which are identified by the "200" and "300" series grades; 304 and 316 stainless steels are austenitic grades. A 416 steel is a martensitic grade, and 430 is a ferritic grade. One of the common PH stainless steels is a 17-4 PH grade. A popular duplex grade is AL-6XN.

As might be expected, the weldability of these grades varies significantly. The austenitic grades are very weldable with today's available filler metal compositions. These grades can be subject to hot short cracking, which occurs when the metal is very hot. This problem is solved by controlling the composition of the base and filler metals to promote the formation of a "delta ferrite" phase which helps eliminate the hot short cracking problem.

Typically, cracking will be avoided in the austenitic stainless steels by selecting filler metals with a delta ferrite percent of 4–10%. This percentage is often referred to as the *Ferrite Number* and can be measured with a magnetic gauge. The delta ferrite can be measured using the magnetic gauge since delta ferrite is BCC and magnetic, while the primary phase, austenite, is FCC and non-magnetic.

The ferritic grades are also considered weldable with the proper filler metals. The martensitic grades are the most difficult to weld, and often require special preheating and post weld heat treatment. Procedures have been developed to weld these materials and must be followed carefully to avoid cracking problems and maintain the mechanical properties of the base metals. The PH and duplex stainless steels are also weldable, but attention must be given to the changes in mechanical properties caused by welding.

One of the common problems found when welding the austenitic grades is referred to as *carbide precipitation*, or *sensitization*. When heated to the welding temperatures, a portion of the base metal reaches temperatures in the 800°–1600°F range, and within this temperature range, the chromium and carbon present in the metal combine to form chromium carbides. The most severe temperature for this formation is about 1250°F, and this temperature is passed through twice on each welding operation cycle, once on heating to weld and again on cooling to room temperature.

These chromium carbides typically are found along the grain boundaries of the structure. The result of their formation is the reduction of the chromium content within the grain itself adjacent to the grain boundary, called *chromium depletion*, resulting in reducing the chromium content below that required for resisting corrosion. The final result of this chromium depletion of the grain is a reduced corrosion resistance of the grain itself due to its reduced chromium content. In certain corrosion environments, the edges of the grains corrode at a high rate, and is called *intergranular corrosion attack*, or IGA (see Figure 8.17).

Sensitization of austenitic stainless steels during welding can be prevented by several methods. The first method involves reheat treating the complete structure by heating to 1950°–2000°F. This solution annealing breaks up the chromium carbides, permitting the carbon to be redissolved into the structure. However, this heat treatment can cause severe distortion of welded structures. Following heat treatment the structure must be rapidly quenched in water to avoid reformation of the chromium carbides (see Figure 8.18).

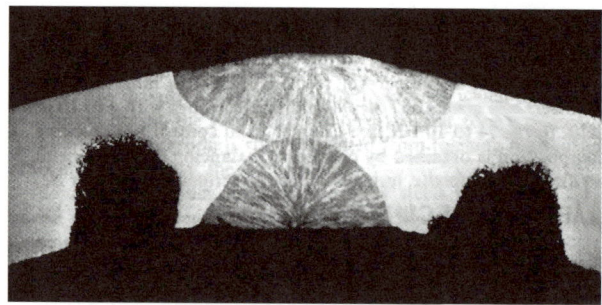

Figure 8.17—Intergranular Corrosion Attack on Austenitic Stainless Steel Caused by Sensitization During Welding

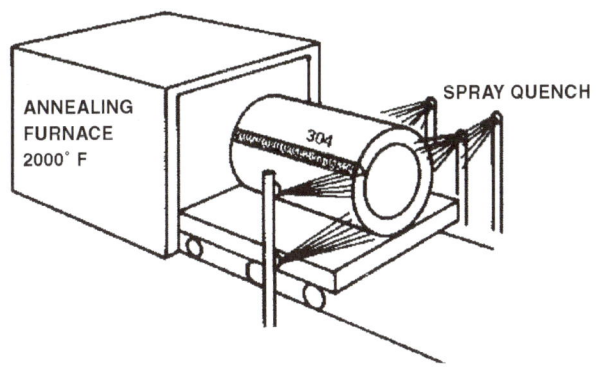

Figure 8.18—Quenching of Austenitic Stainless Steel to Maximize Corrosion Resistance

A second method is the addition of stabilizers to the base and filler metals. The two most common examples for stabilization are the addition of titanium or niobium (columbium) to the 300 series alloys in amounts equal to 8 or 10 times the carbon content. These alloying stabilizers preferentially combine with the carbon and reduce the amount of carbon available for chromium carbide formation, maintaining the grains' chromium content and corrosion resistance. When titanium is added, we have the austenitic stainless alloy 321; when niobium is added, we have the 347 grade (see Figure 8.19).

The third method is the reduction of the carbon content in the base and filler metals. Initially, these low carbon austenitic stainless steels were referred to as *Extra Low Carbon*, or ELC for abbreviation. Today, they are referred to by the letter "L" meaning the carbon content is less than 0.03%. (The standard grades contain up to 0.08% carbon.) By reducing the carbon content in the alloy, less carbon is available to combine with the chromium, and sensitization is reduced during welding (Figure 8.19). These low carbon grades have slightly reduced mechanical properties because of their lower carbon content, and this must be considered when selecting these alloys, especially for high temperature use.

Aluminum and its Alloys

Aluminum alloys have a very tenacious oxide film on their surfaces which forms very rapidly when the bare aluminum is exposed to air, and these oxide films give protection in corrosive environments. These same oxides on the surface interfere with the joining processes. To braze or solder these alloys, fluxes are used to break down the oxide film so the parts can be joined. When welding, alternating current is used which results in breaking down of the oxides by the current reversal of AC welding, and reformation of the oxide film is avoided by shielding with helium or argon gas. The AC welding method is sometimes referred to as a *surface cleansing technique*.

The metallurgy of aluminum and its alloys is very complex, especially regarding the great number of alloy types and heat treatments. The proper filler metals for most weldable grades and heat treat conditions can be found in AWS A5.10, *Specification for Bare Aluminum and Aluminum Alloy Welding Electrodes and Rods*.

Copper and its Alloys

Pure copper and many of its alloys cannot be hardened by a quench and temper heat treatment as steel can. These alloys are usually hardened and made stronger by the amount of "cold work" introduced when forming them into the various shapes. The act of welding softens the cold worked material and must be considered before welding on work-hardened copper alloys. There is a series of copper alloys that are strengthened by "aging," a treatment similar to the precipitation hardening used on the PH stainless steels. When welding on these alloys, a postweld heat treatment is usually specified to restore the original mechanical properties.

One of the major problems with welding of copper and its alloys is due to their relative low melting point and very high thermal conductivity. Considerable heat must be applied to the metal to overcome its loss through conductivity, and the relatively low melting point often results in the metal melting earlier than expected and flowing out of the weld joint. Most copper alloys are weldable with proper technique and practice.

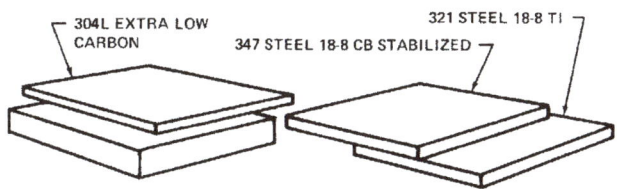

Figure 8.19—Prevention of Sensitization in Austenitic Stainless Steel

Summary

The metallurgy of welding is an important consideration for any welded component because the metallurgical changes which occur can have very significant effects on the resulting mechanical properties of the weld and base metals. A number of welding procedure requirements exist because the metallurgist or welding engineer desires to control the resulting mechanical properties of the weldment. Consequently, the welding inspector may be asked to monitor some of these requirements to assure that the resulting fabrication is satisfactory.

Changes in the metal properties will occur based on the amount of heat which is applied as well as the rate at which the heat is removed from the metal. You have seen how these factors cause changes in the metal properties. Consequently, it is now easier to understand which welding variables are important and why they need to be controlled during the welding operation.

Key Terms and Definitions

alpha iron—a BCC solid solution of carbon in iron, and stable at room temperature. Also named ferrite.

annealing—a heat treatment designed to place the metal in its softest, lowest strength condition.

austenite—an FCC solid solution of carbon in iron which forms upon heating above the A3 transformation line. Also, the room temperature stable phase of the 300 series of stainless steels. Also called gamma iron.

autogenous—in welding, a weld made without filler metal, melting portions of the base metal for filler.

bainite—a phase of iron, as contrasted with pearlite, which forms on cooling. A very fine particle sized structure difficult to resolve on the optical microscope.

BCC—body centered cubic.

BCT—body centered tetragonal.

carbon equivalent—a number calculated by one of several different formulae which aids in determining the required preheat temperature.

carburizing—a case hardening process that diffuses carbon into a solid ferrous alloy by heating the metal in contact with a carbonaceous material (one containing carbon). There are two types of carburizing; pack carburizing and gaseous carburizing.

cementite—iron carbide, Fe_3C.

conduction—in heat transfer, the transmission of heat from particle to particle.

convection—in heat transfer, the transmission of heat by the mass movement of the heated particles.

crystal—or grain; the individual unit formed on solidification, separated from other crystals by grain boundaries.

delta ferrite—a phase of stainless steel alloys which resists cracking at high temperatures.

deoxidizers—elements or compounds which preferentially combine with oxygen to keep it from reacting with heated base or weld metal.

dewpoint—the temperature at which any moisture present will condense; a measure of gas "dryness."

diffusion—movement of atoms within a solution, be it a solid, liquid, or a gas.

discontinuity—any interruption in the normal homogeneous pattern or arrangement of a metal.

duplex—in metals, a type of stainless steel containing approximately 50% ferrite and 50% austenite.

eutectoid—in steel, the alloy with a carbon content of exactly 0.8%.

ferrite—a BCC solid solution of carbon in gamma iron; also named alpha iron.

FCC—face centered cubic.

gamma iron—an FCC solid solution of carbon in iron; also named austenite.

grain—in metals, the individual crystal formed on solidification; see crystal.

HAZ—heat-affected zone; the metal adjacent to a weld that does not melt but is affected by the heat of welding.

HCP—hexagonal close packed.

hot short—the formation of intergranular hot cracks as a result of iron sulfide contained in the grain boundaries at elevated temperatures (1800°F).

hypereutectoid—a steel alloy with more than 0.8% carbon.

hypoeutectoid—a steel alloy with less than 0.8% carbon.

IGA—intergranular corrosion attack; caused by sensitization of stainless steels.

joule—the metric unit for work or heat (energy).

ksi—one thousand pounds per square inch.

lamellar—layered, or plate-like.

martensite—an unstable constituent of iron, formed without diffusion by rapid quenching from the austenite phase above the transformation line, A3.

metallurgical bond—the special type of atomic bonding holding metal atoms together.

molecule—a group of chemically bonded atoms.

nascent—a single atom, as in a single (nascent) hydrogen atom as opposed to molecular hydrogen which is composed of two hydrogen atoms as H_2; all gases are molecular in their natural state.

nitriding—a case hardening process that introduces nitrogen into the surface of a ferrous material at elevated temperatures in the presence of ammonia or nitrogen.

normalizing—a heat treatment whereby a steel is heated into the austenite range and cooled in still air.

notch toughness—the ability of a metal to absorb an impact load (energy) with the existence of surface notches.

organic—materials made up primarily of the elements of carbon, oxygen and hydrogen.

pearlite—a layered, or lamellar, structure composed of ferrite and cementite (iron carbide).

peening—severe mechanical deformation of a metal.

phase transformation—in metals, a change in the atomic structure.

postheat—a thermal treatment given to a weldment after welding is completed.

precipitation hardening—a hardening mechanism, different from quenching and tempering, which relies on the formation of a precipitate during the heat treating cycle for increasing strength and hardness.

preheat—a thermal treatment given to a joint prior to welding.

psi—pounds per square inch.

purging—the secondary application of an inert or unreactive gas to protect the back side of weldments during welding.

quenching—in heat treatment, very rapid cooling from elevated temperatures.

residual stress—stress remaining immediately after a welding or forming operation.

safety factor—a multiplier used in design making the structure stronger than actually required; typically 3 or 4 for pressure vessels and 5 or more for bridges and castings.

segregation—in alloying, the separation, or non-homogeneity, of two or more elements or phases.

sensitization—or carbide precipitation; the formation of chromium carbides resulting in depletion of the chromium from the individual grains and reducing the metal's corrosion resistance to intergranular corrosion (IGA).

shielding—the primary protection from atmospheric gases during the welding operation; obtained from fluxes, electrode coatings, or inert or reactive gases.

slag—the material formed when welding fluxes or electrode coatings combine with atmospheric gases or contaminants during welding.

solid solubility—the ability of metals to dissolve within each other when in a solid form through diffusion mechanisms.

stainless steels—alloys containing a minimum of 12% chromium selected for their corrosion resistance.

stress relieving—a heat treatment which relieves a metal's residual stress by heating, holding at temperature, and cooling per a prescribed cycle.

stress riser—or stress raiser; a surface notch or geometry which multiplies the applied stress to increase the actual stress in a component.

tempering—a heat treatment which reduces the strength and hardness of as-quenched steels and restores ductility and toughness.

thermal expansion—the expansion, or growth, of a material upon being heated.

toughness—the ability of a material to absorb energy.

unit cell—a symmetrical shape with the smallest number of atoms that completely describes a unique structure of a metal or phase.

Module 9
Weld and Base Metal Discontinuities

Contents

Introduction and Background 9-2

Discussion of Discontinuities 9-4

Summary 9-25

Key Terms and Definitions 9-26

Module 9—Weld and Base Metal Discontinuities

Introduction and Background

One of the most important parts of the welding inspector's job is the evaluation of welds to determine their suitability for an intended service. During the various stages of this evaluation, the inspector will be looking for irregularities in the weld or weldment. We commonly refer to these irregularities as discontinuities.

In general, a discontinuity is described as any interruption in the uniform nature of an item. Therefore, a bump in a highway could be considered to be a type of discontinuity because it interrupts the smooth, uniform surface of the pavement. In welding, the types of discontinuities with which we are concerned are such things as cracks, porosity, undercut, incomplete fusion, etc.

Knowledge of these discontinuities is important to the welding inspector for a number of reasons. First, the inspector will be asked to visually inspect welds to determine the presence of any of these discontinuities. If any are discovered the welding inspector must then be capable of describing their nature, location, and extent. This information will be required to determine whether or not that discontinuity requires repair, as described in the applicable job specifications.

If additional treatment is deemed necessary, the welding inspector must be capable of accurately describing the discontinuity to the extent that it can be satisfactorily corrected by production personnel.

Before describing these discontinuities, it is extremely important to understand the difference between a discontinuity and a defect. Too often, people mistakenly use the two terms interchangeably. As a welding inspector, you should strive to realize the distinction between the terms *discontinuity* and *defect*.

While a *discontinuity* is some feature which introduces an irregularity in an otherwise uniform structure, a *defect* is a specific discontinuity which can impair the suitability of that structure for its intended purpose. That is, a defect is a discontinuity of a certain type, and one which occurs in an amount great enough to render that particular object or structure unsuitable for its intended service based on criteria in the applicable code.

In order to determine if a particular discontinuity is actually a defect, there must be some standard which defines the acceptable limits of that discontinuity. When its size or concentration exceeds these limits, it is deemed a defect. We can therefore think of a defect as a "rejectable discontinuity." So, if we refer to some feature as a defect, we are implying that it is rejectable and requires some further treatment to bring it into acceptable limits to a particular code.

Depending on the intended service of the part in question, an existing discontinuity may or may not be considered to be a defect. Consequently, each industry uses specific codes or standards which describe the acceptable limits for those discontinuities which could affect the successful performance of various components.

Therefore, the following discussion of weld discontinuities will deal with their characteristics, causes, and cures without specific reference to their acceptability. Only after their evaluation in accordance with an applicable standard can a judgment be made as to whether they are acceptable discontinuities or rejectable defects.

We can, however, talk in general about the effect or criticality of certain discontinuities. Such a discussion will help you understand why certain discontinuities are unacceptable, regardless of their size or extent, while the presence of minor amounts of others is considered to be acceptable.

One way in which this can be explained relates to the specific configuration of that discontinuity. Configurations of discontinuities can be separated into two general groups: linear and non-linear. Linear discontinuities exhibit lengths which are much greater than their widths. Non-linear discontinuities, on the other hand, have length and width dimensions which are essentially the same. When present in a direction perpendicular to the applied stress, a linear discontinuity usually represents a far more critical situation than does a nonlinear type because it is more likely to propagate and cause a failure.

Another way in which the shape of a discontinuity relates to its criticality, or effect on the integrity of a structure, is its end condition. By end condition, we are referring to its specific sharpness at its extremities. In general, the sharper the end of the discontinuity, the more critical it becomes. This is because a sharper discontinuity is more likely to propagate. Again, this is also dependent on its orientation with respect to the applied stress. We most often associate linear discontinuities with sharp end condition. So, if there is a linear discontinuity having a sharp end condition lying transverse to the applied stress, this represents the most detrimental situation with respect to the ability of that member to carry an applied load.

If we were to rate some of the more common discontinuities with respect to the sharpness of their end conditions, starting with the sharpest, we would generally have crack, incomplete fusion, incomplete joint penetration, slag inclusion, and porosity. This order coincides with the amounts of these discontinuities permitted by most codes. There are only a few instances in which any amount of cracking is allowed. Incomplete fusion may also be forbidden or at least limited to minor amounts. Most codes will permit the presence of small amounts of incomplete joint penetration and slag, and some porosity. Depending on the industry and the intended service these amounts will vary, but in general the sharper the discontinuity the more its presence is restricted.

To further explain the importance of the end condition on the severity of a discontinuity, let's review an example of how a crack's propagation could be stopped using a technique which you may have seen performed. The technique referred to is the placement of a drilled hole at the end of a crack in a component. While this does not correct the cracking, it may stop its further propagation. This is accomplished because the sharp ends of the crack are rounded sufficiently by the radius of the drilled hole to reduce the stress concentration to the point that the material can withstand the applied load without further crack propagation.

A final way in which the criticality of a discontinuity is judged relates to the way in which a part or structure will be loaded during service. For example, if a weld forms a part of a pressure boundary, those weld discontinuities constituting a significantly large percentage of the wall thickness will usually be most damaging. In the case of a structure which will be loaded in fatigue (i.e., cyclic loading), those discontinuities forming sharp notches on the surfaces of the structure will generally cause failure more readily than those beneath the surface. These surface notches act as stress risers which tend to concentrate, or amplify, the stresses at that notch point. Such a stress concentration can result in a localized overload condition even though the stress applied to the full cross section may be low. Stress risers can amplify the applied stress by factors as high as ten in the case of a sharp surface crack.

This can be shown by the example of a piece of welding wire which you would like to break. One way of accomplishing this would be to bend the wire back and forth until it finally broke. However, it may take many cycles to produce this failure. If you were to take a similar piece of welding wire, place it on a hard surface, and strike it with the sharp edge of a chipping hammer, you would produce a sharp notch on the wire's surface. Now, only one or two bends would be necessary to result in the failure of that wire because the notch represents a significant stress concentration of the applied bending load.

So, for a structure which must withstand fatigue loading, the surfaces should be free of those discontinuities which provide sharp notches. Consequently, parts subjected to fatigue loading in service are often required to have their surfaces machined to very smooth finishes. Abrupt changes in contour or geometry are also avoided.

For these types of components, one of the most effective methods of inspection is visual. Therefore, you, as a welding inspector, can play an extremely important role in determining how well these components will behave in service. The suitability of these structures for their intended service can be judged on the presence or sharpness of any surface discontinuities.

Discussion of Discontinuities

Having provided this background on discontinuities in a general sense, let's now discuss some of the more common weld and base metal discontinuities found during normal inspection activities. Those with which we will concern ourselves are listed, and the definitions for each can be found in the AWS standard A3.0, *Standard Welding Terms and Definitions*, or in the Key Terms and Definitions section at the end of this module.

- crack
- incomplete fusion
- incomplete joint penetration
- inclusion
- slag inclusion
- tungsten inclusion
- porosity
- undercut
- underfill
- overlap
- convexity
- weld reinforcement
- arc strike
- spatter
- lamination
- lamellar tear
- seam/lap
- dimensional

Cracks

The first of the discontinuities to be discussed is the crack, the most critical discontinuity. This criticality is due to cracks being characterized as linear, as well as exhibiting very sharp end conditions. Since the ends of cracks are extremely sharp, there is a tendency for the crack to grow, or propagate, if a stress is applied.

Cracks are initiated when the load, or stress, applied to a member exceeds its tensile strength. In other words, there was an overload condition which caused the crack. The stress can occur during welding, immediately after, or when a load is applied. While the applied load may not exceed the load carrying ability of a member, the presence of a notch, or stress riser, could cause the localized stress at the tip of the stress riser to exceed the tensile strength of the material. In such a case, cracking could occur at this stress concentration. Therefore, you commonly see cracking associated with both surface and subsurface discontinuities which provide such a stress riser in addition to those associated with the welding operation itself.

We can categorize cracks in several different ways. One way of grouping cracks is by characterizing them as either hot or cold cracks. These terms are an indication of the metal temperature at which the fracture occurred. This is often a way in which we can decide exactly why a particular crack resulted, since some types of cracks are characteristically either hot or cold cracks.

Hot cracks usually occur as the metal solidifies, at some elevated temperature. The propagation of these cracks is considered to be intergranular; that is, the cracks occur between individual grains. If we observe the fracture surfaces of a hot crack, we may see various "temper" colors on the fracture faces indicating the presence of that crack at an elevated temperature. *Cold cracks* occur after the metal has cooled to ambient temperature. Those cracks resulting from service conditions would be considered cold cracks. Delayed, or underbead, cracks resulting from entrapped hydrogen would also be categorized as cold cracks. The propagation of cold cracks can be either intergranular or transgranular; that is, either between or through the individual grains, respectively.

Cracks can also be described by their direction with respect to the longitudinal axis of the weld. Those lying in a direction parallel to the longitudinal axis are referred to as *longitudinal* cracks. Similarly, those cracks lying perpendicular to the weld's longitudinal axis are called *transverse* cracks. These directional references apply to cracks occurring in either the weld or base metals. Longitudinal cracks can result from transverse shrinkage stresses of welding or stresses associated with service conditions. Figure 9.1 shows a longitudinal crack in the center of a groove weld. The weld also contains surface porosity which may have contributed to the crack's propagation.

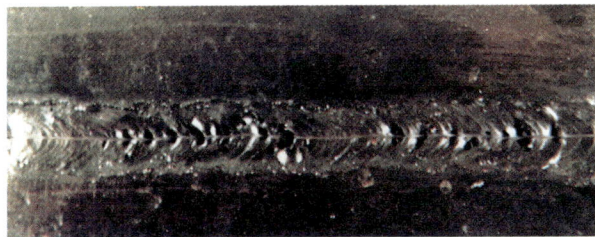

Figure 9.1—Longitudinal Crack

Transverse cracks are generally caused by the longitudinal shrinkage stresses of welding acting on welds or base metals of low ductility. Figure 9.2 shows two transverse cracks occurring in a GMAW deposit on HY-130 steel and propagating into the base metal. Figure 9.3 illustrates throat cracks in fillet welds.

Finally, we can further differentiate between various types of cracks by giving a description of their exact locations with respect to the various parts of the weld. These descriptions include throat, root, toe, crater, underbead, heat-affected zone, and base metal cracks.

Throat cracks are so named because they extend through the weld along the weld throat, or the shortest path through the weld's cross section. These are longitudinal cracks and are generally considered to be hot cracks. A throat crack can be observed visually on the weld face. Consequently, the term *centerline* crack is often used to describe this condition.

Joints exhibiting high restraint transverse to the weld axis are susceptible to throat cracking, especially in situations where the weld cross section is small. So, such things as thin root passes and concave fillet welds could result in a throat crack because their reduced cross sections may not be sufficient to withstand the transverse weld shrinkage stresses. Figure 9.4 is an example of a throat crack in a fillet weld.

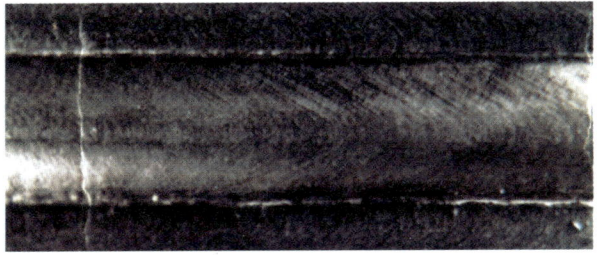

Figure 9.2—Transverse Cracks

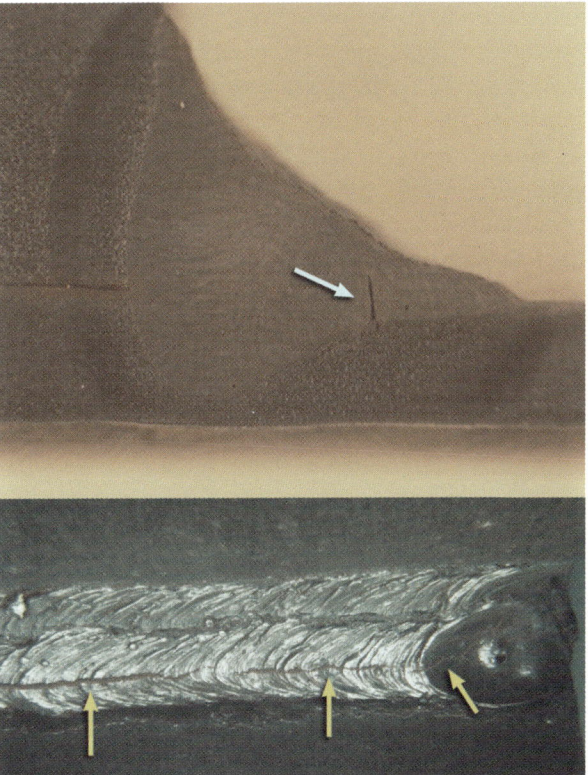

Figure 9.3—Throat Cracks in Fillet Welds

Figure 9.4—Throat Crack

Root cracks are usually longitudinal; however, their propagation may be in either the weld or base metal. They are referred to as root cracks because they initiate at the weld root or the root surface of the weld. Like throat cracks, they are generally related to the existence of shrinkage stresses from welding. Therefore, they are usually considered to be hot cracks. Root cracks often result when joints are improperly fitted or prepared. Large root openings, for example, may result in a stress concentration which produces root cracks.

Toe cracks are base metal cracks which propagate from the toes of welds. Weld configurations exhibiting weld reinforcement or convexity may provide a stress riser at the welds' toes. This, combined with a less ductile microstructure in the heat-affected zone, increases the susceptibility of the weldment to toe cracks. Toe cracks are generally considered to be cold cracks. The stress causing the occurrence of toe cracks could be the result of either the transverse shrinkage stresses of welding, some applied service stresses, or a combination of the two. Toe cracks occurring in service are often the result of fatigue loading of welded components. Typical toe cracks are shown in Figure 9.5.

Crater cracks occur at the termination point of individual weld passes. If the technique used by the welder to terminate the arc does not provide for complete filling of the molten weld puddle, the result could be a shallow region, or crater, at that location. The presence of this thinned area, combined with the shrinkage stresses from welding, may cause individual crater cracks or networks of cracks radiating from the center of the crater. When there is a radial array of crater cracks, they are commonly referred to as star cracks.

Since crater cracks occur during the solidification of the molten puddle, they are considered to be forms of hot cracks. Crater cracks occurring in a GTAW bead on aluminum are shown in Figure 9.6. Crater cracks can be extremely detrimental because there is a tendency for the crack to propagate further, as shown in Figure 9.7.

Although the primary cause of crater cracks relates to the technique used by the welder to terminate a weld pass, these cracks can also result from the use of filler metals having flow characteristics which produce concave profiles when solidified. An example of this phenomena is the use of those stainless steel covered electrodes bearing designations ending with "-16" (i.e., E308-16, E309-16, E316-16, etc.). This ending designates a titania type coating which will produce a characteristically flat or slightly concave weld profile. Consequently, when these electrodes are used, the welder must

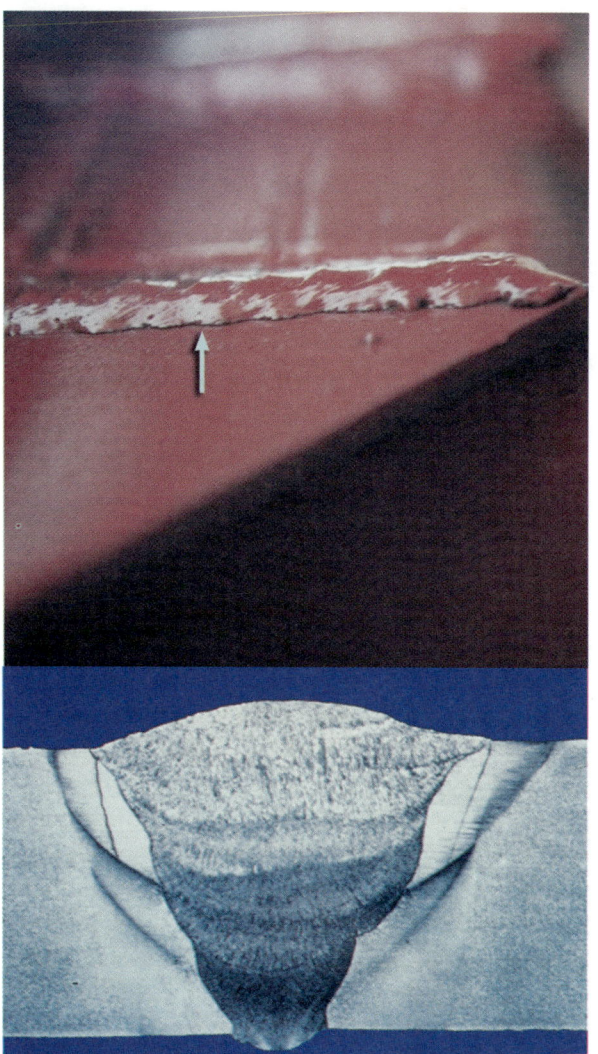

Figure 9.5—Toe Cracks

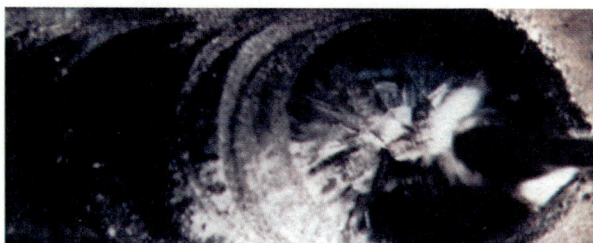

Figure 9.6—Closeup of Crater Cracks in Aluminum Weld

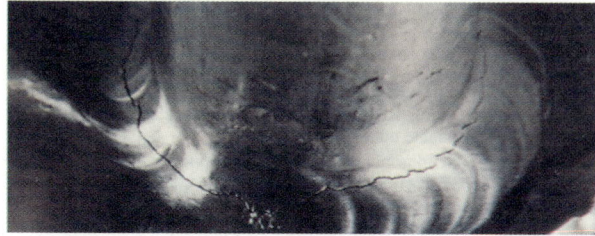

Figure 9.7—Propagation of Crater Crack in Aluminum Weld

take extra precautions and fill the craters sufficiently to prevent crater cracks.

The next category of crack is the underbead crack. Although related to the welding operation, the underbead crack is located in the heat-affected zone instead of the weld metal. As the name implies, it will characteristically lie directly adjacent to the weld fusion line in the heat-affected zone. When cross-sectioned, underbead cracks will often appear to run directly parallel to the fusion line of a weld bead. Figure 9.8 illustrates the typical configurations of underbead cracks. These are typically subsurface and difficult to detect. However, they may propagate to the surface, which allows for their discovery during visual inspection.

Underbead cracking is a particularly detrimental type of crack because it may not propagate until many hours after welding has been completed. For this reason, underbead cracks are sometimes referred to as delayed cracks. Consequently, for those materials which are more susceptible to this type of cracking, final inspection should not be performed until 48 to 72 hours after the weld has cooled to an ambient temperature. High strength steels are particularly susceptible to this type of cracking (e.g., ASTM A 514).

Underbead cracks result from the presence of hydrogen in the weld zone. The hydrogen could come from the filler metal, base metal, surrounding atmosphere, or organic surface contamination. If there is some source of hydrogen present during the actual welding operation, it may be absorbed by the heat-affected tone. When hot, the metal can hold a great deal of this atomic, or nascent, hydrogen, designated as the hydrogen ion, H^+.

However, once solidified, the metal has much less capacity for the hydrogen. The tendency of the hydrogen ions is to move through the metal structure to grain boundaries in the heat-affected zone. At this point, individual atoms of hydrogen may combine to form hydrogen molecules (H_2). This gaseous form of hydrogen requires more volume and is now too large to move through the metal structure. These molecules are now trapped. If the surrounding metal does not exhibit sufficient ductility, the internal pressure created by the trapped hydrogen molecules can result in underbead cracking.

As the welding inspector, you should be aware of this potential problem and take precautions to prevent its occurrence. The best technique for the prevention of underbead cracking is to eliminate sources of hydrogen when welding susceptible materials. With SMAW, for example, low hydrogen electrodes may be used. When specified, they should be properly stored in an appropriate holding oven to maintain this low moisture level. If allowed to remain in the atmosphere for prolonged periods, they may pick up enough moisture to cause cracking. Parts to be welded should be cleaned adequately to eliminate any surface sources of hydrogen. Preheat may also be prescribed to help eliminate this cracking problem.

Since the heat-affected zone is typically less ductile than the surrounding weld and base metal, cracking may also occur there without the presence of hydrogen. In situations of high restraint, shrinkage stresses may be sufficient to result in heat-affected zone cracking, especially in the case of brittle materials such as cast iron. A particular type of heat-affected zone crack which has already been discussed is the toe crack.

Cracking may also be present in the base metal itself. These types of cracks may or may not be associated with the weld. Quite often, base metal

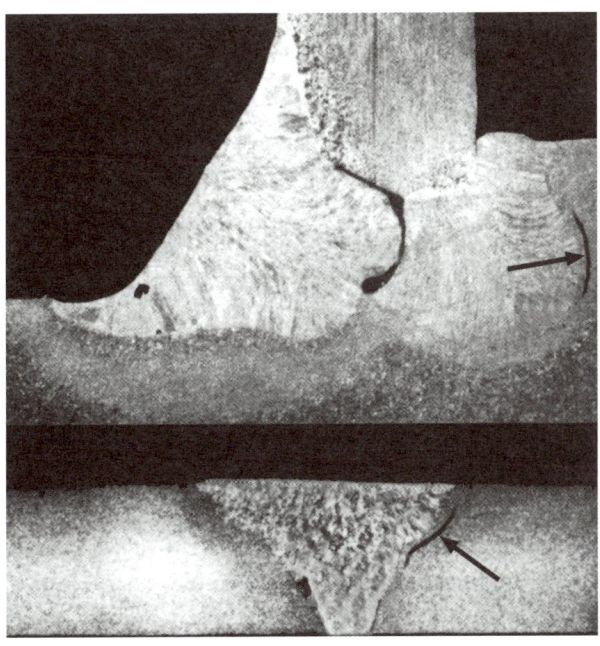

Figure 9.8—Underbead Cracks

cracks are associated with stress risers which result in cracking once the part has been placed in service.

Radiographically, cracks appear as fine, rather well defined dark lines which can be differentiated from other linear discontinuities because their propagation path is not perfectly straight, but tends to wander as the crack follows the path of least resistance through the material's cross section. Figure 9.9 shows a radiograph of a typical longitudinal crack which is probably associated with the weld root. Figure 9.10 illustrates how typical transverse cracks may appear on a radiograph.

Incomplete Fusion

By definition, *incomplete fusion* is described as a "weld discontinuity in which fusion did not occur between weld metal and fusion faces or adjoining weld beads." That is, the fusion is less than that specified for a particular weld. Due to its linearity and relatively sharp end condition, incomplete fusion represents a significant weld discontinuity. It can occur at numerous locations within the weld zone. Figure 9.11 shows some of these various locations for incomplete fusion.

Figure 9.11(A) shows the occurrence of incomplete fusion at the original groove face as well as between individual weld beads. Quite often, incomplete fusion also has slag inclusions associated with it. In fact, the presence of slag due to insufficient cleaning may prevent the fusion from occurring.

We most often think of incomplete fusion as being some internal weld flaw. However, it can occur at the surface of the weld as well. This is shown in Figure 9.11(B) and displayed pictorially in Figure 9.12.

Another non-standard term for incomplete fusion is *cold lap*. This term is often, and incorrectly, used to describe incomplete fusion between the

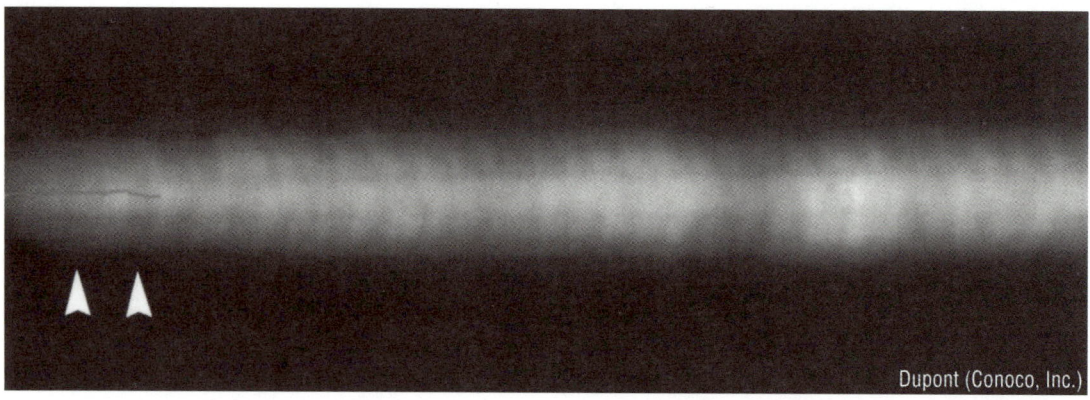

Figure 9.9—Radiograph of Longitudinal Crack

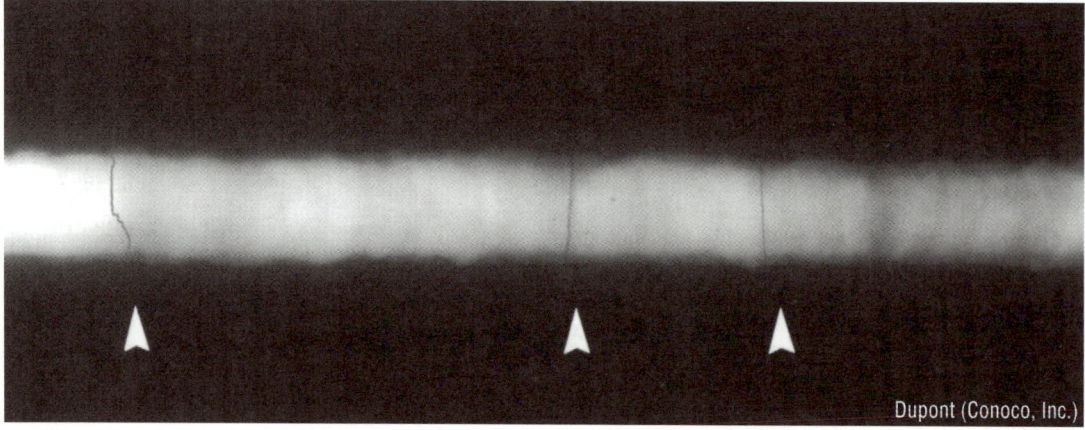

Figure 9.10—Radiograph of Transverse Cracks

weld and base metal or between individual weld beads, especially when using gas metal arc welding. Figures 9.13 and 9.14 show incomplete fusion (cold lap) occurring between the weld and base metals.

Incomplete fusion can result from a number of conditions or problems. Probably the most common cause of this discontinuity is the improper manipulation of the welding electrode by the welder. Some processes are more prone to this problem because there is not enough concentrated heat to adequately melt and fuse the metals. For example, when using short circuiting transfer GMAW, the welder must concentrate on directing the welding arc at every location of the weld joint where fusion is to be obtained. Otherwise, there will be areas which do not exhibit the proper amount of melting, and therefore fusion.

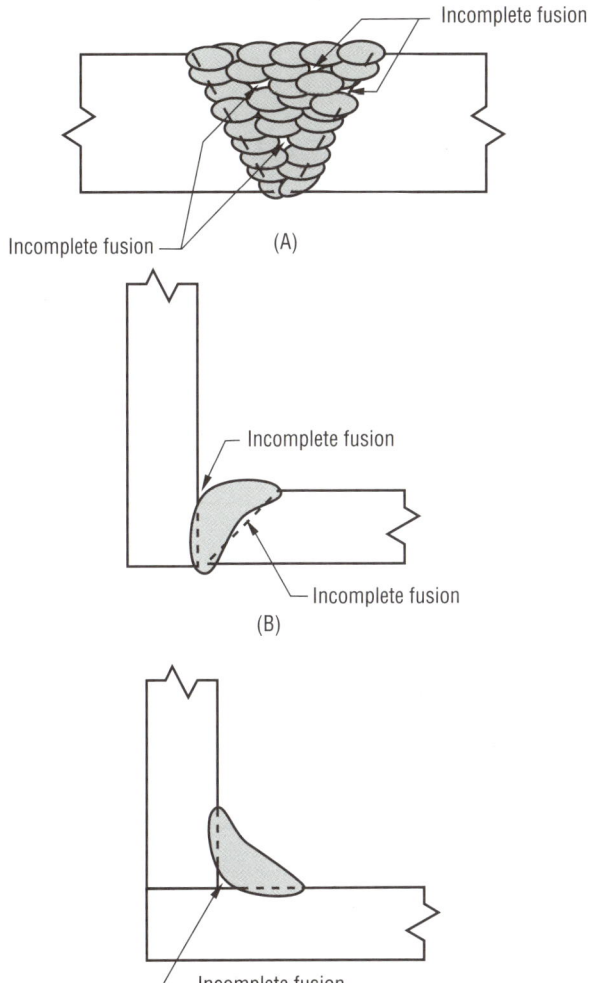

Figure 9.11—Various Locations of Incomplete Fusion

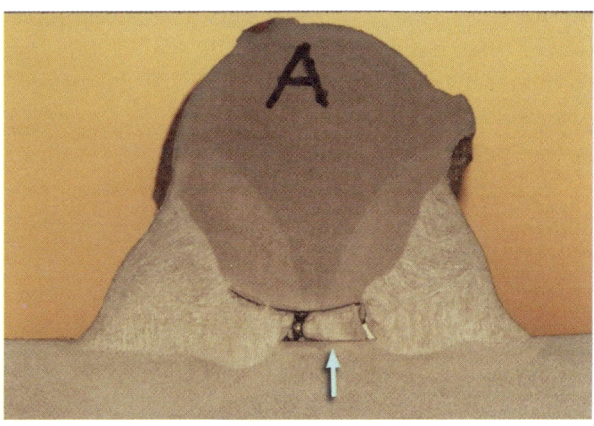

Figure 9.13—Incomplete Fusion of Flare-Bevel Groove Welds

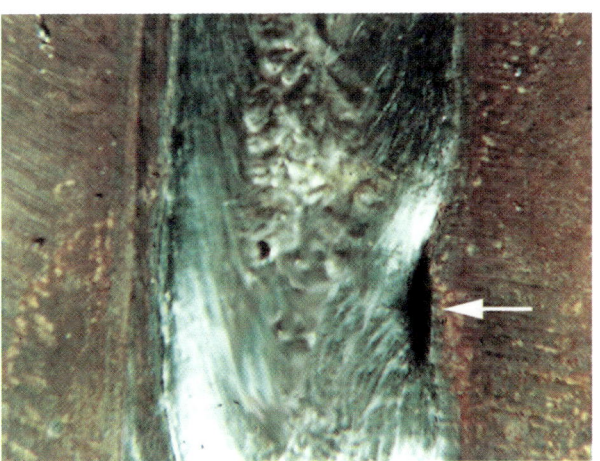

Figure 9.12—Incomplete Fusion at Weld Face

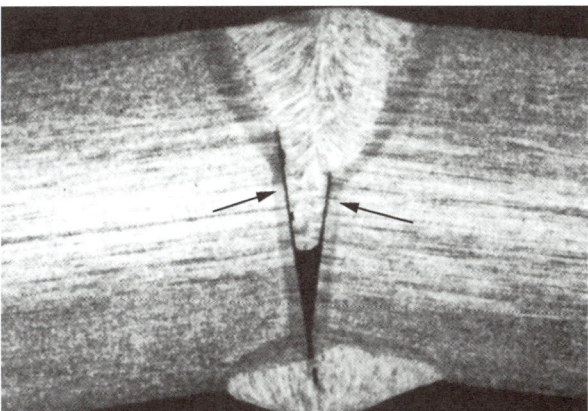

Figure 9.14—Incomplete Fusion Between Weld and Base Metal

In other situations, the actual configuration of the weld joint may limit the amount of fusion which can be attained. An example is the use of a weld groove with an insufficient groove angle for the process and electrode diameter being used. Finally, extreme contamination, including mill scale and tenacious oxide layers, could also prevent the attainment of complete fusion.

Incomplete fusion is very difficult to detect with radiography unless the angle of radiation is oriented properly. Typically, the incomplete fusion is adjacent to the original groove face and has very little width and volume, making it very difficult to resolve radiographically unless the radiation path is parallel and in line with the discontinuity.

If incomplete fusion is radiographically visible, it will usually appear on the film as darker density lines which are generally straighter than the images of either cracks or elongated slag. The lateral position of these indications on the film will be a hint as to their actual depth. For example, in a single-V groove weld, incomplete fusion near the root will appear near the weld centerline while the presence of incomplete fusion closer to the weld face will appear on the radiograph as an image positioned closer to the weld toe.

Figure 9.15 is a radiograph representing linear images likely to be produced by incomplete fusion along the groove faces of the original weld joint.

Incomplete Joint Penetration

Incomplete joint penetration, unlike incomplete fusion, is a discontinuity associated only with groove welds. It is a condition where the weld metal does not extend entirely through the joint thickness when complete penetration is required by a specification. Its location is always adjacent to the weld root. Figure 9.16 illustrates several examples of incomplete joint penetration. Most codes place limits on the amount and degree of incomplete joint penetration permissible, and several codes do not accept any incomplete joint penetration.

There is another name which can be correctly applied to the conditions shown in Figure 9.16 if the welds meet the designer's requirements. They could be called *partial joint penetration*; that is, they were not intended to be complete penetration welds. For example, in a joint where the design requirements specify partial joint penetration welds, and this is commonly done, then the weld examples shown would be acceptable if the weld sizes are adequate. However, in a joint where the weld is required to have complete joint penetration, the presence of incomplete joint penetration is cause for rejection.

It should be noted that previously the condition now correctly referred to as incomplete joint penetration has been called by several non-standard terms. Some of these were inadequate penetration, lack of penetration, IP, and LP. For groove welds, the correct term is incomplete joint penetration, and should be used instead of the other terms. Figure 9.17 shows a photograph of this condition at the root of a groove weld, and Figure 9.18 shows the radiographic image of incomplete joint penetration.

Incomplete joint penetration can be caused by the same conditions which result in incomplete fusion; that is, improper technique, improper joint configuration, or excessive contamination.

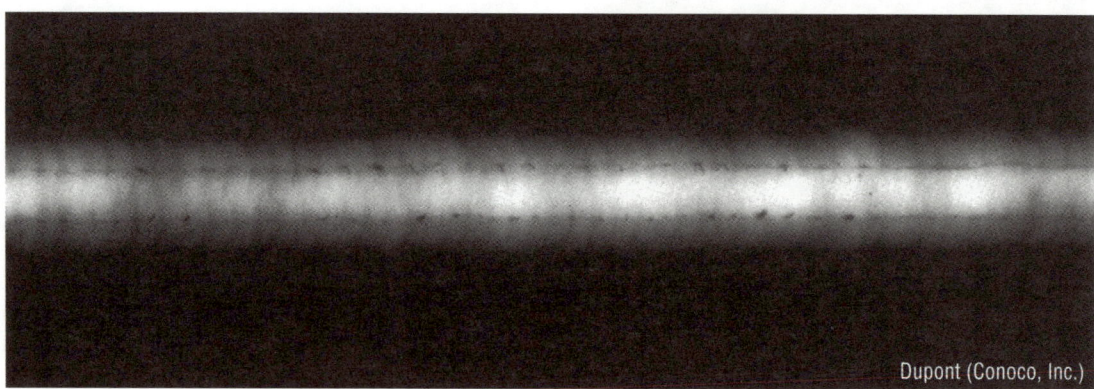

Figure 9.15—Radiograph of Side-Wall Incomplete Fusion

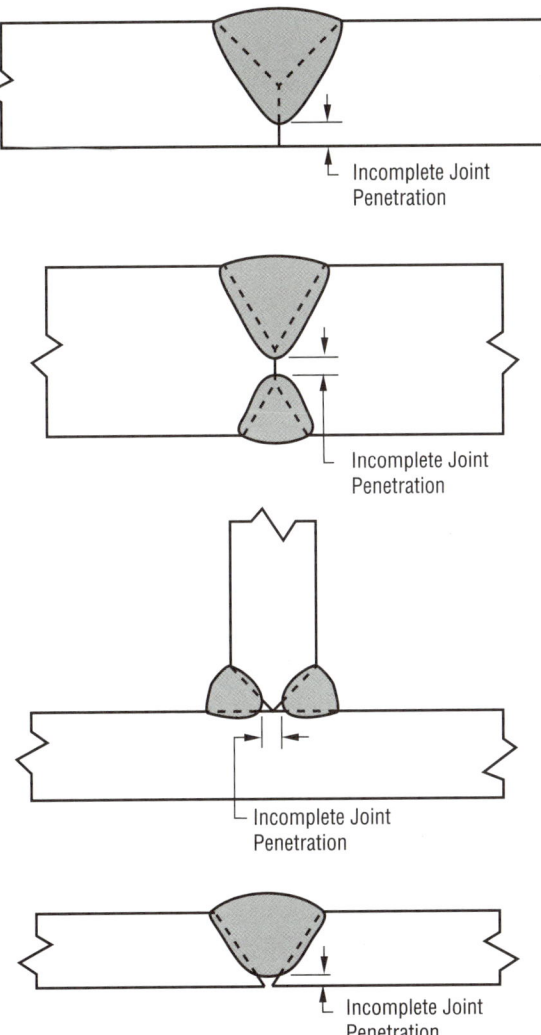

Figure 9.16—Examples of Incomplete Joint Penetration In Groove Welds

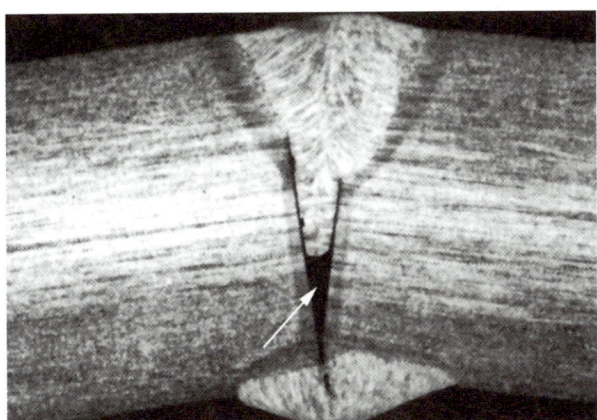

Figure 9.17—Incomplete Joint Penetration

The radiographic image caused by incomplete penetrations will typically be a dark, straight line. It will usually be much straighter than incomplete fusion because it is associated with the original weld preparation at the root. It will be centered in the width of the weld for those groove welds in which both members are prepared.

Inclusions

The definition of *inclusion* is entrapped foreign solid material, such as slag, flux, tungsten, or oxide. Thus, the term inclusion can include both metallic and nonmetallic categories. Slag inclusions, as the name implies, are regions within the weld cross section or at the weld surface where the molten flux used to protect the molten metal is mechanically trapped within the solidified metal. This solidified flux, or slag, represents a portion of the weld's cross section where the metal is not fused to itself. This can result in a weakened condition which can impair the serviceability of the component. Although we normally think of slag inclusions as being totally contained within the weld cross section, we sometimes observe them at the surface of the weld as well. Figure 9.19 shows an example of such a surface slag inclusion.

Like incomplete fusion, slag inclusions can occur between the weld and base metal or between individual weld beads. In fact, slag inclusions are often associated with incomplete fusion. Slag inclusions can only result when the welding process uses some type of flux shielding. They are most often caused by improper techniques used by the welder. Such things as improper manipulation of the welding electrode and insufficient cleaning between passes can result in the presence of slag inclusions. Often, the improper manipulation of the electrode or incorrect welding parameters could result in undesirable weld profiles which could then hinder cleaning of the slag between passes. Subsequent welding would then cover the trapped slag to produce slag inclusions.

Figure 9.20 has the appearance of slag inclusions but is actually penetration faults at the start/stop areas of the weld.

Since the density of slag is usually much less than that of metals, slag inclusions will generally appear on a radiograph as relatively dark indications, having rather irregular shapes, as shown in Figures

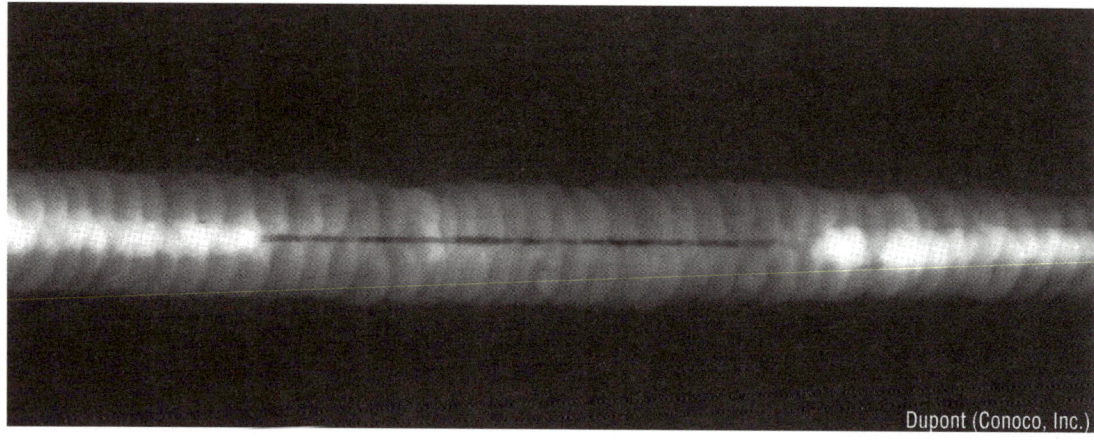

Figure 9.18—Radiographic Image of Incomplete Joint Penetration

Figure 9.19—Surface Slag Inclusion

9.20 and 9.21. However, there are covered welding electrodes whose slag is about the same density as the metal, and as you might expect, slag inclusions stemming from the use of these electrodes are very difficult to determine radiographically.

Tungsten inclusions are almost always associated with the GTAW process, which uses a tungsten electrode to produce an arc. If the tungsten electrode makes contact with the molten weld puddle, the arc can extinguish and the molten metal can solidify around the tip of the electrode. Upon removal, the tip of the electrode will very likely break off and remain embedded in the final weld if not removed by grinding.

Tungsten inclusions can also result when the welding current being used for GTAW is in excess of that recommended for a particular diameter of electrode. In such a case, the current density may be great enough that the electrode starts to decompose and pieces may be deposited in the weld metal. This can also occur if the welder does not properly grind the point on the tungsten electrode. If the grinding marks are oriented such that they form rings around the electrode instead of being aligned with its axis, they could form stress risers which might cause the tip of the electrode to break off preferentially. Other reasons for the occurrence of tungsten inclusions include:

(1) contact of filler metal with hot tip of electrode;
(2) contamination of the electrode tip by spatter;
(3) extension of electrodes beyond their normal distances from the collet, resulting in overheating of the electrode;
(4) inadequate tightening of the collet;
(5) inadequate shielding gas flow rates or excessive wind drafts resulting in oxidation of the electrode tip;
(6) use of improper shielding gas;
(7) defects such as splits or cracks in the electrode;
(8) use of excessive current for a given size electrode;
(9) improper grinding of the electrode; or
(10) use of too small an electrode.

Tungsten inclusions are seldom found on the surface of the weld unless the welding inspector has the opportunity to look at an intermediate pass after a piece of tungsten has been deposited. The primary way in which tungsten inclusions are revealed is through the use of radiography. Since tungsten has a much greater density than steel or aluminum, it will show up as a definite light area on the radiographic film. This is shown in Figure 9.22.

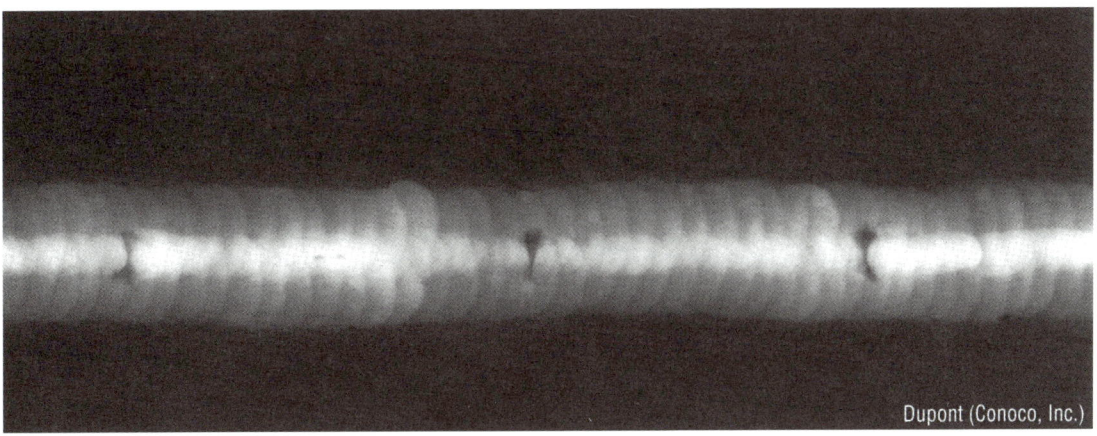

Figure 9.20—Radiographic Image of Penetration Faults at the Starts and Stops of Individual Root Beads

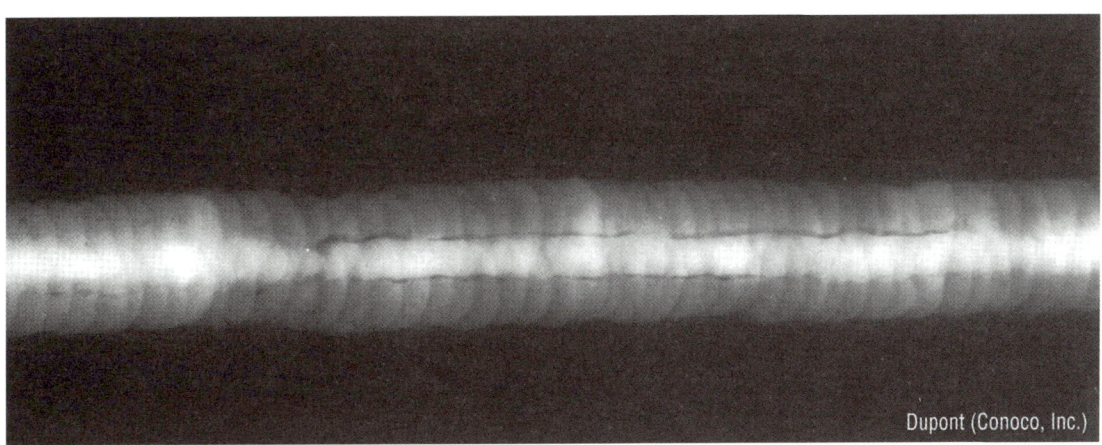

Figure 9.21—Radiographic Image of Elongated Slag Inclusions

Figure 9.22—Radiographic Image of Tungsten Inclusions

Porosity

AWS A3.0 defines porosity as "cavity-type discontinuities formed by gas entrapment during solidification." Therefore, we can think of porosity as being voids or gas pockets within the solidified weld metal. Due to its characteristically spherical shape, porosity is normally considered to be the least detrimental discontinuity. However, in cases where a weld must form some pressure boundary to contain a gas or liquid, porosity might then be considered to be more damaging. This is due to the possibility of the porosity providing a leak path.

Like cracking, there are several different names given to specific types of porosity. They refer, in general, to the relative location or the specific shape of the individual porosity pockets. Therefore, such names as uniformly scattered porosity, cluster porosity, linear porosity, and piping porosity are used to better define the occurrence of porosity. A single cavity is also referred to as porosity or a cavity.

Uniformly scattered porosity refers to numerous cavities which occur throughout the weld in no particular pattern. Cluster porosity and linear porosity, however, refer to specific patterns of several cavities. Cluster porosity describes a number of cavities grouped together while the term linear porosity refers to a number of cavities which are grouped in a straight line.

With these types, the cavities, or gas pockets, are usually spherical in shape. Figure 9.23 shows an example of uniformly scattered porosity on the weld surface. Figure 9.24 illustrates linear porosity with an accompanying crack, and Figure 9.25 shows the presence of some isolated pockets of porosity at the weld surface.

There are some types of porosity where the individual gas pockets are not spherical but are elongated. Figure 9.26 is an example of elongated porosity occurring at the weld surface. This is often referred to as *wormhole* porosity. Such a surface condition can occur when gases are trapped between the molten metal and solidified slag. One situation in which this phenomenon can occur is when the depth of granular flux used for SAW is excessive. When this occurs the weight of the flux may be too great to permit the gas to escape properly.

Another form of elongated porosity is piping porosity. Piping-type porosity usually represents the

Figure 9.23—Uniformly Scattered Surface Porosity

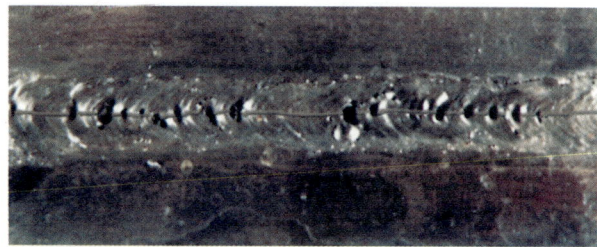

Figure 9.24—Linear Surface Porosity with Connecting Crack

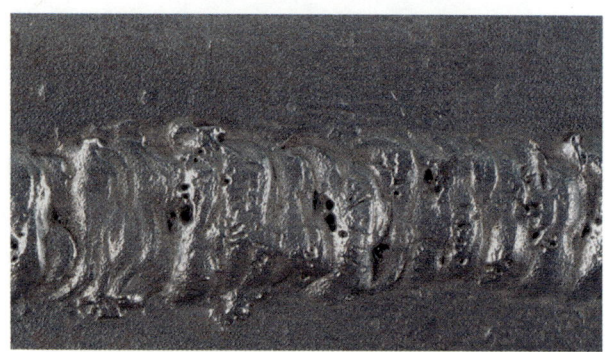

Figure 9.25—Isolated Surface Porosity

Figure 9.26—Elongated Surface Porosity

most detrimental condition if liquid or gas containment is the primary function of the weld because it represents a more significant leak path possibility (see Figure 9.27).

Porosity is normally caused by the presence of contaminants or moisture in the weld zone which decompose due to the welding heat and form gases. This contamination or moisture can come from the electrode, the base metal, the shielding gas, or the surrounding atmosphere. However, variations in the welding technique could also cause this porosity. An example would be the use of an excessively long arc during SMAW with a low hydrogen type electrode. Another example would be the use of excessively high travel speeds with SAW resulting in piping porosity. Therefore, when porosity is encountered, it is a signal that some aspect of the welding operation is out of control. It is then time to investigate further to determine what factor, or factors, are responsible for the presence of this weld discontinuity.

When porosity is shown on a weld radiograph, it will appear as a well-defined dark region because it represents a significant loss of material density. It will normally appear as a round indication except in the case of piping porosity. This type of porosity will have a tail associated with the rounded indication.

Figure 9.28 illustrates a radiograph showing cluster porosity, and an example of linear porosity is shown in Figure 9.29.

Undercut

Undercut is a surface discontinuity which occurs in the base metal directly adjacent to the weld. It is a condition in which the base metal has been melted away during the welding operation and there was insufficient filler metal deposited to adequately fill the resulting depression. The result is a linear groove in the base metal which may have a relatively sharp configuration. Since it is a surface condition, it is particularly detrimental for those structures which will be subject to fatigue loading. Figure 9.30 shows the typical appearance of undercut in both fillet and groove welds. It is interesting to note that for groove welds, the undercut may occur at either the face or root surface of the weld.

Figure 9.31 illustrates the typical visual appearance of undercut in a fillet weld. This picture is evidence of how undercut is best revealed visually. That is, there is a definite shadow produced by the undercut when the lighting is properly positioned. Experienced welding inspectors understand this phenomenon and use techniques such as laying a flashlight on the base metal surface to result in a

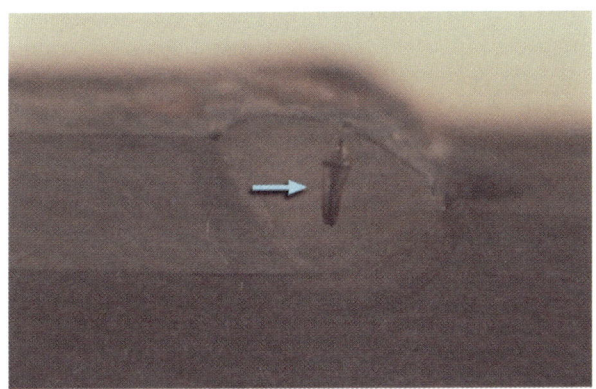

Figure 9.27—Fillet Weld Piping Porosity

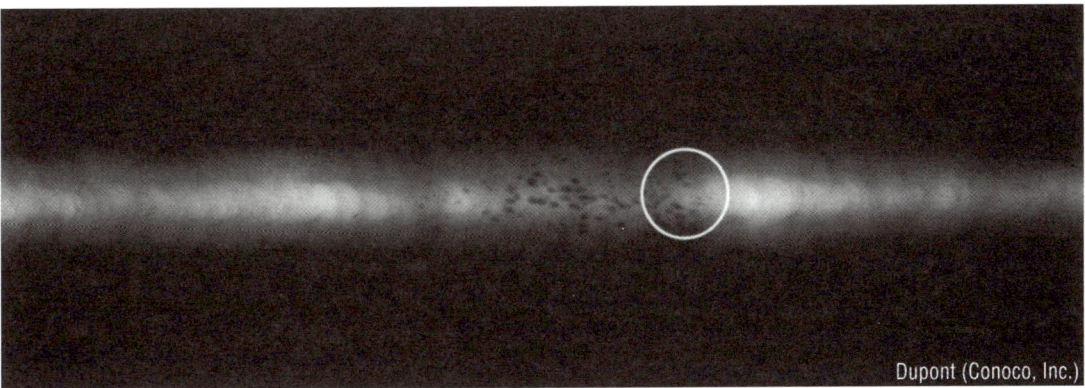

Figure 9.28—Radiograph of Cluster Porosity

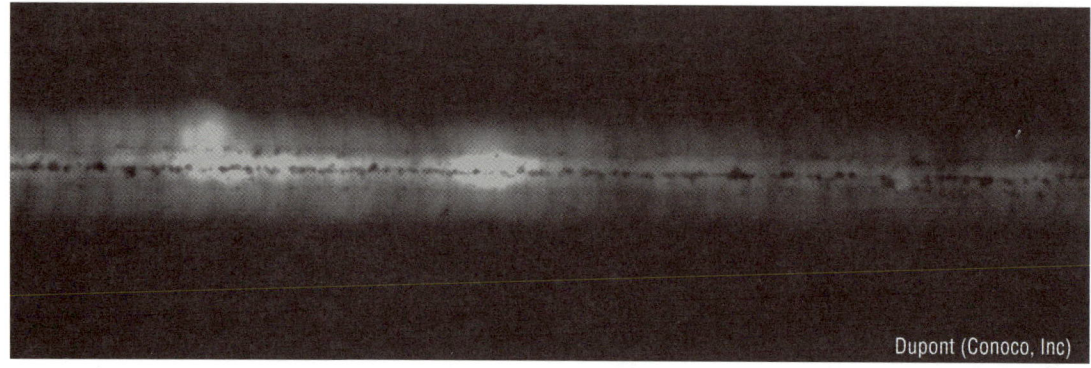

Figure 9.29—Radiograph of Linear or Aligned Porosity Near the Root of the Weld

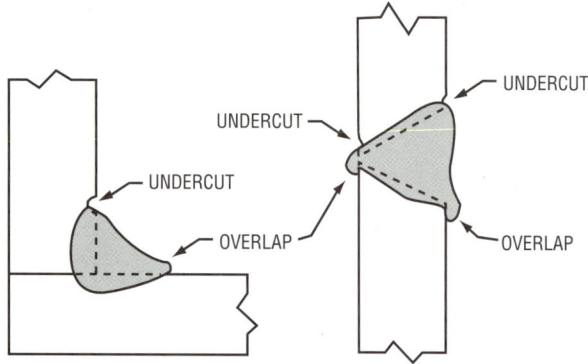

Figure 9.30—Typical Appearance of Undercut and Overlap in Groove and Fillet Welds

Figure 9.31—Undercut Adjacent to Fillet Weld

shadow being cast in any location where undercut exists.

Another technique is to perform final visual inspection of the weldment after painting, especially when the paint being used is a light color such as white or yellow. When viewed under normal lighting, the shadows cast by the presence of undercut are much more pronounced. The only problem with this technique is that the paint must then be removed from the undercut area prior to any repair welding to prevent the occurrence of other discontinuities such as porosity. Of course, the part would then require repainting after any weld repairs are completed.

Undercut is normally the result of improper welding technique. More specifically, if the weld travel speed is excessive, there may not be sufficient filler metal deposited to adequately fill depressions caused by the melting of the base metal adjacent to the weld. Undercut could also result when the welding heat is too high, causing excessive melting of the base metal, or when electrode manipulation is incorrect.

When found on a radiograph, and it is not being suggested this technique be used for its discovery, undercut will appear as a dark, fuzzy indication at the toe or root of the weld reinforcement, as shown in Figure 9.32. It must be noted that radiographic detection of surface undercut is a definite waste of time, money, and resources. Surface undercut is found by careful visual inspection; once found, it should then be repaired if necessary prior to radiographic inspection.

Underfill

Underfill, like undercut, is a surface discontinuity which results in a loss of material cross section. However, underfill occurs in the weld metal of a groove weld whereas undercut is found in the base metal adjacent to the weld. In simple terms, underfill results when there is not sufficient filler metal deposited to adequately fill the weld joint. When discovered, it usually means the welder has not finished making the weld, or has not understood the welding requirements. Figure 9.33 shows the appearance of underfill in a groove weld configuration.

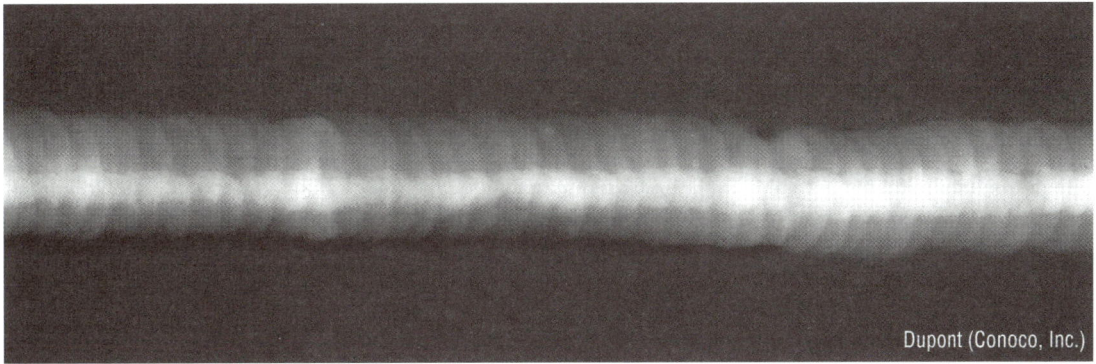

Figure 9.32—Radiographic Image of External Surface Undercut

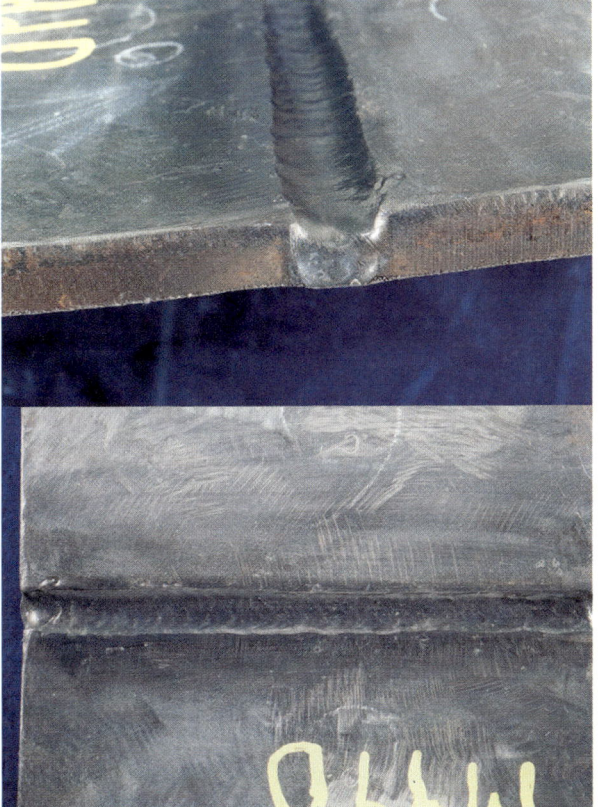

Figure 9.33—Underfill in Groove Welds

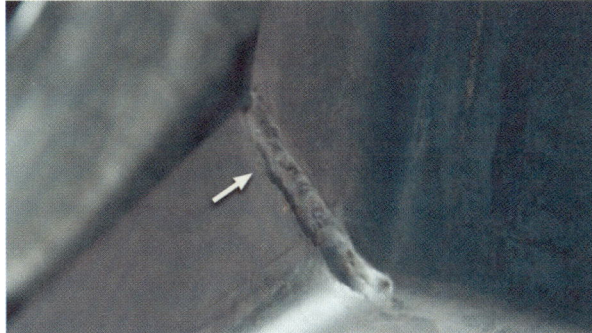

Figure 9.34—Underfill

undercut, when the lighting is properly oriented, there is a shadow produced because of the surface depression.

The primary cause of underfill is the technique employed by the welder. Excessive travel speeds do not allow sufficient filler metal to be melted and deposited to fill the weld zone to the level of the base metal surface.

Overlap

Another surface discontinuity which can result from improper welding techniques is overlap. Overlap is described as the protrusion of weld metal beyond the weld toe or weld root. It appears as though the weld metal has overflowed the joint and is laying on the adjacent base metal surface. Due to its characteristic appearance, overlap is sometimes referred to as "rollover." Rollover is a non-standard term and should not be used.

Figure 9.35 shows how overlap may appear for both fillet and groove welds. As was the case for both undercut and underfill, overlap can occur at

Like undercut, underfill can occur at both the face and root surfaces of the weld. Underfill at the weld root of pipe welds is sometimes referred to as *internal concavity* or the slang term, "suckback." It can be caused by excessive heating and melting of the root pass during deposition of the second pass.

Figure 9.34 illustrates the visual appearance of underfill at the face of a groove weld. As with

Figure 9.35—Overlap in Fillet and Groove Welds

either the weld face or weld root of groove welds (see Figure 9.30). Once again, there is a definite shadow cast when the lighting is properly oriented.

Overlap is considered to be a significant discontinuity since it can result in a sharp notch at the surface of the weldment. Further, if the amount of overlap is great enough, it can hide a crack which may propagate from this stress riser. The occurrence of overlap is normally due to an improper technique used by the welder. That is, if the welding travel speed is too slow, the amount of filler metal melted will be in excess of that amount required to sufficiently fill the joint. The result is that this excessive metal spills over and lays on the base metal surface without fusing. Some types of filler metals are more prone to this type of discontinuity since, when molten, they are too fluid to resist the forces of gravity. Therefore, they may only be used in positions in which gravity will tend to hold the molten metal in the joint. Overlap and undercut often occur when welding in the horizontal position, as illustrated in Figure 9.30.

Convexity

This particular weld discontinuity applies only to fillet welds. Convexity refers to the amount of weld metal buildup on the face of the fillet weld beyond what would be considered flush. By definition, it is the maximum distance from the face of a convex fillet weld perpendicular to a line joining the weld toes. Figure 9.36 illustrates a fillet weld with convexity.

Within certain limits, convexity is not damaging. In fact, a slight amount of convexity is desirable to insure that concavity is not present which can reduce the size and strength of a fillet weld. However, when the amount of convexity exceeds some limit, this discontinuity becomes a significant flaw. The fact that additional weld metal is present is not the real problem unless one considers the economics of depositing more filler metal than is absolutely necessary. The real problem created by the existence of excess convexity is that the resulting fillet weld profile now has sharp notches present at the weld toes. These notches can produce stress risers which could weaken the structure, especially when that structure is loaded in fatigue. Therefore, excessive convexity should be avoided during welding, or corrected by depositing additional weld metal at the weld toes to provide a smoother transition between the weld and base metals.

Convexity results when welding travel speeds are too slow, when too little heat is used, or when the electrode manipulation is incorrect. The result is that excess filler metal is deposited and it does not

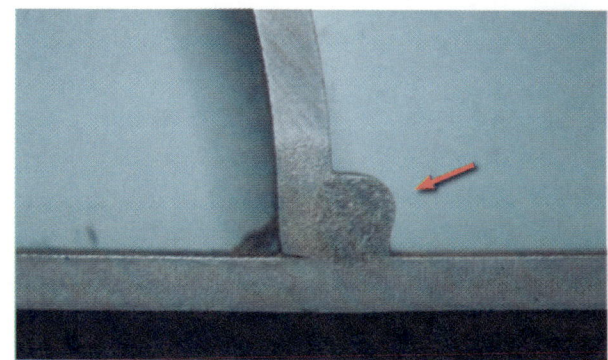

Figure 9.36—Fillet Weld Convexity

properly wet the base metal surfaces. The presence of contamination on the base metal surface or the use of shielding gases which do not adequately clean away these contaminants can also result in this undesirable fillet weld profile.

Weld Reinforcement

Weld reinforcement is similar to convexity except that it describes a condition which can only be present in a groove weld. Weld reinforcement is described as that weld metal in excess of the amount required to fill a joint. Two other terms, face reinforcement and root reinforcement, are specific terms which describe the presence of this reinforcement on a particular side of the welded joint. As the names imply, face reinforcement occurs on the side of the joint from which welding was done, and root reinforcement occurs on the opposite side of the joint.

Figure 9.37 shows the face and root reinforcement for a weld joint welded from one side. For a weld joint welded from both sides, the reinforcement on both sides is described as the face reinforcement, which is shown in Figure 9.38.

Like convexity, the problem associated with excessive reinforcement lies with the sharp notches that may be created at each weld toe instead of the fact that there is more weld metal present than is necessary. The greater the amount of weld reinforcement, the more severe the notches. The graph shown in Figure 9.39 illustrates the effect of the amount of reinforcement on the fatigue strength of the weld joint.

Figure 9.38—Face Reinforcement on Both Sides of Joint

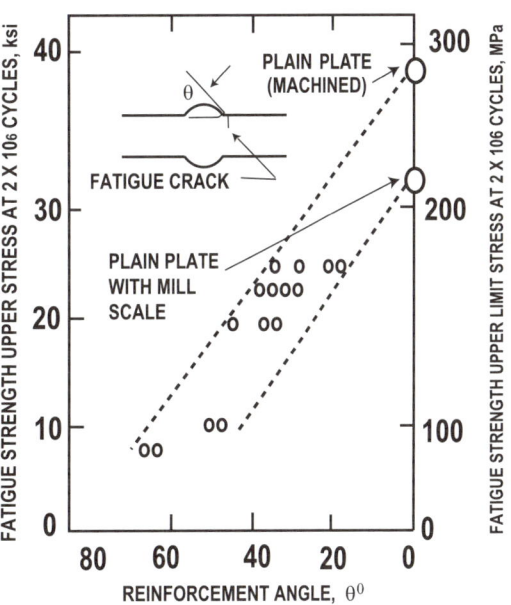

Figure 9.39—Effect of Weld Reinforcement Angle on Fatigue Strength

Looking at this graph, it is obvious that as the reinforcement angle increases (caused by an increase in the amount of weld reinforcement), there is a significant decrease in the fatigue resistance of the weld joint. Most codes prescribe maximum limits for the amount of weld reinforcement permitted. However, reducing the amount of weld reinforcement does not really improve the situation, as shown in Figure 9.40.

As the illustrations show, only after performing blend grinding to increase the weld reinforcement

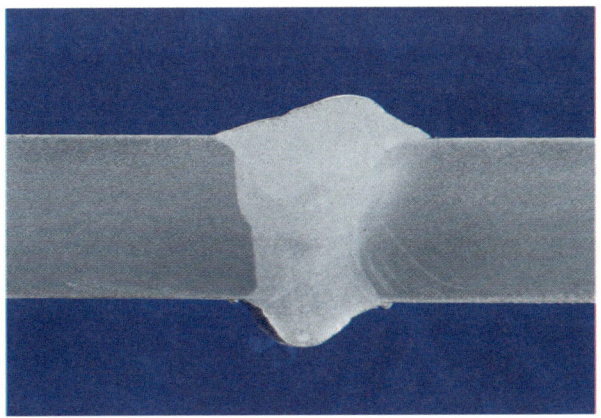

Figure 9.37—Face and Root Reinforcement

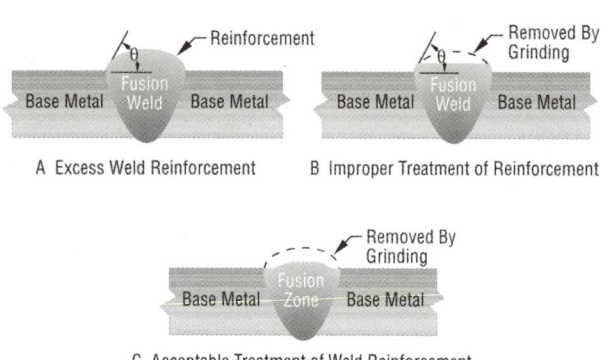

Figure 9.40—Unacceptable and Acceptable Treatment of Excess Weld Reinforcement

angle and increase the "notch" radius is the situation really improved. Grinding to remove the top of the weld reinforcement does nothing to decrease the sharpness of the notches at the weld toes. Reinforcement height is diminished with grinding to possibly meet the code requirements, but the condition of concern remains. Excessive weld reinforcement results from the same reasons as given for convexity, with the welding technique being the predominant cause.

Arc Strikes

The presence of an arc strike can be a very detrimental base metal discontinuity, especially on the low alloy, high strength steels. Arc strikes result when the arc is initiated on the base metal surface away from the weld joint, either intentionally or accidentally. When this occurs, there is a localized area of the base metal surface which is melted and then rapidly cooled due to the massive heat sink created by the surrounding base metal. On certain materials, especially high strength steels, this can produce a localized heat-affected zone which may contain martensite. If this hard, brittle microstructure is produced, the tendency for cracking can be great. Numerous failures of structures and pressure vessels can be traced back to the presence of a welding arc strike which provided a crack initiation site resulting in a catastrophic failure.

Figure 9.41 is a photomicrograph showing an arc strike on the surface of a boiler tube. The darkened microstructure formed is martensite. This particular arc strike provided a crack initiation site which resulted in the ultimate failure of this boiler tube.

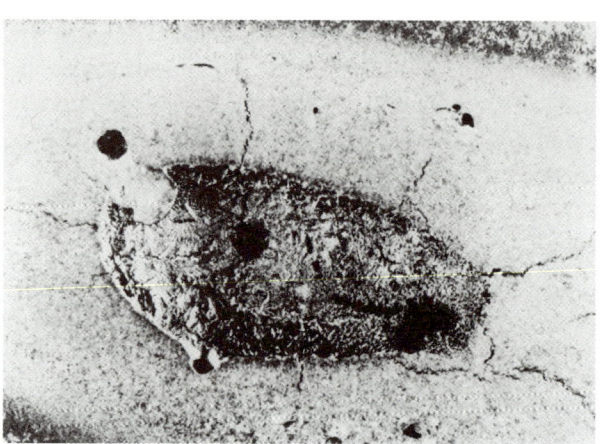

Figure 9.41—Photomicrograph of Martensitic Structure Produced by an Arc Strike

Arc strikes are normally caused by improper welding techniques. Welders should be informed of the potential damage caused by arc strikes. Due to the potential damage they represent, arc strikes should never be permitted. The welder should not be performing production welding if he or she insists on initiating the welding arc outside the weld joint. So, it becomes a matter of discipline and work attitude. Improper connection of the work clamp to the work can also result in the production of arc strikes.

Another important note applies to the inspection of welds using the "prod" type magnetic particle testing method. Since this method relies on the conduction of electricity through the part to produce the magnetic field, the possibility exists that arc strikes can be produced during the inspection if there is not adequate contact between the prods and the metal surface. Although not as severe as welding arc strikes, these arc burns could also produce harmful effects.

Spatter

AWS A3.0 describes spatter as, "metal particles expelled during fusion welding that do not form a part of the weld." We more commonly think of those particles which are actually attached to the base metal adjacent to the weld. However, particles which are thrown away from the weld and base metal are also considered to be spatter. For that reason, another definition might be those particles of metal which comprise the difference between the amount of metal melted and the amount of metal actually deposited in the weld joint.

In terms of criticality, spatter may not be a great concern in many applications. However, large globules of spatter may have sufficient heat to cause a localized heat-affected zone on the base metal surface similar to the effect of an arc strike. Also, the presence of spatter on the base metal surface could provide a local stress riser which could cause problems during service. An example of such a situation is shown in Figure 9.42, where a crack formed at a globule of spatter which stuck to the base metal. The presence of this stress concentration along with a corrosive environment resulted in a form of stress corrosion cracking known as caustic embrittlement. When spatter is present, however, it does detract from the otherwise pleasing appearance of a satisfactory weld. Such a condition is illustrated in Figure 9.43.

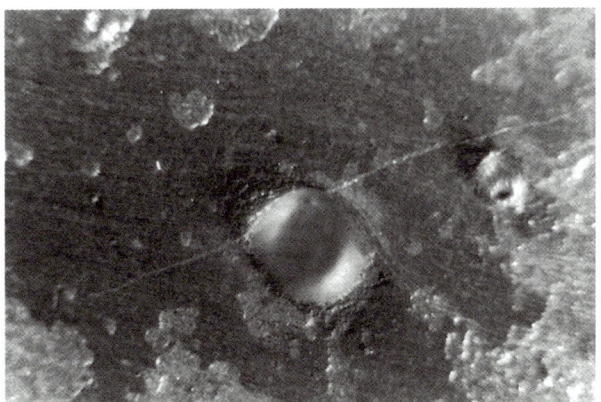

Figure 9.42—Crack Formed at Weld Spatter on Base Metal Surface

Figure 9.43—Weld Spatter on Base Metal

Another feature of spatter which could result in problems has to do with the irregular surface which is produced. During inspection of the weld using various nondestructive methods, the presence of spatter could either prevent the performance of a valid test or produce irrelevant indications which could mask real weld flaws. For example, the presence of spatter adjacent to a weld may prevent adequate coupling of the transducer during ultrasonic testing. Also, spatter could cause problems for both the performance and interpretation of magnetic particle and penetrant testing. And spatter can certainly cause problems if the surfaces are to be painted; spatter can cause premature failure of applied coatings.

Spatter can result from the use of high welding currents which can cause excessive turbulence in the weld zone. Some welding processes are considered to characteristically produce greater amounts of spatter than others. For example, short circuiting and globular transfer GMAW tend to produce more spatter than the use of spray transfer. Another aspect which will help control the amount of spatter produced is the type of shielding gas used for GMAW and FCAW. The use of argon mixtures will reduce the amount of spatter compared to the amount produced when straight CO_2 shielding gas is used. Spatter can also be reduced by applying anti-spatter compounds to the adjacent areas.

Lamination

This particular discontinuity is a base metal flaw. Laminations result from the presence of nonmetallic inclusions which occur in steel when it is being produced. These inclusions are normally forms of oxides which are produced when the steel is still molten. During subsequent rolling operations, these inclusions become elongated to form stringers. If these stringers are particularly large, and take a planar shape, they are referred to as *laminations*. The most massive form of lamination arises from a condition referred to as *pipe*, which develops in the upper part of the steel ingot during the final stages of solidification. Sometimes, on infrequent occasions, this pipe is not completely cropped off the ingot prior to rolling into plate or bar. The pipe cavity usually contains some complex oxides, which are rolled out within the plate to form laminations. Another term mistakenly used interchangeably with the term lamination is delamination. The AWS standard B1.10, *Guide for the Nondestructive Examination of Welds,* defines the two words differently. B1.10 refers to *delamination* as the separation of a lamination under stress. Thus, according to the AWS standard, the primary difference between the two terms is only the degree of separation of the plate sections.

The heat of fusion welding may be sufficient to remelt the stringers in the lamination zone immediately adjacent to the weld, and the ends of the stringers may either fuse or they may open up.

Laminations may also show up during thermal cutting, where the heat of the cutting operation may be sufficient to open the planar stringers to the point that they can be visually observed. Laminations may or may not present a detrimental situation, depending on the way in which the structure is loaded. If the stresses are acting on the material in a direction perpendicular to the lamination, it will severely weaken the structure. However, laminations oriented parallel to the applied stress may not cause any concern.

If a lamination is present on the surface of a weld preparation, it can cause further problems during welding. In such a case, weld metal cracks can propagate from the laminations due to a concentration of stress. An example of this phenomenon is shown in Figure 9.44.

Another problem related to the presence of laminations open to the groove face is that they are prime sites for the accumulation of hydrogen. During welding, hydrogen could be dissolved in the molten metal and provide a necessary element for the occurrence of underbead cracking.

Since laminations are the result of the steelmaking process, there is little that can be done to prevent their occurrence other than adequate ingot cropping. Purchasing steels having low levels of contaminants will drastically reduce the tendency toward the presence of laminations. However, the welder and welding inspector can do nothing to prevent their occurrence. About all that can be done is to perform an adequate visual and/or nondestructive examination to reveal the presence of laminations before a piece of laminated material is included in a weldment.

The best method for the discovery of laminations other than visual inspection is the use of ultrasonic

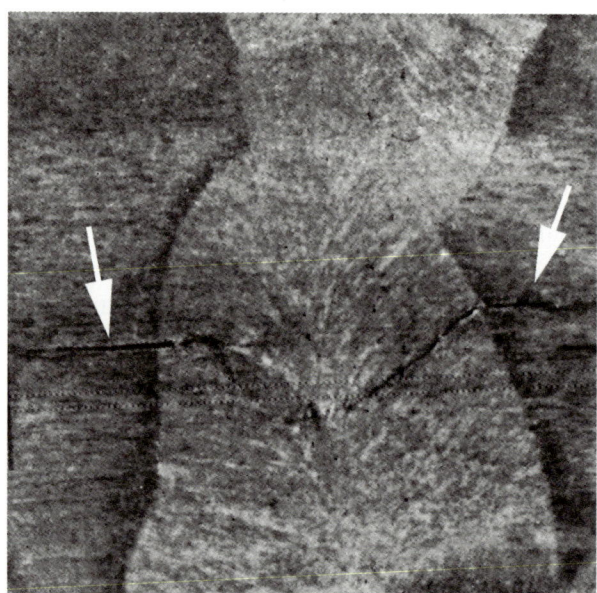

Figure 9.44—Weld Metal Cracking Due to Presence of Lamination

testing. Radiography will not reveal laminations because there is no change in the radiographic density of a metal even if it is laminated. To illustrate this, imagine the radiography of two 1/4-in. plates placed one on top of the other compared to a single 1/2-in. plate. Review of the film for each would reveal no difference in film density, because the radiation is still passing through the same total thickness of metal.

Lamellar Tear

Another base metal discontinuity of importance is the lamellar tear. It is described as a terrace-like fracture in the base metal with a basic orientation parallel to the rolled surface. Lamellar tears occur when there are high stresses in the Z direction, or through-thickness direction, often resulting from welding shrinkage. The tearing always lies within the base metal, usually outside the heat-affected zone and generally parallel to the weld fusion boundary. Figure 9.45 shows some typical configurations in which lamellar tearing may occur.

Lamellar tearing is a discontinuity directly related to the actual configuration of the joint. Therefore, those joint configurations in which the shrinkage stresses from welding are applied in a direction which tends to pull the rolled material in its Z direction, or through-thickness direction, will

Figure 9.45—Weld Configurations Which May Cause Lamellar Tearing

Other factors affecting a material's susceptibility to lamellar tearing are its thickness and the degree of contaminants present. The thicker the material and the higher the inclusion content, the greater the possibility of experiencing lamellar tearing.

For the onset of lamellar tearing, three conditions must exist simultaneously. They are: stress in the through-thickness direction, susceptible joint configuration, and material having a high inclusion content. So, to prevent the occurrence of lamellar tearing, any one of these elements must be eliminated. Generally, the problem is solved by using cleaner steels.

Seams and Laps

Seams and laps are additional base metal discontinuities related to the steel-making process. They differ from laminations in that they are open to the rolled surface of the metal instead of the edge. If cross-sectioned, they may run parallel to the rolled surface for some distance and then tail off toward that surface. Seams are described as straightline longitudinal crevices or openings that may appear on the surface of steel. Seams are primarily caused by imperfections in the steel ingot, by improper handling after pouring, or by variations in heating and rolling practice. Laps are the result of overfilling in the rolling mill passes that causes fins or projections which turn down as the material rolls through succeeding stands in the mill train.

Figures 9.46 and 9.47 show examples of a deep seam and a cluster of shallow seams, respectively. An example of a lap is shown in Figure 9.48. Since seams and laps result from improper processing at the steel mill like laminations, the welding inspector has little control over their occurrence other than detecting them if they appear in material to be used in fabrication. They are best revealed using either visual, magnetic particle, penetrant, ultrasonic or eddy current testing.

Dimensional

Up to this point, all of the discontinuities discussed could be classified as structural type flaws. However, there is another group of discontinuities which can be classified as dimensional irregularities. Dimensional discontinuities are size and/or shape imperfections. These irregularities can occur in the welds themselves or in the overall welded

be more susceptible to lamellar tearing. As we learned in Module 6, when a metal is rolled, it will characteristically exhibit lower strength and ductility in this Z direction as compared to its properties in the longitudinal and transverse directions.

Figure 9.46—Deep Seam on the Surface of a Semi-Finished Rolled Product

Figure 9.47—Clustered Seams on the Surface of a Semi-Finished Rolled Product

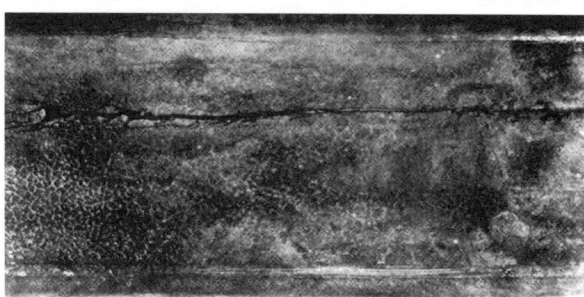

Figure 9.48—Lap on the Surface of Rolled Steel Billet

entire weldment to assure that the heat of welding has not caused excessive distortion or warpage.

Discontinuities in Laser and Electron Beam Welding

Most of the discontinuities previously discussed can also occur in electron and laser beam welds. In addition, there are several discontinuities which are unique to these welding processes. This is due to the narrow and deep weld profiles produced by these processes as well as the high travel speeds used.

The first of these is a *missed joint*, where the beam is deflected off the weld joint as shown in Figure 9.49. This can be caused if the axis of the small diameter laser or electron beam is not aligned with the joint root (especially for a thick part) or along the length of the joint. For electron beam welding, even when the beam is properly aligned, magnetic forces on the beam can cause deflection off the joint. Electron or laser beam welding of metals with dissimilar magnetic properties can also produce this deflection (as shown in Figure 9.49).

If the beam misses the joint along the top surface, visual inspection should easily detect it. However, if the "missed" joint occurs subsurface (as in Figure 9.49), or where the face of the weld is wider (a "nail head" shape), this is difficult to detect

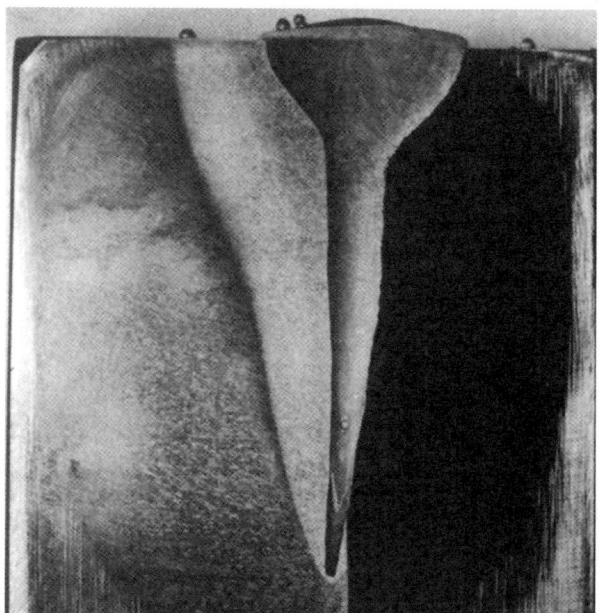

Figure 9.49—Beam Deflection (Missed Joint)

structure. Since dimensional discontinuities could render a structure unsuitable for its intended service, they must be considered and checked by the welding inspector.

This inspection could consist of the measurement of weld sizes and lengths to assure that there is sufficient weld metal to transmit the applied loads. Other measurements should be made of the

by visual inspection. A missed joint may have only a small fraction of the expected and required strength.

Another discontinuity in electron and laser beam welds is formation of voids at the bottom of the weld, which is often referred to as *root porosity*. These are typically caused by gases which form in the weld metal that do not have sufficient time to escape up through the deep weld metal. Because electron beam welding is done under a vacuum, porosity can easily become entrapped. Voids or cold shuts also form where the molten metal does not completely fill the cavity produced by the beam (during welding). The deeper the weld, the worse the problem can be.

Shrinkage voids and microfissures or hot cracks can also form near the weld centerline of electron and laser beam welds. This is because the weld solidifies from the two sides toward the centerline. This produces tremendous shrinkage stresses along the weld centerline which can produce these discontinuities.

Another problem with electron or laser beam welding is inconsistent penetration or *spiking*. This occurs mainly in partial joint penetration welds, but also can occur near the root of complete joint penetration welds. Figure 9.50 shows spiking.

For electron beam welding, this is caused by variations in the power density of the beam, by the vaporization of elements during welding, and by the turbulent weld pool. It occurs mainly with higher power, deeper penetrating welds. In laser beam welding, variable penetration can also be caused by coupling and decoupling of the laser beam. These are caused by reflection of the beam off the top surface of the components and interaction with the plume of vapors above the weld, and are affected by the type of laser.

In complete joint penetration laser, and especially in electron beam, welds there is a tendency for *over penetration*. When this occurs, there is a tendency for liquid metal to be expelled from the root of the weld in the form of spatter. This can be a problem when welding on a vessel or container, where the spatter may adhere to the opposite surface, or become a loose particle inside the vessel.

Spiking is often accompanied by incomplete fusion along the root or sides of the weld, as shown in Figure 9.50. This is often caused by a very nar-

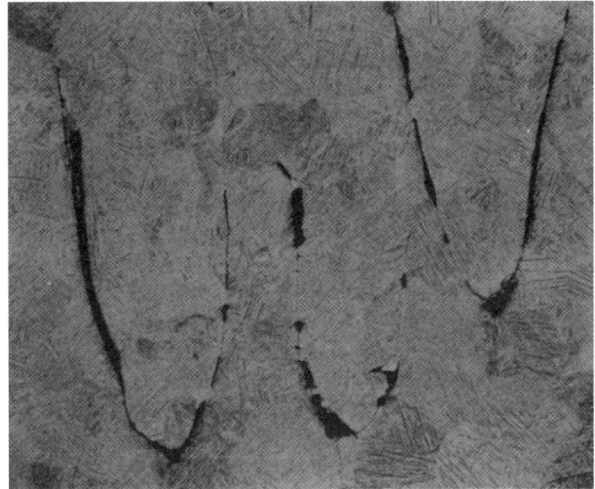

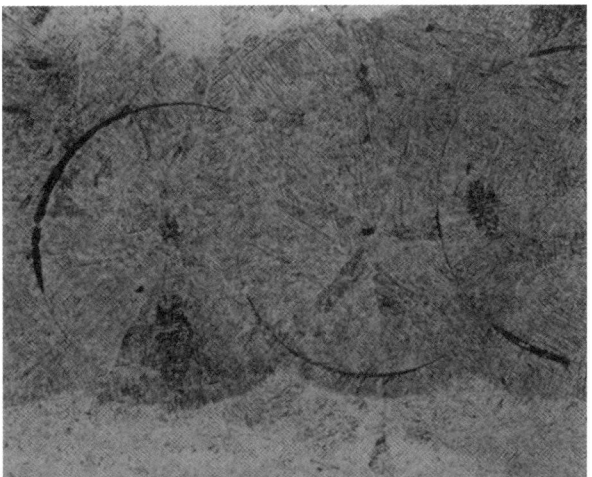

Figure 9.50—Inconsistent Joint Penetration ('Spiking') Along with Incomplete Fusion in Vertical (Top) and Horizontal (Bottom) Section Through an Electron Beam Weld in a Titanium Alloy

row beam, by improper or inconsistent joint fitup, and by fast travel speeds.

Summary

Imperfections may exist in the weld and/or base metal; they are generally described as discontinuities. If a certain discontinuity is of sufficient size, it may render a structure unfit for its intended service. Codes normally dictate the permissible limits for discontinuities. Those greater than these limits are termed defects. Defects are discontinuities which require some corrective action.

Discontinuity severity is based on a number of factors, including whether it is linear or nonlinear, the sharpness of its ends, and whether or not it is open to the surface.

Discontinuities exist in a number of different forms, including cracks, incomplete fusion, incomplete joint penetration, inclusions, porosity, undercut, underfill, overlap, convexity, weld reinforcement, arc strikes, spatter, laminations, lamellar tears, seams/laps, and dimensional types.

By knowing how these discontinuities can form, the welding inspector may be successful at spotting these causes and preventing problems from occurring.

Key Terms and Definitions

arc strike—a discontinuity resulting from an arc, consisting of any localized remelted metal, heat-affected metal, or change in the surface profile of any metal object.

atomic hydrogen—the ionic form of hydrogen, denoted as H^+ as opposed to molecular hydrogen which contains two atoms of hydrogen and is denoted as H_2. A synonym for atomic hydrogen is nascent hydrogen.

collet—in welding terms, a part of a welding torch forming a shroud.

convexity—the maximum distance from the face of a convex fillet weld perpendicular to a line joining the weld toes.

crack—a fracture type discontinuity characterized by a sharp tip and high ratio of length and width to opening displacement.

crater crack—a crack forming at the termination of a weld.

defect—a discontinuity which exceeds the permissible limit of a code; a rejectable discontinuity requiring repair or replacement.

delamination—the separation of a lamination under stress.

density—the ratio of the mass of an object to its volume, usually in terms of grams per cubic centimeter or pounds per cubic foot; also refers to the darkness of radiographic film; the darker areas are noted as having a higher density.

discontinuity—any irregularity in the normal pattern of a material; any interruption of the uniform nature of an item.

inclusion—entrapped foreign solid material, such as slag, flux, tungsten, or oxide.

incomplete fusion—a weld discontinuity in which fusion did not occur between weld metal and fusion faces or adjoining weld beads.

incomplete joint penetration—a joint root condition in a groove weld in which weld metal does not extend through the joint thickness.

intergranular—referring to conditions which occur at or follow the grain boundaries of a metal. An intergranular crack would initiate and propagate along a metal's grain boundaries.

lamellar tear—a subsurface terrace and step-like crack in the base metal with a basic orientation parallel to the wrought surface caused by tensile stresses in the through-thickness direction of the base metals weakened by the presence of small dispersed, planar shaped, nonmetallic inclusions parallel to the metal surfaces.

lamination—a type of discontinuity with separation or weakness generally aligned parallel to the worked surface of a metal.

nascent hydrogen—see Atomic Hydrogen.

overlap—in fusion welding, the protrusion of weld metal beyond the weld toe or weld root.

pipe—in metal ingot casting, the severe shrinkage occurring at the top center portion of the ingot, usually containing oxides.

planar—of or pertaining to a plane; lying in a plane.

porosity—cavity-type discontinuities formed by gas entrapment during solidification or in a thermal spray deposit.

propagate—growth, or continuation of growth; to get larger.

protrusion—a projection outward; a jutting out.

radiograph—a film made by passing X- or gamma radiation through an object to determine the quality of its internal structure.

safe-ending—the practice of drilling a small hole at each end of a crack to increase the crack end radius and stop further crack propagation.

seam/lap—longitudinal base metal surface discontinuities on wrought products.

shielding gas—protective gas used to prevent or reduce atmospheric contamination, as of a molten weld metal.

slag inclusion—an inclusion of slag.

spatter—the metal particles expelled during fusion welding that do not form a part of the weld.

stress risers—conditions such as notches, cracks, or geometry which increase the applied stress by factors of 2 to 10.

stringer—in metallurgy, an elongated oxide or nonmetallic inclusion within the metal.

titania—a titanium oxide; a coating type for covered electrodes in welding.

transgranular—referring to conditions which cross or pass through the metal's grains. A transgranular crack has a path across the grains as opposed to intergranular cracking which follows a path along the grain boundaries.

transverse—lying, situated, placed across; having a path from side to side.

tungsten inclusion—an inclusion of tungsten.

undercut—a groove melted into the base metal adjacent to the weld toe or weld root and left unfilled by weld metal.

underfill—a condition in which the weld face or root surface extends below the adjacent surface of the base metal.

weld reinforcement—weld metal in excess of the quantity required to fill a joint; at the face or root.

wrought—the term applied to the working or forming of metal while it is solid to form shapes, as opposed to a cast product which forms directly from the molten state.

Module 10
Visual Inspection and Other NDE Methods and Symbols

Contents

Introduction 10-2

Visual Inspection (VT) 10-3

Nondestructive Testing (NDT) 10-12

NDE Symbols 10-27

Summary 10-29

Key Terms and Definitions 10-30

Module 10—Visual Inspection and Other NDE Methods and Symbols

Introduction

In any effective welding quality control program, visual inspection provides the basic element for evaluation of the structures or components being fabricated. In order to gain some assurance as to the suitability of the welding for its intended service, codes and standards will always stipulate the performance of visual inspection as the minimum level of acceptance/rejection evaluation. Even when other nondestructive or destructive methods are specified, they are actually meant to supplement, or reinforce, the basic visual inspection. When we consider the various other methods used to evaluate welds, they can really be thought of as simply visual enhancement techniques, because the final evaluation of the test or test results will be accomplished visually.

It has been proven in numerous situations that an effective program of visual inspection will result in the discovery of the vast majority of those defects which would be found later using some other more expensive nondestructive test method. It is important to realize, however, this is possible only when the visual inspection is accomplished before, during, and after welding, by a trained and qualified inspector. Simply looking at the finished weld without the benefit of seeing those preceding fabrication steps can only provide a limited assurance of a weld's suitability.

The primary limitation of the visual inspection method is that it will only reveal those discontinuities which are apparent on the surface. Therefore, it is important for the welding inspector to look at many of the initial and intermediate surfaces of the joint and weld.

Due to its relative simplicity as well as the minimal amount of equipment required, visual inspection is a very cost effective quality control tool. This cost effectiveness is further enhanced when visual inspection reveals a defect soon after it occurs, and it can then be corrected immediately and most economically. An example of this would be the discovery of a root pass crack. If discovered prior to deposition of additional passes, the repair is relatively simple compared to the cost if not discovered until after the weld is completed. Many times these additional costs involve more than simply the higher cost of a more extensive repair. Often the major concern is in the additional time required to accomplish that repair. When a defect is discovered right after it occurs, the time for repair is minimal so there is little impact on the overall job schedule.

While visual inspection is considered to be a relatively simple evaluation method, do not be deceived into thinking that it can be performed by just any individual. The American Welding Society has recognized the importance of using only those individuals having at least a minimum level of knowledge and experience for the performance of visual inspection. To answer these needs, the Certified Welding Inspector program has been developed to judge the suitability of an individual for a position as a welding inspector. When an individual satisfies the experience requirements and successfully passes a series of examinations, he or she is considered to be capable of effectively performing visual inspection of welds and weldments.

While visual inspection is usually considered to be less complicated than some of the other nondestructive test methods, that does not imply that just anyone can effectively perform this operation. By simply reviewing the preceding nine sections, it becomes evident that whoever performs visual inspection must be knowledgeable in numerous areas of expertise. It takes years of experience and training to become familiar with all of these various aspects of welding inspection. In essence, the welding inspector must be familiar with all of the techniques used to produce welds as well as those methods used to evaluate the finished product.

This final section will deal with the application of visual weld inspection as the basic element of a quality control program, and cover those additional NDE techniques which effectively supplement visual testing. Within the constraints of this presentation, it will be impossible to accurately describe the responsibilities of every welding inspector in every industry. Each individual situation will have associated with it particular practices and procedures which will not apply to some other situation. However, this discussion will attempt to describe, in general, many of the responsibilities which may be part of a welding inspector's job. So, in essence, the information included will serve to summarize how each of the elements discussed in the preceding nine sections will actually be applied by the welding inspector during the performance of his or her daily duties.

Visual Inspection (VT)

Since the welding inspector's responsibilities can be extensive and will occur at various stages of the fabrication sequence, a helpful aid is an "inspection checklist." Such a document will help the welding inspector organize the inspection effort and assure each specific task has been performed. An example of such a listing is shown in Figure 10.1.

Also, the various tools used by the visual welding inspector will be reviewed. While the visual inspection method is characterized as requiring a minimum of tools, there are certain devices which can help the welding inspector perform more easily and effectively. Figure 10.2 illustrates some of these tools which might be used by the welding inspector to aid in evaluating welds and weldments.

It has been mentioned that the only way in which visual inspection can be considered to effectively evaluate the quality of welds is to apply that inspection at every step of the fabrication process. Unless there is an ongoing program, certain discontinuities could be missed. Further, the main reason to perform the inspection on a continuing basis is to discover problems as soon after they occur as pos-

Before Welding

___ Review applicable documentation
___ Check welding procedures
___ Check individual welder qualifications
___ Establish hold points
___ Develop inspection plan
___ Develop plan for recording inspection results and maintaining those records
___ Develop system for identification of rejects
___ Check condition of welding equipment
___ Check quality and condition of base and filler materials to be used
___ Check weld preparations
___ Check joint fitup
___ Check adequacy of alignment devices
___ Check weld joint cleanliness
___ Check preheat, when required

During Welding

___ Check welding variables for compliance with welding procedure
___ Check quality of individual weld passes
___ Check interpass cleaning
___ Check interpass temperature
___ Check placement and sequencing of individual weld passes
___ Check backgouged surfaces
___ Monitor in-process NDT, if required

After Welding

___ Check finished weld appearance
___ Check weld size
___ Check weld length
___ Check dimensional accuracy of weldment
___ Monitor additional NDT, if required
___ Monitor postweld heat treatment, if required
___ Prepare inspection reports

Figure 10.1—Sample Welding Inspection Checklist

sible so they can be corrected most efficiently. Therefore, this discussion of the welding inspector's "visual inspection" duties will be organized in terms of those tasks which are performed before, during, and after welding.

In some respects the responsibilities of the welding inspector prior to the start of welding may be the most important. It can at least be said that unless

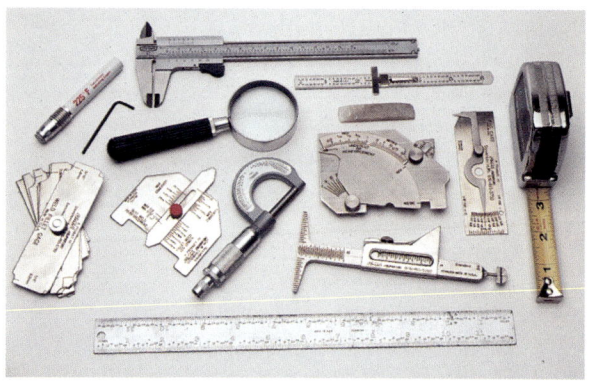

Figure 10.2—Visual Inspection Tools

this aspect of the inspection job is performed satisfactorily, there may be problems encountered later in the fabrication process. Many of these tasks apply to the organization of the inspection which will follow, including becoming familiar with welding requirements, determining when inspections are to be performed, and developing systems for reporting and maintaining inspection information.

One of the first duties of the welding inspector at the onset of a new job is to review all of the documentation relating to the actual welding which is to be performed. Some of the documents which may be reviewed include drawings, codes, specifications, procedures, etc. These documents contain information which is very valuable to the welding inspector. In essence, they describe what, when, where, and how inspection is to be performed. Therefore, they provide ground rules for all of the inspections which will follow. This will help the welding inspector plan how to proceed in evaluating the welding to assure that it complies with job requirements.

Some of the information gained from this review of documents refers to the materials to be used for the welded fabrication. Depending upon the type of material specified, there may be special requirements for its fabrication. For example, if a quenched and tempered steel is specified, there will usually be a need for welding heat input control. So, the welding inspector will be required to monitor the welding with this in mind.

Another preliminary step related to the materials being used is to check whether or not there are welding procedures which cover the required welding. In addition to the types of materials being welded, the welding inspector must check if the qualified welding procedures provide adequate coverage with regard to welding process(es), technique, filler metal type, position, etc. If some aspect of the upcoming production welding is not properly covered by the existing welding procedures, then new procedures must be developed and qualified in accordance with the appropriate code. The welding inspector may also be responsible for the monitoring, testing, evaluation, and recording of those welding procedure qualifications.

Once all of the appropriate welding procedures have been qualified, it is then necessary to review the certification papers of the individual welders to assure they are considered to be qualified and certified to perform the production welding in accordance with the approved welding procedures. Some of the specific limitations relating to the individual welder qualification include materials being welded, process(es), position, technique, joint configuration, etc. Those welders who do not have the proper qualification and certification must then be tested to assure that they are capable of performing production welding in accordance with the applicable procedures.

It is often helpful for the welding inspector if there is a listing of all of the production welders showing those procedures which they are considered qualified to perform. Further, some codes require that the welders permanently identify all of the production welds they have made. If this is the case, there should be an accompanying log showing the appropriate identification stamp of each welder. There may also be a code requirement relating to the period of effectiveness of a welder's qualification. In such cases, a running log should be maintained and available for the welding inspector to review to determine if an individual welder has used a particular procedure within the specified time period. If not, the welder may require requalification.

Once the welding inspector has reviewed the appropriate documents related to the specific inspection job, he or she may elect to establish hold points. These are simply preselected steps in the fabrication sequence when the work must stop until the inspector has a chance to review the work completed to that point. Fabrication cannot continue until the welding inspector has approved the work up to that stage of the operation. This allows the work to be accepted on a step-by-step basis instead of waiting until the entire structure or component is completed. In that way, problems can be located and corrected with relatively little effect on the fabrication schedule. This also reduces the possibility that some minor problem which occurs at some early step will result in a major defect later in the fabrication sequence.

Another important preliminary step for the welding inspector is to develop a suitable plan for performing the inspections and recording and maintaining the results. Through experience, the welding inspector becomes more aware of how important this step can be. The inspector must know when a particular inspection task is to be performed and how it will be accomplished. There must be a plan so that important aspects of the fabrication process are not left uninspected. In general, the inspector can base this system on the basic steps of the fabrication process, so the inspection plan might simply use the production schedule as an outline for when a particular inspection step will be performed.

Once an inspection has been performed, there must be a suitable system established for the recording of the inspection results. This system should include provisions for the types and contents of reports, the distribution of those reports, as well as some method of logically maintaining the records so that they can be retrieved and reviewed by others familiar with the job. Basically, the reports, and the system developed to maintain those reports, should be as simple as possible while still providing adequate information which is understandable to all personnel involved at some future review.

Another related matter involves the identification and treatment of rejects. At the beginning of every job, the welding inspector must establish some system whereby a rejected weld can be reported and identified. This system should include provisions for marking the location of a reject so that the production personnel understand the nature and location of the defect to enable them to easily find and repair the existing problem. There should also be some established convention regarding the reporting of that reject so that all involved individuals know that a defect exists and must be corrected. The marking used to indicate the presence and location of a defect should be some unique type or color so that it is clearly visible and descriptive to both quality control and production personnel. Finally, the system should describe how the reinspection after repair will be initiated and performed. Once performed, the method of reporting the results should be established so that the original rejection report is accompanied by a subsequent acceptance report.

The condition of the welding equipment to be used will also have an effect on the resulting weld quality. Consequently, the welding inspector should evaluate the performance and condition of the equipment. This includes welding power sources, wire feeding equipment, ground cables and clamps, flux and electrode storage devices, gas shielding hoses and accessories, etc. When evaluating the welding power sources, the accuracy of the equipment meters should be checked using calibrated voltmeters and ammeters so the welding parameters can be accurately determined during the production welding. Due to the inherent inaccuracies of some of these equipment meters, this can be an important step toward alleviating welding problems.

Once all of these tasks have been performed, it is now time to perform some preweld inspection of the materials and their configurations. One of these steps is to evaluate the quality of the base materials and welding filler materials. If problems exist in either of these items, they will likely create additional problems later in the fabrication sequence. If not discovered early enough, a material problem can be extremely costly when one considers the costs associated with the application of additional fabrication steps. So, it is extremely important that these problems are found before a great deal of time and materials have been applied. An example would be the presence of a lamination in a structural member. If not discovered until all the cutting, drilling, punching, and welding have been performed, the costs of those operations cannot usually be recovered. The supplier may have to simply replace the defective member, and the fabrication begins again from the start.

The inspection of the base materials will vary from simple visual inspection of the metal's surface to an elaborate combination of various nondestructive test methods to evaluate both the surface and subsurface quality of the material. The criticality of the structure or component will dictate to a certain degree the extent of inspection required.

Inspection of the welding filler materials to be used is also very important. Moisture or contamination present in the flux or on the electrode surface can result in serious weld quality problems. For example, if low hydrogen electrodes are required, problems such as underbead cracking and porosity can result if they are not properly protected from the atmosphere. Therefore, the welding inspector should be aware of how they will be stored and handled to prevent the pickup of excessive moisture or contamination.

After the inspection of all the materials to be used, the next step is to evaluate the quality and accuracy of the weld joint preparations. In the case of groove welds, such items as bevel angle, depth of preparation, root face dimension, and groove radius (for J and U grooves) should be checked visually. This inspection may require the use of additional tools such as a scale, a tape measure, or devices to measure the bevel angles and radii. Examples of some of these measuring tools are shown in Figure 10.3.

After the weld joint preparations have been checked and approved, the welding inspector should then evaluate the weld joint fitup. That is, he or she should check the alignment and relative placement of the two members being joined. If the dimensional accuracy of the component or structure is improper at this stage, it is unlikely that subsequent welding will improve the situation sufficiently. Items to be checked at this phase include root opening, angular alignment, planar alignment (high-low), groove angle, etc. In situations where distortion is expected, there may be a specified preset dimension with the idea that the initially improper alignment will be corrected by the resulting distortion from welding. Figure 10.4 illustrates some examples where the members are purposely misaligned or bent to allow for the expected distortion.

Devices such as those used for the evaluation of the weld joint preparation may also be used during this aspect of the inspection process. In some instances, it may also be helpful to use specially-made templates or gauges to check these dimensional features when the configuration is a common one for a particular job, or the shape is to be repeated many times.

The accuracy of the joint fitup will have an effect on the final dimensions of the weldment. In addition, variations in the fitup could have a direct bearing on the resulting weld quality. For example, if the groove angle or root opening is insufficient, the welder may not be able to properly fuse the weld metal to the groove face. Excessive groove angles or root openings may require additional welding which in turn could result in excessive

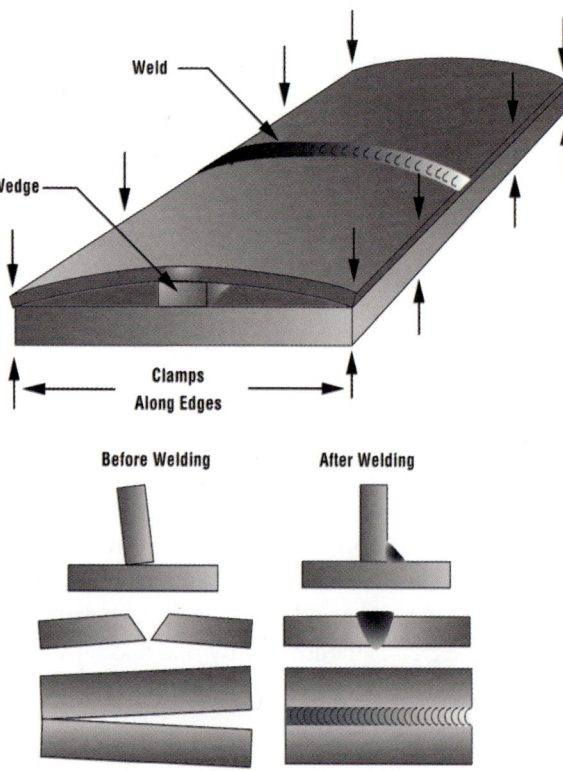

Figure 10.3—Typical Measuring Devices

Figure 10.4—Prebending and Presetting Pieces to Allow for Distortion

distortion. In the case of a fillet weld, if a root opening is present, deposition of a weld of the specified size will produce an effective throat less than the theoretical throat required by the designer. This is illustrated in Figure 10.5.

In a case like this, the actual weld size should normally be increased by the amount of the root opening which is present to provide the necessary weld cross section. Therefore, the welding inspector should note any root opening which is present during fillet weld fitups so the resulting welds can be accurately sized when completed.

If any jigs, fixtures, or other alignment devices are used, the welding inspector should check to assure that they provide the proper alignment and that they are massive enough to maintain that alignment during the welding operation. If tack welds are added to assist in this alignment, they should be inspected to assure that they are not defective. Cracked tack welds should be removed and redeposited prior to final welding; if not corrected, the crack will remain and possibly grow to produce a detrimental situation which will require a major repair effort if discovered later in the fabrication sequence.

During the inspection of the weld joint fitup, it is also important that the welding inspector carefully check the cleanliness of the weld zone. The presence of contaminants and moisture could significantly affect the resulting weld quality. Such things as moisture, oil, grease, paint, rust, mill scale, galvanizing, etc. can introduce levels of contamination which may not be tolerated by the welding process. The result could be the presence of porosity, cracking, or incomplete fusion in the completed weld.

One final feature which should be checked prior to the commencement of welding is the preheat, when required. The welding procedure will indicate the requirements for this preheat, and it may be stated as either a minimum, a maximum, or both. This specified preheat should be checked slightly away from the weld joint instead of on the groove face itself. In fact, all base metal within a distance equal to the thickness of the members, but not less than 3 in., should be raised to the appropriate preheat temperature. This preheat can be checked using a variety of methods, including temperature indicating crayons, surface pyrometers, thermocouples, or surface thermometers. Examples of some temperature indicating crayons are shown in Figure 10.6.

In order to continue the ongoing welding quality control, the welding inspector also has numerous things to check as the welding is actually being performed. As was the case for those inspections performed prior to welding, these checks can hopefully detect problems when they occur so they can be more easily corrected. During this phase of the fabrication process, the inspector's knowledge of welding will be extremely beneficial, since part of the inspection will involve the evaluation of the actual

Figure 10.5—Fillet Weld on T-Joint with Root Opening

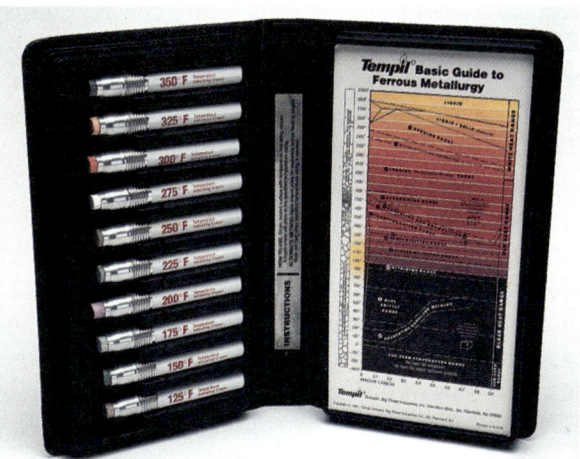

Figure 10.6—Some Typical Temperature-Indicating Crayons

welding technique as well as the resulting weld quality. It is realized that it is unrealistic to think that the welding inspector can observe the deposition of each and every weld pass. Therefore, the experienced welding inspector should be able to select those aspects of the welding sequence which are considered to be critical enough to warrant his or her presence.

When conducting welding inspection during the welding operation, the welding inspector must rely on the welding procedure to provide a basis for inspection. This document will specify all of those important aspects of the welding operation, including welding process, materials, specific technique, preheat and interpass temperature, plus any additional information which describes how the production welding should be performed.

So, the welding inspector's job will essentially consist of monitoring the production welding to assure that it is being conducted in accordance with the appropriate procedure. This also implies that any problems with the procedure can be discovered and corrected so that satisfactory welds can be produced.

One of the parts of the welding inspection which occurs during welding is the visual examination of the individual weld passes as they are deposited. At that time, any surface discontinuities can be detected and corrected, if necessary. It is also important to note any weld profile irregularities which may hinder subsequent welding. An example of such a situation may occur during the welding of a multipass groove weld. If one of the intermediate passes is deposited such that it exhibits a very convex profile which creates a deep notch at its toe, that configuration may prevent a subsequent pass from properly fusing at that location. If noted, the welding inspector could ask that some grinding be done to assure that thorough fusion can be attained.

Checking the in-process quality is especially critical in the case of the root pass or root layer. In most situations, this portion of the weld cross section represents the most difficult welding condition, especially in the case of an open root configuration. In conditions of high restraint, the shrinkage stresses from welding may be sufficient to fracture the root pass if it is not large enough to resist those stresses. The welding inspector should be aware of these problems and thoroughly check the root pass prior to any additional welding so that irregularities can be found and corrected as they occur.

Another feature which should be evaluated during the welding operation relates to the interpass cleaning. If the welder fails to thoroughly clean the weld deposit between individual passes, there is a great possibility that slag inclusions or incomplete fusion could result. This is especially critical when using a welding process which uses a flux for protective shielding. However, careful interpass cleaning is still recommended for those processes using gas shielding. Proper cleaning may be hindered when the deposited weld exhibits a convex profile which prevents sufficient access to the slag coating. As indicated above, it may then be necessary to perform additional grinding to remove the objectionable profile and facilitate proper cleaning.

The interpass cleaning of welds can be accomplished by any method which yields the appropriate result, including use of such tools as manual chipping hammers, pneumatic chipping hammers, grinders, manual wire brushes, power wire brushes, etc. When using some of these tools on softer materials, it is important that the action is not so aggressive that the weld is cracked or otherwise damaged. It is also possible during cleaning operations to deform the material to such a degree that existing discontinuities are "masked" and remain undetected. Care should also be taken to prevent the deforming or marring of the base metal adjacent to the weld. If a weld requires a harsh treatment to remove the slag, quite possibly the real problem is associated with the welding process or technique.

For those welding procedures requiring interpass temperature control, the welding inspector may need to monitor this aspect of the process as well. Just as with the preheat, the interpass temperature could be specified as a minimum, maximum, or both. The interpass temperature should also be measured on the base metal surface near the weld zone and not in the weld joint itself. Figure 10.7 shows a digital pyrometer which is very effective for measuring interpass temperatures.

During the welding operation, the welding inspector may also check the placement of individual weld passes for multipass welds. Improper placement of individual passes may make deposition of subsequent passes more difficult or even impossible. Figure 10.8 shows an example of how

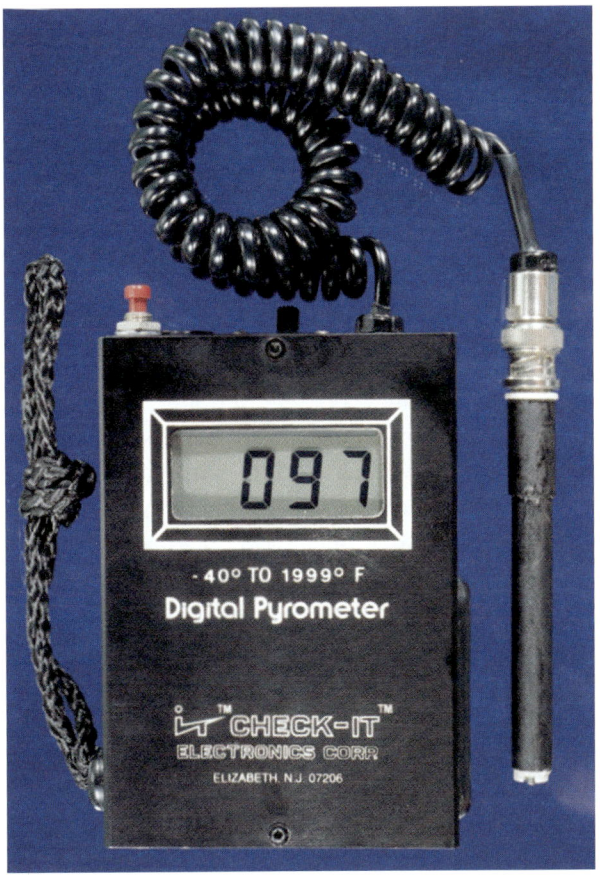

Figure 10.7—Digital Pyrometer for Temperature Measurement

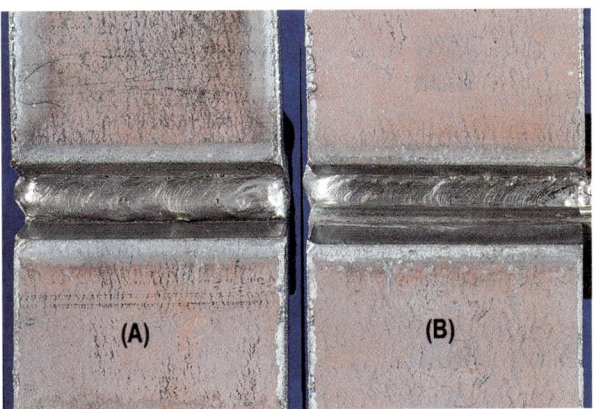

Figure 10.8—Placement of Root Passes

between the first pass and the groove face. To correct this problem, the welding inspector might ask that the welder do some grinding to open that gap slightly, as shown in Figure 10.8(B). Of course, the proper position of this root bead would have been for it to have fused to both members of the joint with a single bead.

Figure 10.9 illustrates both the incorrect and correct methods for placement of passes in a multipass fillet weld. In Figure 10.9(A), the initial pass only fuses to one of the members and leaves a tight gap at the joint root. The second pass cannot properly fuse in this area. Figure 10.9(B) shows the proper way to place the two passes.

In addition to checking the placement of weld passes, the welding inspector may also be asked to monitor the sequencing and placement of individual segments of welds. This is usually a concern for those situations where excessive distortion may result from welding too much in one area. This sequencing may require that the welder first deposit passes on one side of the joint and then move to the opposite side to reduce the amount of angular distortion which could result from welding from one side only. Figure 10.10 illustrates this technique on

improper placement of a root pass makes successful deposition of the next pass extremely difficult.

Looking at Figure 10.8(A), you can imagine that it will be very hard to deposit a second pass and obtain adequate fusion in the tight gap that remains

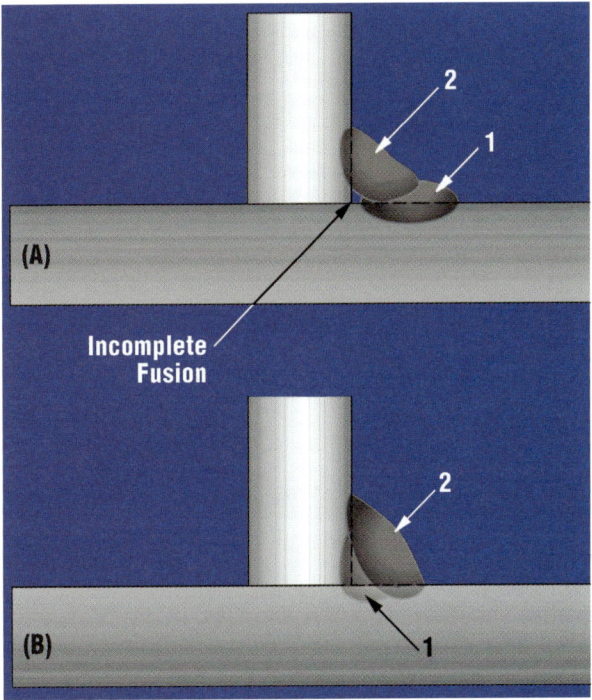

Figure 10.9—Placement of Passes for Multipass Fillet Weld on T-Joint

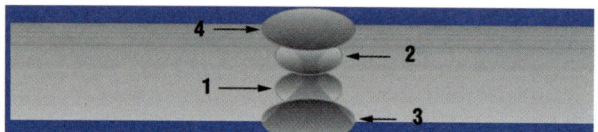

Figure 10.10—Weld Sequencing in Double V-Groove Weld

a double V-groove weld. Figure 10.11 shows how double fillet welded joints can be sequenced to reduce distortion.

In some cases, the method used to reduce distortion is to deposit the individual weld passes using a backstepping technique. With this method, the direction of travel for the individual passes is opposite that for the general progression of welding along the weld axis. So, each subsequent weld pass starts ahead of the previous pass and progresses toward it. This is illustrated in Figure 10.12.

When full penetration groove welds are designed to be welded from both sides, there must be some method of gouging the weld root of the first side prior to welding the second side. The welding inspector should examine that backgouged surface prior to the welding of the second side. If

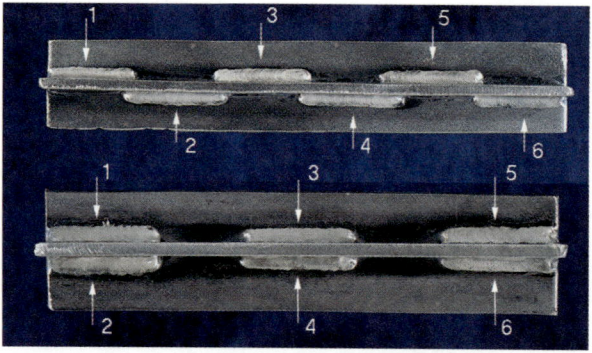

Figure 10.11—Weld Sequencing of Double Fillet T-Joints

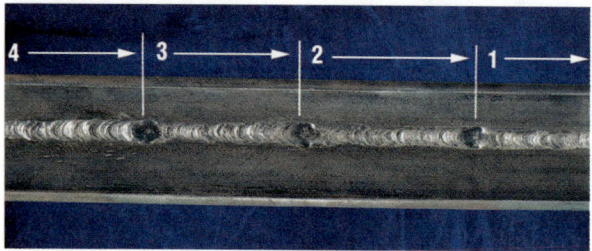

Figure 10.12—Backstep Technique

this is not done, there is a possibility that slag inclusions or other discontinuities won't be removed and would then be included in the finished weld. Not only should the welding inspector be sure that all discontinuities have been removed by this backgouging operation, but also that the configuration of that backgouge is adequate to assure that the opening is sufficient to allow for successful deposition of additional passes. Problems occurring at this location can be easily corrected at this time in the operation compared to the relative difficulty expected if excavation is required on the finished weld.

While most of these items monitored during welding are really the responsibility of the welder, it is still important that the welding inspector check to assure that the welder understands the welding requirements and follows instructions adequately. The welding inspector usually has a better grasp of the overall quality expected, so he or she can more easily spot problems and initiate corrective measures.

Once a weld has been completed, the welding inspector must then examine the finished product to assure that all of the preceding steps have been performed successfully to produce a quality weld. If all of the preliminary steps have been performed as required, then the post-weld inspection should simply confirm that the weld is of sufficient quality. However, the codes specify the required attributes of the finished weld, so the welding inspector must examine the weld visually to determine if those requirements have been met.

In general, visual inspection after welding consists of looking at the appearance of the finished weld. This visual examination will detect surface discontinuities in the weld and base metal. Of special importance during this aspect of the welding inspection is the evaluation of the weld's profile. Sharp surface irregularities can result in premature failures of that component during service. These visual features are evaluated in accordance with the applicable code which will describe the permissible amount of a certain type of discontinuity.

Included in this visual examination of the weld is the measurement of the weld to determine if its size is correct as required by the drawing. For a groove weld, you are primarily concerned if the weld groove is filled flush with the base metal surfaces without excessive reinforcement. Any underfill must be corrected by depositing additional weld metal.

In the case of fillet welds, the size determination is normally accomplished with the aid of a fillet weld gauge. There are numerous different types of fillet weld gauges which can be used, including gauges or templates which are specially made for use on a particular fillet weld configuration. There are also several types of fillet weld gauges which are manufactured for use in the measurement of general fillet welds.

One common type of fillet weld gauge consists of a series of sheet metal templates which have been machined to produce two different types of cutouts. The individual templates are selected based on whether the weld is concave or convex as well as the size of fillet weld required. The welding inspector selects the gauge of the proper type and size and compares the existing weld size with that gauge.

Since fillet weld sizes are designated as nominal dimensions, there should realistically be a tolerance applied to this measurement. Since commercially-available gauges are typically graduated in 1/16 in. increments, it would seem reasonable to gauge fillet weld sizes to the closest 1/32 in. Conditions warranting such an approach include the difficulty in positioning your eyes properly to view the gauge, the fact that weld sizes cannot be thought of in terms of typical machining precision, gauge imprecision, base and weld metal surface irregularities, and the difficulty in determining the exact location of the toe of a convex fillet weld. Figure 10.13 illustrates the template type gauge being used to measure a fillet weld; this is the type of gauge used on the CWI Practical Examination.

When measuring a fillet weld, the weld size is determined by the size of the largest isosceles right triangle which can be totally contained within the weld's cross section. So, for a convex profile, the leg and size dimensions are the same. However, a fillet weld exhibiting a concave profile will be sized based on the throat dimension. The welding inspector must first decide what the apparent fillet weld profile actually is, convex or concave. If it is not readily apparent, both the leg and throat dimensions should be measured using both type templates to determine if the weld's size is sufficient. In the case of an unequal leg fillet weld, the weld size will be governed by the shorter of the two legs.

As mentioned above, when using the sheet metal template type gauges, the two different shapes of cutouts will be used depending upon whether the fillet weld profile is convex or concave. Once the inspector decides which profile is present, he or she selects that shaped template for the weld size which is specified. If the weld is convex, the proper gauge shape will actually be measuring the leg dimension. Similarly, for the concave fillet weld profile, the proper gauge shape will be measuring the existing throat dimension. Regardless of the shape of the template, the size indicated will be related to the required size of the theoretical triangle which is inscribed in the existing fillet weld cross section. Use of this type of fillet weld gauge is illustrated in Figure 10.14 for various fillet weld configurations.

When a weld has been measured to determine if it is of sufficient size, the welding inspector must then evaluate its length to assure that there has been enough weld metal deposited to satisfy the drawing requirements. This is of special importance where intermittent fillet welds have been specified. Here each segment must be measured as well as their center-to-center, or pitch, distances. For continuous groove or fillet welds, they are only considered to be of sufficient length if they are filled to their required cross section for the entire length of the shorter of the two members being joined.

Other measurements are required to evaluate the overall dimensional accuracy of the weldment. This is important since the shrinkage stresses from welding may have caused the size of the part to change. For example, a weld deposited around the outside of

Figure 10.13—Use of Template Type Fillet Gauge

Figure 10.14—Methods for Measuring Fillet Size

a machined bore will probably cause the diameter of that bore to be distorted, necessitating further machining to provide the appropriate bore size. Some of this dimensional evaluation will be to determine if any distortion resulted from welding. The localized heat of welding can cause members to be distorted or misaligned with respect to other parts of the weldment. These measurements will determine if the amount of distortion which is present is enough to cause the part to be rejectable.

Some welds must also be evaluated using other nondestructive test methods in addition to the visual inspection. You may also perform this testing if you are certified in the required technique, or it could be done by some other nondestructive testing specialist. If someone else performs the testing, you may be required to monitor that operation. Perhaps your only involvement relates to the review of the test personnel certification records and the inspection report, which is created to assure that the findings are in accordance with the applicable code or standard. You may also be responsible for the maintenance of those test records.

There may also be requirements relating to the post-weld stress relief or other heat treatments which are specified to modify the as-welded properties of the weldment. The welding inspector may be responsible for the monitoring of these thermal treatments as well. If so, the operation must be performed in accordance with written procedures or code requirements.

Once all of these visual inspection steps have been completed, reports must be created to explain all aspects of the evaluations which were performed. These reports should specify various features of the inspection, including what was inspected, when it was inspected, who performed the inspection, the applicable acceptance criteria, and the results of the inspection. As mentioned earlier, these reports should be as simple (and legible) as possible while still providing enough information so that others can understand what was done and what were the findings.

As has been discussed, visual inspection comprises the basic element of any welding quality control program. Although quite simple, this method is capable of finding most of the discontinuities which result from the welding operation. However, visual inspection is limited to the discovery of surface irregularities. Therefore, it must be done at all phases of the fabrication sequence to provide adequate coverage. In general, there are certain responsibilities of the welding inspector which are to be performed before, during, and after the welding operation. When properly applied, visual inspection is able to detect problems when they occur, which greatly reduces the costs associated with the correction of those defects. With this background on visual testing completed, we must look at the next phase of weld inspection.

Nondestructive Testing (NDT)

One of the purposes of an effective quality control program is to determine the suitability of a given base metal or a weld to perform its intended service. One way to judge that suitability is to subject the base metal or weld to destructive tests which would provide information about the performance of that test object. The major disadvantage of such an approach is that, as the name implies, the test object is destroyed in the process. Therefore, a number of tests have been developed to provide an indication of the acceptability of the test object without rendering it unusable for service.

These various tests are referred to as *nondestructive tests*, because they permit the nondestructive evaluation (NDE) of the metal or component. Even the destructive testing of a given percentage of parts can be expensive and assumes that the untested parts are of the same quality as those tested. Nondestructive tests yield indirect, yet still valid, results and, by definition, leave the test object unchanged and ready to be placed in service if acceptable.

As mentioned, there are numerous nondestructive tests used to evaluate the base metals to be joined as well as the completed welds. The most common test methods will be discussed, noting both their advantages and limitations, and applications. However, all of these nondestructive test methods share several common elements. These essential elements are summarized below:

(1) A source of probing energy or medium
(2) A discontinuity must cause a change or alteration of the probing energy
(3) A means of detecting this change
(4) A means of indicating this change
(5) A means of observing or recording this indication so that an interpretation can be made

The suitability of a particular nondestructive test for a given application will be determined by considering each of these factors. The source of the probing energy or probing medium must be suitable for the test object and the discontinuity of concern. If present, a discontinuity must then be capable of somehow modifying or changing the probing medium. Once changed, there must be some way of detecting these changes. These changes to the probing medium by the discontinuity must create an indication or otherwise be recorded. Finally, this indication must be preserved somehow so that it can be interpreted.

As each of these nondestructive test methods are discussed, it is important to understand how they each provide the essential elements. This will aid in deciding which nondestructive test method is best suited for a particular application.

Over the years, numerous nondestructive test methods have been developed. Each one has associated with it various advantages and limitations making it more or less appropriate for a given application. With the number of test methods available, it is important to select that method which will provide the necessary results. In many cases, several different tests may be applied to provide adequate assurance that the material or component is satisfactory. Since so many tests exist, it would be difficult to mention each one in the context of this course.

Therefore, we will concentrate on those more common nondestructive test methods which are commonly used for the evaluation of base metals and weldments. Those test methods to be discussed are noted, with their abbreviations in parentheses.

(1) Penetrant (PT)
(2) Magnetic Particle (MT)
(3) Radiographic (RT)
(4) Ultrasonic (UT)
(5) Eddy Current (ET)

While the welding inspector is usually not called upon to perform these tests, other than visual testing covered previously, it is important that he or she has a basic understanding of these other tests for several reasons. First, the welding inspector should be aware of the advantages and limitations of these methods. This will assist in deciding which test might be used to provide some additional information about the apparent quality of a material or weld. In that way, visual evaluation can be further substantiated by some additional testing. Knowledge of the advantages and limitations will also help in determining if the nondestructive testing specialist doing the actual testing is applying the test properly. Since the welding inspector may be called on to monitor the performance of, or maintain records about these tests, this knowledge should aid in understanding the results.

As each of the various tests is discussed, there will be a description of the advantages and limitations, as well as the basic principles of operation. The necessary equipment for each test will also be discussed, and there will also be mention of some of the typical applications of each of the methods.

Penetrant Testing (PT)

In general terms, penetrant testing reveals surface discontinuities by the bleedout of a penetrating medium against a contrasting colored background. This is accomplished by applying a penetrant (usually liquid) to the cleaned surface of the test piece. Once this penetrant is allowed to remain on the surface for a prescribed time (dwell time), it will be drawn into any surface opening by capillary action. Subsequent removal of excess penetrant and application of a developer draws remaining penetrant from discontinuities. The resultant indications are shown in high contrast and magnify the presence of the discontinuity so it can be visually interpreted.

There are two primary ways in which penetrant materials are grouped: specifically, the type of indication produced, and the method of excess penetrant removal. The two penetrant indications are visible and fluorescent. The visible dye (usually red) produces a vivid red indication against a white developer background when viewed under white light. The fluorescent penetrant produces a greenish, fluorescent indication against a light background when observed under ultraviolet (black) light. Since the human eye can more readily perceive a fluorescent indication than a visible indication, use of fluorescent penetrant can result in a more sensitive test.

The second way in which penetrants are categorized refers to the method by which excess penetrant is removed from the test surface. They can be water washable, solvent removable, or post-emulsifiable. Water washable penetrants contain an emulsifier which allows the oily penetrant to be rinsed off the surface with a low pressure water spray. Solvent removable penetrants require solvent to remove the surface penetrant from the test object. Post-emulsifiable penetrants are removed by adding an emulsifier after the dwell time. The application of this emulsifier to the penetrant on the test surface permits it to be removed with water in the same manner as the water washable type. By combining the characteristics of these two classifications, six different types of penetrants can be produced:

(1) Visible/Water Washable
(2) Visible/Solvent Removable
(3) Visible/Post-Emulsifiable
(4) Fluorescent/Water Washable
(5) Fluorescent/Solvent Removable
(6) Fluorescent/Post-Emulsifiable

With any of these types, the basic steps are essentially the same, except for post-emulsifiable penetrants which require an additional step to apply the emulsifier. So, with any of the methods, there are four general steps to follow, making the test a relatively simple test to perform. However, even as simple as it appears, it is important that each of the steps be carefully performed in its proper sequence; otherwise, the test results will not be reliable.

The first step involved in performing penetrant testing is to thoroughly clean the surface of the test object. Since penetrant testing is used to reveal surface discontinuities, this step is extremely important. If anything is blocking the surface opening of a discontinuity, it will prevent the penetrant from entering the opening; consequently, the discontinuity will not be revealed. The test object should be free of oil, dirt, rust, paint, etc. When cleaning softer materials, such as copper or aluminum, care must be taken if the surface is cleaned using some mechanical technique like wire brushing or blast cleaning. An aggressive mechanical cleaning operation might tend to smear the surface metal to cover an existing surface opening and prevent its discovery. Figure 10.15 depicts a cleaned test surface.

Figure 10.15—Cleaned Test Surface

Once the surface is suitably clean and has been allowed to dry, the penetrant is applied. On small parts this can be done by dipping them into a bath of penetrant. On larger parts, the penetrant can be applied by spraying or brushing. The penetrant is allowed to remain on the test surface for a time period between 5 and 30 minutes, and this time is referred to as the dwell time. The exact length of this dwell time depends upon the penetrant manufacturer's recommendations, the temperature of the part, and the size of the discontinuities of concern. The test surface must be kept moistened with penetrant during this entire time so that the penetrant can flow into surface openings. Figure 10.16 shows the penetrant being applied to the test surface.

The penetrant is drawn into tight cracks by an action referred to as capillary action; this phenomena was previously discussed with regard to brazing filler metal being drawn into the braze joint. Capillary action causes liquids to be drawn into tight clearances. Following the prescribed dwell time, the surface of the test object is thoroughly and carefully cleaned of the excess penetrant. Care must be taken to clean the surface sufficiently to prevent the occurrence of excessive penetrant background and other nonrelevant indications which may mask real indications and cause them to be missed. However, this cleaning operation should not be so aggressive that it washes penetrant out of shallow discontinuities. Figure 10.17 shows the proper procedure for removing excess penetrant.

Once the excess penetrant has been removed, a developer is applied. It can be a dry powder or a

Figure 10.16—Penetrant on Test Surface and In Crack

Figure 10.17—Excess Penetrant Removed

powder suspended in a volatile liquid which readily evaporates, leaving the powder on the surface. It is important that the developer be applied in a thin, uniform layer. In fact, a good technique is to apply the developer in several very thin layers, allowing a couple of minutes between successive developer applications to avoid excessive developer buildup. A thick layer of developer can mask very small indications.

The sensitivity of the penetrant test is dependent upon the size of the developer powder particles as well as the thickness of the developer layer on the test surface. Large particle size and thick layers of developer will tend to decrease the sensitivity of the penetrant test. This developer draws penetrant out of any surface discontinuities to create a contrasting indication in much the same manner that an absorbent material soaks up liquid. This "bleedout" magnifies tiny discontinuities to provide indications which can be easily seen. The discontinuity indication can then be evaluated as to whether it is considered to be detrimental. When using a visible dye penetrant, the evaluation is done under white light whereas use of fluorescent penetrants will require that the evaluation be performed using an ultraviolet (black) light in a darkened area. Figure 10.18 illustrates how the visible indication is produced by the bleedout of the penetrant through the layer of developer.

There are numerous advantages which can be gained when using penetrant testing. First, the use of penetrants is not limited to metallic test objects. Any nonporous material can be tested for surface discontinuities in this manner. It is also well suited for evaluating weld or braze joints between dissim-

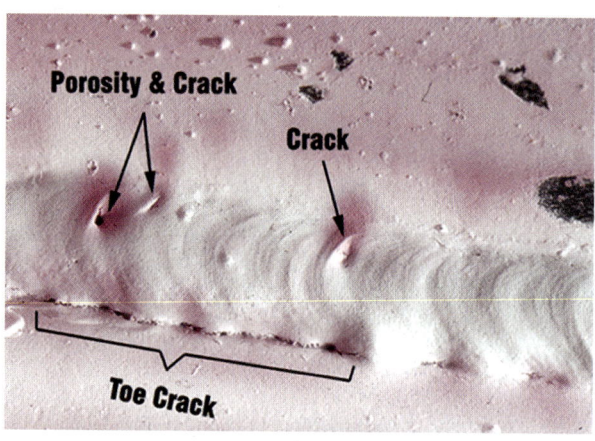

Figure 10.18—Visible Indication After Application of Developer

ilar metals, which may present a problem for other test methods. And it can be applied to nonmagnetic metals when other techniques are not applicable. The process is also quite portable, especially the solvent removable type. For this method, there are convenient aerosol cans of penetrant, developer, and cleaner which can be taken to virtually any test location. Depending on the type of penetrant system being used, the equipment required may be very minimal, allowing the user to penetrant test without a significant capital expense as compared to other test methods.

Among the limitations of penetrant testing is the most prominent one—the fact that it will not detect subsurface discontinuities. Also it is hindered because it is a relatively time-consuming test when compared to magnetic particle testing. The surface condition of the part has a significant effect on the reliability of the test, so the cleaning required for certain applications may be extensive. There is also a need for cleaning the test part after the testing is completed. When testing rough, irregular surfaces, which are often present as the result of welding, the presence of nonrelevant indications may make interpretation difficult.

The equipment required to perform penetrant testing is relatively simple and may consist only of a penetrant, cleaner, lint-free rags, developer and, if required, an emulsifier. A good white light source is required for visible dye penetrants and a good ultraviolet light source is required for fluorescent types. In addition, fluorescent penetrant testing requires a darkened area to monitor cleaning and interpretation of test results. A magnifying glass can also prove useful when very minute discontinuities are being evaluated.

Once an indication has been discovered, it can be permanently recorded using photography or sketches. The indication can also be lifted off the test surface and transferred to a test report form using a transparent plastic tape.

When using the PT method, it is imperative to remove all testing materials including excess penetrant, cleaner, and developer prior to welding. Striking an arc on a surface containing these materials not only affects weld quality, but it can also result in the formation of noxious or even hazardous fumes which can create a serious safety hazard for personnel.

Magnetic Particle Testing (MT)

This particular nondestructive test method is used primarily to discover surface discontinuities in ferromagnetic materials. While indications can be observed from subsurface discontinuities very near the surface, they are very difficult to interpret, and often require testing by other methods. Other NDE techniques are usually required for subsurface discontinuity detection and interpretation. However, surface discontinuities present in a magnetized part will cause the applied magnetic field to create "poles" of opposite sign on either side of the discontinuity, creating a very attractive force for iron particles. If iron particles, which are "magnetic particles" since they can become magnetized, are sprinkled on this surface, they will be held in place by this attractive field to produce an accumulation of iron particles and a visual indication of the discontinuity.

While several different types of magnetic particle tests exist, they all rely on this same general principle. Therefore, all of these tests will be conducted by creating a magnetic field in a part and applying the iron particles onto the test surface.

To understand magnetic particle testing, it is necessary to have some basic knowledge of magnetism; therefore, it is appropriate to describe some of its important characteristics. To begin this discussion, refer to Figure 10.19 which shows a diagram of the magnetic field associated with a bar magnet.

Looking at this diagram, there are several principles of magnetism which are demonstrated. First,

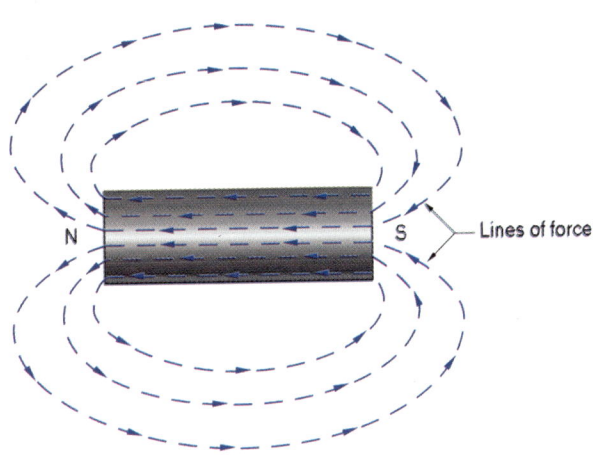

Figure 10.19—Magnetic Field Around a Bar Magnet

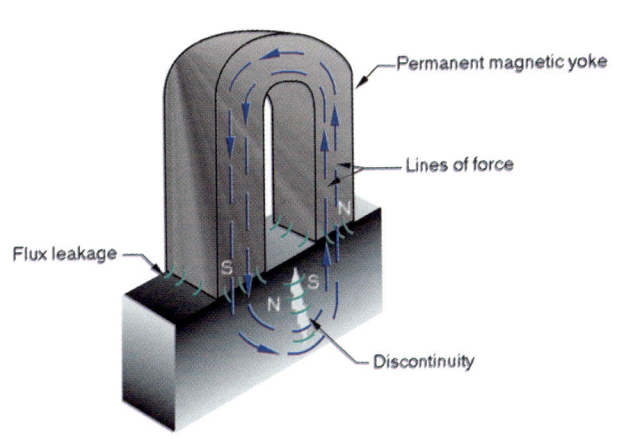

Figure 10.20—U-shaped Magnet in Contact with a Ferromagnetic Material Containing a Discontinuity

there are magnetic lines of force, or magnetic flux lines, which tend to travel from one end (or pole) of the magnet to the opposite end (pole). These poles are designated as the north and south poles. The magnetic flux lines form continuous loops which travel from one pole to the other in a single direction. These lines always remain virtually parallel to one another and will never cross each other. Finally, the force of these flux lines (and therefore the intensity of the resulting magnetic field) is greatest when they are totally contained within a ferrous or magnetic material. Although they will travel across some air gap, their intensity is reduced significantly as the length of the air gap is increased.

Figure 10.20 shows a configuration in which a bar magnet similar to the one in Figure 10.19 has been bent into a U-shape and is in contact with a magnetic material containing a discontinuity. There are still magnetic lines of force traveling in continuous loops from one pole to the other. However, now the piece of steel has been placed across the ends of the magnet to provide a continuous magnetic path for the lines of force. While there is some flux leakage present at the slight air gaps between the ends of the magnet and the piece of steel, the magnetic field remains relatively strong because of the continuity of the magnetic path.

Now consider the discontinuity which is present in the steel bar; in the vicinity of that discontinuity, there are magnetic poles of opposite sign created on either side of the air gap present at the discontinuity. These poles of opposite sign have a strong attractive force between them, and if the area is sprinkled with iron particles, those particles will be attracted and held in place at the discontinuity.

Therefore, to perform magnetic particle testing, there must be some means of generating a magnetic field in the test piece. Once the part has been magnetized, iron particles are sprinkled on the surface. If discontinuities are present, these particles will be attracted and held in place to provide a visual indication. The examples discussed so far have depicted permanent magnets. However, use of permanent magnets for magnetic particle testing is done infrequently; most magnetic particle testing uses electromagnetic equipment. An electromagnet relies on the principle that there is a magnetic field associated with any electrical conductor, as shown in Figure 10.21.

When electricity is passed through a conductor, the magnetic field which is developed is oriented perpendicular to the direction of the electricity. There are two general types of magnetic fields which are created in test objects using electro-

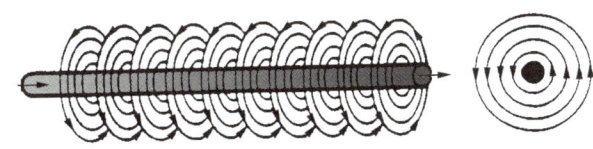

Figure 10.21—Magnetic Field Around an Electrical Conductor

magnetism: longitudinal and circular. The types refer to the direction of the magnetic field which is generated in the part. When the magnetic field is oriented along the axis of the part, it is referred to as longitudinal magnetism. Similarly, when the direction of the magnetic field is perpendicular to the axis of the part, it is called circular magnetism. There are several ways in which these two types of magnetism can be created in a test part.

Figure 10.22 illustrates a typical longitudinal magnetic field created by surrounding the part with a coiled electrical conductor. When using a stationary magnetic particle testing machine, this would be referred to as a "coil shot." When electricity passes through this conductor, a magnetic field is created as shown.

With this magnetic field, those flaws lying perpendicular to the lines of force will be easily revealed. Those lying at 45° to the magnetic field will also be shown, but if a flaw lies essentially parallel to the induced magnetic field, it will not be revealed.

The other type of magnetic field is referred to as circular magnetism. To create this type of field, the part to be tested becomes the electric conductor so that the induced magnetic field tends to surround the part perpendicular to its longitudinal axis. On a stationary testing machine, this would be called a "head shot." This is illustrated in Figure 10.23.

With circular magnetism, longitudinal flaws will be revealed while those lying transverse will not. Those at approximately 45° will also be shown. An important aspect of the circular magnetic field is that the magnetism is totally contained within the ferromagnetic material whereas the longitudinal magnetic field is induced in the part by the electric conductor which surrounds it. For this reason, the circular magnetic field is generally considered to be somewhat more powerful, making circular magnetism more sensitive for a given amount of electric current. When trying to determine the orientation of discontinuities which are likely to form indications, start by determining the direction of the electric current, then consider the direction of the induced magnetic field, and then determine the discontinuity orientation which will give optimum sensitivity.

Both types of magnetic fields can also be generated in a part using portable equipment. A longitudinal field results when the "yoke" method is used, as shown in Figure 10.24. A yoke unit is an electromagnet, and is made by winding a coil around a soft magnetic material core. Current flowing through the coil induces a magnetic field which flows across the test object between the ends of the yoke.

To produce a circular magnetic field with a portable unit, the "prod" technique is used. Use of this method for weld testing is illustrated in Figure 10.25. Either alternating (AC) or direct current (DC) can be used to induce a magnetic field. The magnetic field created by alternating current is strongest at the surface of the test object. AC current will also provide greater particle mobility on the surface of the part allowing the particles to move about more freely which aids flaw detection, even when the surface of the part may be rough and irregular.

Direct current induces magnetic fields which have greater penetrating power and can be used to

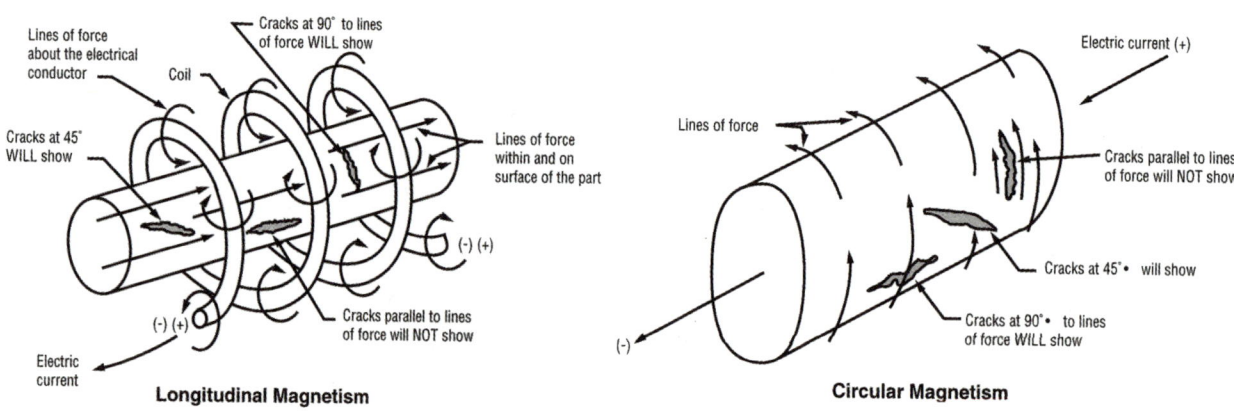

Figure 10.22—Longitudinal Magnetism

Figure 10.23—Circular Magnetism

Figure 10.24—Yoke Method

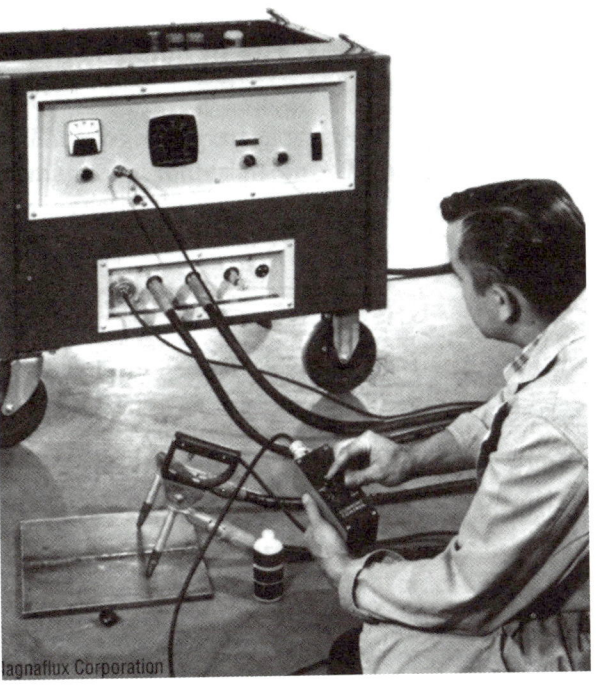

Figure 10.25—Prod Method

detect near-surface discontinuities. However, these indications are very difficult to interpret. A third type of electric current is referred to as half wave rectified AC and can be thought of as a combination of both AC and DC. With this type of power usage, benefits of both types of current can be achieved.

It has been stated that magnetic particle testing is most sensitive to discontinuities perpendicular to the magnetic lines of flux and that discontinuities parallel to the lines of flux might not be detected at all. At angles between these extremes a grey area exists. In general, if the acute angle formed between the lines of flux and the long axis of the discontinuity is greater than 45°, the discontinuity will form an indication. At angles less than 45° the discontinuity might not be detected. Therefore, to provide complete evaluation of a part to locate flaws lying in all directions, it is necessary to apply the magnetic field in two directions 90° apart.

Applications of magnetic particle inspection include the evaluation of materials which are considered to be magnetic at the test temperature. Such materials include steel, cast iron, some of the stainless steels (not the austenitic stainless steels), and nickel. It cannot be used for testing aluminum, copper, or other materials which cannot be magnetized. Properly applied, this test method can detect extremely fine surface discontinuities and will give "fuzzy" indications of larger, near-surface flaws.

Equipment used with this test method varies in size, portability and expense. Lightweight AC yoke units are extremely portable and useful for inspection of objects too large to test otherwise. Such objects might include buildings, bridges, tanks, vessels, or large weldments. Less-portable equipment includes prods and coils. Both typically require a special power source and may have limited mobility. Stationary equipment usually includes mechanisms for both head and coil shots. Parts inspected in stationary units might well be small with extremely high inspection rates or surprisingly large with correspondingly lower inspection rates. The stationary units include demagnetization mechanisms.

The iron particles used are very small and are often dyed to provide a vivid color contrast with that of the test object. Colors commonly available include gray, white, red, yellow, blue, and black. These are called visible particles and are used under a strong visible light source. Iron particles can also be obtained that are fluorescent under black light, and their test sensitivity is greater.

These magnetic particles are applied as a dry powder with a low velocity air stream or are flowed over the part as a suspension in inhibited water or light oil. The dry method is called dry magnetic particle testing, and the oil or water suspension method is called wet magnetic particle testing. Both methods are frequently used, but the wet fluorescent method has higher sensitivity and has become the method of choice for many field and shop applications. The advantages of MT are rapid testing speed and low cost. The method can be made extremely portable and is very good for the detection of surface discontinuities. Testing can be done through thin paint coatings.

The major limitation of magnetic particle testing is that it can only be used on materials that can be magnetized. Other limitations are that most parts require demagnetization after testing and that very thick coatings may mask detrimental indications. Demagnetization is usually done by the AC method and is done by either removing the part from the magnetizing field slowly or reducing the induced magnetizing current applied to the part to zero. Electricity is required for most applications; this may limit portability. Rough surfaces such as those seen on welds or castings can make evaluation more difficult.

Results of magnetic particle testing may be recorded by sketching, photographing or by placing adhesive cellophane tape over the indication and then transferring the tape to a clean piece of white paper.

Radiographic Testing (RT)

Radiography is a nondestructive test method based on the principle of preferential radiation transmission, or absorption. Areas of reduced thickness or lower density transmit more, and therefore absorb less, radiation. The radiation which passes through a test object will form a contrasting image on a film receiving the radiation.

Areas of high radiation transmission, or low absorption, appear as dark areas on the developed film. Areas of low radiation transmission, or high absorption, appear as light areas on the developed film. Figure 10.26 shows the effect of thickness on film darkness. The thinnest area of the test object produces the darkest area on the film because more radiation is transmitted to the film. The thickest area of the test object produces the lightest area on the film because more radiation is absorbed and thus less is transmitted. Figure 10.27 shows the effect of the material density on film darkness.

Of the metals shown in Figure 10.27, lead has the highest density (11.34 g/cc), followed in order by copper (8.96 g/cc), steel (7.87 g/cc), and then

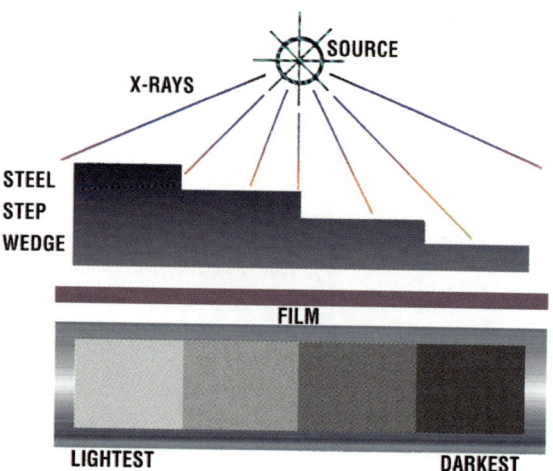

Figure 10.26—Effect of Part Thickness on Radiation Transmission (Absorption)

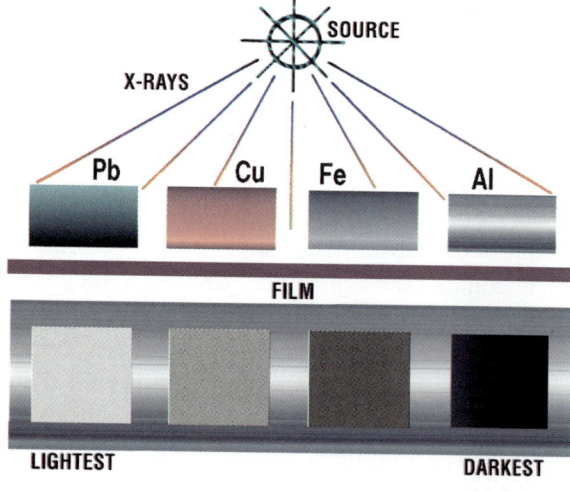

Figure 10.27—Effect of Material Density on Radiation Transmission (Absorption)

aluminum (2.70 g/cc). With the highest density (weight per unit volume), lead absorbs the most radiation, transmits the least radiation, and thus produces the lightest film.

Lower energy, non-particulate radiation is in the form of either gamma radiation or X-rays. Gamma rays are the result of the decay of radioactive materials; common radioactive sources include Iridium 192, Cesium 137, and Cobalt 60. These sources are constantly emitting radiation and must be kept in a shielded storage container, referred to as a "gamma camera," when not in use. These containers usually employ lead and steel shielding.

X-rays are man-made; they are produced when electrons, traveling at high speed, collide with matter. The conversion of electrical energy to X-radiation is achieved in an evacuated (vacuum) tube. A low current is passed through an incandescent filament to produce electrons. Application of a high potential (voltage) between the filament and a target metal accelerates electrons across this voltage differential. The action of an electron stream striking the target produces X-rays. Radiation is produced only while voltage is applied to the X-ray tube. Whether using gamma or X-ray sources, the test object is not radioactive following the test.

Subsurface discontinuities which are readily detected by this method are those having different densities than the material being radiated. This includes voids, metallic and nonmetallic inclusions, and favorably aligned incomplete fusion and cracks. Voids, such as porosity, produce dark areas on the film, because they represent a significant loss of material density. Metallic inclusions produce light areas on the film if their density is greater than that of the test object. If the inclusion density is less than the metal, it shows as a dark area on the film.

For example, tungsten inclusions in aluminum welds, produced by improper gas tungsten arc welding techniques, appear as very light areas on the film; the density of tungsten is 19.3 g/cc. Nonmetallic inclusions, such as slag, usually produce dark areas on the film. However some electrode coatings produce slag having a density very similar to the deposited weld metal, and slag produced from them is very difficult to find and interpret. Cracks and incomplete fusion must be aligned such that the depth of the discontinuities are nearly parallel to the radiation beam for detection. Surface discontinuities will also show on the film; however, using radiation testing to find these types is not recommended since visual inspection is much more economical. Some of these surface discontinuities include undercut, excessive reinforcement, incomplete fusion, and melt-through. Radiographic testing is very versatile and can be used to inspect all common engineering materials.

The equipment required to perform radiographic testing begins with a source of radiation; this source can be either an X-ray machine, which requires electrical input, or a radioactive isotope which produces gamma radiation. The isotopes usually offer increased portability. Either radiation type requires film, a light-tight film holder, and lead letters are used to identify the test object. Because of the high density of lead and the local increased thickness, these letters form light areas on the developed film. Image Quality Indicators (IQI), or penetrameters (pennys) are used to verify the resolution sensitivity of the test. These IQIs are usually one of two types: *hole* or *wire*. They are both specified as to material type; in addition, the hole type will have a specified thickness and hole sizes, while the wire type will have specified wire diameters. Sensitivity is verified by the ability to detect a given difference in density due to the penetrameter thickness and hole diameter, or wire diameter. Figure 10.28 shows both types of the IQI or penetrameters. Figure 10.29 shows the placement of the hole type IQI on a plate weld prior to radiography.

Figure 10.28—Shim and Wire Type Image Quality Indicators (Penetrameters)

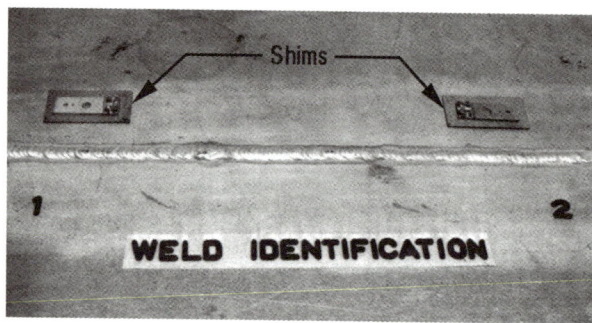

Figure 10.29—Placement of Penetrameters and Weld Identification. Note Shims Beneath IQIs.

Hole penetrameters vary in thickness and hole diameters depending on the metal thickness being radiographed. Figure 10.30 shows the essential features of a #25 IQI used by the ASME Code; its thickness and hole dimensions will be noted for illustration. Here, the penetrameter thickness is 0.025 in., hence the designation of #25 for the IQI thickness in mils (a #10 is 0.010 in. thick, a #50 is 0.050 in. thick, etc.). The hole diameters and positions are specified and are noted in terms of multipliers of the individual IQI thickness. The largest hole in a #25 IQI is 0.100 in. and is called the "4T" hole, referring to the fact that it is equal to four times the IQI thickness, and it is placed nearest the IQI lead number. A "2T" hole (0.050 in.) is positioned furthest away from the lead number 25 and is equal to two times the IQI thickness. The smallest hole between the 4T and 2T hole is referred to as the "1T" hole and is exactly equal to the IQI thickness, 0.025 in. These holes are used to verify film resolution sensitivity, which is usually specified to be 2% of the weld thickness. However, a 1% sensitivity can also be specified, but is more difficult to attain (these are specified in the codes).

Film processing equipment is required to develop the exposed film, and a special film viewer with variable high intensity lighting is best for interpretation of the film. Because of the potential dangers of radiation exposure to humans, radiation monitoring equipment is always required.

The major advantage of this test method is that it can detect subsurface discontinuities in all common engineering materials. A further advantage is that the developed film serves as an excellent permanent record of the test if the film is stored properly away from excessive heat and light.

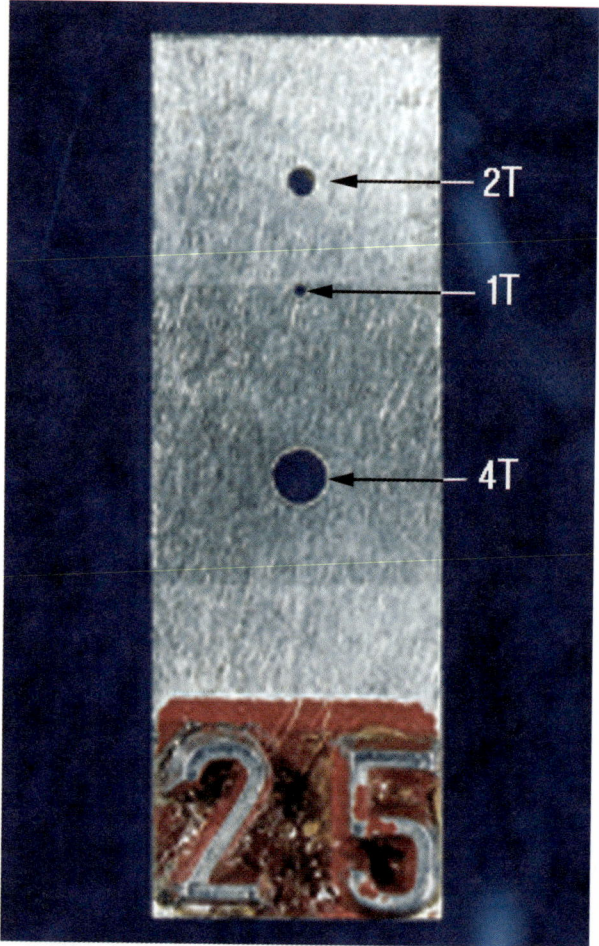

Figure 10.30—Features of a Hole IQI

Along with these advantages are several disadvantages. One of those is the hazard posed to humans by excessive radiation exposure. Many hours of training in radiation safety are required to assure the safety of both the radiographic test personnel and other personnel in the testing vicinity. For that reason, the testing may be performed only after the test area has been evacuated, which may present scheduling problems. Radiographic testing equipment can also be very expensive, and the training periods required to produce competent operators and interpreters are somewhat lengthy. Interpretation of film should always be done by those currently certified to a minimum Level II per ASNT's SNT TC-1A or to AWS RI. Another limitation of this test method is the need for access to both sides of the test object (one side for the source and the opposite for the film), which is shown in Figure 10.31.

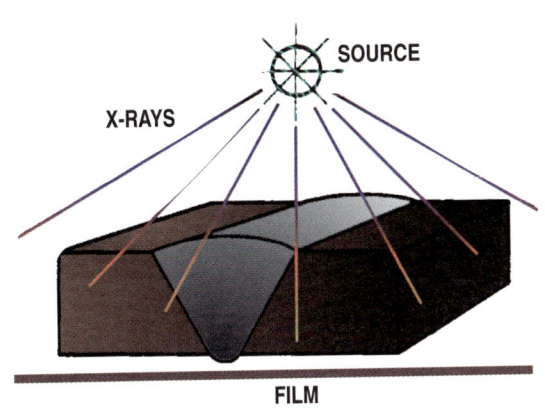

Figure 10.31—Orientation of Radiation Source, Test Plate and Radiographic Film

Another disadvantage of radiographic testing is that it may not detect those flaws which are considered to be more critical (e.g., cracks and incomplete fusion) unless the radiation source is preferentially oriented with respect to the flaw direction. Further, certain test object configurations (e.g., branch or fillet welds) can make both the performance of the testing and interpretation of results more difficult. However, experienced test personnel can obtain radiographs of these more difficult geometries and interpret them with a high degree of accuracy.

Ultrasonic Testing (UT)

Ultrasonic testing (UT) is an inspection method which uses high frequency sound waves, well above the range of human hearing, to measure geometric and physical properties in materials. Sound waves travel at different speeds in different materials. However, the speed of sound propagation in a given material is a constant value for that material. There are several ways that sound travels through a material, but that distinction is not of importance for a discussion at this level. One type of sound wave, called longitudinal, travels about 1100 feet per second in air, about 19,000 feet per second in steel, and about 20,000 feet per second in aluminum.

Ultrasonic testing uses electrical energy in the form of an applied voltage, and this voltage is converted by a transducer to mechanical energy in the form of sound waves. The transducer accomplishes this energy conversion due to a phenomenon referred to as the "piezoelectric" effect. This occurs in several materials, both naturally-occurring and man-made; quartz and barium titanate are examples of piezoelectric materials of each type. A piezoelectric material will produce a mechanical change in dimension when excited with an electric pulse. Similarly, this same material will also produce an electric pulse when acted upon mechanically. An example of the common use of piezoelectric materials is found in the electronic lighters available for starting gas ranges, gas grills, cigarette lighters, etc. In these examples, the piezoelectric crystal is squeezed and released suddenly, resulting in the generation of an electric spark which jumps across a gap to ignite the gas.

To perform ultrasonic testing, the transducer is attached to an electronic base unit. Following a prescribed startup sequence and calibration procedure, the base unit acts as an electronic measuring device. This machine will generate precise electronic pulses which are transmitted through a coaxial cable to the transducer which has been placed in acoustic contact with the test object. These pulses are of very short duration and high frequency (typically 1 to 10 million Hz, or cycles per second). This high frequency sound has the ability to be directed precisely, much like the beam from a flashlight.

When excited by the electronic pulses, the transducer responds with mechanical vibration and creates a sound wave that is transmitted through the test object at whatever speed is typical for that material. A similar phenomenon can be heard when a metal is struck with a hammer to provide a "ringing." This ringing is a sonic (lower frequency) sound wave which travels through the metal. You may have experienced a case where a defective piece of metal is found because of the dull "thud" which results when it is struck.

The generated sound wave will continue to travel through the material at a given speed and return to the transducer when it encounters some reflector, such as a change in density, and is reflected. If that reflector is properly oriented, it will bounce the sound back to the transducer at the same speed and contact the transducer. When struck by this returning sound wave, the piezoelectric crystal will convert that sound energy back into an electronic pulse which is amplified and can be displayed on a cathode ray tube (CRT) as a visual indication to be interpreted by the operator.

By using calibration blocks having specific density, dimensions, and shapes, the ultrasonic unit can be calibrated to measure the time taken by the sound in its travel path and convert this time to part dimensions. Thus the ultrasonic equipment allows the operator to measure how long it takes for the sound to travel through a material to a reflector and back to the transducer, from which dimensional data can be generated as to the reflector's distance below the surface and its size.

Figure 10.32 illustrates a typical calibration sequence on a steel step wedge for a longitudinal beam transducer used for thickness determinations. The transducer is placed on the various known thicknesses of the calibration block and the instrument is adjusted to provide the corresponding screen presentations. Once this operation is complete, the operator can then read the dimension of a test piece directly from the screen by noting where the indication rises vertically along the horizontal axis. With single transducers and multiple reflections, very accurate measurements can be made using a "peak to peak" method rather than the rise off the horizontal line. This technique takes the dimensions between several peaks and averages the data for a thickness measurement.

In general, the screen presentation provides the operator with two types of information. First, indications will appear at various locations along the horizontal axis of the screen. (There will always be an initial indication, called a *main bang*, which will be located near the left side of the screen.) When the sound enters a part and bounces off a reflector returning to the transducer, its return is indicated by a signal rising vertically from the horizontal line. Second, the signal height can be measured and gives a relative measurement of the amount of sound reflected. Once an instrument has been calibrated, the location of the indication reflector on the horizontal axis can be related to the physical distance which the sound has traveled in the part to reach the reflector. The height of that signal on the screen is a relative indication of the size of the reflector. Using this information, the experienced operator can usually determine the nature and size of the reflector and relate it back to a code or specification for acceptance or rejection.

There are two basic types of ultrasonic transducers. (1) Longitudinal waves, or straight beam transducers are used to determine material thicknesses or the depth of a discontinuity below the material surface. These transducers transmit the sound into the part perpendicular to the surface of the part, as shown in Figure 10.32. (2) Shear waves, or angle beam transducers are used extensively for weld evaluation because they send the sound into the part at an angle, allowing testing to be accomplished without the need for removal of the rough weld reinforcement (Figure 10.33). Quite often a longitudinal beam transducer is attached to a plastic wedge which provides the necessary angle. Figure 10.34 shows how the sound propagates through a material when an angle beam is used.

There are two general types of ultrasonic testing, contact and immersion. In contact testing, the transducer is actually placed against the surface of the part. Since the high frequency sound is not readily transmitted through air, a liquid is placed between the test object and the transducer for improved contact. This liquid is referred to as the *couplant*. In immersion testing, the part to be evaluated is placed underwater and the sound is transmitted from the transducer and into the part through the water. Contact testing has the advantage of being portable while immersion is more convenient for production testing of small or irregularly shaped parts.

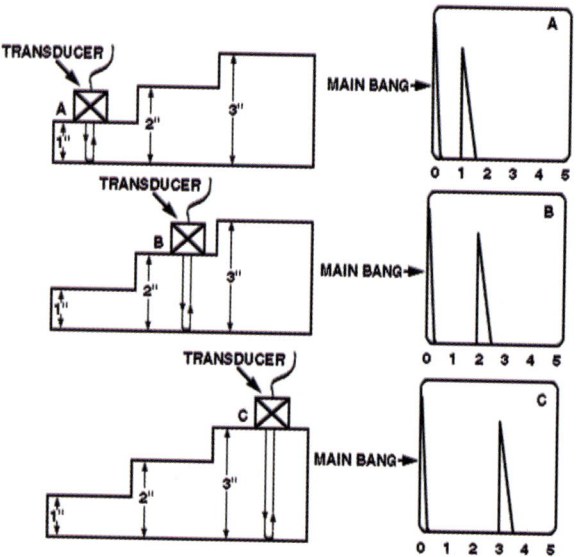

Figure 10.32—Calibration Sequence for Longitudinal Beam Transducer

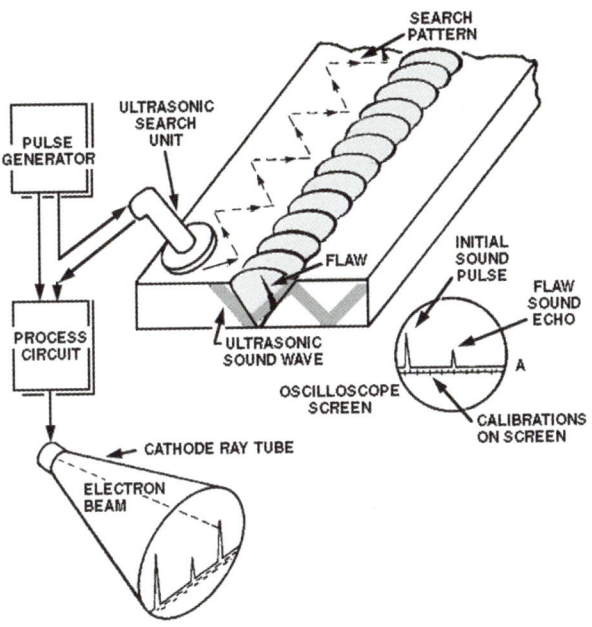

Figure 10.33—Sound Reflection from a Discontinuity Using Shear Wave UT

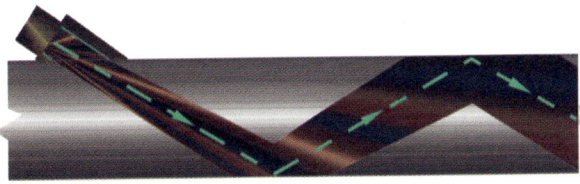

Figure 10.34—Angle Beam Propagation

The applications of ultrasonic testing include both surface and subsurface flaw detection. This method is most sensitive to planar discontinuities, especially those which are oriented perpendicular to the sound beam. Laminations, cracks, incomplete fusion, inclusions, and voids in most materials can all be detected by this method. Along with soundness determinations, thickness measurements can also be made.

The equipment required for ultrasonic testing includes an electronic instrument with either a CRT or digital display. Using an instrument with a CRT, an ultrasonic operator can determine the location, size, and type of many discontinuities. Instruments with digital displays are usually limited to dimensional measurements such as metal thickness. However, when measuring corroded materials for wall thickness, it is best to use an instrument having a scope output for greatest accuracy.

A suitable transducer couplant will also be necessary for ultrasonic testing. Many different materials are used as couplants; some commonly used couplants are oil, grease, glycerine, water, and cellulose powder or corn starch mixed with water. Transducers are available in a wide variety of sizes and styles. Many transducers are mounted on plexiglass wedges which allow the sound to enter the test object at various angles for shear wave testing.

The last equipment requirement is a calibration standard. For material thickness measurements, the calibration standards should be of the same material as the test object and must have known and accurate dimensions. For flaw detection, the calibration standards should meet the above requirements plus contain a machined "flaw," such as a side-drilled hole, a flat-bottomed hole, or a groove. The location and size of this "flaw" must be known and accurate. Signals from discontinuities in the test object are compared with the signals from the calibration standard "flaw" to determine their acceptability. For angle beam testing used in weld testing, one calibration standard is the IIW Block that provides for beam exit and shear wave verification. As noted, the calibration standard should be of the same material; when this is not practical, another material may be substituted and a correction curve, based on the difference in sound velocities of the two materials, is developed for correcting the actual data.

One of the primary benefits of ultrasonic testing is that it is considered to be truly a volumetric test. That is, it is capable of determining not only the length and lateral location of a discontinuity, but it will also provide the operator with a determination of the depth of that flaw beneath the surface. Another major advantage of ultrasonic testing is that it only requires access to one side of the material to be tested. This is a big advantage in the inspection of vessels, tanks, and piping systems. Another important advantage is that ultrasonic testing will best detect those more critical planar discontinuities such as cracking and incomplete fusion. Ultrasonic testing is most sensitive to discontinuities which lie perpendicular to the sound beam. Because various beam angles can be achieved with plexiglass wedges, ultrasonic testing can detect laminations, incomplete fusion, and cracks that are oriented such that detection with radiographic testing would be very difficult.

Ultrasonic testing has deep penetration ability, up to 200 in. in steel, and can be very accurate. Modern ultrasonic testing equipment is very lightweight and often battery powered making this equipment quite portable. The newer instruments have data storage built into the units, which are hand held and only weigh one or two pounds. Stored data can be transferred to a computer for trend analysis and permanent storage.

The major limitation of this test method is that it requires a highly skilled and experienced operator because interpretation can be difficult. Also, the test object surface must be fairly smooth, and couplant is required for contact testing. Reference standards are required, and this test method for weld inspection is generally limited to groove welds in materials that are thicker than 1/4 in.

Eddy Current Testing (ET)

When a coil carrying AC is brought near a metal specimen, eddy currents are induced in the metal by electromagnetic induction. The magnitude of the induced eddy currents depends on many factors, and the test coil is affected by the magnitude and direction of these induced eddy currents. When the test coil is calibrated to known standards, the eddy current method can be used to characterize many test object conditions. Figure 10.35 is a schematic presentation of the eddy currents induced in a test object when the test coil is placed near the surface.

Eddy current testing is a highly versatile test method. It can be used to measure the thickness of thin sections, electrical conductivity, magnetic permeability, hardness, and the heat treatment condition of test objects. This test method can also be used to sort dissimilar metals and to measure the thickness of nonconductive coatings on electrically conductive test objects. In addition, this method can be used to detect cracks, seams, laps, voids, and inclusions near the test object's surface.

The equipment required for eddy current testing includes an electronic instrument with either a CRT or meter display and a coil probe consisting of one or more electrical turns. The test coil can be a probe type for evaluating a surface, a cylindrical coil which surrounds a circular or tubular part, or an inside diameter coil which is passed inside a tube or hole. The calibration standards depend on the desired information. Thickness measurements

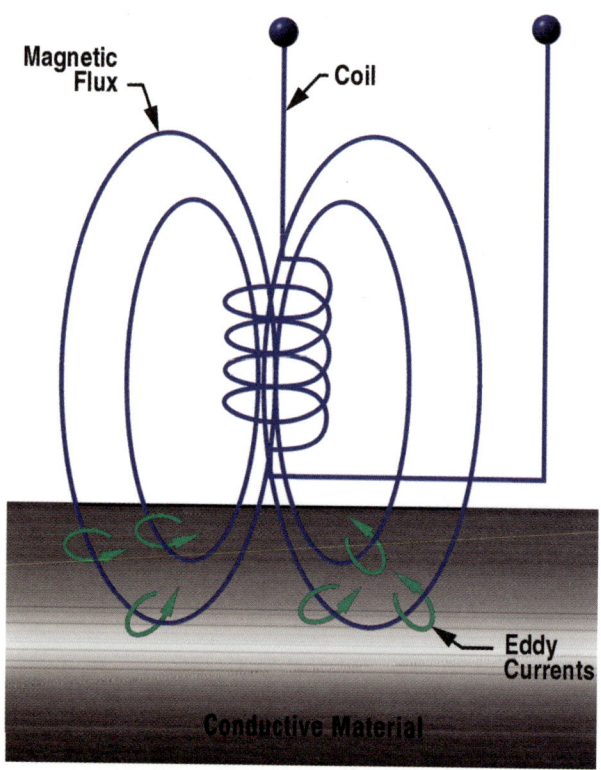

Figure 10.35—Induced Eddy Current in Test Object

require calibration standards of the same material and of known and accurate thicknesses. Heat treatment determination requires standards of the same material with known heat treatment histories.

Figure 10.36 illustrates some typical CRT displays for various types of eddy current evaluations, including metal sorting by conductivity, corrosion thinning, flaw detection, and determination of coating thickness.

One of the major advantages of eddy current testing is that it can be readily automated. The probe need not touch the test object, no couplant is required, and the method is expedient, all of which makes "assembly line" inspection relatively easy. Because testing does not require that the probe contact the part, inspection of hot parts is facilitated. Finally, eddy current testing can be used for the inspection of any electrically conductive material, whether magnetic or nonmagnetic.

The major limitation of eddy current testing is that highly skilled operators are required to calibrate the equipment and interpret results. It is limited to the testing of electrically conductive materials

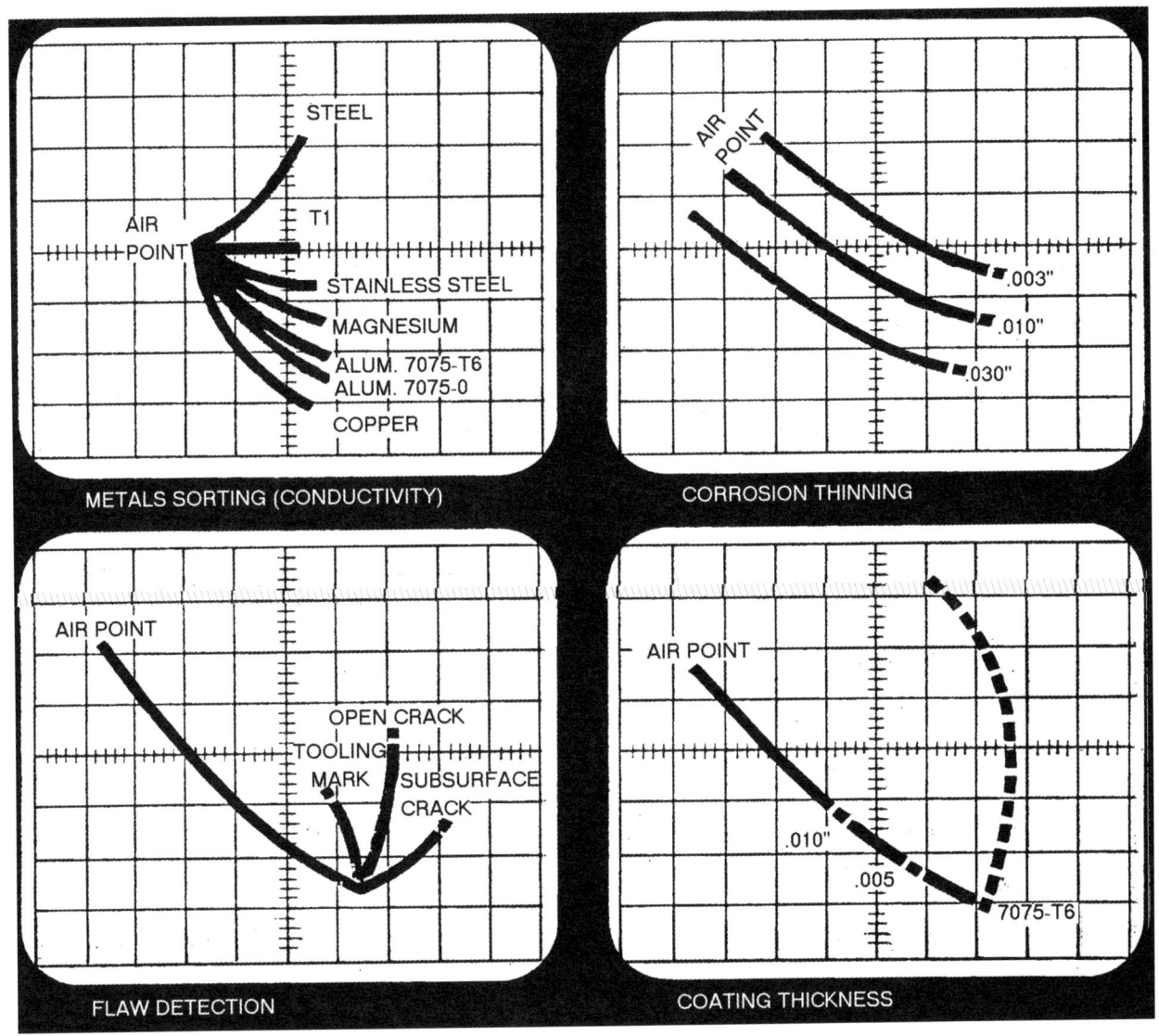

Figure 10.36—Typical CRT Displays for Eddy Current Testing

and its maximum penetration is shallow (typically 3/16 in. or less). The reference standards required for eddy current testing can be quite elaborate and numerous. Surface dirt or contamination that is magnetic or electrically conductive may affect test results and must be removed. And, eddy current testing of magnetic materials may require special probes and techniques.

A major application for eddy current testing is the evaluation of tubing such as that found in heat exchangers. By passing an inside diameter test coil through the inside of the tube, a vast amount of information can be gained regarding corrosion, cracking, pitting etc.

NDE Symbols

Just as we have welding symbols to aid in specifying exactly how the welds are to be done, NDE symbols provide similar information for our inspection and testing work. Once joined, it will usually be necessary to inspect those welds to determine if the applicable quality requirements have been satisfied. When required, tests can be specified through the use of nondestructive testing symbols which are constructed in much the same manner as the welding symbols described earlier. Figure 10.37 shows the general arrangement of the basic elements of the nondestructive testing symbol. As is the case for the

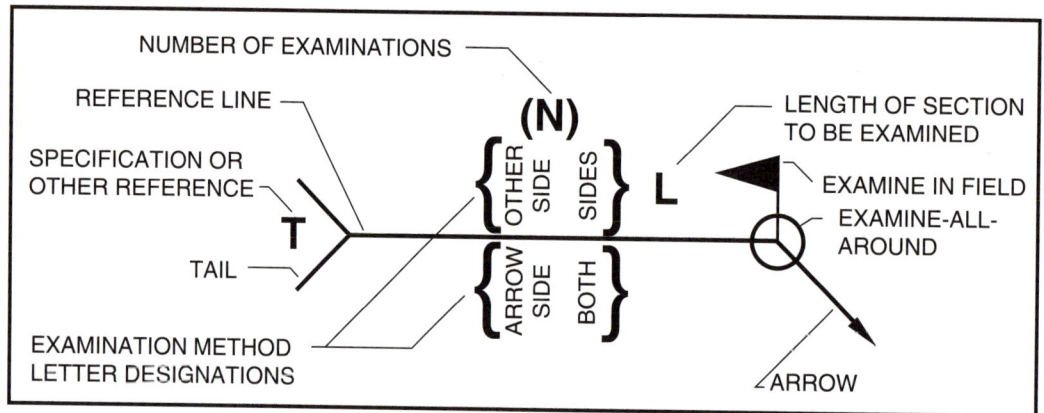

Figure 10.37—Standard Location of Elements for Nondestructive Examination Symbols

welding symbol, information below the reference line refers to the testing operation performed on the arrow side of the joint, and information above the line describes the treatment for the other side. Instead of weld symbols, there are basic NDE testing symbols which are letter designations for the various testing processes. These are shown below:

Type of Test	Symbol
Acoustic Emission	AET
Eddy Current	ET
Leak	LT
Magnetic Particle	MT
Neutron Radiographic	NRT
Penetrant	PT
Proof	PRT
Radiographic	RT
Ultrasonic	UT
Visual	VT

Figures 10.38, 10.39, and 10.40 show testing symbols applied to the arrow side, other side, and both sides, respectively. If it is not significant which side is to be tested, the test symbol can be centered on the reference line, as shown in Figure 10.41. There is also a convention for describing the extent of testing required. A number to the right of the test symbol refers to the length of weld to be tested, as shown in Figure 10.42.

If no dimension exists to the right of the test symbol, it implies that the entire length of the joint is to be tested, which is similar to the convention for welding symbols. Other ways to describe the extent of testing are to specify a percentage of the weld

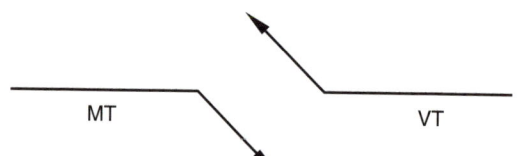

Figure 10.38—Nondestructive Testing on Arrow Side

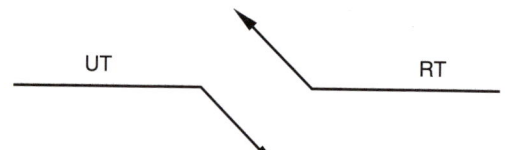

Figure 10.39—Nondestructive Testing on Other Side

Figure 10.40—Nondestructive Testing on Both Sides

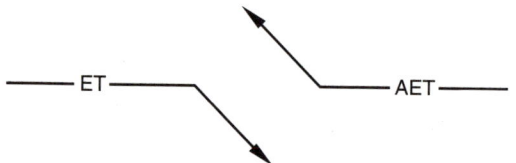

Figure 10.41—Nondestructive Testing Where There is No Side Significance

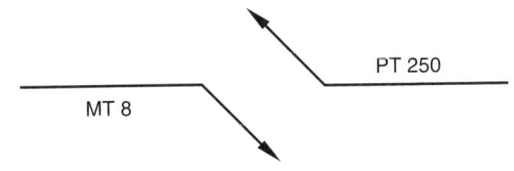

A. Length Shown

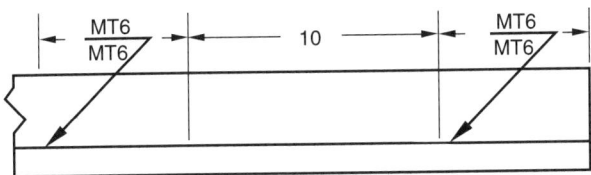

B. Location Shown

Figure 10.42—Designations for Length and Location of Weld to Be Tested

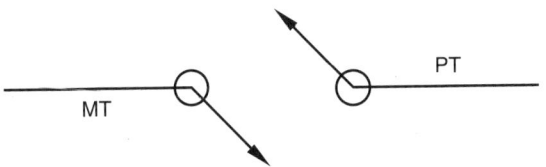

Figure 10.45—Use of Test-All-Around Symbol

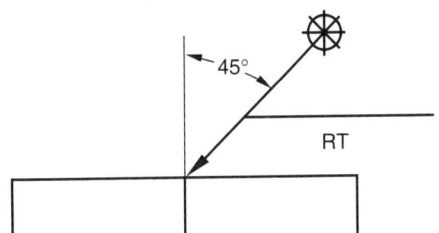

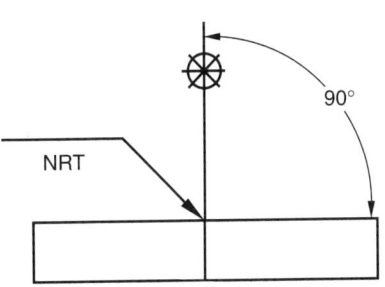

Figure 10.46—Symbols Showing Orientation of Radiation Source

length or the number of pieces to be tested. Figure 10.43 illustrates the application of a percentage to describe partial testing, and Figure 10.44 shows how to specify the number of tests, in parentheses, to be performed. If testing is to be performed entirely around a joint, the test all around symbol can be applied as shown in Figure 10.45.

In the case of radiographic or neutron radiographic testing, it may be helpful to describe the placement of the radiation source to optimize the information received from these tests. If desired, the orientation of the radiation source can be symbolized as illustrated in Figure 10.46.

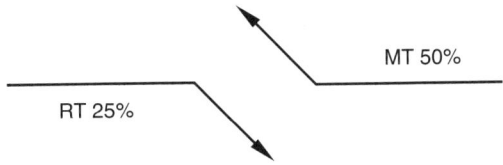

Figure 10.43—Designation of Percentage of Weld Length to Be Tested

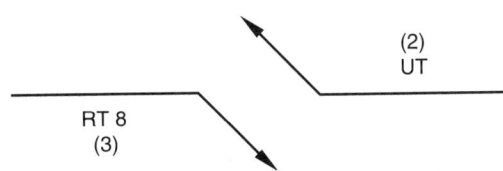

Figure 10.44—Designation of Number of Tests to Be Performed on a Joint or at Random Locations on the Welds

These testing symbols can also be combined with welding symbols as shown in Figure 10.47.

Summary

There are numerous nondestructive test methods available since no single test method is considered to provide a complete evaluation of the properties or soundness of a material. As a welding inspector, it may be necessary to determine which test is best suited for a particular application. Consequently, the inspector must understand how the various tests are conducted, but, more importantly, be capable of deciding which test would be best suited for providing the necessary information to support the visual inspection.

As an AWS certified welding inspector, it may be your job to see that the inspections are done by qualified personnel and that proper records are

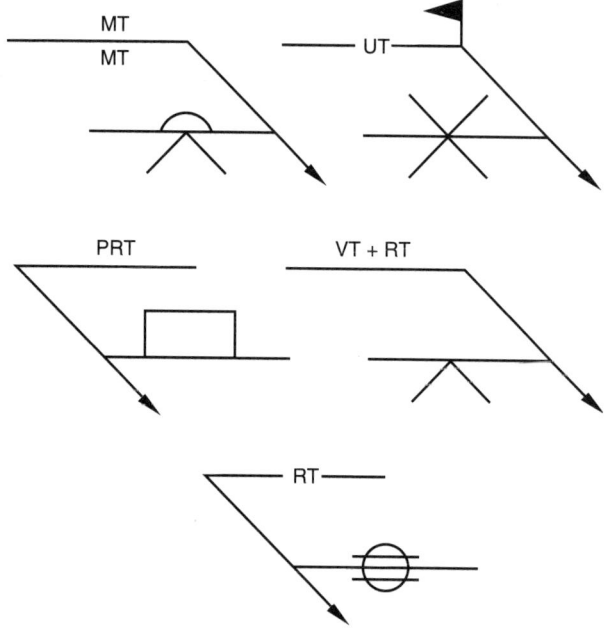

Figure 10.47—Combination of Welding and Testing Symbols

prepared and maintained. While other nondestructive tests may be specified, the requirement for visual inspection should be automatic and completed prior to any other test method.

Also, the welding inspector spends a great deal of time communicating with others involved in the welded fabrication of various structures and components. The use of welding and testing symbols is an important part of that communication process, because this is the "shorthand" used to convey information from the designer to those involved in the production and inspection of that product. The welding inspector is expected to understand the many features of these symbols so that weld and inspection requirements can be determined.

Key Terms and Definitions

backstep—in welding, a technique where the direction of travel for individual passes is opposite that for the general progression of welding along the weld axis.

bleedout—in penetrant testing, the "wicking" action of the developer to draw the penetrant out of a discontinuity to the surface of the part being tested; the surface indication caused by the penetrant after application of the developer.

capillary action—the effect of the surface tension of liquids causing them to be drawn into tight clearances.

couplant—in ultrasonic testing, the liquid applied to a test object to improve transducer contact.

CRT—Cathode Ray Tube; a scope used for displaying electrical signals.

density—in metals, density refers to a weight per unit volume, such as grams per cubic centimeter or pounds per cubic foot. In radiographic testing, film density refers to the darkness of the film; a low density film is light and a high density film is dark.

developer—in penetrant testing, a dry powder or a solution of fine absorbent particles to be applied to a surface, usually by spraying, to absorb penetrant contained within a discontinuity and magnify its presence.

dwell time—in penetrant testing, the time the penetrant is permitted to remain on the test surfaces to permit its being drawn into any surface discontinuities.

eddy currents—small induced currents in conductive materials caused by the proximity of a current carrying coil.

excess penetrant—in penetrant testing, the penetrant remaining on the surface after a portion of it has been drawn into the discontinuity by capillary action.

ferromagnetic—referring to ferrous metals, iron based, which can be magnetized.

flaw—in NDT, a synonym for a discontinuity. A flaw must be evaluated per a code to determine its acceptance or rejection.

fluorescence—the property of a substance to produce light when acted upon by radiant energy, such as ultraviolet light.

flux—in magnetism, the term referring to the magnetic field or force.

galvanizing—adding a thin coating of zinc to the surfaces of a carbon or low alloy steel for corrosion protection.

gamma rays—radiation emitted from a radioactive isotope such as Iridium 192.

hertz—in engineering, the term denoting cycles per second.

hold points—preselected steps in the fabrication process where work must be stopped to permit inspection.

hole IQI—an IQI consisting of a thin piece of material of specified thickness in mils, containing hole diameters based on IQI thickness. See IQI.

IQI—Image Quality Indicator, a device used to determine test resolution sensitivity for RT testing; also called a penetrameter, or penny.

isosceles triangle—A triangle with two equal sides.

main bang—in UT, the term referring to the signal on the CRT generated by the transducer in contact with air.

NDE—Nondestructive Examination.

NDI—Nondestructive Inspection.

NDT—Nondestructive Testing.

parameter—a quantity or constant whose value varies with the circumstances of its application.

penetrameter—see IQI.

penny—see IQI.

piezoelectric—a property of some materials to convert mechanical energy to electrical energy and vice versa.

pole—in magnetism, the term referring to the polarity of the two ends of a magnet; a magnet has a north and a south pole.

prod—in magnetic particle testing, the conductive test electrodes used to induce magnetism in a part.

right triangle—designating a triangle with one angle equal to 90 degrees.

ultrasonic—sound frequencies greater than the range of normal hearing; usually 1 to 10 megahertz.

wire IQI—an IQI consisting of several wires of varying diameters. See IQI.

X-rays—radiation emitted from an electrical device.